Signals and Communication Technology

More information about this series at http://www.springer.com/series/4748

Artem Boriskin · Ronan Sauleau
Editors

Aperture Antennas for Millimeter and Sub-Millimeter Wave Applications

Springer

Editors
Artem Boriskin
Technicolor R&D France
Cesson-Sévigné
France

Ronan Sauleau
IETR
Université de Rennes 1
Rennes
France

ISSN 1860-4862 ISSN 1860-4870 (electronic)
Signals and Communication Technology
ISBN 978-3-319-62772-4 ISBN 978-3-319-62773-1 (eBook)
DOI 10.1007/978-3-319-62773-1

Library of Congress Control Number: 2017946656

Printed on acid-free paper

This Springer imprint is published by Springer Nature
The registered company is Springer International Publishing AG
The registered company address is: Gewerbestrasse 11, 6330 Cham, Switzerland

Contents

Introduction

This book is about challenges and solutions: the fundamental challenges associated with the exploration of the millimeter (mm) wave band and the antenna solutions introduced to overcome these challenges.

The rapid development of the wireless communication and radar technologies that took place in the second half of the twentieth century, thanks to the growth of the semiconductor industry, resulted in a fast overload of the available frequency bandwidth at microwaves.

A number of technical solutions were explored to answer this challenge, including multiple-input and multiple-output (MIMO) methods, frequency and spatial multiplexing, and finally increase of the operational frequency towards the mm- and submm-wave ranges that are capable of providing higher data rates compatible with the needs of today and future applications.

Although the increase of the operational frequencies promised a natural solution for the limited bandwidth problem, the exploration of the mm-wave band raised a number of fundamental issues that handicapped development of mm-wave antenna technologies. This includes the low radiation efficiency of planar antennas, significant material loss intrinsic to both metals and dielectric materials, and high atmospheric absorption in some portions of the mm-wave band. These fundamental problems together with the lack of powerful mm-wave solid state generators required development of electrically large radiating structures, whose total efficiency and physical dimensions were not always compatible with specifications of the target applications, such as high-data-rate communications, intelligent transport systems, and security imaging and sensing applications.

These challenges have determined the main trends in the research and development of mm-wave antennas. Getting a high radiation efficient became of paramount importance to compensate for the high losses and low output power of mm-wave generators. For the same purpose, a point-to-point communication concept based on high gain antennas was proposed instead of an omnidirectional broadcasting used at microwaves. This research direction naturally leads to exploration of various types of focusing and collimating systems, including metallic reflectors and dielectric lenses that can be considered as two milestone technologies that created a base for a majority of the antenna solutions existing at the mm-wave band until now.

However, being comparatively simple to design and fabricate, reflector and lens antennas were often too heavy and bulky. In the original designs, the primary feed of such antennas should have been placed at a certain distance from the phase correcting element (i.e., shaped (or not) reflector acting in the reflection mode or a lens acting in the transmission mode) that required additional volume and structure complexity.

A solution for this problem was found in the introduction of the so-called integrated lens antennas (ILAs) that assume a direct coupling between the feed and dielectric lens. This solution was first introduced in the late 1980s and became one of the most reliable antenna technologies used from lower mm-wave to infrared ranges. In particular, it has been widely studied for point-to-point communications, automotive radars, wireless backhauling, and imaging systems.

Although being quite effective in terms of aperture efficiency and directivity, ILAs often suffer from a significant material loss and large dimensions. These weak points are often considered as a bottleneck for the use of ILAs for mobile and on-board applications requiring lower profile and reduced size/weight solutions. Implementation of multi-beam and scanning ILAs was also associated with certain difficulties due to the limited field of view (FoV), which is intrinsically restricted by the focusing capabilities of extended hemielliptical lenses typically used in ILAs.

A solution for the latter problem (i.e., limited FoV) was found by using radially symmetrical lenses that were already well-know at that time in the optical domain. The aperture of such lenses, typically having the shape of a multi-shell sphere, can be easily shared between several primary feeds, each generating an independent beam in the far zone. In such a way, an angular coverage of up to $\pm 90°$ and even more can be achieved, being limited only by an aperture blockage by other feeds and elements of the lens supports. However, the size of such lenses, limited number of standard low-loss dielectric materials with the desired dielectric permittivity values, and difficulties associated with the fabrication and assembly of multi-shell lenses affected the attractivity of radially symmetrical lens antennas (RSLA). Because of this, their usage has been largely limited to multi-beam automotive radars and base stations for wireless communication systems requiring a point-to-multi-point broadcasting.

A solution for the oversize and overweight problems of lens antennas was also found in optics in the form of Fresnel-type diffractive lens antennas (DLAs). The idea of using diffractive-type zone plates, capable of producing a required phase shift for a portion of the wave front of an incident electromagnetic wave, dates back to the nineteenth century. This concept widely used in optics was successfully scaled down to the mm-wave range. Thanks to their planar topology and simple fabrication, DLAs became an attractive solution for applications sensitive to the weight and size of the antenna system, but still requiring a high directivity, such as security imaging and space communication. More advanced configurations of DLAs were successfully integrated in conformal dielectric radomes. However, the operational principle of DLAs naturally restricts their total efficiency, which may not be sufficient for some applications.

As a next step in the evolution, two complementary antenna technologies were proposed, namely the reflect array (RA) and transmit array (TA) antennas, whose operational principles rely on the use of discrete arrays of phase correcting elements operating in reflection or transmission, respectively. These antennas assumed illumination of the entire array by a single feed, same as it was in the original designs of reflector and lens antennas. However, the discrete structure of the reflecting/transmitting phase correcting elements enabled a new functionality that was not possible in any of earlier configurations, namely the reconfigurability (e.g., variation of the antenna radiation pattern) that may be enabled via electronic tuning of the phase delay provided by each element (unit cell) of the array. Unlike active phased arrays, which are widely used at lower frequencies but less effective at mm-wave range due to the low radiation efficiency, RA and TA antennas are capable of providing a twofold advantage, namely a higher radiation efficiency and reconfigurability. Thanks to their advanced performance characteristics and reduced weight, RA and TA antennas are widely used nowadays for on-board and space applications.

Another antenna solution that was introduced aiming at producing low-profile antennas with a simultaneous increase of its directivity was the so-called Fabry–Perot antenna (FPA). Its operational principles rely on the use of a cavity covered with a semi-transparent mirror that causes multiple reflections inside the cavity, thus allowing to increase the electrical size of the radiating aperture over a limited bandwidth, whose bound mainly depend on the antenna directivity. Although the operational concepts of FPAs were introduced in the microwave community in the 1950s, research on FPA really started in the mid-1990s. Many high-performance designs have been proposed, most of them below 100 GHz, with the main goal to enlarge the bandwidth and provide multiband capabilities.

Of course, the progress in the development of mm-wave antennas was not solely associated with the evolution of the focusing/collimating systems. It was also supported by the advances in the primary feed design. Indeed, the mismatch between the operational bandwidth of primary feeds, whose function was originally limited to matching a waveguiding system with a free space in order to minimize the reflection loss, and quasi-optical focusing systems aimed at correcting the phase front of the radiated wave remained (and still remains) a bottleneck for design of all types of mm-wave antennas. To handle this fundamental problem, significant efforts were devoted to development of advanced feeds, having different topologies and operational principles and, thus, capable of providing a required radiation efficiency, bandwidth, and illumination conditions for the phase correcting elements (e.g., lenses and reflectors) used to increase the directivity and/or shape of radiation pattern of a primary feed antenna.

In some cases, the research into the primary feeds resulted in the development of antennas that do not require an additional focusing/collimation element. Among those are flat horn antennas (FHA), whose operational principle relies on the excitation of leaky waves (LW) that propagate along the corrugated flanges of the horn, thus contributing to the formation of the total radiation pattern. Optimization of corrugated surface parameters enables one to support desired LWs that, in its turn, defines the phase and amplitude distribution along the antenna aperture. In the limit, one can open the mouth of a corrugated horn antenna to get a totally flat profile. Such flat horn antennas can be fed by standard waveguide feeds as well as narrow slots. The latter family is often referred as the bulls-eye antennas, thanks to the shape of the annular grooves surrounding the central slot. Its operational principle is linked to the extraordinary transmission through sub-wavelength apertures. The low-profile corrugated horn antennas were recognized as being of potential interest for indoor communication systems.

In recent years, the idea of using functional surfaces, capable of supporting different types of waves (e.g., surface and leaky waves) or, instead, preventing their propagation, has been attracting a growing attention that was boosted by the advances in the metamaterial domain. Indeed, the possibility of changing locally the surface impedance of the functional surface and thus manipulating with the phase and amplitude of the fields in the entire aperture of a metasurface antenna (MSA), opened the door for development of a new class of antennas with very attractive performance characteristics, including shaped-beam patterns and near-field focusing capabilities. Moreover, an active control over the surface impedance can potentially enable development of reconfigurable and scanning MSAs capable of competing with alternative approaches, namely the RA and TA antennas. Such a functionality may be achieved, for instance, thanks to the use of advanced electrically and/or optically tunable materials, like graphene.

In this book, the aforementioned antenna technologies are reviewed from the point of view of their maturity and potential for future development. Practical examples are provided in order to illustrate the evolution, current trends, and specific features intrinsic to each antenna technology. The cross-links are provided between the chapters to facilitate benchmarking of different antenna solutions. The last two chapters are devoted to the challenges associated with antenna measurement at mm-wave bands. The pitfalls in the design and characterization of different types of mm-wave aperture antennas are identified in order to help readers in the selection of an appropriate antenna solution and its effective implementation.

In such a way, the book represents a survey of the state-of-the-art and recent developments in the field of the mm- and submm-wave antennas technologies being in the focus of interest of the 5-year ESF RNP Newfocus program (2010–2015). The authors gratefully acknowledge the financial support provided by the European Science Foundation (ESF) through the RNP Newfocus program, which enabled 76 research projects in between 43 research teams from 19 countries worldwide, whose outcomes constitute the core of the current book.

Artem Boriskin
Ronan Sauleau

Part I
Antenna Systems

Chapter 1
Integrated Lens Antennas

Artem V. Boriskin, Ronan Sauleau, Jorge R. Costa and Carlos Fernandes

Abstract This chapter presents the Integrated Lens Antenna (ILA) technology as it evolved since its introduction aiming to respond to the needs of emerging applications such as high-data-rate communication, intelligent transport, and mm-wave imaging. The topics covered include the ILA design concepts as well as the electromagnetic phenomena intrinsic to dielectric lenses that may affect ILA performance. The aspects of the ILA technology related to selection of the primary feeds, lens materials, and fabrication methods are also revised. A few practical examples are provided to illustrate the current and future trends of this technology.

Keywords Integrated lens antennas · Geometrical optics · Physical optics · Lens feeds · Dielectric materials · Lens manufacturing · Lens profile design · Optimization

A.V. Boriskin (✉)
Research & Innovation Division, Technicolor R&D France,
975 avenue des Champs Blancs, 35576 Cesson-Sévigné, France
e-mail: artem.boriskin@technicolor.com

R. Sauleau
Université de Rennes 1, Institut d'Electronique et de Télécommunications
de Rennes (IETR), UMR CNRS 6164, Campus de Beaulieu, 263 Avenue
du Général Leclerc - CS 74205, 35042 Rennes Cedex, France
e-mail: Ronan.Sauleau@univ-rennes1.fr

J.R. Costa
Instituto de Telecomunicações, Instituto Universitário de Lisboa
(ISCTE-IUL), Lisbon, Portugal
e-mail: jorge.costa@lx.it.pt

C. Fernandes
Instituto de Telecomunicações, Instituto Superior Técnico,
Universidade de Lisboa, Lisbon, Portugal
e-mail: carlos.fernandes@lx.it.pt

A. Boriskin and R. Sauleau (eds.), *Aperture Antennas for Millimeter and Sub-Millimeter Wave Applications*, Signals and Communication Technology,
DOI 10.1007/978-3-319-62773-1_1

1.1 Introduction

Integrated lens antennas (ILAs) were originally introduced in the late 1980s for high-resolution imaging and sensing applications to compensate the low directivity and low radiation efficiency of printed microstrip antennas at the millimeter (mm) wave bands.

A partial remedy for the low-radiation efficiency problem was found in a direct integration between printed antennas and dielectric lenses. This approach helped one to minimize losses associated with modes supported by a dielectric substrate, which is an essential counterpart of a printed antenna. Such modes (i.e., both first TE and TM modes of a dielectric slab) have no cutoff frequencies and, thus, can be excited even for very thin substrates. In case of a groundless microstrip antenna, the presence of the lens on top of the planar feed also improved distribution of the radiated power between the upper and lower half spaces. As demonstrated in [1], the ratio of power radiated by a microstrip antenna printed at the interface of two dielectric media towards the upper and lower half space is defined by the index ratio at this interface:

$$P_1/P_2 = (\varepsilon_1/\varepsilon_2)^{3/2}, \tag{1.1}$$

where P_1 and P_2 is the total power radiated towards the upper and lower half space filled in with the dielectric material having relative permittivity ε_1 and ε_2, respectively.

This physical phenomenon suggests the use of electrically large dielectric lenses made of high permittivity materials capable of imitating unbounded dielectric medium. In some embodiments typically used at upper mm and THz ranges, the primary feeds were printed directly on the flat base of a plano-convex lens. Such lenses were often referred as “substrate lens” [2].

Optimization of the lens shape and structure provides additional degrees of freedom in optimizing ILA performance characteristics such as directivity and shape of the radiation pattern, phase center position, as well as radiation and aperture efficiencies. In the 90s, the ILA concept has resurfaced as an effective configuration to produce highly shaped radiation patterns for mm-wave mobile applications [3,4, 60–62]: for these applications the objective was no longer enhancing the directivity by producing a collimated beam but instead reshaping the output beam to match demanding amplitude templates. For instance, shaped ILAs were used to obtain isoflux illumination of areas having specific shapes, like a rectangle, which is a typical scenario for mm-wave wireless LANs and mm-wave mobile communications [4, 5].

Collimated beams lenses may by of primary importance for mm waves point-to-point links due to the need to compensate the high material and propagation losses to achieve a desired power link budget. To reach the highest possible directivity, dielectric lenses with elliptical profiles were proposed [6, 7]. This shape was suggested by the ray-tracing theory, which predicts that a spherical wave radiated by a point source can be transformed into a locally plane wave (or vice versa) by an elliptical lens, provided that the source is located in one of the ellipse foci and the ellipse eccentricity is properly related to the lens material permittivity

$$e = \frac{f}{a} = \varepsilon^{-1/2}, \tag{1.2}$$

where $f = b(\varepsilon - 1)^{-1/2}$ is the focal length of the ellipse and b is its minor semi-axis (Fig. 1.1a).

As the correction occurs only for the portion of the wave front impinging on the front part of the lens, it was suggested to cut the ellipse through its rear focus and attach the resulting flat-bottom lens directly on top of a primary feed, as a way to solve simultaneously the radiation efficiency, beamforming, and assembly problems. In another embodiment, a hemielliptical lens was extended using a cylindrical

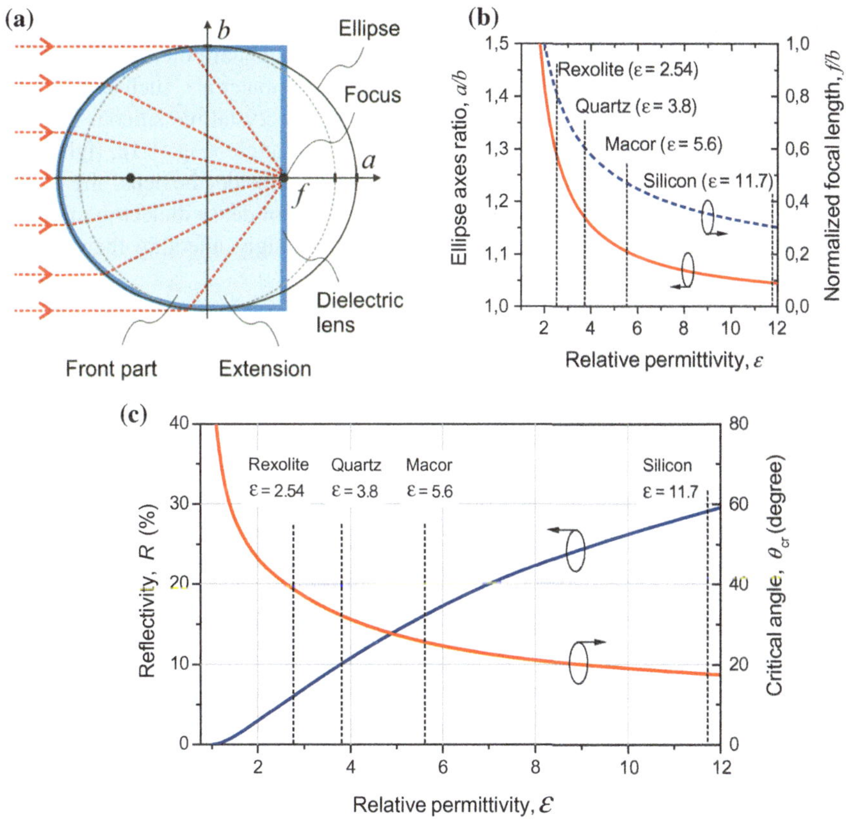

Fig. 1.1 **a** Topology and notation of an extended hemielliptical lens, **b** parameters of the elliptical lens versus relative permittivity of the lens material, **c** reflectance at normal incidence and critical angle for plane wave incident on an air-dielectric interface from the denser medium side

extension, whose height was selected so that the primary feed occurs close to the rear focus of the ellipse. Dimensions of the ellipse semi-axes and its extension length suggested by Eq. (1.2) are presented in Fig. 1.1b for some standard dielectric materials. In other embodiments, the elliptical shape of the lens was approximated by a hemispherical one, whereas the difference in the height of the hemiellipse and hemisphere was again compensated by the height of the cylindrical extension.

As follows from Eq. (1.2), the lens material may be selected rather arbitrarily, whereas Eq. (1.1) suggests the use of dielectrics with the highest possible permittivity. A larger electrical size of the lens aperture is also favorable because it may provide a higher directivity value for the same physical size of the lens. On the other hand, high-index materials usually have larger density, which may cause increase of the lens weight. In some cases, it is partly compensated by a smaller height of the lens (see Fig. 1.1b) but may still be critical for some applications. What is more important, an abrupt change of the refractive index value at the air-dielectric boundary may cause a strong backreflection and excitation of multiple internal reflections inside the lens that may deteriorate impedance matching of the primary feed and the ILA radiation pattern. As shown in Fig. 1.1c, reflectivity value for the plane wave normally incident on the surface of standard dielectric materials may vary from about 6% observed for low-permittivity materials, such as Rexolite, up to about 30% observed for dense materials, such as crystalline silicon. Because of this, and also due to a reduced value of the critical angle (Fig. 1.1c, right axis) that causes a larger amount of energy being trapped inside the lens, the modal features in the electromagnetic behavior of lenses made of dense dielectric materials may dominate over the ray-tracing focusing thus strongly affecting their performance characteristics [8, 9].

A solution for the latter problem related to internal reflections was found in using one or more matching layers in the form of quarter-wavelength layer made of a material with a lower permittivity [10] or in a form of corrugations, acting as a layer with an effective permittivity lower than that of the lens material [11]. These solutions have been widely used in optics for decades and are well established nowadays at upper mm wave and THz ranges. At the same time, they are less popular at lower frequencies mostly because of the fabrication and assembly difficulties associated with multi-shell structures. This makes the use of low-permittivity-shaped lenses to be a preferable solution for low mm-wave bands.

Regarding the optimal lens dimensions and choice of materials, there is no universal guideline derived so far. This selection is usually made based on a trade-off approach with respect to a technical specification derived for a certain application. The factors typically taken into account are the loss tangent of the dielectric material, its density, its homogeneity, permittivity, ease and cost of manufacturing, including the cost of the supports and matching layer(s), if applicable.

As follows from above, the hemielliptical (or hemispherical) shape of the lens suggested by ray optics takes into account neither the actual size of the feed aperture, nor its radiation pattern. Thus, it should be considered only as an initial guess for the optimization process that may include optimization of the lens shape, size, material, as well as the primary feed radiation pattern and position. Some hints on how to reach the optimal match between the feed and lens parameters have been provided in [12], based on the analysis of the lens edge illumination conditions.

All of the previous issues related to lens size, energy coupling from the feed to the lens, material permittivity choice and internal reflections also apply for arbitrarily shaped ILAs. These kinds of lenses can be used to compensate for the effects associated with finite size of the lens and nonnegligible size of the primary feed aperture, as well as a possibly nonuniform and nonsymmetrical pattern of the primary feed [13, 14]. ILAs can also be used to provide highly "shaped-beam" radiation patterns according to given amplitude templates.

While enabling an effective control over the ILA radiation pattern at a certain frequency, in certain designs the use of shaped lenses may restrict the operation bandwidth of ILAs, especially if it is defined with respect to the shape of the radiation pattern or position of the ILA phase center. To overcome this problem, the use of multi-material shaped lenses has been recommended. In [15] it was demonstrated that a double-layer shaped lens can provide a higher radiation efficiency as well as a more stable radiation pattern and phase center position achieved thanks to the lower contrast at the air-dielectric boundaries and additional degree of freedom in optimizing the antenna characteristics provided by the additional interfaces between different shells of the lens. The same conclusions have been reached in [16], where a complementary focusing structure in a form of a multi-shell lens-like dome has been reported.

Another trend in the design of ILAs is related to the reduction of the lens weight and dimensions. Solutions proposed in this context include the use of lenses with air cavities enabling reduction of the lens focal distance [17]. For some applications, requiring fan-type beams with different angular width in the azimuthal and elevation planes, the use of planar slab lenses was suggested [18].

There is also a number of mm-wave applications, which require multi-beam or scanning performance. Among those are point-to-point and point-to-multi-point communications, imaging, and intelligent transport systems. The sought multi-beam functionality was enabled using beam-switching arrays with each feed producing an independent beam in a given direction [19, 20]. This approach provided a good reliability and fast scanning rates with a field of view (FoV) of about $\pm 25°$. For larger scan angles, a rapid degradation of the antenna directivity occurs that is explained by a shift of the feed position from the focal arc of the lens observed for the off-axis feeds as well as a mismatch in the orientation of the primary feed (radiating at broadside) and the desired scan angle [21].

At present, the enlargement of the ILA scan angle is probably one of the most critical challenges associated with ILAs, which handicap further advances of this technology. Solutions proposed in this context include multi-objective optimization over the lens and feed parameters [22, 23], mechanical scanning [24], and the use of shaped bifocal lenses [25] and double lens focusing systems [26, 27].

Another available solution is based on the use of multi-shell spherical [28] and hemispherical [29] lenses, however, this solution is not always effective at upper mm waves because of the fabrication and assembly difficulties associated with the multi-shell structure of such lenses [30].

There exists another solution for resolving the limited FoV problem that has been developed by nature as a result of the million-year evolution, i.e., the insect compound eye, which comprises plurality of low-resolution sensors arranged along a curvilinear surface [31]. Technical implementation of such a device at mm waves is difficult but, potentially, feasible using soft polymer materials.

Of course, the progress of ILA technology was not only associated with the design of focusing elements but was also tightly linked with the advances in primary feed designs. The role of a primary feed of any aperture antenna is twofold: impedance matching between a primary feed and its feeding system (e.g., hollow metallic waveguide or microstrip line) and the optimal illumination conditions of the focusing and/or beamforming element (e.g., a lens). The types of feeds used as counterparts of ILAs include various types of microstrip patch antennas printed on a PCB board [26] or directly at the bottom of the lens [1, 2, 23], as well as waveguide feeds [4, 32, 16]. The latter typically provides a wider bandwidth but require additional space for the waveguide feeding structure or a microstip-to-waveguide transition that may spoil the total ILA bandwidth. The matching problem between a waveguide feed and a lens is typically solved using dielectric tapers integrated inside the waveguide. However, the bandwidth provided by these solutions is not sufficient for some applications, especially at upper mm and submm-wave ranges. Solutions for ultra-wideband ILAs have been provided in [33] using a crossed exponential tapered slot antenna configuration, or in [34] using non-dispersive leaky-wave planar feeds, both cases intended for imaging systems.

Finally, the use of primary feeds in a form of active arrays enabled another interesting functionality not possible with a single-feed ILA, namely a reconfigurable radiation pattern, whose shape (e.g., half power beam width) can be changed by switching between different combinations of array elements [35]. Such ILAs with a reconfigurable FoV were recently proposed for automotive radars requiring switching between long-range and short-range scanning regimes.

Looking back at the evolution of the ILA technology, we can see that after about 30 years of development the ILA technology has reached a certain level of maturity and became an essential part of many established and emerging mm wave and THz applications. At the same time, there are still some functionalities that have not been enabled due to certain fundamental issues intrinsic to the technology, such as a wide FoV and reconfigurability. Being of high practical importance, these constrains may define future trends in the development of the ILA technology.

In a summary for this overview of the ILA technology evolution, it is worth mentioning various numerical techniques that have been developed for the analysis and synthesis of ILAs, among those are: high frequency techniques based on the geometrical optics (GO) and physical optics (PO) [13, 4, 36, 37, 5], frequency domain techniques based on two-dimensional and three-dimensional boundary integral equation (BIE) methods [8, 38], and time domain techniques, based on the two-dimensional, two-dimensional-BoR, and three-dimensional FDTD methods [39]. The GO/PO techniques were widely used for fast analysis and synthesis of elliptical and shaped ILAs. The integral equation methods enabled accurate analysis of the interplay between the geometrical optics focusing and modal (resonant) features of compact-size dielectric lenses. In particular, use of Muller's BIEs enabled one, for the first time, to investigate the drastic influence of the so-called half-bowtie resonances on the performance of elliptical ILAs [40, 9] as well as to assess efficiency of GO/PO [8] and FDTD [41] methods in characterization of such resonances. Finally, the FDTD techniques were effectively used for wideband analysis of body-of-revolution (BoR) and three-dimensional single- and multi-material structures metal-dielectric structures, e.g., [42, 16]. Nowadays, these complementary techniques are implemented in a number of in-house [43] and commercial software.

In the following sections, we present general guidelines for the design and implementation of ILAs intended for various mm-wave applications. The topics covered include: ILA design steps, lens materials, primary feeds and ILA fabricating methods. The discussion is followed by a few practical examples of ILAs illustrating recent advances in the ILA technology.

1.2 Lens Antenna Design

Figure 1.2 presents the generic lens geometry that of an ILA. It consists of a single- or double-shell dielectric lens, shaped in the elevation coordinate and eventually also in azimuth coordinate. It may be fed by a single feed or by an array of feed elements attached or not to the lens base. In the general case, feed elements are either on-axis (x = 0; y = 0; z = 0) or off-axis. Different options exist for the feed element, as will be discussed in the following section, and of course its configuration and radiation characteristics influence the final lens shape.

The adopted design procedure for shaped lenses is summarized in Fig. 1.3. Two alternative approaches are presented: one provides a fast first guess of the lens shape, based on closed-form analytical expressions—labeled (a); the second more elaborate approach—labeled (b)—combines lens performance analysis tools with an optimization algorithm that successively generates and evaluates meaningful populations of test lenses until a specified figure of merit is reached.

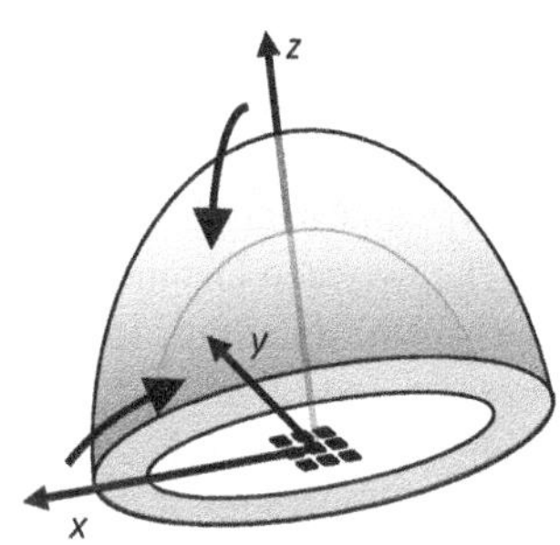

Fig. 1.2 Generic geometry of a substrate integrated lens antennas (ILA)

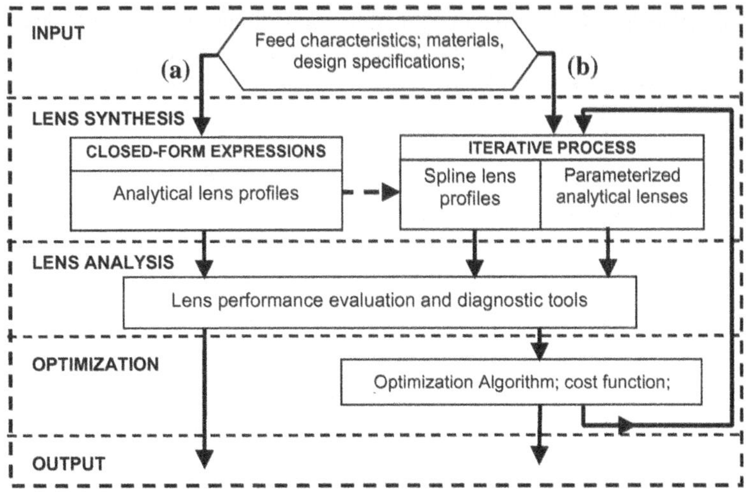

Fig. 1.3 Block diagram of lens design steps

Closed-form lens synthesis expressions can be obtained as a first guess for some classes of target specifications, based upon Geometrical Optics (GO) formulation. The lens' overall dimensions and the principal radii of curvature at every point on the surface are required to be much larger than the wavelength to ensure the GO validity. The feed is characterized only by its phase center and by its far-field pattern in an unbounded medium with the same permittivity of the lens material. The GO lens designs are inherently broadband as long as the lens dimensions are large compared to the wavelength, the material dispersion is negligible, and the feed radiation pattern remains constant versus frequency.

The lens shape is directly determined from the solution of a set of partial differential equations corresponding to Snell laws at the lens interfaces and additional conditions related to the output requirements. The calculation time is generally in the order of seconds, which makes it attractive as a first guess and as a seed for a subsequent refinement optimization loop if required. Alternatively, the GO formulation can include shaping parameters that enable running the closed-form expressions in the optimization loop (b) of Fig. 1.3 to match an appropriate cost function [43]. This narrows down the search space to physically meaningful lenses.

The GO-based formulation is inherently independent from frequency. An appropriate analysis method is required to evaluate the lens actual performance taking into account diffraction effects, frequency dependence of the feed radiation pattern, and material dispersion.

The two-step Geometrical Optics/Physical Optics (GO/PO) method [6] is widely used in the literature to determine the radiation pattern of lens antennas (also of reflector and other aperture antennas). It offers a very good compromise between quality of results and computation time, making it attractive for integration into lens design optimization loops. It takes as input the lens shape, the material permittivity, the feed position, and the far-field pattern from the feed in the lens material. The GO is used for evaluating the field distribution over the lens/reflector surface, and PO is used to calculate its contribution to the near- or far-field radiation. GO/PO is especially useful for very large structures in terms of wavelength and there is no upper size limit for its applicability. On the contrary, GO fails for small lenses [8, 42], where the feed can no longer be replaced by its far-field radiation pattern and the ray propagation approach is no longer valid.

An alternative formulation was developed to circumvent some of the indicated limitations of GO/PO in the characterization of small lenses [44]. It is a PO-based method that uses point-like sources as basis functions to decompose the pre-calculated feed aperture fields at the lens base and calculate the resulting incident near-field at the lens surface. The proposed method was implemented only for axial-symmetric lenses with on-axis feed.

Another limitation of the GO approach is that it fails to characterize completely the total internal reflection phenomenon [37]. However, analysis showed that this effect is more important in collimating lenses with canonical shapes, like the elliptical lens; it is imperceptible in most of the shaped lenses.

Full-wave numerical solvers like the CST Microwave Studio™ transient solver [45] can be used to analyze the final lens design prior to fabrication. Such tool takes into account all the effects that are handled with some approximations in the previously discussed methods but, due to the involved computation time, it is not viable option for integration in lens design optimization loops.

Genetic Algorithms (GA) optimization method has long been used because of its flexibility, robustness, global search character, inherent capability to avoid trapping in local minima and complete absence of restrictions on the behavior of the function to be optimized, e.g., [22, 14]. The drawback is its relatively slow convergence, and hence two different approaches were used to reduce the number of shape parameters that model the unknown lens surfaces (parallel boxes in the (b) path of Fig. 1.3):

- Splines-based representation of the lens surface;
- Parametric optimization of selected parameters from the GO closed-form design expressions.

The splines-based representation is the most flexible of the two solutions, allowing for a limitless search space. As few as 4–5 nodal points np_n of the lens surface are used as the target for optimization (see Fig. 1.4) while a cubic spline approximation is used to represent the lens surface between consecutive pairs of these wide-spaced nodal points. Especially with double-shell lenses, this strategy often leads to the presence of caustics near the outer lens surface. These lenses are automatically eliminated from the test population but this may significantly reduce the number of "individuals" in the test population, with a negative impact on the convergence of the optimization method.

The alternative strategy was to construct the test population based on lens shapes obtained directly from the previously described GO closed-form expressions, where the shaping parameters are incorporated. These shaping parameters were used as the

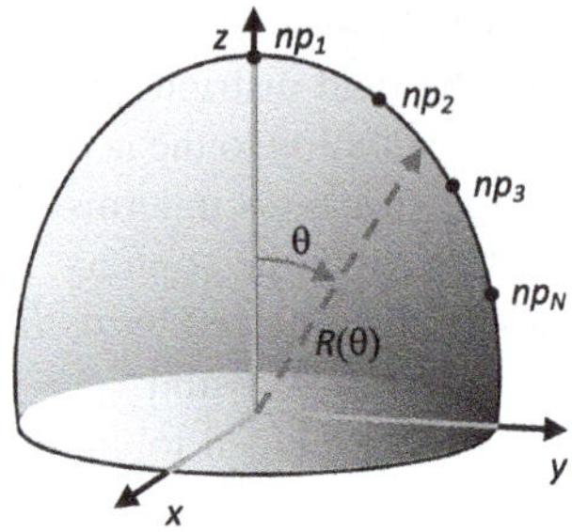

Fig. 1.4 Geometry for spline parameterization of a generic lens interface

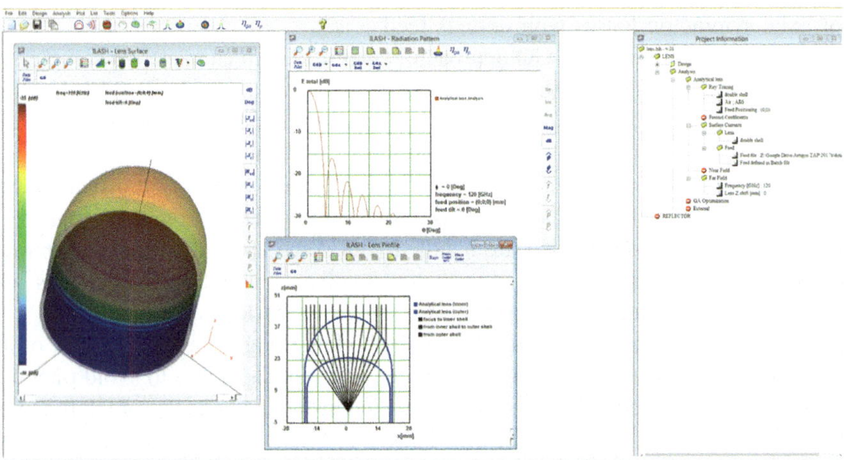

Fig. 1.5 Screenshot of ILASH software tool [43]

target for optimization, thus ensuring that all the generated lenses are necessarily viable and none needs to be eliminated a priori. This favors convergence.

In order to quantify the merit of a lens in the optimization procedure, versatile cost functions were defined combining different target conditions like: far-field amplitude template, maximum power transfer, main beam power transfer, Gaussicity, phase center position or beam scanning. For broadband and for scanning lens design, multi-frequency and/or multi-feed position optimization runs were required.

A software tool was developed ILASH [43] that implements the previously described methods for the design, analysis, and optimization of axial-symmetric multi-shell lens antennas with homogenous refractive index. Its interface, presented in Fig. 1.5, provides the user with a simple way to interact with the kernel, to generate and manipulate lens design data, to fully characterize the lens performance and to export and import results. It is also possible to monitor in real time several aspects of the optimization process like the cost function value, lens parameter evolution, and convergence.

1.3 Lens Antenna Implementation

1.3.1 Feeds

For any lens output specification, at least two items need to be carefully addressed prior to the lens synthesis process: the lens feeding structure and the lens dielectric materials.

Like the feeds for reflectors, lens feeds have specific requirements in terms of radiation pattern, polarization, and phase center depending on the application. The shape of an integrated lens is tailored to obtain a prescribed output radiation pattern, and thus it obviously depends on the feed radiation pattern inside the lens body. Three main technologies are the most common for the lens integrated feeds: waveguide, coaxial, and planar printed feeds.

Figure 1.6a shows one example of a dedicated V-band (50–75 GHz) waveguide feed [32] that is used in Sect. 1.4. Near its open end, the waveguide is loaded with a high permittivity dielectric wedge and tapered down to a narrow 1.4 mm square aperture designed to present a good power transfer to the lens. The square aperture produces a circularly symmetric radiation pattern inside the lens body, equivalent to a printed double-slot radiation pattern in the same conditions [6], with linear polarization and about 8 dBi directivity (depending on lens material permittivity). This directivity is adequate to illuminate most part of the useable aperture from our shaped integrated lenses. The reflection coefficient at the waveguide port is lower than −10 dB.

However, some lenses have much wider bandwidth than this V-band waveguide feed. A similar feed was fabricated for the Q-band (30–50 GHz) with the same external dimensions, Fig. 1.6b, producing exactly the same symmetric radiation pattern and the same directivity [32] to allow testing wideband lenses by separate sub-bands.

The outer wall of the developed waveguide feeds is threaded and screwed into a matching hole opened at an extended part of the lens base. This strategy enables

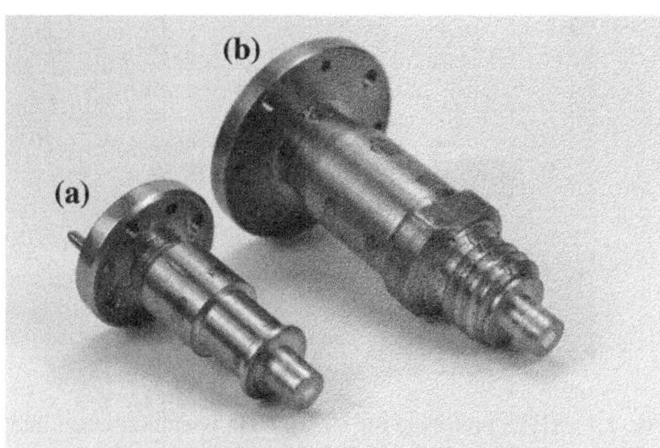

Fig. 1.6 Photograph of the fabricated waveguide feeds [32]: **a** for 62.5 GHz operation; **b** for 43 GHz operation

simple and reliable exchange of different test feeds in the same lens, or the reuse of the same test feed in different lenses without compromising alignment and measurement repeatability in lab evaluation tests.

However, an actual ultra-wideband low profile feed is also required for the end-application, compatible with simple integration of ultra-wideband bolometer detectors used, for example, in radio astronomy applications [46, 47]. Planar antenna structures printed directly at the lens base [48, 49] are the preferred solution for submillimeter wave applications where waveguides are difficult to manufacture. However, once the feed is printed or glued to the lens, it is not possible to reuse it in a different lens, making this type of feeds less attractive for laboratory systematic testing. A very common integrated lens feed is the double-slot [6], however its bandwidth is limited. Wideband self-complementary log-periodic printed antennas have been used as integrated lens feeds in the literature, but its polarization is not sufficiently stable versus frequency [48].

A dedicated broadband printed antenna was developed, which is especially adequate for integration with bolometers or mixer diodes. It is based on a new crossed exponentially tapered slot configuration combined with an intersecting square slot, which is onwards referred to as the XETS antenna. A fabricated prototype for 30–70 GHz operation attached to a MACOR™ lens [33] is shown in Fig. 1.7. The feed is developed in two layers of metallization [33]. The slot layer is in direct contact with the lens base; the other layer (Fig. 1.7b) contains a replica of two opposing central petals from the front face, printed at the same position in order to provide a capacitive coupling with the front layer. These petals are used to mount a Schottky diode (used as a mixer). The Intermediate Frequency (IF) retrieving circuit with soldering pads for the coaxial cable is also shown in Fig. 1.7b. Depending on the lens permittivity, more than 60% impedance bandwidth can be achieved with reasonably stable radiation pattern and almost pure linear polarization across the whole bandwidth; cross-polarization level is in the order of −15 dB.

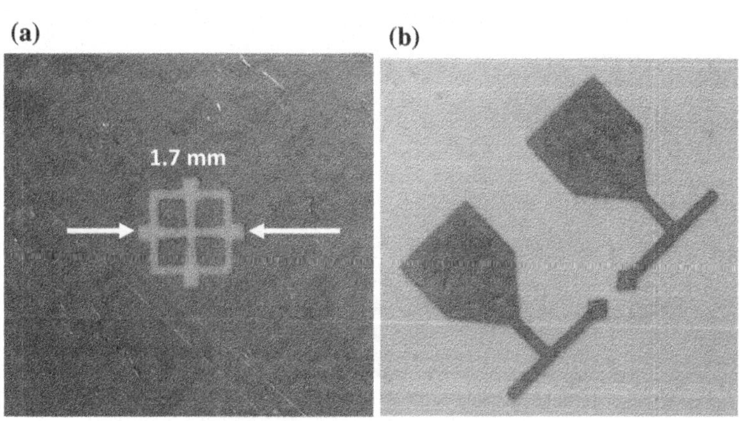

Fig. 1.7 XETS prototype for 30–70 GHz operation when integrated with a MACOR™ lens [33]: **a** photograph of the front face; **b** back face

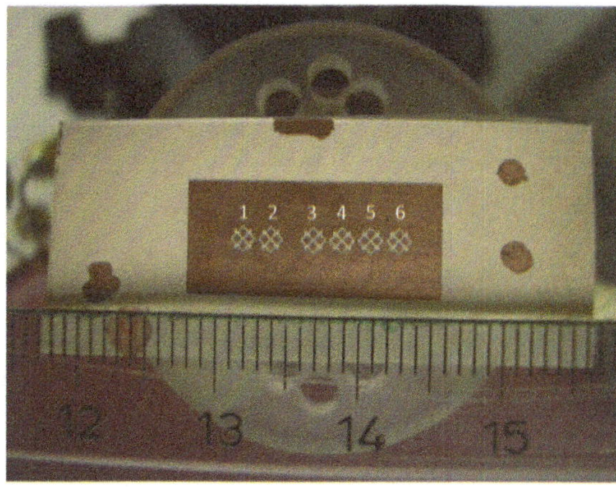

Fig. 1.8 Front side of the multiple XETS prototype [50]

This is a clear advantage over self-complementary log-periodic printed feeds [49]. The phase center of this radiating element is very stable across the whole frequency band and lies in the antenna plane.

These broadband XETS elements can be closely packed for scanning lens applications, see Fig. 1.8. The multiple XETS elements are aligned along the feed H-Plane [50]. Simulations have shown that a separation between XETS centers of only 2 mm in this plane ensures better than 18 dB isolation between adjacent feeds across the operating bandwidth, despite the presence of the very close IF circuits on the backface.

1.3.2 Lens Materials

The second issue to be addressed prior to the lens design is the selection and characterization of the dielectric material. Common materials are PTFE, polystyrene, fused quartz, ceramics, silicon, or alumina. In most cases, the electromagnetic characteristics of candidate materials are specified by the manufacturer for microwave frequencies only, with tolerances that are incompatible with an accurate lens design in the most stringent applications. Thus, the dielectric complex permittivity must be evaluated in-house to ensure a good agreement between predictions and measurements. Besides the electrical properties, the homogeneity and the isotropy of the raw material must also be checked.

Two most used methods for permittivity measurement are: the waveguide method [51] and the open resonator Fabry–Perot method [52, 53]. The waveguide method is very convenient for low to moderate permittivity (up to 5) and loss tangent (in the 10^{-4} to 10^{-2} range). Parallelepiped samples of the dielectric material are precisely cut to fill completely the cross section of the rectangular waveguide short-circuited sample holder, see Fig. 1.9. Under single-mode TE_{10} conditions, the

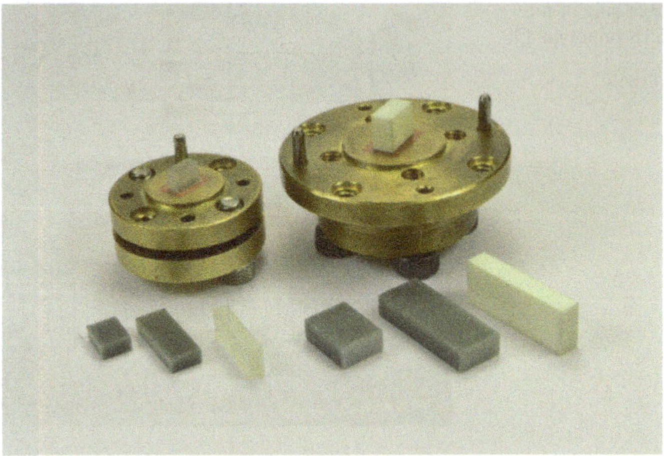

Fig. 1.9 Example of waveguide short-circuited sample holder for V-band and a few dielectric samples [51]

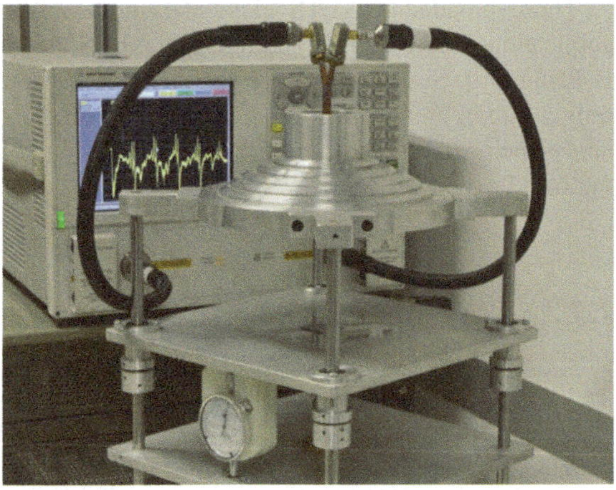

Fig. 1.10 Fabry-Perot open resonator for V-band complex permittivity measurements [53]

complex permittivity is calculated from the measured s_{11} frequency responses for the (i) dielectric filled and (ii) empty sample holder. The method is quite accurate, provided that the air gaps between the sample and waveguide walls are less than 50 μm in the V-band. Because the required samples are very small (typically $3.9 \times 1.9 \times 5.0\ \text{mm}^3$ in V-band), material non-homogeneity or anisotropy can be evaluated by cutting samples from different parts and different orientations of the raw material block.

Excitation of higher order modes in the dielectric filled sample holder for relative permittivity values typically larger than 5 tends to introduce significant error in the complex permittivity determination, especially in the loss tangent. The Fabry–Pérot open resonator method [52] does not have this limitation. In Fig. 1.10, a

Table 1.1 Measured permittivity values of different materials at 60 GHz

Material	Dielectric constant	Loss tangent
ABS-M30 (three-dimensional printed)	2.48	0.008
Acrylic glass	2.5	0.0118
Alumina	9.3	0.0013
Fused quartz	3.8	0.0015
MACOR	5.5	0.0118
PLA (three-dimensional printed)	2.85	0.014
Polyethylene	2.3	0.0003
Polypropylene	2.2	0.0005
Polystyrene (Rexolite)	2.5	0.0004
Teflon	2.2	0.0002

plane-concave configuration was selected, since it involves the fabrication of only one spherical mirror while allowing precise positioning of the dielectric samples directly on the planar mirror, without any perturbation from sample holders. The resonator dimensions were chosen to produce at the planar mirror a 6 mm beam waist fundamental Gaussian mode with well-defined linear polarization. The quality factor of the empty Fabry–Perot resonator is 10^6. The complex permittivity of the sample dielectric material disk is calculated from the S_{21} frequency response measured between the feeding and pick-up waveguide probes [53] for the empty and for the dielectric loaded resonator. The measurement is sensitive to polarization, and thus, by rotating the sample disk about the cavity z-axis, it is possible to clearly identify possible anisotropies of the material sample and calculate the complex permittivity along the anisotropy axes [53].

The measured permittivity for typical dielectrics is shown in Table 1.1. High permittivity values of the lens material allow more flexibility to comply with beam shaping requirements and better power coupling from bidirectional planar feeds to the lens. But this comes at the cost of higher lens internal reflection loss detuning of the feed input impedance [54]. Matching layers [10] or multiple shell lenses can cope with these opposite dependencies on material permittivity.

1.3.3 Fabrication

Besides the electromagnetic properties, the choice of the dielectric material may be limited by the selected process for lens fabrication. Several techniques exist to manufacture a dielectric lens antenna. The most common approaches use either computer numerical control milling machine (CNC), moulding or three-dimensional additive manufacturing. Milling is a common approach, but it may not be appropriate for hard materials, typically with high permittivity values like alumina.

Fig. 1.11 Lens fabrication by CNC milling machine technique

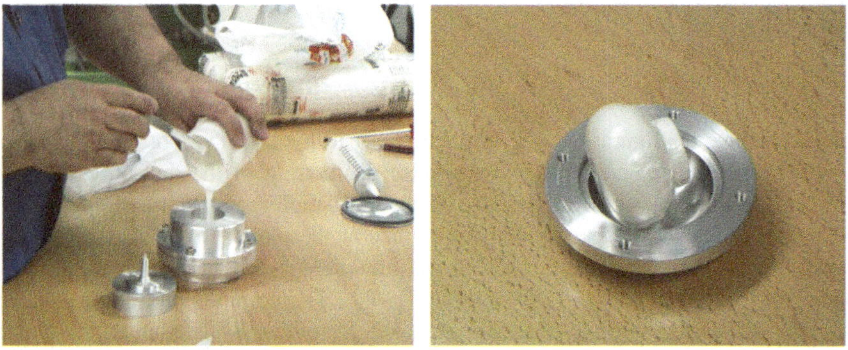

Fig. 1.12 Lens fabrication by molding technique

CNC enables to produce physical objects from its digitized three-dimensional description, by automatically excavating the object shape out from a block of raw material, Fig. 1.11. This type of fabrication process is especially appropriate for antennas with complex shapes, tight dimensional tolerances and good surface finish as happens for instance with shaped lens antennas at millimeter waves. It is more indicated for small series production or for lab prototyping as opposed for instance to the injection molding process, which is cost effective for mass production. CNC milling may, however, be used for mold production for the latter technology.

There are no special limitations on the materials that can be used for CNC milling, except that they must be machineable. Some examples of commercially available machineable dielectrics are listed in Table 1.1. The appropriate cutting tool, its rotation speed, the cutting speed, feed rate, and cutting depth must be appropriate for the selected material. The specific values for these cutting parameters can be found in manufacturers' material data sheets. Depending on the dielectric material and CNC milling machine characteristics, 50 μm fabrication accuracies can be achieved. For some type of materials it is possible to fabricate the lens using a moulding technique, Fig. 1.12. In this case, a mold with the shape of the lens has to be manufactured where the dielectric, in liquid phase, is poured or injected into. Usually, the assembly must be left for a few hours before the mold can be removed. Depending on the material, the lens dimensions may change marginally after material curing; this has to be considered when designing the mold. Moulding is generally considered a cost-effective technique for mass-production, but it is of limited advantage for lens lab prototyping due to the mold manufacturing cost. Furthermore, it is not easy to avoid trapped gas bubbles or to control the homogeneity of the mixture as it becomes solid, without an appropriate facility.

In the case of multilayer lenses particular attention has to be given to the matching and alignment of the layers produced by milling or molding technologies. Air gaps between the layers should be avoided when assembling the lens. The presence of air gaps as small as 0.03 wavelengths in multilayer lenses can reduce the antenna gain and increase side lobe level of the radiation pattern. This effect is more intense with the increase of the number of layers [55] or with higher permittivity materials [56]. A glue can be used to mitigate the effect of the air gaps as long as it has approximately the same dielectric constant of one of the adjacent layers.

Three-dimensional printing is becoming an attractive technique to fabricate elaborate dielectric prototypes at relatively low cost. As with CNC, it starts with a three-dimensional digitized model of the object, but the rendering is based on an additive process where successive layers of a material are laid down dot by dot to construct the desired shape. Various materials can be considered, but thermoplastic based three-dimensional printers are the most popular. Although the commonly used materials tend to have moderately high losses and the fabrication tolerance of low-end printers is presently of the order of 200 μm, the viability of a three-dimensional printed lens antenna made of ABS ($\tan\,\delta = 0.008$) for short-range indoor wireless link at 60 GHz has been recently demonstrated [57, 58]. A similar technique can be used to manufacture lenses using higher permittivity lower loss materials like Alumina [11].

1.4 Lens Design Examples

This section presents selected lens examples for different applications. These and other examples are discussed in full detail in the indicated references.

1.4.1 Constant Flux and Flat-Top ILAs for mm-Wave WLAN

The cellular Mobile Broadband Systems (MBS) concept was introduced in 1992 and developed throughout the 90s [59]. It was intended to transport, in a transparent way, Asynchronous Transfer Mode cells (ATM) over the air interface at data rates up to 155 Mbit/s, withstanding mobile velocities up to 100 km/h, while maintaining

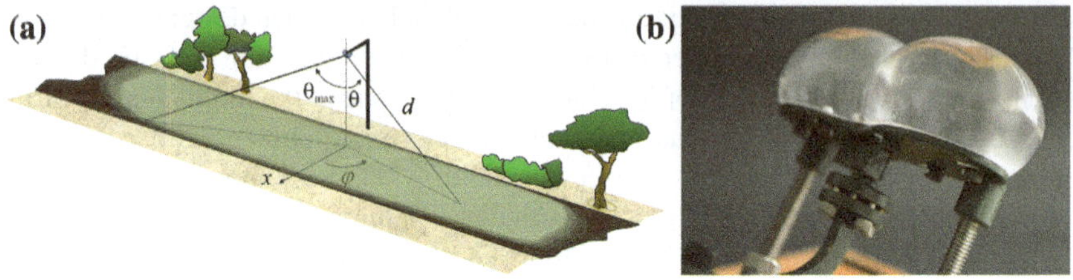

Fig. 1.13 **a** Geometry of a street sectored cell; **b** photograph of a shaped lens antenna for a constant flux illumination of the street at 60 GHz [60]

the quality of service of the fixed network. Provisional sub-bands were allocated for these systems at 39.5–40.5, 42.5–43.5 GHz, and 62–63, 65–66 GHz.

At the base station, a shaped beam was required to approximate a constant flux illumination of the cell to favor the link budget, with reasonably sharp drop at the cell edge to control cell spillover (see Fig. 1.13a). This characteristic corresponds to cosec^2 type gain contribution from the antennas to the link budget within the cell, which ideally compensates for the free space loss at each observation direction. An acrylic three-dimensional lens was designed using a modification of GO closed-form expressions to produce constant flux illumination across a long street section when hanging from a lamp post located 10 m above the street level at the center of the cell [60]. The lens was fed by the embedded open-end aperture of a dielectric loaded rectangular waveguide operating in the TE_{10} mode. The lens prototype is shown in Fig. 1.13b [60]. The dimensions are 10 cm × 4 cm × 2 cm and the lens produces a bidirectional radiation pattern with 13.2 dBi gain at 62.5 GHz at the maximum distance direction (θ_{max}, $\varphi = \pm 90°$), with linear polarization. The lens was successfully used in an outdoor system demonstrator operating at 32 Mbit/s gross bit rate. In a follow-up project, these lenses were redesigned for the 40 GHz band with the same specification but using a lower loss dielectric material, polystyrene. A similar radiation pattern was obtained in this case, but with slightly higher gain of 14 dBi. The received power over the street cell showed that by using a single-shaped lens antenna at the center of the cell it is possible to produce a rectangular-type of coverage above the threshold level, almost confined to the cell limits [61]. No similar type of performance has been presented before for any antenna. In a different configuration, a three-dimensional lens was developed to produce a constant flux illumination with sharp square boundaries, appropriate for indoor room coverage with controlled wall illumination [62], Fig. 1.14a. The lens was made of acrylic with the size 84 mm × 32 mm (diagonal × depth) for operation at 62.5 GHz. The lens was fed by the circularly polarized TE_{11} mode of a circular waveguide immersed in the lens body and mounted near the ceiling. Figure 1.14b shows the received power distribution beneath the lens, measured while the receiving lens antenna, with a flat-topped radiation pattern, traveled on a kart to scan a 4 m × 8 m floor area. The difference in height between base station and the mobile terminal antenna was 0.5 m. The figure shows a constant received power in the illuminated region and a sharp drop at

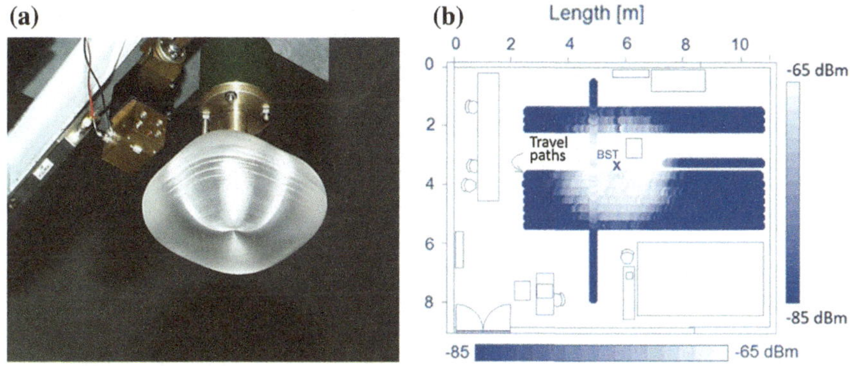

Fig. 1.14 **a** Dielectric lens antenna for a square cell fed by a circular waveguide; **b** received power distribution measured in the laboratory at 62.5 GHz [62]

the cell edge as required. The $\sec^2$ radiation pattern has the interesting property that the cell dimensions can be scaled just by changing the height difference between the fixed and the mobile antennas. This topic is further developed in reference [62], where the calculated and measured radiation patterns, and also the resulting channel time dispersion characteristic, are given.

1.4.2 Broadband ILAs as a Feed for Reflectors

The most common reflector feeds are based on horn antennas. Despite many variants that horns may have, they all share the same feature of aperture antennas, which is a radiation pattern that increases its directivity with the frequency. When large bandwidths are involved, this may represent an important reduction of the reflector aperture illumination efficiency. This problem is often circumvented by splitting the wideband into multiple sub-bands, each with its own feed horn; but then a complex set-up is required to minimize aberrations due to off-axis positioning of the multiple horns.

A challenging specification for a reflector feed is the ability to maintain a constant radiation pattern beam width with stable phase center over a 1:3 bandwidth. A double-shell shaped dielectric lens can be designed to meet such specification [15]. The lens is intended to feed a 90° offset reflector, Fig. 1.15a. The lens output beam produces a virtual focus located far below the lens and close to the corresponding reflector focal point. A 20-λ diameter lens prototype was designed and fabricated using MACOR™ and acrylic (Fig. 1.15b). The optical transparency of the acrylic reveals the inner shell, which appears distorted due to refraction effects. The lens is fed by the waveguide device described in Sect. 1.3.1 (Fig. 1.6).

The measured lens radiation patterns are shown in Fig. 1.16a for two distinct frequencies in the Q- and V-band (40 and 62.5 GHz). The measured lens radiation

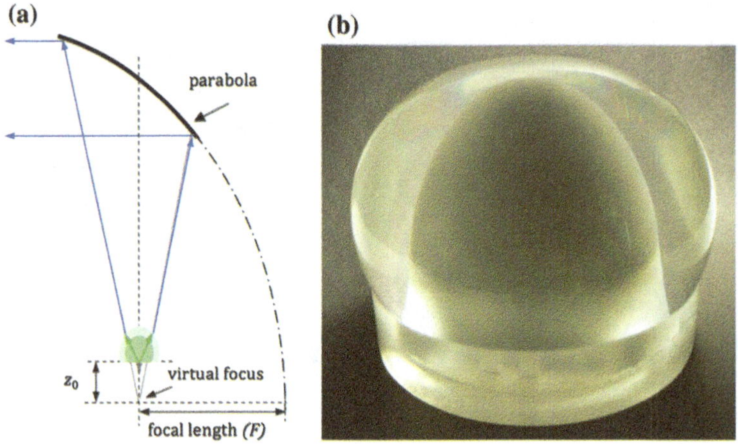

Fig. 1.15 **a** Problem geometry; **b** fabricated MACOR™/acrylic lens prototype [15]

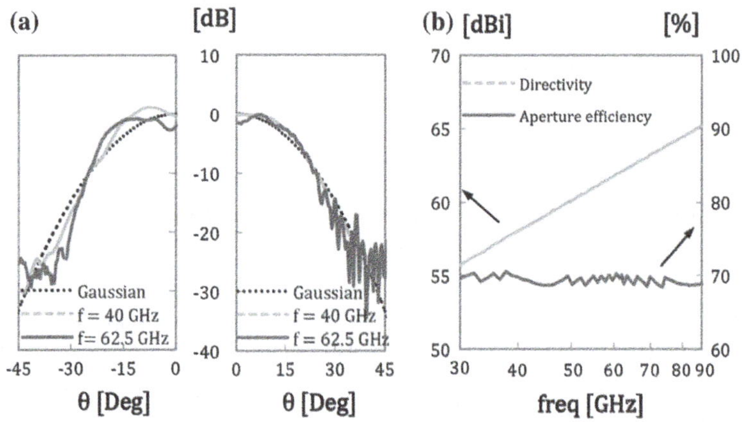

Fig. 1.16 **a** Measured radiation patterns of the lens at 40 and 62.5 GHz; **b** calculated directivity and aperture efficiency of the reflector with frequency plotted on log scale [15]

pattern agrees very well with the Gaussian template with $\alpha_0 = 23°$ Gaussian width (dashed curve). It is possible to see that the shape and beam width of the lens radiation pattern are identical in both bands as desired. The same behavior was demonstrated by full-wave simulation between 30 and 90 GHz [15]. The lens output beam Gaussicity is better than 94% over the band. The measured phase center position of the lens agrees with the value imposed during the design ($z_0 = 68.7$ mm) and it is identical in both frequencies. Full-wave simulations confirm phase center stability over the 1:3 bandwidth. The performance of the lens-reflector assembly was simulated using the ILASH software [43]. Figure 1.16b shows that the aperture efficiency is practically constant versus frequency as

Fig. 1.17 Fabricated shaped double-shell lens prototype [63]

required and, in agreement with that property, the reflector directivity in dB increases almost linearly with logarithm of the frequency.

In a second example, a shaped double-shell dielectric lens (Fig. 1.17) has been designed as a primary feed for a multi-beam single reflector system operating in the Ka-band satellite up-link (30 GHz) and down-link (20 GHz) bands, complying with gain and edge-of-coverage directivity (EoC) requirements [63]. An assembly of dual-band printed feeds is integrated at the base of a single lens, each feed producing a virtual focus far behind the lens base and coincident with the reflector focal arch.

The lens is axial-symmetric, formed by two embedded shells of Polystyrene (inner shell) and MACOR^{TM} (outer shell). The used double-shell lens approach, instead of a single-material lens, allows an extra degree of freedom to accommodate an aberration mitigation condition. In fact, two refracting surfaces are required to impose two design conditions, one associated with the phase center and the other with low aberration. Further, it was verified that selecting a higher permittivity for the outer shell material favors lower beam tilt for the off-axis feeds (lower spillover losses) while allowing more space between adjacent feeds at the lens base. Therefore, this lens based primary feed system is proposed as a low-complexity solution to enable fitting more beams per solid angle than conventional single-feed-per-beam systems based on a cluster of focal-plane horns. The proof of concept lens prototype with 87 mm diameter and 62 mm height, fed by a linear arrangement of five dual-band printed feeds was fabricated and tested at the Ka-band [63]. The lens measured radiation patterns were post-processed to evaluate the combined performance of the lens with an offset F/D $\approx$ 1 reflector system designed for 45 dBi EoC directivity, see Fig. 1.18. It is shown that it duplicates the reflector aperture efficiency compared to horn fed systems with same feed separation.

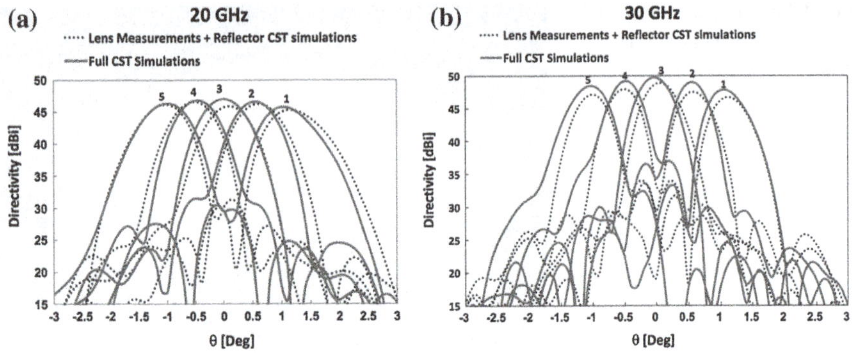

Fig. 1.18 Radiation patterns of an F/D = 0.95 reflector obtained by illuminating a simulated CST reflector with the post-processed measured lens radiation patterns and lens simulated radiation patterns: **a** at 20 GHz; **b** at 30 GHz [63]

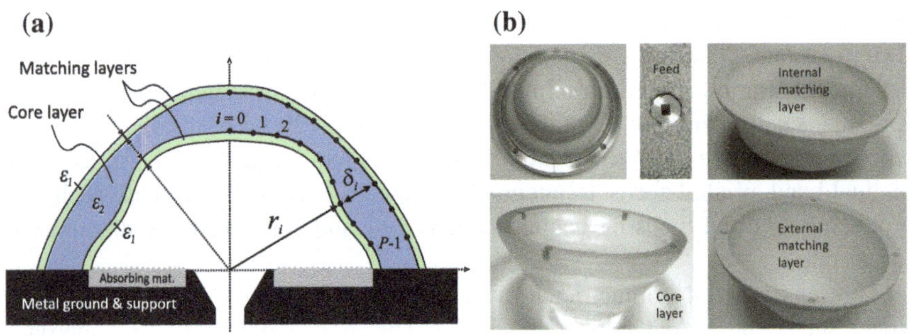

Fig. 1.19 Multi-shell lens-like dome: **a** topology in the E-plane cross-section, **b** prototype

1.4.3 Broadband Lens-Like Dome Antenna

Another example of a broadband ILA is presented in Fig. 1.19. Unlike earlier examples, this antenna comprised a multilayer lens-like dome, whose profile was optimized using a genetic algorithm aiming at producing a stable Gaussian-type radiation pattern across the entire V-band (50–75 GHz). The desired functionality was achieved thanks to the simultaneous optimization of the shape for both internal and external surfaces of the dome [16]. The backreflection from both air-dielectric interfaces was minimized using quarter-wavelength matching layers. The dome optimization method was similar to the one represented in Fig. 1.3 (loop b) with a number of cubic spline nodes uniformly distributed along both surfaces (Fig. 1.19a).

The antenna was fed with E-plane sectoral horn having a rectangular aperture and a nearly symmetrical radiation pattern that justified the use of a rotationally symmetrical dome.

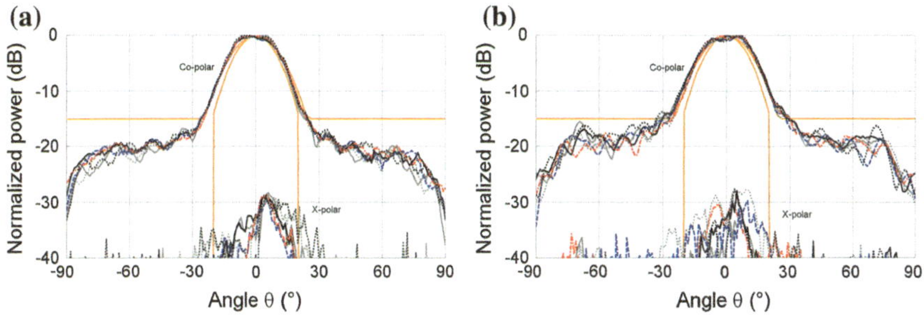

Fig. 1.20 Measured radiation patterns of the optimized LLDA with two matching layers: **a** E-plane, **b** H-plane. The families of 6 curves corresponds to different frequencies: ——— 50 GHz; ·········· 55 GHz; ——— 60 GHz; -------- 65 GHz; ▪▪▪▪▪▪ 70 GHz; ·········· 75 GHz (borrowed from [16] with permission from IEEE)

The dome core layer was made of Rexolite ($\varepsilon = 2.53$), with two matching layers made of a low-permittivity foam ($\varepsilon = 1.59$). All counterparts of the shaped dome have been fabrication via micro-machining (Fig. 1.19b). The feasibility of a single-step fabrication of each shell and their assembly was guaranteed by the constrains to the dome feasible shape that have been implemented in the optimization routine.

The measured radiation patterns in both planes presented in Fig. 1.20 for five frequency points reveal successful achievement of the optimization goal. The main advantages of the lens-like dome antenna include wideband performance and light weight, which may be important for on-board applications. Note that a certain constrain on the shape of external surface can be easily added in the optimization routine. For instance, it may be optimized to provide optimal aerodynamic characteristics. Possible weak points include complexity of mechanical assembly of the multilayer structure and rigidity of the foam matching layers, which may prevent scaling this design to upper mm-wave bands.

1.4.4 Multi-beam ILAs

The first example refers to an integrated multi-beam lens antenna intended to produce 3 dB beamwidths of the order of 6° at the 60 GHz band, with high Gaussicity, gain scan loss lower than 1 dB over ±20° scan angle and complying with a maximum diameter of 70 mm [64]. The challenge was to minimize internal reflections and to ensure a reasonably stable Gaussian beam shape across the scanned angular interval. A double-shell configuration was adopted to allow imposing both a beam collimation condition and the Abbe sine condition to minimize aberration of the output beam. The selected lens materials were the MACOR^{TM}/acrylic combination (permittivity 5.5/2.53), and the feed was the especially designed waveguide aperture described in Sect. 1.3.1 for 62.5 GHz

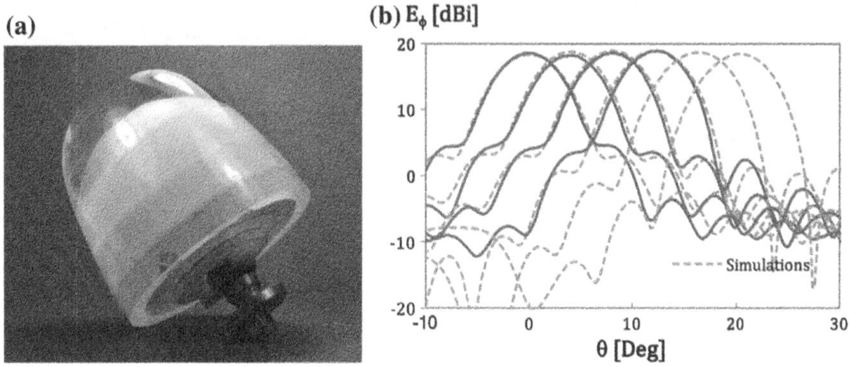

Fig. 1.21 **a** Fabricated MACOR™/acrylic lens prototype, showing the attached 60 GHz band waveguide feed; **b** simulated and measured gain radiation patterns at 62.5 GHz, in the H-plane, with feed at 0, 1.1, 2.2, 3.3, 4.4, and 5.5 mm from the lens axis [64]

operation. Figure 1.21a shows the lens antenna prototype which was designed using ILASH [43]. The lens dimensions are 60 mm diameter at the base and 37 mm height; the outer shell material is transparent, and thus it reveals the inner shell, which appears distorted due to refraction effects.

Figure 1.21b shows the measured and simulated radiation pattern results for different feed positions at 1.1 mm steps along the lens base in the H-plane. This produces overlapping of consecutive beams near −1.5 dB level with respect to maximum. Beam overlap at −3 dB occurs for wider separation between consecutive feeds, compatible with the used feed aperture size (not shown). The figure shows excellent agreement between ILASH predictions and measurements and confirms the effectiveness of the proposed double-shell lens concept regarding beam scan linearity, internal reflections, beam shape and Gaussicity preservation with extremely low scan loss up to 20° elevation. The achieved gain was of the order of 18–19 dBi and the scan loss better than 1 dB. As a consequence of the GO-based lens design, the scan angle dependence on the feed off-axis position was practically independent of frequency and the beam Gaussicity was better than 95%.

The second example of a multi-beam ILA concerns an antenna with a reconfigurable FoV (Fig. 1.22). Such a functionality may be of particular interest of automotive radar systems requiring switching between short-range (SR) and long-range (LR) regimes. The presented antenna relies on the concept proposed in [35] but, instead of using a standard extended cylindrical lens, it is based on a shaped lens, whose profile was optimized in three cut-planes to provide a desired beam width in both horizontal and elevation planes and both operational regimes (Fig. 1.22). The lens was fabrication via micro-machining (Fig. 1.11c) in Rexolite and excited by a five-element non-equally-spaced linear array (in H-plane) of aperture-coupled microstrip patch antennas. The array is symmetric (the central patch is located at the origin of the coordinate system) and all patches are excited in phase. Modification of the lens surface illumination laws is provided by a selective

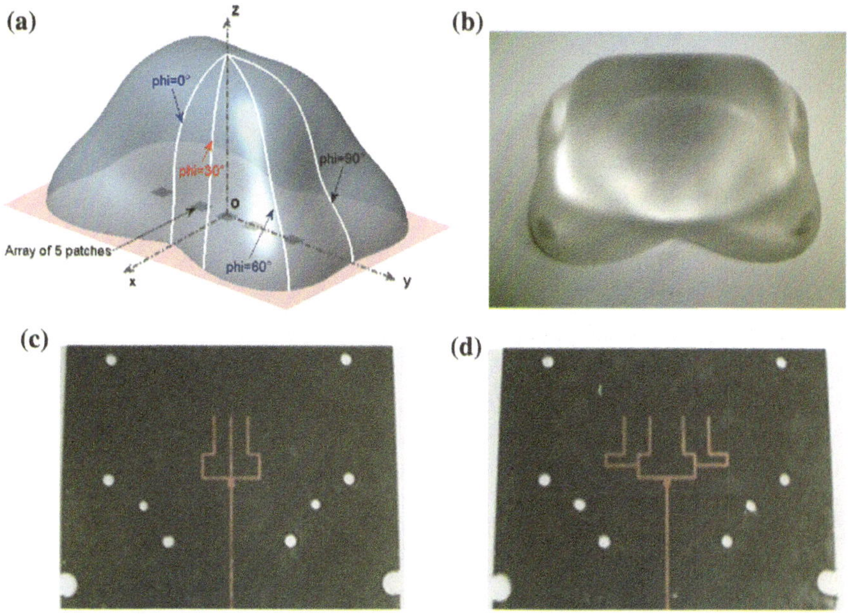

Fig. 1.22 Shape ILA with a reconfigurable field of view: **a** CAD model, **b** shaped lens micromachined in Rexolite material, **c**, **d** back view of the BFM networks used to excite the primary feed array for SR and LR regimes, respectively

excitation of the array elements. To simplify the architecture of the beamforming network (BFN), the excitation amplitude of the array elements was limited to two states ON or OFF. The lens shape was optimized to transform a sub-array pattern (corresponding to a specific state of the BFN) into a prescribed far-field beam. The design goal is to enable switching between two beams with half power beam widths of $10° \times 10°$ and $10° \times 40°$, respectively.

To facilitate fabrication, the antenna was designed to operate at 24 GHz with two operational regimes implemented using two different BFN configurations comprising three central ones for the SR regime (Fig. 1.22c) and four offset ones for the LR regime (Fig. 1.22d), respectively. The simulated and measured patterns for both regimes are represented in Fig. 1.23 and Fig. 1.24, respectively. The slight asymmetry of the measured curves was attributed to the manual alignment tolerances in the assembly of the multilayer BFN. Nevertheless, good agreement between the simulated (FDTD) and measured curves was obtained with the desired beam width values achieved in both regimes. The other radiation characteristics of the prototype operating in the wide and narrow-beam regimes were determined to be the following: measured gain is 12.6 and 16.6 dB with the operational bandwidth $S_{11} < -10$ dB is 14.6% (22.3–25.8 GHz) and 5% (24.7–25.5 GHz),

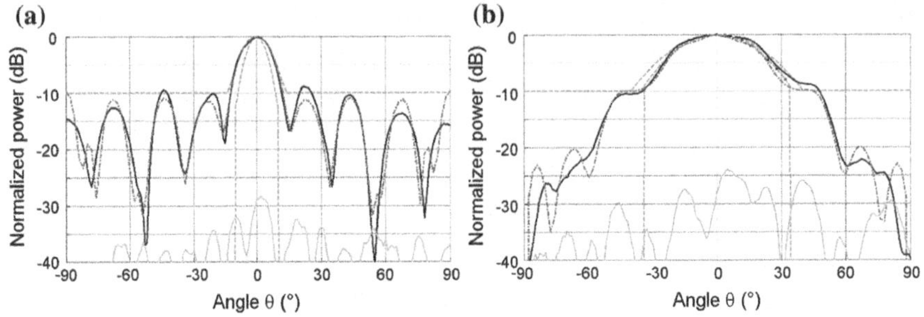

Fig. 1.23 Radiation patterns of the ILA operating at 24 GHz in the wide-beam regime: **a** elevation E-plane, **b** azimuth H-plane. Co-polar measured: ——, co-polar simulated: ··········, X-polar measured: ——, template: -----

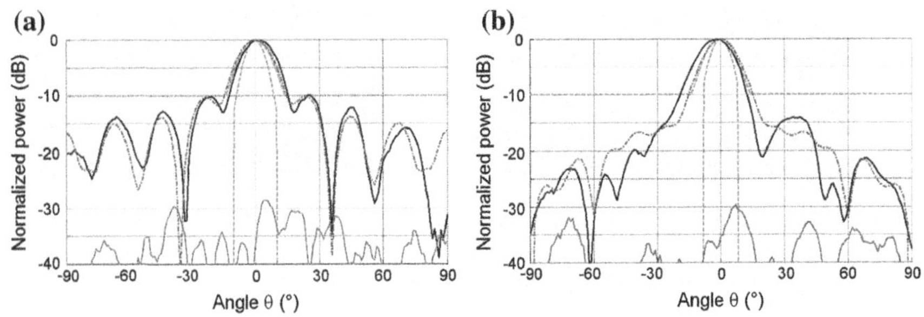

Fig. 1.24 Same as in Fig. 1.23 for the ILA operating in the narrow-beam regime

respectively. The matching difficulties in the narrow-beam regime were associated with fabrication intolerances, and thus, in principle, can be avoided. The radiation patterns fit well the target template within the entire operational bandwidth. The realized antenna radiation efficiency was estimated as 58 and 53% for the wide- and narrow-beam regimes, respectively. A quite high insertion loss was attributed to the material losses associated with the lens and BFN.

This example is reported as a proof of concept for ILAs with a highly reconfigurable FoV, which can be achieved thanks to a combination of a reconfigurable primary feed array and shaped dielectric lens. This concept can be generalized for other frequency bands and types of primary feeds. It can also be implemented to support a double-band functionality with the same shaped lens enabling two different beamforming functions for different excitation conditions. Such antennas may find application in a number of application domains including intelligent transport.

1.4.5 Beam-Steering ILAs

Beam-steering ILAs are attractive solutions for automotive radars and point-to-point (or point-to-multipoint) communication systems. An established solution for such antennas comprises an extended hemielliptical (or hemispherical lens) fed by a focal array (Fig. 1.2). The weak point of this design is a rapid drop of the directivity value for scan angles beyond about 20° [6, 20]. The following examples illustrate three possible alternatives to overcome this limitation.

The first example is a mechanical scanning ILA proposed in Ref. [65] for Low Earth Orbit satellite data link communications at 26 GHz. The lens was required to produce a sector beam with shaping in elevation, and to enable a simple mechanical azimuth scanning. The target gain versus elevation angle θ is given approximately by $\sec(k\theta)$. With an appropriate choice of the k shaping parameter, this elevation radiation pattern compensates path loss attenuation taking into account the Earth curvature.

The designed shaped lens is axial-symmetric but a spherical air cavity is excavated through its base at a calculated offset distance with respect to the lens symmetry axis, Fig. 1.25a. The feed aperture is at the center of this spherical air cavity. The asymmetry introduced by the feed position produces an adequate directive shaped beam. This is an extremely simple solution where the feed is fixed and the lens rotates about the feed axis to provide the required beam scanning. The feed is not in physical contact with the lens, thus avoiding the need for fault prone rotary joints in long continuous usage. The assembly is very compact and lightweight (lens diameter is less than 5 wavelengths and height is less than 3 wavelengths) complying with mass and volume restrictions for satellites.

The same concept of a moving lens with fixed feed was extended for azimuth plus elevation mechanical scanning in a different application context. Wireless indoor transmission of uncompressed HD video signal between a video device like

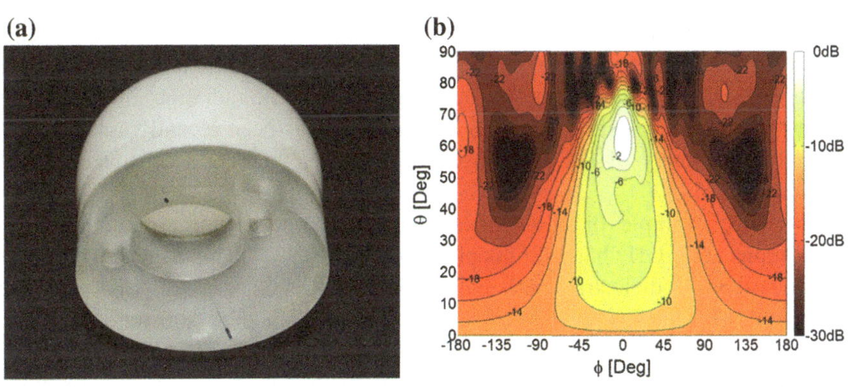

Fig. 1.25 **a** MACOR^{TM} axial-symmetric 50 mm diameter shaped lens; **b** radiation performance (in polar coordinates) of the lens computed with the GO/PO method [65]

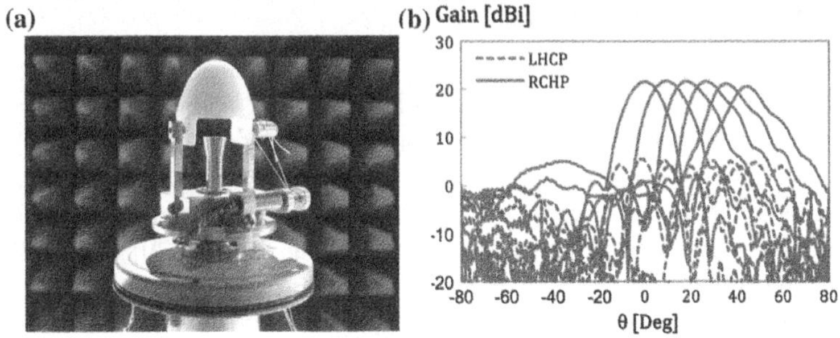

Fig. 1.26 **a** Photograph of the manufactured polyethylene lens plus horn feed; **b** measured radiation patterns of the lens antenna for several lens tilt angles [24]

a camcorder and a HD TV Display may soon become a requirement for mass market home applications. It proposes to use the unlicensed spectrum from 57 to 66 GHz for such applications. For mass consumer applications, the radio link must use inexpensive low power sources and hence high gain antennas (>20 dBi) are required to favor the link budget. The narrow beam needs to be steered to allow for the user mobility. A new configuration of a mechanical steerable beam antenna has been developed where a dielectric lens pivots in front of a single stationary moderate gain feed (see Fig. 1.26a) [24]. Again the feed is not in physical contact with the lens. As discussed previously, this eliminates the need for rotary joints and the associated drawbacks. The designed lens is made of polyethylene and is such that both the input and output lens surfaces are shaped to accommodate a beam collimation condition and a scan angle maximization condition. The lens output beam is aligned with the lens axis for all lens orientations within the operation cone. Thus, pivoting the lens around two of its main axis allows steering the beam in elevation and azimuth. The important difference with respect to conventional scanning lenses is that for all inclinations of the lens the focus is always exactly coincident with the aperture phase center. In this way, aberrations are strongly reduced allowing wider scan angles than conventional solutions. The fabricated lens assembly demonstrated a −45° to +45° elevation scan capability over full azimuth, with 21 dB gain and better than −1.1 dB gain scan loss (see Fig. 1.26b). The radiation efficiency was always above 95%. A simple system demonstrator has been recently assembled to prove the feasibility of the antenna concept, of the tracking mechanism and its algorithms [66].

The third example of a beam-steering ILA proposed for automotive radar applications at 77 GHz band [26] is presented in Fig. 1.27. This antenna comprises two plano-convex dielectric lenses having the same diameter but different radius of curvature: one attached on top of a planar feed (labeled as substrate lens) and another one placed at a certain distance above the first one (labeled as objective lens). As illustrated in Fig. 1.27a via a ray-tracing diagrams, such a topology enables one to obtain a flatter shape of the focal arc than that observed in case of a

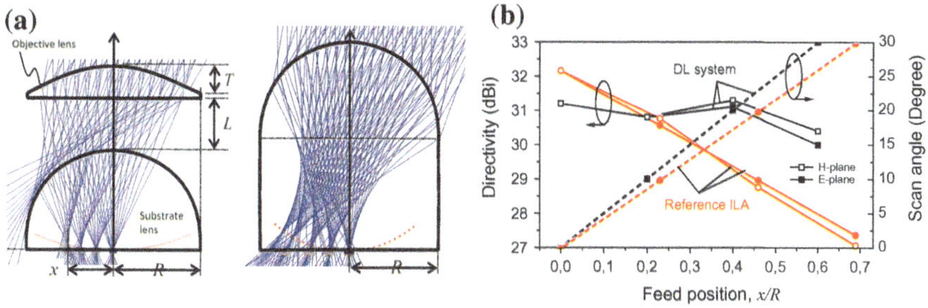

Fig. 1.27 Double-lens ILA: **a** topology and ray-tracing diagrams for the proposed DL-ILA and a standard synthesized elliptical lens considered as a reference solution, **b** simulated directivity in both principal planes and scan angle of the DL-ILA and the reference antenna versus feed position at 77 GHz

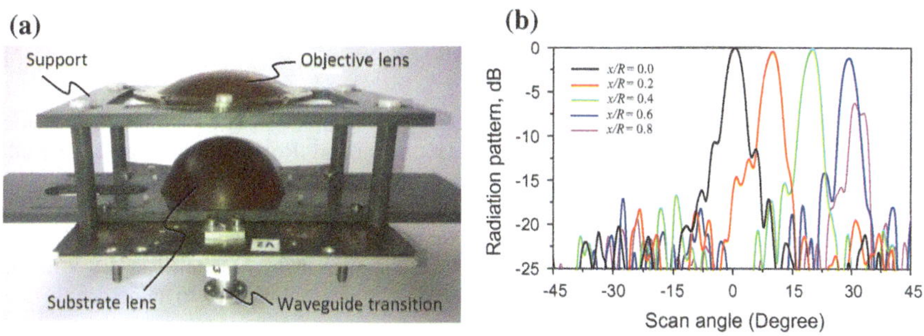

Fig. 1.28 **a** Prototype of the DL-ILA, **b** measured radiation pattern in H-plane at 77 GHz

standard synthesized elliptical lens reported in [6]. Reciprocally, a lower scan loss is achieved for off-axis feed positions (Fig. 1.27b). In this study, both ILAs have diameter $2R = 16 \times \lambda_0$, which corresponds to a maximum directivity of 34 dBi (for a uniform field distribution). In case of the double lens (DL) system, the distance between the lenses and thickness of the objective lens have been optimized aiming at achieving a minimum scan loss for the angular range of ±30°. The optimal design reported in [26] provided 1.1 and 0.9 dB scan loss for the most offset position in the E and H planes, respectively. Comparison with the reference synthesized elliptical ILA [6] antenna shows a 3.9 dB advantage in terms of the directivity scan loss at 30° for the DL system that has been achieved by the expense of 1 dB loss in terms of the broadside directivity.

The DL-ILA design concept was experimentally validated via prototyping (Fig. 1.28). The two 62.4 mm diameter lenses were micromachined in a thermoplastic polyetherimide resin (Ultem™, $\varepsilon = 3.01$). The off-axis characteristics of the antenna were measured by changing the feed position with respect to optical axis of the DL focusing system. For this purpose, a special support and positioning system was used, enabling synchronized displacement of the lenses with respect to the primary feed position. The antenna was fed by an aperture-coupled patch antenna excited through a waveguide-to-microstrip transition, coupled ($S_{11} < -10$ dB) from 74.5 to 80 GHz.

The radiation pattern of the antenna was measured at three frequency points, i.e., 75, 77, and 79 GHz. The realized scan loss in both principal planes and all frequency points does not exceed 1.2 and 1.0 dB for the E and H-planes, respectively. For brevity, only the 77-GHz pattern in presented in Fig. 1.28b.

The obtained results confirmed a strong potential of the DL concept for ILAs with a wide FoV and/or low scan loss performance. Additional advantages of the DL-ILA revealed by this example include a smaller total height of the focusing system and, consequently, reduced weight and smaller material loss. Analysis showed that the latter advantage may fully compensate the additional scattering loss associated with the objective lens. A potential weak point of DL-ILA is complexity with assembly and alignment of the DL focusing system. A possible extension of this concept may include integration of a substrate ILA with a shaped dome acting as an objective lens. This approach may help further improve FoV as well as enlarge the operational bandwidth.

Acknowledgements The authors acknowledge contributions of the former students and postdocs at IETR (France), namely: Barbara Barès, Gaël Godi, Anthony Rolland, Tinh Ngyen, and others and at IT (Portugal): Eduardo Lima, Carla Medeiros, Catarina Cruz, and others. The authors also thank Laurent Le Coq for antenna measurements. The authors acknowledge the collaboration from Vasco Fred, Jorge Farinha, and Carlos Brito for prototype construction, and António Almeida and Jorge Silva for prototype measurements. The authors also thank Maarten van der Vorst for discussions on some aspects of the lens design. The authors are also indebted to numerous students that actively contributed to this work in the framework of their theses.

References

1. M.R. Rebeiz, Millimeter-wave and terahertz integrated circuit antennas. Proc. IEEE **80**(11), 1748–1770 (1992)
2. J.V. Rudd, D.M. Mittleman, Influence of substrate-lens design in terahertz time-domain spectroscopy. J. Opt. Soc. Am. B **19**(2), 319–329 (2002)
3. C.A. Fernandes, P.O. Francês, A. Barbosa, Shaped coverage of elongated cells at millimetre waves using a dielectric lens antenna, in *European Microwave Conference (EMC), 1995, Bolonha, Italy, September 1995*, vol. 1 (1995), pp. 66–70
4. C.A. Fernandes, Shaped-beam antennas, in *Handbook of Antennas in Wireless Communications, Chapter 15*, ed. by L.C. Godara (CRC, New York, 2002)
5. C. Salema, C.A. Fernandes, R.K. Jha, *Solid Dielectric Horn Antennas* (Artech House, Boston, 1998)
6. D.F. Filipovic, S.S. Gearhart, G.M. Rebeiz, Double-slot antennas on extended hemispherical and elliptical silicon dielectric lenses. IEEE Trans. Microw. Theory Tech. **41**(10), 1738–1749 (1993)
7. D.B. Rutledge, M.S. Muha, Imaging antenna arrays. IEEE Trans. Antennas Propag. **AP-30** (4), 535–540 (1982)

8. A.V. Boriskin, G. Godi, R. Sauleau, A.I. Nosich, Small hemielliptic dielectric lens antenna analysis in 2-D: boundary integral equations versus geometrical and physical optics. IEEE Trans. Antennas Propag. **56**(2), 485–492 (2008)
9. A.V. Boriskin, R. Sauleau, Drastic influence of the half-bowtie resonances on the focusing and collimating capabilities of 2-D extended hemielliptical and hemispherical dielectric lenses. J. Opt. Soc. Am. A **27**(11), 2442–2449 (2010)
10. N.T. Nguyen, R. Sauleau, C.J. Martínez Pérez, Very broadband extended hemispherical lenses: role of matching layers for bandwidth enlargement. IEEE Trans. Antennas Propag. **57** (7), 1907–1913 (2009)
11. N.T. Nguyen, N. Delhote, M. Ettorre, D. Baillargeat, L. Le Coq, R. Sauleau, Design and characterization of 60-GHz integrated lens antennas fabricated through ceramic stereolithography. IEEE Trans. Antennas Propag. **58**(8), 2757–2762 (2010)
12. A.V. Boriskin, R. Sauleau, A.I. Nosich, Performance of hemielliptic dielectric lens antennas with optimal edge illumination. IEEE Trans. Antennas Propag. **57**(7), 2193–2198 (2009)
13. B. Barès, R. Sauleau, L. Le Coq, K. Mahdjoubi, A new accurate design method for millimeter-wave homogeneous dielectric substrate lens antennas of arbitrary shape. IEEE Trans. Antennas Propag. **53**(3), 1069–1082 (2005)
14. G. Godi, R. Sauleau, L. Le Coq, D. Thouroude, Design and optimization of three-dimensional integrated lens antennas with genetic algorithm. IEEE Trans. Antennas Propag. **55**(3), 770–775 (2007)
15. C.A. Fernandes, E.B. Lima, J.R. Costa, Broadband integrated lens for illuminating reflector antenna with constant aperture efficiency. IEEE Trans. Antennas Propag. **58**(12), 3805–3813 (2010). doi:10.1109/TAP.2010.2078463
16. N.T. Nguyen, A.V. Boriskin, A. Rolland, L. Le Coq, R. Sauleau, Shaped lens-like dome for UWB antennas with a Gaussian-like radiation pattern. IEEE Trans. Antennas Propag. **61**(4), 1658–1664 (2013)
17. N.T. Nguyen, A.V. Boriskin, A. Rolland, L. Le Coq, R. Sauleau, Size and weight reduction of integrated lens antennas using a cylindrical air cavity. IEEE Trans. Antennas Propag. **60**(12), 5993–5998 (2012)
18. L. Xue, V. Fusco, Patch fed planar dielectric slab extended hemi-elliptical lens antenna. IEEE Trans. Antennas Propag. **56**(3), 661–666 (2008)
19. D.F. Filipovic, G.P. Gauthier, S. Raman, G.M. Rebeiz, Off-axis properties of silicon and quartz dielectric lens antennas. IEEE Trans. Antennas Propag. **45**(5), 760–766 (1997)
20. X. Wu, G.V. Eleftheriades, T.E. van Deventer-Perkins, Design and characterization of single- and multiple-beam mm-wave circularly polarized substrate lens antennas for wireless communications. IEEE Trans. Microw. Theory Tech. **49**(3), 431–441 (2001)
21. A. Karttunen, J. Ala-Laurinaho, R. Sauleau, A.V. Räisänen, Extended hemispherical integrated lens antenna with feeds on a spherical surface, in *Proceedings European Conference Antennas Propagation (EuCAP), Gothenburg, Sweden* (2013), pp. 2539–2543
22. A.V. Boriskin, R. Sauleau, Numerical investigation into the design of shaped dielectric lens antennas with improved angular characteristics. Prog. Electromagn. Res. B **30**, 279–292 (2011)
23. G.C. Trichopoulos, G. Mumcu, K. Sertel, L.H. Mosbacker, P. Smith, A novel approach for improving off-axis pixel performance of Terahertz focal plane arrays. IEEE Trans. Microw. Theory Tech. **58**(7), 2014–2021 (2010)
24. J.R. Costa, E.B. Lima, C.A. Fernandes, Compact beam-steerable lens antenna for 60-GHz wireless communications. IEEE Trans. Antennas Propag. **57**(10), 2926–2933 (2009). doi:10.1109/TAP.2009.2029288
25. A.L. Peebles, A dielectric bifocal lens for multibeam antenna applications. IEEE Trans. Antennas Propag. **31**(5), 599–606 (1988)
26. N.T. Nguyen, A.V. Boriskin, L. Le Coq, R. Sauleau, Improvement of scanning performance of integrated lens antenna using a double lens focusing system. IEEE Trans. Antennas Propag. **64**(8), 3698–3702 (2016)

27. X. Wu, G.V. Eleftheriades, Two-lens and lens-fed reflector antenna systems for MM-wave wireless communications, in *Proceedings of the IEEE Antennas Propagation Symposium (APS), Salt Lake City, UT* (2000), pp. 660–663
28. B. Schoenlinner, X. Wu, J.P. Ebling, G.V. Eleftheriades, G.M. Rebeiz, Wide-scan spherical-lens antennas for automotive radars. IEEE Trans. Microw. Theory Tech. **50**(9), 2166–2175 (2002)
29. J. Thornton, S. Gregson, D. Gray, Aperture blockage and truncation in scanning lens-reflector antennas. IET Microw. Antennas Propag. **4**(7), 828–836 (2010). doi:10.1049/iet-map.2009.0283
30. Z. Sipus, et al., This book/Chapter 2
31. K.H. Jeong, J. Kim, L.P. Lee, Biologically inspired artificial compound eyes. Science **312**, 557–561 (2006)
32. C.A. Fernandes, E.B. Lima, J.R. Costa, Tapered waveguide feed for integrated dielectric lens antenna performance tests, in *EUROCON—International Conference on Computer as a Tool (EUROCON), 27–29 April 2011* (IEEE, 2011), pp. 1–4
33. J.R. Costa, C.A. Fernandes, Broadband slot feed for integrated lens antennas. IEEE Antennas Wirel. Propag. Lett. **6**, 396–400 (2007). doi:10.1109/lawp.2007.900954
34. A. Neto, UWB, Non dispersive radiation from the planarly fed leaky lens antenna—Part 1: Theory and design. IEEE Trans. Antennas Propag. **58**(7), 2238–2247 (2010)
35. N.T. Nguyen, R. Sauleau, M. Ettorre, L. Le Coq, Focal array fed dielectric lenses: an attractive solution for beam reconfiguration at millimeter waves. IEEE Trans. Antennas Propag. **59**(6), 2152–2159 (2011)
36. D. Lemaire, C.A. Fernandes, P. Sobieski, C.A. Barbosa, A method to overcome the limitations of G.O. in the design of axis-symmetrical lenses. Int. J. Infrared Millimetrew. **17**(8), 1377–1390 (1996)
37. D. Pasqualini, S. Maci, High-frequency analysis of integrated dielectric lens antennas. IEEE Trans. Antennas Propag. **52**(3), 840–847 (2004)
38. P. Otero, G.V. Eleftheriades, J.R. Mosig, Integrated modified rectangular loop slot antenna on substrate lenses for millimeter- and submillimeter-wave frequencies mixer applications. IEEE Trans. Antennas Propag. **46**(10), 1489–1497 (1998)
39. H. Hoteit, R. Sauleau, B. Philippe, Ph Coquet, J.-P. Daniel, Vector and parallel implementations for the FDTD analysis of millimeter wave planar antennas. Int. J. High Speed Comput. **10**(2), 209–234 (1999)
40. A.V. Boriskin, A.I. Nosich, S.V. Boriskina, T.B. Benson, P. Sewell, A. Altintas, Lens or resonator? Electromagnetic behavior of an extended hemielliptical lens for a sub-mm wave receiver. Microw. Opt. Technol. Lett. **43**(6), 515–518 (2004)
41. A.V. Boriskin, A. Rolland, R. Sauleau, A.I. Nosich, Assessment of FDTD accuracy in the compact hemielliptic dielectric lens antenna analysis. IEEE Trans. Antennas Propag. **56**(3), 758–764 (2008)
42. G. Godi, R. Sauleau, D. Thouroude, Performance of reduced size substrate lens antennas for millimeter-wave communications. IEEE Trans. Antennas Propag. **53**(4), 1278–1286 (2005)
43. E.B. Lima, J.R. Costa, M.G. Silveirinha, C.A. Fernandes, ILASH—software tool for the design of integrated lens antennas, in *2008 IEEE Antennas and Propagation Society International Symposium*, vols. 1–9 (IEEE, New York, 2008), pp. 863–866
44. J.R. Costa, C.A. Fernandes, Analysis of small integrated lens antennas using multiple point source physical optics discretization of the feed aperture field. Journees Internationales de Nice sur les Antennes—JINA **1**, 466–467 (2004)
45. Computer Simulation Technology, CST, http://www.cst.com/
46. D.S. Alexei, N.G.T. Gregory, S. Roman, Hot-electron effect in superconductors and its applications for radiation sensors. Supercond. Sci. Technol. **15**(4), R1 (2002)
47. P.H. Siegel, THz instruments for space. IEEE Trans. Antennas Propag. **55**(11), 2957–2965 (2007). doi:10.1109/TAP.2007.908557

48. L. Liu, H. Xu, R.R. Percy, D.L. Herald, A.W. Lichtenberger, J.L. Hesler, R.M. Weikle, Development of integrated terahertz broadband detectors utilizing superconducting hot-electron bolometers. IEEE Trans. Appl. Supercond. **19**(3), 282–286 (2009). doi:10.1109/TASC.2009.2018268
49. A.D. Semenov, H. Richter, H.W. Hubers, B. Gunther, A. Smirnov, K.S. Il'in, M. Siegel, J. P. Karamarkovic, Terahertz performance of integrated lens antennas with a hot-electron bolometer. IEEE Trans. Microw. Theory Tech. **55**(2), 239–247 (2007). doi:10.1109/TMTT.2006.889153
50. J.R. Costa, C.A. Fernandes, Integrated imaging lens antenna with broadband feeds, in *The Second European Conference on Antennas and Propagation, 2007, EuCAP 2007* (2007), pp. 1–6
51. M.G. Silveirinha, C.A. Fernandes, J.R. Costa, A graphical aid for the complex permittivity measurement at microwave and millimeter wavelengths. IEEE Microw. Wirel. Compon. Lett. **24**(6), 421–423 (2014). doi:10.1109/LMWC.2014.2310470
52. M.N. Afsar, X. Li, H. Chi, An automated 60 GHz open resonator system for precision dielectric measurements. IEEE Trans. Microw. Theory Tech. **38**(12), 1845–1853 (1990). doi:10.1109/22.64565
53. C.A. Fernandes, J.R. Costa, Permittivity measurement and anisotropy evaluation of dielectric materials at millimeter-waves, in *XIX IMEKO World Congress: Fundamental and Applied Metrology, Proceedings, Budapest 2009* (IMEKO, 2009), pp. 673–677
54. M.J.M. van der Vorst, P.J.I. de Maagt, A. Neto, A.L. Reynolds, R.M. Heeres, W. Luinge, M. H.A.J. Herben, Effect of internal reflections on the radiation properties and input impedance of integrated lens antennas-comparison between theory and measurements. IEEE Trans. Microw. Theory Tech. **49**(6), 1118–1125 (2001)
55. K.W. Kim, Y. Rahmat-Samii, Spherical Luneburg lens antennas: engineering characterizations including air gap effects, in *Antennas and Propagation Society International Symposium, 21–26 June 1998*, vol. 2064 (IEEE, 1998), pp. 2062–2065
56. N.T. Nguyen, R. Sauleau, C.J. Martinez Perez, M. Ettorre, Finite-difference time-domain simulations of the effects of air gaps in double-shell extended hemispherical lenses. IET Microw. Antennas Propag. **4**(1), 35–42 (2010). doi:10.1049/iet-map.2008.0255
57. A. Bisognin, D. Titz, F. Ferrero, G. Jacquemod, R. Pilard, F. Gianesello, D. Gloria, D. Lugara, E.B. Lima, J.R. Costa, C.A. Fernandes, C. Luxey, Noncollimating MMW polyethylene lens mitigating dual-source offset from a Tx/Rx WiGig module. IEEE Trans. Antennas Propag. **63**(12), 5908–5913 (2015). doi:10.1109/TAP.2015.2484420
58. A. Bisognin, A. Cihangir, C. Luxey, G. Jacquemod, R. Pilard, F. Gianesello, J.R. Costa, C.A. Fernandes, E.B. Lima, C.J. Panagamuwa, W.G. Whittow, Ball grid array-module with integrated shaped lens for wigig applications in eyewear devices. IEEE Trans. Antennas Propag. **64**(3), 872–882 (2016). doi:10.1109/TAP.2016.2517667
59. L. Fernandes, Developing a system concept and technologies for mobile broadband communications. IEEE Pers. Commun. **2**(1), 54–59 (1995). doi:10.1109/98.350865
60. C.A. Fernandes, Design of shaped lenses for non-symmetric cells in MBS, in *Antennas and Propagation Society International Symposium, 1999, 11–16 July 1999*, vol. 2444 (IEEE, 1999), pp. 2440–2443
61. C.A. Fernandes, Shaped dielectric lenses for wireless millimeter-wave communications. IEEE Antennas Propag. Mag. **41**(5), 141–150 (1999). doi:10.1109/74.801527
62. C.A. Fernandes, L.M. Anunciada, Constant flux illumination of square cells for millimeter-wave wireless communications. IEEE Trans. Microw. Theory Tech. **49**(11), 2137–2141 (2001). doi:10.1109/22.963149
63. C.C. Cruz, J.R. Costa, C.A. Fernandes, S.A. Matos, Focal-plane multi-beam dual-band dielectric lens for Ka-band. IEEE Antennas Wirel. Propag. Lett. **16**, 432–436 (2017). doi:10.1109/LAWP.2016.2582263
64. J.R. Costa, M.G. Silveirinha, C.A. Fernandes, Evaluation of a double-shell integrated scanning lens antenna. IEEE Antennas Wirel. Propag. Lett. **7**, 781–784 (2008). doi:10.1109/lawp.2008.2008403

65. J.R. Costa, C.A. Fernandes, G. Godi, R. Sauleau, L. Le Coq, H. Legay, Compact Ka-band lens antennas for LEO satellites. IEEE Trans. Antennas Propag. **56**(5), 1251–1258 (2008). doi:10.1109/tap.2008.922690
66. J.R. Costa, E.B. Lima, C.R. Medeiros, T. Radil, R.C. Martins, P.M. Ramos, C.A. Fernandes, Development of an indoor wireless personal area network based on mechanically steered millimeter-wave lens antenna, in *Instrumentation and Measurement Technology Conference (I2MTC), 3–6 May 2010* (IEEE, 2010), pp. 1202–1206

Chapter 2
Multi-shell Radially Symmetrical Lens Antennas

Zvonimir Sipus and Tin Komljenovic

Abstract In this chapter, we review several aspects of the analysis and the design of multi-shell spherical and circular cylindrical lens antennas. Such lens antennas are attractive for implementation in communication and radar systems, in particular in the millimeter-wave frequency band, due to their broadband behavior, excellent focusing properties, possibility of beam scanning, and the ability to form multiple beams. In order to develop an efficient analysis tool, needed for successful design of multi-shell lens antennas, we first demonstrate the principles of the analysis algorithm for calculating the EM field distribution in general multilayer structures (i.e., inside a structure with an arbitrary number of layers). This algorithm is designed for spherical and circular-cylindrical geometries with elementary excitation. To model realistic lens antennas, we introduce additional flexibility that allows the analysis of actual feed antennas that usually do not follow the symmetry properties of the lens. Finally, by connecting the algorithm with an optimization subroutine, a powerful analysis and design tool is created. All the aspects of the proposed analysis approach are explained and illustrated with examples. Furthermore, some practical problems which are encountered in the design of these types of lens antennas are highlighted and common solutions are presented and compared to the ideal situations.

Z. Sipus (✉)
Faculty of Electrical Engineering and Computing, University of Zagreb,
Unska 3, HR-10000 Zagreb, Croatia
e-mail: zvonimir.sipus@fer.hr

T. Komljenovic
Department of Electrical and Computer Engineering, University of California,
Santa Barbara, CA 93106, USA

A. Boriskin and R. Sauleau (eds.), *Aperture Antennas for Millimeter and Sub-Millimeter Wave Applications*, Signals and Communication Technology,
DOI 10.1007/978-3-319-62773-1_2

2.1 Introduction

In recent years, there is a growing interest for antennas in communication and radar systems at higher microwave and millimeter-wave frequencies. Different kinds of spherical and cylindrical lens antennas are attractive solutions for such applications due to their focusing properties, capability of beam scanning, ability to form multiple beams, and broadband behavior. Furthermore, the typical problems with microwave lenses in the past, such as their bulkiness and weight are no longer an obstacle for practical implementation since by moving to higher frequencies and by using novel materials the size and weight have been considerably reduced. This added degree of freedom in terms of weight and size has in many cases led to the increase in the electrical size of the lenses in order for them to meet requirements for more demanding applications. In addition, the feed can be any type of antenna (horn, dipole, microstrip patch), and typically it is a part of the mechanical system that points the main beam in the desired direction. Altogether, the lens antenna system is a complex electromagnetic design problem that requires efficient and accurate analysis tools that can take into account various practical problems and demands.

The inhomogeneous lenses have first been analysed and designed using geometrical optics and ray-tracing techniques [1–4]. However, these are approximate methods valid for structures with all the dimensions large compared to the wavelength. Furthermore, they cannot take into account internal reflections. The general numerical methods, such as the finite difference time domain [5] or the finite element method in [6], can be applied to the lens analysis. However, such methods are suitable for structures having limited electric dimensions, usually not more than 20 free space wavelengths, and are time- and memory-consuming. Therefore, there is a need for specialized software in which the geometrical properties of the considered structures are analytically taken into account, and which can be efficiently connected with an optimization routine.

The analytical solution for spherical and circular-cylindrical multilayer structures is derived from the solution of the homogeneous Helmholtz equation, known as Mie series. It has been extensively used and reported for plane wave scattering by a conducting and dielectric sphere [7, 8]. This approach has been generalized for multilayered dielectric lens using different formulations: Mie series augmented matrix solution [9, 10], scalar potentials [11], mode matching technique [12–15] or dyadic Green functions [16–19]. The similar analysis approach can be applied for analyzing other types of structures (e.g., conformal microstrip antennas [20–23]).

2.2 Analysis of Spherical and Circular Cylindrical Multilayer Structures

For many structures and technical problems, one requires the knowledge of the field distribution scattered and/or inside multilayer spherical or cylindrical structures. Examples would be lens antennas, radomes, optical fibers, geophysical probing, scattering reduction structures (cloaking), etc. In all the considered cases we would like to determine the EM field distribution inside and around the multilayer structure while the position and/or existence of the source can be arbitrary. For example, in scattering problems the sources are faraway from the structure, in waveguide problems we search for eigenmode solutions, while in the antenna problems the sources are a part of the structure. In general, all of these problems can be described using the Green's function approach, i.e., we can consider the physical or equivalent sources as a part of the structure (the waveguide problems can be treated as poles of the Green's functions). The knowledge of the most appropriate Green's function is often essential for implementing the numerical method we would like to apply. For example, planar, cylindrical, and spherical patch antennas and periodic surfaces are frequently analysed by means of the electric field integral equation and the moment method. There the kernel of the integral operator is a Green's function, which takes into account the considered curved multilayer structure.

2.2.1 Green's Functions for Layered Media

In principle, there are two basic approaches for calculating the Green's function of a general multilayer structure: either to analytically derive an expression for it and then code this expression, or to develop a numerical routine for the complete calculation. The analytic approach requires less computer resources than the numerical approach; however, it is very laborious to analytically determine the Green's functions for structures with more than two layers. In such cases, it is more convenient to use a numerical algorithm which determines the Green's function directly. Furthermore, the analytic approach often requires a new derivation of the Green's functions for practically every slightly different problem, such as for different number of layers or for different source locations inside the layers.

In most cases, lenses follow spherical or cylindrical geometry, or a slight variation of these. Therefore, it is reasonable to focus our attention on these geometries and use the inherent properties of these structures to our advantage. Ideal planar, circular cylindrical, and spherical multilayer structures have one property in common: the structure is homogeneous in two dimensions, and varies in the third dimension. For example, the spherical structure in Fig. 2.1 varies in r direction and is homogeneous in θ and ϕ directions. Thus, planar, cylindrical and spherical structures can be referred to as one-dimensional structures since they vary only in

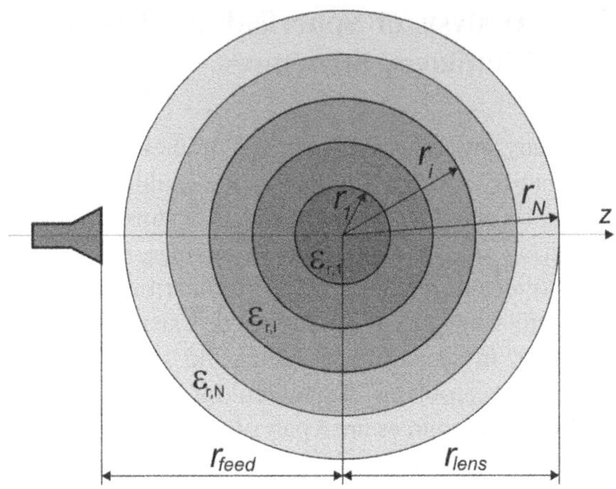

Fig. 2.1 Spherical lens antenna illuminated by a horn antenna

one dimension [24]. We can simplify the problem of determining the field distribution Green's functions for one-dimensional structures if we perform the two-dimensional (2D) Fourier transformation in the coordinates for which the structure is homogeneous. As a result, our original three-dimensional problem is transformed into a series of one-dimensional problems, which are much easier to solve. Instead of using Fourier transformation one can search for modal representation of the EM field, which at the end gives the same EM field expression.

When determining the Green's functions of multilayer one-dimensional structures one can select two paths. The first is to select one component of electric and magnetic potential and then to express the field components at the boundaries to determine the considered potentials. In practice, the algorithm determines the reflection and transmission coefficients of TE and TM waves which are connected (simply by multiplication) in a final matrix that describes the structure (see, e.g., [16]). For example, in the spherical case we can use the Debye potentials ψ_e and ψ_m to characterize the TM and TE waves (note that $\mathbf{r} = r\hat{r}$)

$$\mathbf{E}^{TE} = \nabla \times \mathbf{r}\psi_m \tag{2.1a}$$

$$\mathbf{H}^{TM} = \nabla \times \mathbf{r}\psi_e \tag{2.1b}$$

The Debye potentials ψ_e and ψ_m are determined by solving the scalar Helmholtz differential equation (in spherical coordinate system in this case)

$$\psi_{mn} = z_n(kr)P_n^m(\cos\theta)e^{jm\phi} \tag{2.2}$$

Here ψ_n is the elementary solution of the Helmholtz differential equation, i.e., $P_n^m(\cos\theta)$ are the associated Legendre functions of the first kind and z_n denotes spherical Bessel j_n or Hankel $h_n^{(2)}$ functions. Using the following notation:

$$\mathbf{M}_{mn} = \nabla \times \mathbf{r}\psi_{mn} \tag{2.3a}$$

$$\mathbf{N}_{mn} = \frac{1}{k}\nabla \times \mathbf{M}_{mn} \tag{2.3b}$$

we can express the electromagnetic field as

$$\mathbf{E} = -\sum_{n}\sum_{m} a_{mn}\mathbf{M}_{mn} + b_{mn}\mathbf{N}_{mn}. \tag{2.4a}$$

$$\mathbf{H} = -\frac{j}{\eta}\sum_{n}\sum_{m} b_{mn}\mathbf{M}_{mn} + a_{mn}\mathbf{N}_{mn}. \tag{2.4b}$$

The coefficients are determined by fulfilling the boundary conditions for the tangential EM field components (e.g., for electric field)

$$\begin{aligned}\mathbf{E}^{\tan} = &-\sum_{n=0}^{\infty}\sum_{|m|\le n}\left[a_{nm}^{j}j_n(kr) + a_{nm}^{h}h_n^{(2)}(kr)\right]\left(\hat{\theta}\frac{jm}{\sin\theta}P_n^{|m|}(\cos\theta) - \hat{\phi}\frac{d}{d\theta}P_n^{|m|}(\cos\theta)\right)e^{jm\phi}\\ &+\frac{1}{kr}\frac{d}{dr}\left[b_{nm}^{j}rj_n(kr) + b_{nm}^{h}rh_n^{(2)}(kr)\right]\left(\hat{\theta}\frac{d}{d\theta}P_n^{|m|}(\cos\theta) + \hat{\phi}\frac{jm}{\sin\theta}P_n^{|m|}(\cos\theta)\right)e^{jm\phi}\end{aligned} \tag{2.5}$$

By fulfilling the boundary conditions at each boundary (continuity of the tangential components) the reflection coefficients can be determined, see [16] for details.

The second approach is based on determining the tangential components of the EM field (so-called propagator matrix approach [16]). This approach actually corresponds to the implementation of the Huygens's principle or more precisely, the Love's equivalence theorem [8]. Here the original problem is divided into equivalent subproblems, one for each layer. The field inside each layer is then determined from equivalent currents at the subproblem boundaries and from the sources inside that subproblem. As a theoretical background, in parallel to the Love's equivalence theorem, we can modify the curl Maxwell's equation to "eliminate" the normal component of the EM field, i.e., to get the equations that depend only on the tangential field and source components. This modification is given in [25] and the resulting equations are (e.g., for H-field; by duality principle the equivalent equation can be derived for the E-field)

$$\begin{aligned}-\frac{\partial}{\partial n}\mathbf{H}_t &= \frac{j\eta}{k}\left(1 + \frac{\nabla_t\nabla_t}{k^2}\right)\cdot(\hat{\mathrm{n}} \times \mathbf{E}_t) + \hat{\mathrm{n}} \times \mathbf{J}_{te}\\ \mathbf{J}_{te} &= \mathbf{J}_t - \frac{j}{\eta k}\nabla \times \mathbf{M}_n\end{aligned} \tag{2.6}$$

Here subscripts t and n denote tangential and normal directions (related to the considered boundary), $\hat{\mathrm{n}}$ is the unit vector in the normal direction, and ∇_t is the transverse part of the ∇ operator. In other words, in order to determine the EM fields inside the considered multilayer structure we need to know the tangential EM field components at each boundary as well as tangential component of the excitation currents inside the structure. Note that the excitation current in Eq. (2.6) contains two parts: the tangential electric current J_t and the replacement tangential electric current with the origin in the normal component of the excitation magnetic current M_n.

2.2.2 Description of the G1DMULT Algorithm

As indicated, a canonical three-dimensional problem can be efficiently decomposed into one-dimensional subproblems by applying Fourier transformation for the directions where the structure is homogeneous. This so-called spectral domain approach is the core of an algorithm called G1DMULT for calculating the Green's function in the spectral domain (representing a one-dimensional (1D) spatial domain) for planar, circular cylindrical, and spherical multilayer structures. The G1DMULT algorithm calculates the Green's functions in the same way for all three types of geometries and it will be explained by considering the spherical geometry.

The solution procedure makes use of the Fourier transformation technique. Since the problem is described in spherical coordinate system, we use the vector Legendre transformation in θ and ϕ directions, defined by [26, 24].

$$\tilde{\mathbf{J}}(r,n,m) = \frac{1}{\sqrt{2\pi S(n,m)}} \int_{-\pi}^{\pi} \int_{0}^{\pi} \tilde{\tilde{\mathbf{L}}}(n,m,\theta)\mathbf{J}(r,\theta,\phi) \sin\theta\, e^{-jm\phi} d\theta\, d\phi \tag{2.7a}$$

$$\mathbf{J}(r,\theta,\phi) = \sum_{m=-\infty}^{\infty} \sum_{n=|m|}^{\infty} \frac{1}{\sqrt{2\pi S(n,m)}} \tilde{\tilde{\mathbf{L}}}(n,m,\theta)\, \tilde{\mathbf{J}}(r,n,m) e^{jm\phi} \tag{2.7b}$$

$$\tilde{\tilde{\mathbf{L}}}(n,m,\theta) = \begin{bmatrix} P_n^{|m|}(\cos\theta)\sqrt{n(n+1)} & 0 & 0 \\ 0 & \frac{\partial P_n^{|m|}(\cos\theta)}{\partial\theta} & \frac{-jmP_n^{|m|}(\cos\theta)}{\sin\theta} \\ 0 & \frac{jmP_n^{|m|}(\cos\theta)}{\sin\theta} & \frac{\partial P_n^{|m|}(\cos\theta)}{\partial\theta} \end{bmatrix} \tag{2.7c}$$

$$S(n,m) = \frac{2n(n+1)(n+|m|)!}{(2n+1)(n-|m|)!}. \tag{2.7d}$$

By applying the vector Legendre transformation, the three-dimensional excitations are transformed into harmonic current shells. If the source is infinitely thin in r direction, we get one discrete current shell per source, otherwise we get a continuous distribution of current shells in r direction. The E- and H-fields induced by the harmonic current sources have the same harmonic variations in θ and ϕ as the source. Therefore, only the field variation in the direction perpendicular to the boundaries is unknown, and we have a harmonic one-dimensional (1D) field problem. In this way, the spectral domain problem is interpreted as a one-dimensional spatial domain problem consisting of one-dimensional multilayer structure and harmonic one-dimensional sources in the form of current shells. This is visualized in Fig. 2.2, where also all other steps of the algorithm are shown. We describe the steps of the algorithm below.

The harmonic one-dimensional problem is solved by making use of the equivalent problems, one for each layer. The unknowns are the tangential E- and H-fields at layer boundaries. Since the variation of the E- and H-fields in the direction tangential to the boundaries is harmonic with known periodicity, we only need to determine the complex field amplitudes at the interfaces, i.e., we have four unknowns per boundary. For example, the E-field in the layer j is expressed as

$$\tilde{\mathbf{E}}_j = \tilde{\mathbf{G}}_{EJ}^{\text{homo}}\tilde{\mathbf{J}}_{j-1} + \tilde{\mathbf{G}}_{EJ}^{\text{homo}}\tilde{\mathbf{J}}_j + \tilde{\mathbf{G}}_{EM}^{\text{homo}}\tilde{\mathbf{M}}_{j-1} + \tilde{\mathbf{G}}_{EM}^{\text{homo}}\tilde{\mathbf{M}}_j + \tilde{\mathbf{G}}_{EJ}^{\text{homo}}\tilde{\mathbf{J}}_j^{\text{exci}} + \tilde{\mathbf{G}}_{EM}^{\text{homo}}\tilde{\mathbf{M}}_j^{\text{exci}}, \tag{2.8}$$

where $\tilde{J}_j$ and $\tilde{M}_j$ are equivalent electric and magnetic current sheets at boundary j, $\tilde{J}_j^{\text{exci}}$ and $\tilde{M}_j^{\text{exci}}$ are excitation electric and magnetic currents in layer j (if any), and $\tilde{G}_{\text{homo}}$ is the Green's function of the homogeneous problem (given below). By using $\tilde{\mathbf{J}}_j = \hat{n} \times \tilde{\mathbf{H}}_j$ and $\tilde{\mathbf{M}}_j = -\hat{n} \times \tilde{\mathbf{E}}_j$ Eq. (2.8) can be expressed in terms of the unknown EM field amplitudes $\tilde{\mathbf{E}}_j$ and $\tilde{\mathbf{H}}_j$ at the boundary j between layers j and $j + 1$ and the known excitation currents. Since the tangential E- and H-fields are continuous at the layer boundaries we obtain 4 linear equations per boundary.

The algorithm connects all equivalent subproblems into a system of $4N_{boundary}$ linear equations with the same number of unknowns ($N_{boundary}$ denotes the number of boundaries). Once the amplitudes of the tangential fields have been determined, it is easy to determine the field amplitudes anywhere in the multilayer structure by applying the homogeneous region equivalent of the layer inside which we want to determine the field value.

The core subproblem in the formulation is to calculate the E- and H-fields due to a harmonic current shell excitation of radius r_s in a homogeneous region. This is the only part (i.e., the only subroutine) that is different for planar, cylindrical and spherical case. In the spherical case the formulation is as follows:

(a) from $\tilde{J}_\theta(r_s, m, n)$

$$\tilde{E}_\theta(r, n, m) = -\frac{k}{\omega\varepsilon}\frac{r_s}{r}\tilde{J}_\theta(n, m) \cdot \begin{cases} \hat{H}_n^{(2)\prime}(kr_s)\hat{J}_n'(kr) & r \le r_s \\ \hat{J}_n'(kr_s)\hat{H}_n^{(2)\prime}(kr) & r \ge r_s \end{cases} \tag{2.9a}$$

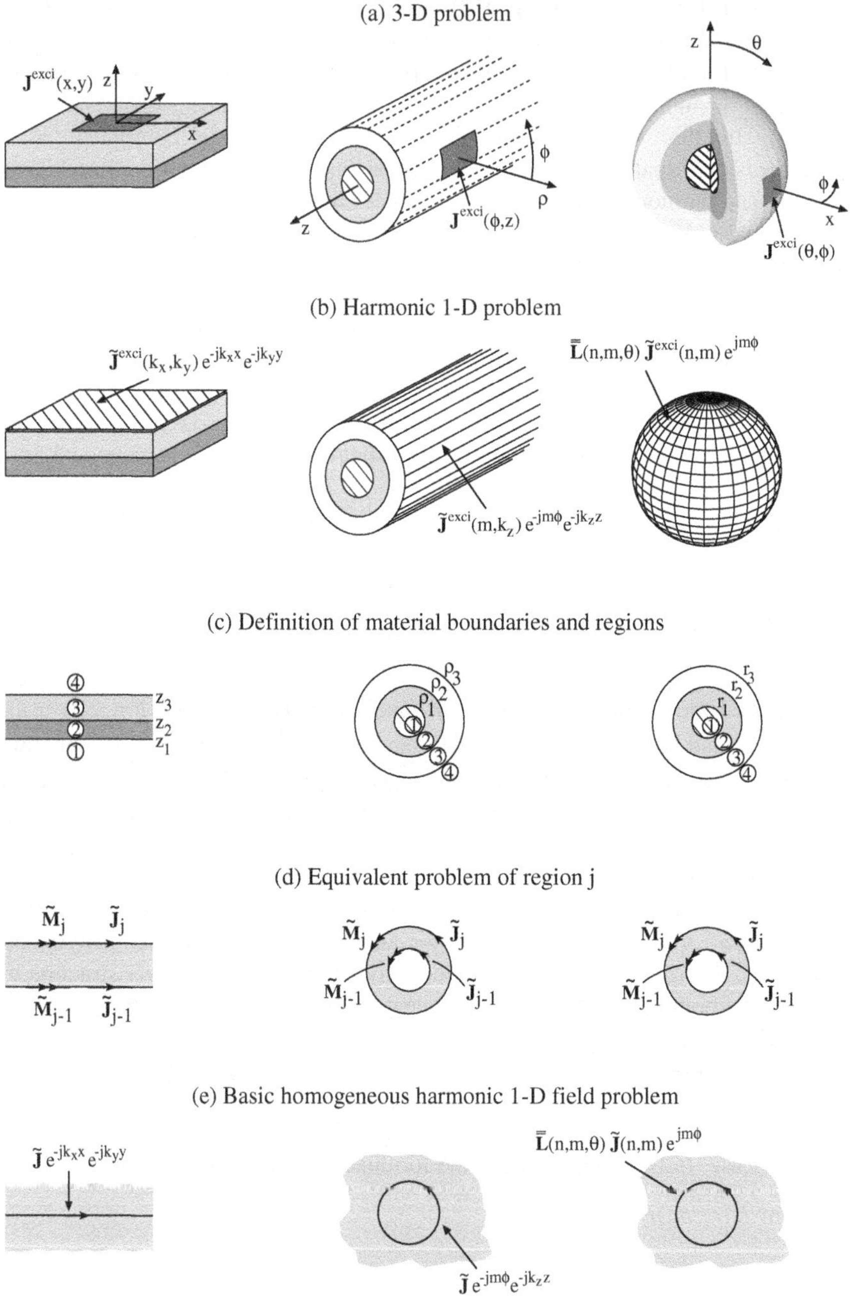

Fig. 2.2 Structure of the G1DMULT algorithm

$$\tilde{E}_r(r,n,m) = -\frac{1}{\omega\varepsilon}\frac{r_s}{r^2}\sqrt{n(n+1)}\,\tilde{J}_\theta(n,m)\cdot\begin{cases}\hat{H}_n^{(2)}(kr_s)\hat{J}_n(kr) & r\leq r_s\\ \hat{J}_n(kr_s)\hat{H}_n^{(2)}(kr) & r\geq r_s\end{cases} \tag{2.9b}$$

$$\tilde{H}_\phi(r,n,m) = \frac{jr_s}{r}\tilde{J}_\theta(n,m)\cdot\begin{cases}\hat{H}_n^{(2)}(kr_s)\hat{J}_n(kr) & r\leq r_s\\ \hat{J}_n(kr_s)\hat{H}_n^{(2)}(kr) & r\geq r_s\end{cases} \tag{2.9c}$$

(b) from $\tilde{J}_\phi(r_s,m,n)$

$$\tilde{E}_\phi(r,n,m) = -\frac{k}{\omega\varepsilon}\frac{r_s}{r}\tilde{J}_\phi(n,m)\cdot\begin{cases}\hat{H}_n^{(2)}(kr_s)\hat{J}_n(kr) & r\leq r_s\\ \hat{J}_n(kr_s)\hat{H}_n^{(2)}(kr) & r\geq r_s\end{cases} \tag{2.10a}$$

$$\tilde{H}_\theta(r,n,m) = \frac{jr_s}{r}\tilde{J}_\phi(n,m)\cdot\begin{cases}\hat{H}_n^{(2)\prime}(kr_s)\hat{J}_n'(kr) & r\leq r_s\\ \hat{J}_n'(kr_s)\hat{H}_n^{(2)\prime}(kr) & r\geq r_s\end{cases} \tag{2.10b}$$

$$\tilde{H}_r(r,n,m) = \frac{jr_s}{kr^2}\sqrt{n(n+1)}\,\tilde{J}_\phi(n,m)\cdot\begin{cases}\hat{H}_n^{(2)}(kr_s)\hat{J}_n(kr) & r\leq r_s\\ \hat{J}_n(kr_s)\hat{H}_n^{(2)}(kr) & r\geq r_s\end{cases} \tag{2.10c}$$

Here $\hat{J}_n(kr)$ and $\hat{H}_n^{(2)}(kr)$ are the Schelkunoff type of spherical Bessel and Hankel functions ($\hat{J}_n(x) = xj_n(x)$ and $\hat{H}_n^{(2)}(x) = xh_n^{(2)}(x)$), and k is the wave number of the considered layer. The fields excited by the magnetic shell can easily be determined by applying the duality concept. The fields caused by the r-directed sources are evaluated using the transverse replacement currents (see [24] for details).

Once we have the fields inside the structure the procedure to obtain the radiation pattern is quite straightforward. If we consider, e.g., the ϕ-component of the electric field in the outermost region with the r-coordinate larger than the r-coordinate of the patch, we have only outward-traveling waves described by $a_{nm}^i\hat{H}_n^{(2)}(k_0 r)$. Therefore, in the outermost region we can connect the ϕ-component of the electric field with different r-coordinates r_1 and r_2 as

$$\tilde{E}_\phi(r_1,n,m) = \tilde{E}_\phi(r_2,n,m)\,\frac{r_2\,\hat{H}_n^{(2)}(k_0r_1)}{r_1\,\hat{H}_n^{(2)}(k_0r_2)} \approx \tilde{E}_\phi(r_2,n,m)\,\frac{j^{n+1}r_2}{\hat{H}_n^{(2)}(k_0r_2)}\,\frac{\exp(-jk_0r_1)}{r_1} \tag{2.11a}$$

$$\tilde{E}_\theta(r_1,n,m) = \tilde{E}_\theta(r_2,n,m)\,\frac{r_2\,\hat{H}_n^{(2)\prime}(k_0r_1)}{r_1\,\hat{H}_n^{(2)\prime}(k_0r_2)} \approx \tilde{E}_\theta(r_2,n,m)\,\frac{j^n r_2}{\hat{H}_n^{(2)\prime}(k_0r_2)}\,\frac{e^{-jk_0r_1}}{r_1} \tag{2.11b}$$

Here r_l represents the r-coordinate of the far-field pattern. This avoids the problem of evaluating Hankel functions with very large argument. Notice that in the lens analysis a slightly larger value than the feed position is chosen for the radius r_2. The final solution is obtained by superposing the spectral solutions, see Eqs. (2.7a–2.7d).

In order to calculate the directivity radiation pattern it is necessary to calculate the radiated power. This is done in spectral domain by using Parseval theorem [8], i.e., the radiated power is equal

$$P^{rad} = r_2^2 \,\mathrm{Re}\left[\sum_{\forall m}\sum_{n\geq|m|} S(n,m)(\tilde{E}_\theta\tilde{H}_\phi^* - \tilde{E}_\phi\tilde{H}_\theta^*)\right]. \tag{2.12}$$

To summarize, the presented modal analysis approach (i.e., the spectral domain approach or the Mie series approach) is completely rigorous (there are no approximations in the formulation) and suitable for analyzing lenses with arbitrary number of layers. However, its application is limited only to canonical geometries (spherical, circular-cylindrical or planar) and one should be aware that numerical problems can appear for electrically very large structures. Two major sources of possible numerical difficulties are connected with the problem of the associated Legendre functions $P_n^m(\cos\theta)$ becoming very large (especially if $m \approx n$, $P_n^{|m|}(\cos\theta) \sim n^{|m-1/2|}$, making the solution numerically unstable), and with the problem of calculating Bessel/Hankel functions of large order. The first numerical difficulty linked to associated Legendre functions can be solved using suitable normalization process [22]. The second numerical problem (Bessel/Hankel functions of large order) can be mitigated by implementing the Debye's asymptotic formulas for Bessel and Hankel functions of large order, by which either the product of Bessel and Hankel function, or the ratio of two Hankel functions, are calculated with extracted exponential parts [22].

Multilayer canonical one-dimensional structures can also be analyzed by expressing the field quantities inside each layer as the superposition of forward and backward propagating waves (as mentioned at the beginning of this chapter). The reflection and transmission coefficients inside each layer can be determined in a recursive way, resulting in the generalized reflection and transmission coefficient of the whole structure (i.e., of the generalized scattering matrix). A detailed description of this method is given in [16]. Although in the presented approach the forward and backward propagating waves are also considered, the way of determining the field quantities is different—the presented G1DMULT algorithm simultaneously determines the tangential field quantities at the interfaces between the layers. Unlike in [16], in the presented method the field solutions to all the subproblems in the procedure are given in terms of Green's functions, i.e., the excitation within each subproblem are equivalent or physical current sources. This makes it easy to modify the code. For example, it is easy to implement the metasurface impedance boundary conditions into the routine, which makes the algorithm suitable for analyzing multilayer metasurface structures (e.g., metasurface lenses).

2.3 Spherical Multi-shell Lenses

Spherical lens antennas in general have many attractive features like broadband behavior, excellent focusing properties, wide scan coverage without scanning losses, possibilities of multiple beams on (in general) different working frequencies, etc. Such properties make lens antennas attractive solution in many applications, in particular in millimeter frequency range where the weight of such antennas is no longer an obstacle for practical implementation due to the increased frequency of operation. Two most common designs follow the spherical Luneburg and half Maxwell fisheye variation of index of refraction (see Fig. 2.3 and [27]), i.e., the respective radial dependency of index of refraction is equal

$$n^2(r) = 2 - r^2 \quad \text{and} \quad n^2(r) = 4\big/(1 + r^2)^2. \tag{2.13}$$

For example, spherical lens antennas can be used for automotive radars [28], radar cross-section (RCS) enhancers [29], satellite communications [30, 31], or for astronomy applications [32].

2.3.1 Modeling of Feed Antennas

As explained in the previous section, spherical lens antennas are usually rigorously analysed using spherical wave decomposition. However, the feed was usually taken only in an approximate way. For example, the Huygens source (approximation of low-gain horns) was considered in [10], dipole sources were considered in [17], and in [12, 13] the feeding horn antenna was modeled by an array of dipoles representing the aperture field distribution of the horn. In more details, in the case of horn feeding, the horn is replaced with equivalent aperture currents by using free space

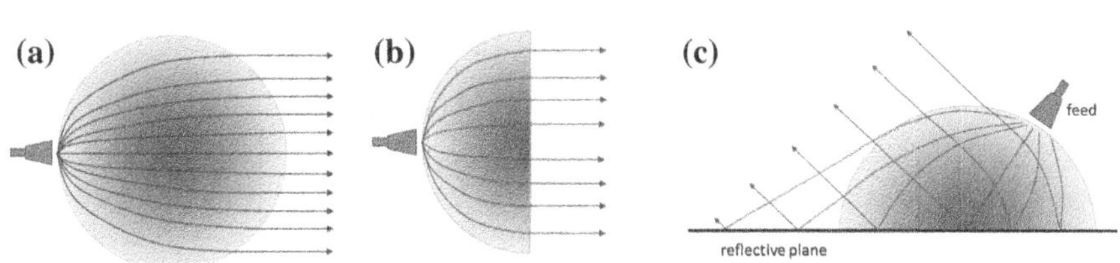

Fig. 2.3 Ray tracing illustration of the properties of **a** Luneburg lens, **b** half Maxwell fisheye lens, **c** hemispherical lens with a ground plane

equivalent principle [8]. For example, if the rectangular horn is used, the field at the horn aperture in the local coordinate system is

$$\mathbf{E}(x,y) = \hat{x}\,E_0 \cos\left(\frac{\pi\,y}{W_y}\right)\, e^{-jk[x^2/\rho_x + y^2/\rho_y]/2} \tag{2.13a}$$

$$\mathbf{H}(x,y) = \hat{y}\,\frac{E_0}{\eta} \cos\left(\frac{\pi\,y}{W_y}\right)\, e^{-jk[x^2/\rho_x + y^2/\rho_y]/2} \tag{2.13b}$$

where W_x and W_y are the width of the horn in x- and y-direction, and ρ_x and ρ_y are distances of the centers of radial waveguides from the horn opening. Therefore, the equivalent currents are

$$\mathbf{J}^{eq}(x,y) = \hat{x}\,\frac{E_0}{\eta} \cos\left(\frac{\pi\,y}{W_y}\right)\, e^{-jk[x^2/\rho_x + y^2/\rho_y]/2} \tag{2.14a}$$

$$\mathbf{M}^{eq}(x,y) = \hat{y}\,E_0 \cos\left(\frac{\pi\,y}{W_y}\right)\, e^{-jk[x^2/\rho_x + y^2/\rho_y]/2} \tag{2.14b}$$

Small horns can be approximated with an elementary area (so-called Huygens source [33]) with E- and H-fields equal to

$$\mathbf{E}^{eq}(r_{feed},\theta,\phi) = \hat{\theta}\frac{\delta(\theta)\,\delta(\phi)\,\delta(r - r_{feed})}{r_{feed}^2 \sin\,\theta} \tag{2.15a}$$

$$\mathbf{H}^{eq}(_{feed},\theta,\phi) = \hat{\phi}\frac{\delta(\theta)\,\delta(\phi)\,\delta(r - r_{feed})}{\eta\, r_{feed}^2 \sin\,\theta}. \tag{2.15b}$$

We can assume that the small horn is located at the $\theta = 0°$ axis with r-coordinate $r = r_{feed.}$ The vector-Legendre transformation of all described models can be found almost completely analytically (for the rectangular horn numerical integration in θ-direction is needed, or alternatively one can consider the array of electric and magnetic dipoles as explained in [13]). For example, the vector-Legendre transformation of the Huygens source is equal

$$\tilde{\mathbf{J}}^{eq}(r_{feed},n,m=\pm 1) = -\frac{1}{8\pi\eta\, r_{feed}^2}\frac{2n+1}{n(n+1)}\begin{bmatrix} 0 \\ 1 \\ jm \end{bmatrix} \tag{2.16a}$$

$$\tilde{\mathbf{M}}^{eq}(r_{feed},n,m=\pm 1) = -\frac{1}{8\pi\, r_{feed}^2}\frac{2n+1}{n(n+1)}\begin{bmatrix} 0 \\ -jm \\ 1 \end{bmatrix} \tag{2.16b}$$

Note that only the terms with $m = \pm 1$ are different from zero.

Rigorous model, based on closing the feed antenna in a small volume and representing the near-field distribution via spherical harmonics, was presented in [15]. The idea is to project the equivalent current distribution (at the interface of a volume that encloses the source) on the spherical wave basis

$$a_{mn} = -\frac{k^2\eta}{\pi(1+\delta_m)} \cdot \frac{1}{S(n,m)} \int_V \left[\mathbf{J}^{eq} \cdot \mathbf{M}_{mn} - \frac{j}{\eta}\mathbf{M}^{eq} \cdot \mathbf{N}_{mn}\right] dV, \tag{2.17a}$$

$$b_{nm} = -\frac{k^2\eta}{\pi(1+\delta_m)} \cdot \frac{1}{S(n,m)} \int_V \left[\mathbf{J}^{eq} \cdot \mathbf{N}_{nm} - \frac{j}{\eta}\mathbf{M}^{eq} \cdot \mathbf{M}_{nm}\right] dV, \tag{2.17b}$$

where δ_m is the Krönecker symbol. The equivalent electric and magnetic currents are determined mostly from computed data. However, in order to implement the proposed method, details of the feeding antenna structure should be known in order to determine the equivalent currents (e.g., by a general EM solver), which is not always the case—one often receives only the far-field measurements from the antenna manufacturer.

In many cases it is beneficial to use a feed model which is based on spherical mode representation of a far-field radiation pattern of the feed antenna and then using an efficient way of connecting fields in local and global coordinate systems. With this a fast and accurate method for analyzing spherical stratified lens antenna systems can be obtained. Note that the arbitrary feed antenna for which we want to find the model can be characterized either by measurements (often performed by the manufacturer) or by analysis using general EM solver. The proposed feed model contains three steps.

First step

The far-field of the feed antenna can be simply represented using vectorial spherical harmonics [7]:

$$\mathbf{E} = -\sum_n \sum_m a_{mn}\mathbf{M}_{mn} + b_{mn}\mathbf{N}_{mn}, \tag{2.18}$$

where the functions $\mathbf{M}_{mn}$ and $\mathbf{N}_{mn}$ are defined in Eqs. (2.3a)–(2.3b).

Second step

The spherical harmonics representation of the feed antenna is given in the local coordinate system, so we need to rewrite it in the global coordinate system. If we make a translation and/or rotation of the coordinate system, a new basis can be defined with respect to the new coordinate system, and any of the considered functions has to be expressed in the new basis. From addition theorems we get expansion coefficients for $\mathbf{M}_{\mathrm{mn}}$ and $\mathbf{N}_{\mathrm{mn}}$ in terms of $\mathbf{M'}_{\mathrm{mn}}$ and $\mathbf{N'}_{\mathrm{mn}}$ which are referenced to the second coordinate system

$$\mathbf{M}_{mn} = \sum_{\nu=0}^{\infty} \sum_{\mu=-\nu}^{\nu} \left(A_{\mu\nu}^{mn} \mathbf{M}'_{\mu\nu} + B_{\mu\nu}^{mn} \mathbf{N}'_{\mu\nu} \right), \tag{2.19a}$$

$$\mathbf{N}_{mn} = \sum_{\nu=0}^{\infty} \sum_{\mu=-\nu}^{\nu} \left(A_{\mu\nu}^{mn} \mathbf{N}'_{\mu\nu} + B_{\mu\nu}^{mn} \mathbf{M}'_{\mu\nu} \right), \tag{2.19b}$$

where $A_{\mu\nu}^{mn}$ and $B_{\mu\nu}^{mn}$ are coefficients that connect the global and local coordinate systems [34, 35]. A way to efficiently calculate them will be discussed at the end of this section.

Third step
The excitation field is rewritten in terms of equivalent currents (suitable for implementation into the code for analyzing spherical multilayer structures [8]):

$$\begin{aligned} \mathbf{J}_{eq} &= (-\hat{\mathbf{r}}) \times \mathbf{H} \\ \mathbf{M}_{eq} &= \mathbf{E} \times (-\hat{\mathbf{r}}) \end{aligned} \tag{2.20}$$

The equivalent currents have the following form in the spectral domain (for example $\tilde{\mathbf{M}}_{eq}$):

$$\begin{aligned} \tilde{\mathbf{M}}_{eq} = -\sum_{n} \sum_{m} \Bigg[& a_{mn} \left(\sum_{\nu=0}^{\infty} \sum_{\mu=-\nu}^{\nu} \sqrt{2\pi S(n,m)} \left(A_{\mu\nu}^{mn} \hat{\phi} + B_{\mu\nu}^{mn} \hat{\theta} \right) \right) \\ & + b_{mn} \left(\sum_{\nu=0}^{\infty} \sum_{\mu=-\nu}^{\nu} \sqrt{2\pi S(n,m)} \left(B_{\mu\nu}^{mn} \hat{\phi} + A_{\mu\nu}^{mn} \hat{\theta} \right) \right) \Bigg] \end{aligned} \tag{2.21}$$

Field representation in local and global coordinate systems.
In order to simplify mathematical description of the electromagnetic field distribution it is desirable to expand field solutions in different coordinate sets which describe the same space. The connection between different representations is made by using addition theorems, i.e., by using formulas which describe representation of a basis set of one coordinate system in terms of a basis set of one other [34]. There is a practical problem with calculating coupling coefficients, in particular with the Gaunt coefficients. This was first achieved by Stein [34] in terms of the Wigner 3jm symbols, and *it turns out to be very complicated and lengthy task* [10] (for this reason no results were shown in [10] that are obtained using transformation from local to global coordinate systems).

Coupling coefficients can be efficiently calculated for special case when the center of the local coordinate system is located at the z-axis (which is our case; multiple sources can be easily treated by rotating the coordinate system [36, 37]). In that case the coupling coefficients in Eqs. (2.19a), and (2.19b) are calculated from expressions

$$A_{mv}^{mn} = (-1)^m j^{v-n} \frac{2v+1}{2v(v+1)} \sum_p i^{-p} [n(n+1) + v(v+1) - p(p+1)] \\ \times a(m, n| - m, v|p)\, z_p(kr_{feed}) \tag{2.22}$$

$$B_{mv}^{mn} = (-1)^m j^{v-n} \frac{2v+1}{2v(v+1)} \sum_p i^{-p} \left[-2jmkr_{feed}\right] \times a(m, n| - m, v|p)\, z_p(kr_{feed}), \tag{2.23}$$

where $a(m,n|\mu,v|p)$ are Gaunt coefficients for which efficient recursion formulas exist for translation in the z-direction (i.e. for $\mu = -m$, see [38] for details).

2.3.2 *Examples of Spherical Multi-shell Lens Antennas*

12-layer Luneburg lens antenna

To demonstrate the results of the presented methodology, let us consider a 12-shell Luneburg lens antenna developed at EPFL [12]. The lens parameters are given in Table 2.1. The operating frequency is 10 GHz. The outer radius of the lens antenna is 15.9 cm, therefore the antenna is approximately $10\lambda_0$ in diameter. As a feed antenna let us consider a rectangular horn for which the radiation pattern is expanded in spherical harmonics. The opening of the rectangular horn is 2.0×3.0 cm and the horn is placed at h = 2.3 cm distance from the lens surface, see [12] for details.

Table 2.1 Lens geometry and electrical characteristics

Shell number	EPFL		Univ. of Rennes		
	Radius (cm)	Permittivity	Shell number	Radius (cm)	Permittivity
1	3.755	1.93	1	8.2	1.92
2	5.52	1.74	2	11.6	1.76
3	7.41	1.71	3	14.2	1.61
4	8.38	1.65	4	16.4	1.46
5	9.415	1.63	5	18.4	1.31
6	9.94	1.56	6	20.0	1.15
7	11.84	1.54			
8	12.5	1.50			
9	13.4	1.40			
10	14.27	1.28			
11	15.15	1.20			
12	15.9	1.10			

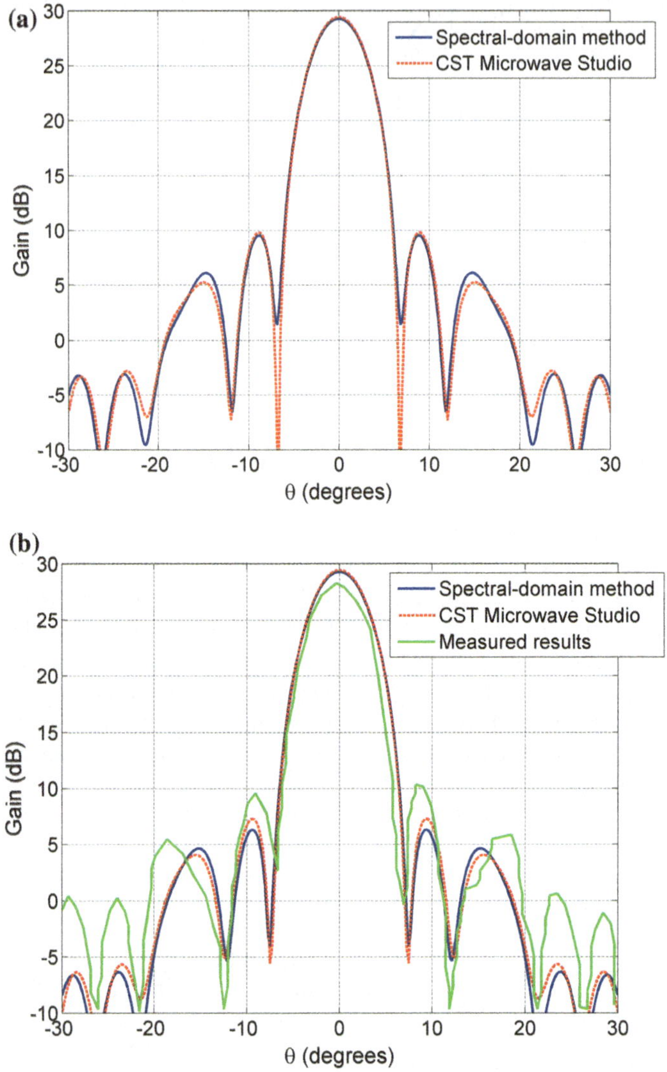

Fig. 2.4 Comparison of the calculated and measured radiation patterns of the 12-layer Luneburg lens antenna developed at EPFL. **a** E-plane, **b** H-plane

The comparison of the calculated radiation pattern of the whole lens antenna with measurements is given in Fig. 2.4. Two analysis methods are compared—the considered analysis method (the presented feed model + G1DMULT algorithm) and results obtained using CST Microwave Studio [39]. There is an excellent agreement between calculated results obtained by two analysis methods. Furthermore, the comparison between calculated results and measurements is also very good. In [12] only the H-plane measurements were provided; this is the reason why there are no measured results in Fig. 2.4a. For this specific antenna, the spherical wave expansion approach is more than four orders of magnitude faster comparing to the time needed for CST calculations. Therefore, the developed program is suitable for coupling with a global optimization algorithm (e.g. with the Particle Swarm Optimization (PSO) algorithm) to allow synthesis of spherically stratified lens antennas [18].

6-layer Luneburg lens antenna

As a second test example, a six-shell Luneburg lens antenna developed at University of Rennes is considered [14, 15]. The lens parameters are also given in Table 2.1. The operating frequency is 6 GHz, i.e., the lens diameter is 8 λ_0. The feeding antenna is simply a waveguide opening with directivity 7.4 dBi placed at distance h = 1.0 cm distance from the lens surface. Figure 2.5 shows comparison of the measured and calculated far-field pattern obtained by our method and by the general electromagnetic solver (CST Microwave Studio [39]). The agreement between CST results and spherical harmonic representation is excellent, and with very good agreement with measurements. Here we have also tested two feed models (the one based on far-field radiation pattern and the rigorous one based on projecting the equivalent currents on the spherical wave basis [15]), and the difference between results obtain by using two feed models is negligible (therefore, the results obtained by both feed models are represented with the same line in Fig. 2.5—spectral domain method).

The analyzed examples of step-index realization of Luneburg lens antenna are two rare examples for which all antenna parameters are given in the open scientific literature (to the best knowledge of the authors, only two additional examples can be found in open scientific literature in [40, 41]). In most of the cases, only a short description of the realized lens is given (e.g., in [42] it is simply stated "*8-inch diameter, ten-step Styrofoam lens at 16.65 GHz*").

2.3.3 Multi-shell Lens Optimization

When designing the lens antenna usually there are a lot of parameters to be determined. Since the proposed algorithms for analyzing the lens antenna are very fast it is practical to connect the developed algorithm with some global optimization routine, for example with the Particle Swarm Optimization (PSO) algorithm. This is an evolutionary algorithm similar to the genetic algorithm and to the simulated annealing, but it operates on a model of social interaction between independent agents and utilizes swarm intelligence to achieve the goal of the optimization problem. It is rather easy and straightforward to implement into the program [43], and its performance is at least comparable to the widely used genetic algorithms [44]. All the optimizations done in this chapter were made using the PSO algorithm.

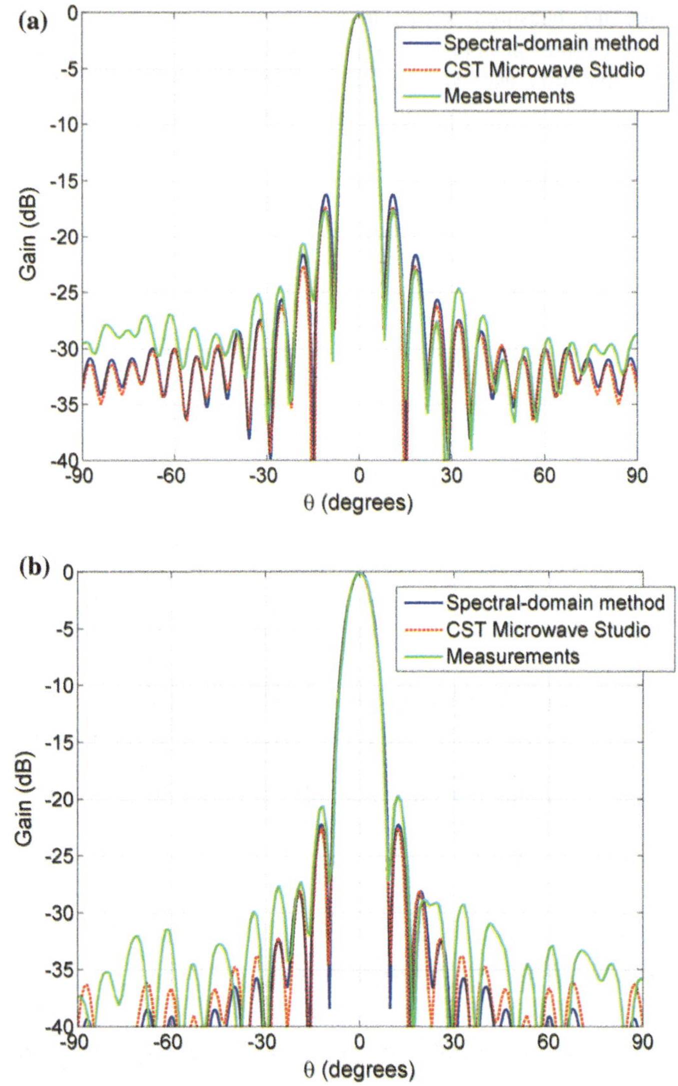

Fig. 2.5 Comparison of the calculated and measured radiation patterns of the 6-layer Luneburg lens antenna developed at University of Rennes. **a** E-plane, **b** H-plane

Characterization of lens and feed antenna

The original Luneburg lens antenna with the permittivity variation $\varepsilon_r(r) = 2 - (r/r_{lens})^2$ has the feed antenna optimally positioned at the lens surface [1]. For practical reasons, it is desirable to move the feed position from the lens surface, i.e., to introduce a gap between the feed and the lens. In order to optimize the lens antenna for new feed position, a modified lens design should be applied [45, 46]

$$\varepsilon_r(r) = \left(1 + \sqrt{1 - \varepsilon_r \left(r/r_{feed}\right)^2}\right) \cdot \exp\left(-\frac{2}{\pi} \int_1^{r_{feed}} \frac{\arcsin(t \cdot r_{lens}/r_{feed})}{\sqrt{t^2 - \varepsilon_r \left(r/r_{lens}\right)^2}} dt\right), \tag{2.24}$$

where r and r_{feed} represents the radial coordinate inside the lens and the position of the feed, respectively. Figure 2.6 shows the optimal permittivities of Luneburg lens as a function of normalized radial coordinate in case if the feed is placed at the lens surface ($r_{\text{feed}} = 1.0 \cdot r_{\text{lens}}$) and if it is placed at $1.1 \cdot r_{\text{lens}}$. Manufactured antenna is always realized from several layers of material with constant permittivity. In other words, the practical realization is just a stepwise approximation of the ideal lens. The theoretical curves are compared to permittivities of two manufactured lenses, with the permittivity value of each layer shown for a radius corresponding to the middle of the considered layer ($(r_{i-1} + r_i)/2$). It can be seen that in both cases there is some uncertainty about the obtained electromagnetic parameters of the realized lens antennas. We address this issue at the end of this section (see Fig. 2.8).

In order to study the effect of feed distance from lens surface and the directivity of feed antenna on total directivity of lens antenna and sidelobe levels, we model an "ideal" type of feed antenna, where the far-field is approximated with a cosine type of pattern

$$\mathbf{E}(r, \theta, \phi) = E_0 \frac{e^{-jkr}}{r} \left[\hat{\theta}(\cos(\theta/2))^{q1} \cos\,\phi - \hat{\phi}(\cos(\theta/2))^{q2} \sin\,\phi\right]. \tag{2.25}$$

With varying factors $q1$ and $q2$ we can model different types of feed antennas. We have selected $(\cos\,(\theta/2))^q$ type of function (instead of the $(\cos\theta)^q$ dependence [47]), since it better approximates the low-directivity antennas which are usually

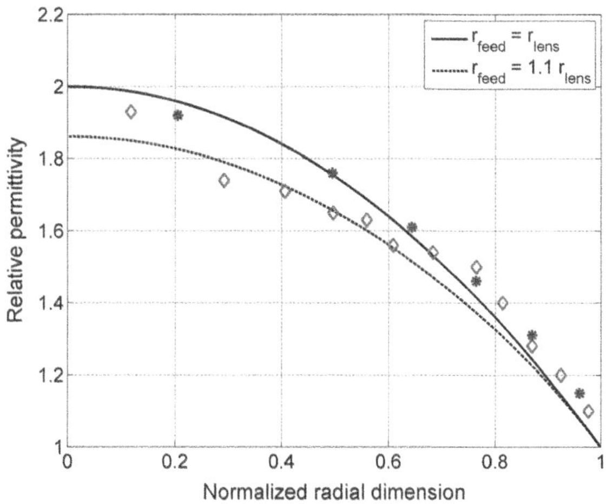

Fig. 2.6 Comparison of permittivity of practical lens realizations and ideal designs for $r_{\text{feed}} = r_{\text{lens}}$ (Luneburg lens) and for $r_{\text{feed}} = 1.1 \cdot r_{\text{lens}}$; diamonds (◇)—lens antenna from EPFL, stars (*)—lens antenna from University of Rennes

used as feeding elements of lens antennas (some authors have applied $(\cos(\theta/2))^q$ dependency for reflector antenna analysis as well [33]). Similar results can also be obtained using a complex Huygens source [48, 49]. First step in antenna system characterization is to determine the optimum directivity of feeding antenna as a function of feed position r_{feed}. We have calculated directivity of modified Luneburg lens antenna (permittivity variation is obtained using Eq. (2.24); the continuous permittivity variation is approximated with 20-layer step-wise distribution) for different q values, i.e., for different directivities of feeding antenna (Fig. 2.7a). For simplicity, we have taken $q1 = q2 = q$. The diameter of the considered lens antenna is 10 λ_0. The optimum directivity of the feed antenna is between 8 and 10 dBi (depending on the designed antenna position r_{feed}), and the directivity of the whole lens antenna does not vary much by moving the feed antenna away from the lens surface (for larger $r_{\text{feed}}/r_{\text{lens}}$ a more directive feed should be used). Therefore, the low-directive horn antennas or waveguide openings are very good elements for feeding the Luneburg lens antennas. In Fig. 2.7b the dependence of sidelobe level and illumination of the edge of the lens antenna is shown for the optimized case. Both, the sidelobe level and edge illumination, do not vary much with the change of feed position; they are around 22 dB and −13 dB, respectively. It is interesting to

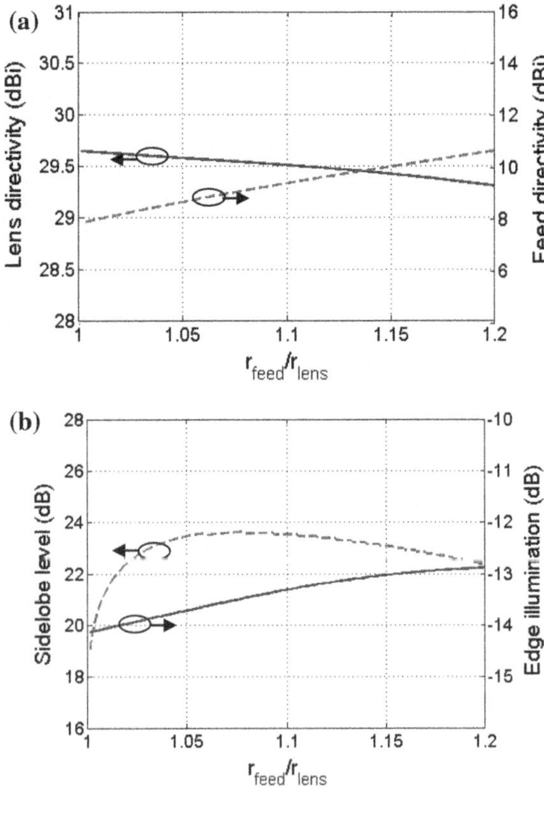

Fig. 2.7 **a** Optimized feed directivity and complete lens antenna directivity as a function of the feed antenna position. **b** Sidelobe level and edge illumination of the optimized case. The diameter of the considered Luneburg lens antenna is 10 λ (reproduced with permission from Ref. [19])

note that the edge illumination is approximately 3 dB lower comparing to the rule-of-thumb value for reflector antennas (–10 dB edge illumination [47]).

To show the effect of feed selection and placement on a practical realization of Luneburg lens we have calculated the total lens antenna directivities as a function of the feed directivity and position for the two considered realizations (see Table 2.1). Due to uncertainties about the obtained electromagnetic parameters of the realized lens antennas (see Fig. 2.6), it is advisable to make such characterization, i.e., to optimize the feed type as well as its position. Results are shown in Fig. 2.8. It can be seen that there is a significant freedom in selecting the feed antenna type and position for optimal performance. This is in particular important since the phase center of the feed antenna is usually located inside the antenna. In the EPFL case the used horn antenna has 9 dBi directivity and the normalized feed position is $r_{\mathrm{feed}} = 1.145 \cdot r_{\mathrm{lens}}$. The obtained sidelobe level and lens edge illumination is 19.5 dB and around −10 dB, respectively, which is comparable with the optimized feed properties. Similar result is also obtained for the antenna developed at University of Rennes. The used waveguide opening has directivity of 7.4 dBi and the normalized feed position is $r_{\mathrm{feed}} = 1.05 \cdot r_{\mathrm{lens}}$. The directivity determined by the spherical wave expansion technique equals 27.6 dBi whereas 27.5 dBi are predicted by CST. The obtained sidelobe level and lens edge illumination is 16.5 dB and between −6.0 dB (E-plane) and −11.3 dB (H-plane), respectively, which is also comparable to the optimized feed properties.

2.3.4 Fabrication and Experimental Results

The challenge of producing gradient index (GRIN) lenses is already mentioned in the articles of James Clerk Maxwell. Spherical Luneburg lenses have been fabricated using various techniques. Crushed Styrofoam was used to build 186 identical wedges—the lens was sliced through the North–South axis and the gradient variation of index of refraction is obtained by compressing Styrofoam pieces [50]. Polystyrene powder, solidified in heated hemispherical molds, was used to build 10-layer step-index Luneburg lens [40]. Ten-layer step-index Luneburg lenses were also built from Styrofoam and Foamed glass [41, 42]. The needed variation of the refractive index can in addition be obtained by drilling holes in the dielectric structures, which can be easily fabricated by using the traditional printed circuit board technique [51, 52]. Recently, a rapid prototyping machine was used to build a Luneburg lens in X-band by changing the size of plastic blocks centered on the junctions of the plastic rod frame [53].

Lens antennas are attractive solutions for many applications due to their focusing properties, capability of beam scanning, ability to form multiple beams and broadband behavior. For example, spherical lens antennas can be used for automotive radars [28], satellite communications [30, 31], for systems that require communication or/and multiple object tracking through a shared aperture [54], radar cross-section (RCS) enhancers [29], or for astronomy applications [32]. One

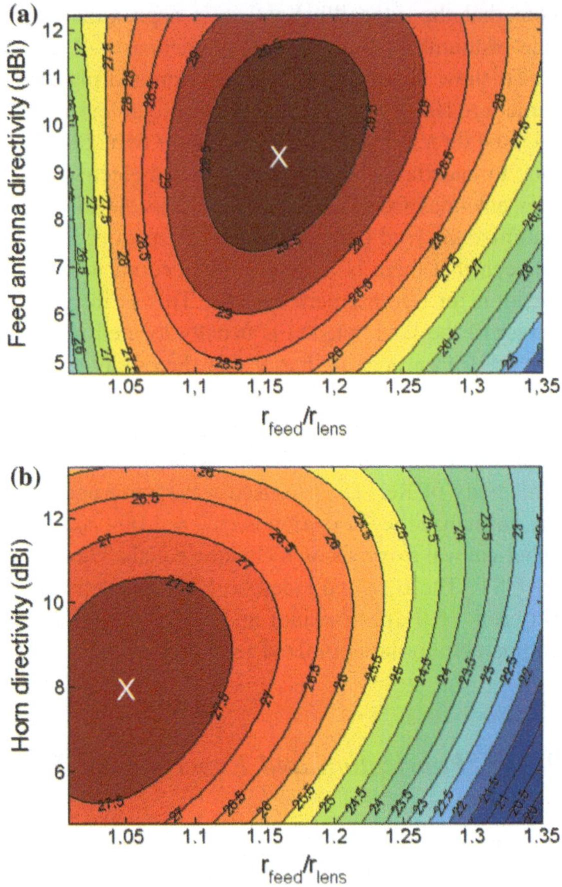

Fig. 2.8 Lens antenna directivity as a function of feed position and directivity. The optimum selection that results with the lens antenna with maximum directivity is marked with X. The parameters of the lens antennas are given in Table 2.1; **a** lens antenna from EPFL, **b** lens antenna from University of Rennes (reproduced with permission from Ref. [19])

of the most attractive applications is satellite communication systems for receiving signals from multiple satellites [30]. With this application in mind, a very useful variation of the Luneburg lens antenna is the hemispherical lens in conjunction with the ground plane. Furthermore, this is also an attractive solution for satellite communications with moving vehicles (e.g., trains, see Fig. 2.3c). If applied on the train, the antenna is mounted on the roof of the train and the mechanical system for positioning the feed antenna enables Internet connection via satellite [31]. Also, an important application of the Luneburg lens is a radar cross-section (RCS) enhancer, realized by placing a reflecting cap on the lens surface [29]. An incoming plane wave is focused at a point on the cap and reflected; it is transmitted in the opposite direction after propagating through the lens the second time. By this, a broadband calibrated scatterer with wide angular coverage is obtained, which is needed for calibrating scattered field measurement systems.

Since the construction of multilayer lenses is not easy, for many applications a constant-n lens or a two–layer lens is proposed that is much easier to construct comparing to Luneburg lenses. For fabrication of optimum constant-index lenses, a material with permittivity of 3.5 is desired, which can be obtained using slip-cast fused silica with small additions of rutile ($\varepsilon_r = 90$) [55]. An investigation on the construction of two layer-lenses with off-the-shell materials was performed in [31], where the considered materials were Rexolite ($\varepsilon_r = 2{,}53$), polyethylene ($\varepsilon_r = 2.28$) and Fused silica ($\varepsilon_r = 3.8$).

Finally, there is a class of two-dimensional Luneburg lenses where the focusing of electromagnetic waves is obtained inside a parallel-plate waveguide. The lens effect can be obtained by changing the thickness of the dielectric plate [56, 57], by drilling holes in a dielectric slab to control the effective permittivity [58], or by printing metallic patterns on dielectric substrate (i.e., variation of index of refraction is obtained using the metasurface concept [59–61]). Recently, the new implementation consisting of two mirrored metallic holey metasurfaces, having ultra-wideband lens property, was introduced [62].

2.4 Cylindrical Multi-shell Lenses

Circular-cylindrical dielectric lenses are attractive antennas because they allow launching multiple fan-beams, each of them originating from one primary feed. Like in the spherical case, the beams can be scanned by moving mechanically the feed around the lens surface or by switching between the feeds. They are also inherently wideband antennas. Such radiation characteristics are of particular interest for many applications at millimeter waves, like Doppler-weather radar, aircraft landing system or imaging systems. These lens antennas have been mostly neglected in the open literature, in contrast to the spherical ones.

In the past, research has been limited to homogeneous cylindrical lenses [63] and cylindrical Luneberg lenses [56, 57] placed between parallel-plate waveguides, i.e., the effect of the finite lens height is not present in this case. There has also been some research regarding antennas with different directivities in two orthogonal planes based on cylindrical Luneberg lens [64], and similar concept has been applied for automotive radars [65]. Cylindrical lenses with special requirements on radiation pattern have also been analyzed as an EM missile launcher, i.e., case when the decay of the energy per unit length is slower than 1/r [66]. Finally, the influence of the whispering-gallery modes on obtained directivity of cylindrical Luneberg lens antennas has been studied in [67]. In this section design of multilayered cylindrical dielectric lens antennas of finite height is discussed (a detailed study made by University of Rennes and University of Zagreb is given in [68]).

2.4.1 Lens Geometry

The generic lens antenna configuration is represented in Fig. 2.9a (the lens antenna geometry will be described using the classical cylindrical coordinate system (r, ϕ, z)). It consists of a multilayer cylindrical dielectric lens illuminated by a feed antenna. The total lens height and radius are labeled h are r_{lens}, respectively. The multilayer lens structure is schematized in Fig. 2.9b. The relative permittivity and thickness of layer #i are labeled $\varepsilon_{r,i}$ and a_i, respectively ($1 \leq i \leq N$). Without loss of generality we have selected a pyramidal horn as a feed antenna. The horn (of length L) is fed by a standard metallic rectangular waveguide linearly polarized along z-axis and d denotes the distance separating the horn aperture ($W_y \times W_z$) from the lens boundary.

The multilayer cylindrical lens can be approximated with an infinite circular cylinder. In that case we can apply the method of analysis described in the previous section, this time for cylindrical structures. The major advantage of this method of analysis is that the determination of the lens radiation performance is extremely fast. As a consequence, the iterative full-wave optimization or synthesis of cylindrical lenses can be made in reasonable amount of time. For lenses of moderate size (around $10 \times \lambda_0$ in diameter), the G1DMULT-based program needs less than one second to compute the three-dimensional far-field radiation patterns, whereas general-purpose commercial software like CST Microwave Studio® [39] requires approximately one hour, depending on the lens height and permittivity of the dielectric layers.

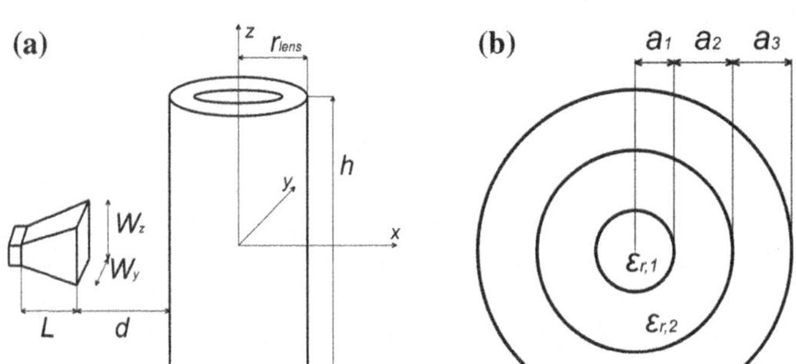

Fig. 2.9 Geometry of the circular-cylindrical dielectric lens antenna. **a** 3-D view. h, r_{lens}, and d denote the lens height, the total lens radius, and the distance between the horn aperture ($W_y \times W_z$) and the lens surface, respectively. L is the length of the horn. **b** Cross-section view of the lens. Definition of the thickness a_i and permittivity $\varepsilon_{r,i}$ of each layer

2.4.2 Range of Validity of Modal Analysis Approach

By essence the modal analysis approach assumes that the cylindrical structures are infinitely long, whereas, in practice, the objective is to minimize the height of the antenna prototype while keeping the lens radiation characteristics close to those of an infinite lens. To determine the range of validity of modal analysis approach, we have carried out an exhaustive parametric study consisting in comparing the far-field radiation patterns of a set of lenses analyzed using both the G1DMULT-based program and CST Microwave Studio®.

An empirical formula has been derived to estimate the minimum cylindrical lens height $h_{\min}$ that would provide, in far-field zone, nearly the same results as for lenses of infinite height

$$h_{\min} = 2(2r_{lens} + d) + W_z \tag{2.26}$$

Here r_{lens} is the total lens radius, d is the distance separating the horn aperture from the lens, and W_z is the height of the feed horn. The formula has been verified for low permittivity materials ($\varepsilon_{r,i} < 3$) as, for our applications, there was no need to use materials with even higher permittivities. The validity of this relation is clearly demonstrated in Fig. 2.10 where the fan-beam pattern of an infinite lens at 30 GHz is compared to the radiation patterns of three finite lenses ($h = h_{\min}$, $h = 0.7 \times h_{\min}$, $h = 0.5 \times h_{\min}$). In cases where the lens height is too small, both bases of the cylinder start to contribute to the antenna pattern. This results in a slight variation of the antenna gain, and in strongly pronounced ripples in E-plane.

2.4.3 Numerical and Experimental Results

The objective of the optimization problems studied here is to maximize the antenna gain while keeping low sidelobe levels. The corresponding fitness function is defined as

$$Fitness = \alpha \cdot Gain - \beta \cdot SL(H) - \gamma \cdot SL(E). \tag{2.27}$$

Here *Gain* is the antenna gain at broadside (i.e., along x-axis), and *SL*(*H*) and *SL* (*E*) are the highest sidelobe levels in the H- and E-planes, respectively. All quantities are expressed in dB. α, β, and γ are weighting coefficients that can be tuned to maximize the gain, or produce fan-beam with low sidelobes. Their typical values are 1, 0.3, and 0.2, respectively. In all cases, the feed and lens parameters are optimized simultaneously since this approach has been shown to be powerful [69].

Influence of the number of layers on the radiation characteristics of small lenses
To the authors' knowledge, in contrast to multilayer spherical lenses (e.g., [14]), there are very few papers dealing with the influence of number of layers on the

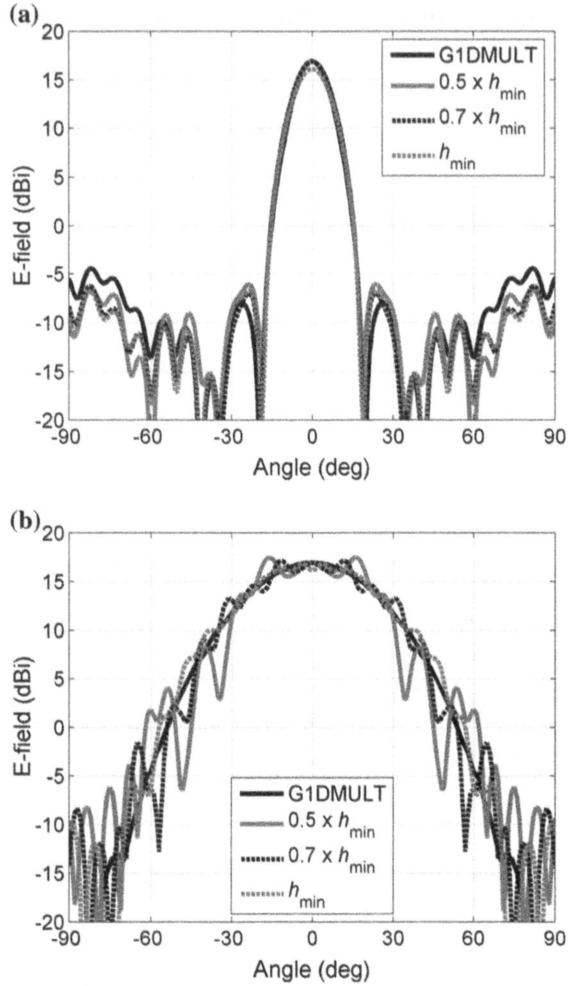

Fig. 2.10 Range of validity of modal analysis approach. Comparison of the radiation patterns computed at 30 GHz with G1DMULT and CST Microwave Studio®. **a** H-plane. **b** E-plane. The lens is made in Teflon ($\varepsilon_{r,1} = 2.1$, $r_{lens} = a_1 = 2.86$ cm) and is illuminated by a pyramidal horn fed by a standard WR-28 waveguide ($L = 2$ cm, $W_y = 2.4$ cm, $W_z = 1$ cm, $d = 0.65$ cm)

radiation characteristics of a cylindrical lens (see, e.g., [67]). The study described here is restricted to lenses with a moderate gain ($\sim$18 dBi). Their total diameter is equal to 5 cm, i.e., $5 \times \lambda_0$, at 30 GHz. The lens and feed horn parameters have been optimized simultaneously, and the number of layers was varied from 1 to 3. The range of variation of the optimization parameters are the following: $0.05\ \text{cm} \le d \le 2\ \text{cm}$, $0.4\ \text{cm} \le W_z \le 2\ \text{cm}$, $0.8\ \text{cm} \le W_y \le 3\ \text{cm}$, $0.5\ \text{cm} \le L \le 3\ \text{cm}$, $1.4 \le \varepsilon_{r,i} \le 5$, $\forall\ i = 1, \ldots, 3$. Both the gain and the sidelobe level were optimized ($\alpha = 1$, $\beta = 0.3$, $\gamma = 0.2$).

Up to seven optimization runs per lens configuration were launched, and the best result was selected based on the value of the fitness function and the visual inspection of the far-field radiation patterns. The geometry of the best designs and their corresponding patterns in H- and E-planes are given in Table 2.2 and

Table 2.2 Influence of the number of layers

	1 layer	2 layers	3 layers
Layer no. 1: thickness (cm)	2.50	0.10	0.73
Layer no. 1: permittivity	1.43	1.56	1.93
Layer no. 2: thickness (cm)	–	2.40	0.72
Layer no. 2: permittivity	–	1.52	1.77
Layer no. 3: thickness (cm)	–	–	1.05
Layer no. 3: permittivity	–	–	1.63
d (cm)	0.82	0.77	0.63
W_z (cm)	1.05	1.03	1.07
W_y (cm)	2.82	2.66	1.57
L (cm)	3.00	2.84	3.00
Gain (dBi)	**17.76**	**17.63**	**17.62**
Sidelobes (dB)	**−22.30**	**−22.95**	**−23.71**

Fig. 2.11, respectively. The radiation characteristics of the best three lens configurations are nearly identical. This demonstrates that there is no need to use more than one layer to optimize the fan-beam characteristics of moderate-sized cylindrical lenses.

Experimental results

To validate experimentally the previous results, a single-layer homogeneous cylindrical lens has been optimized in Ka-band. In this case, the results given in Table 2.2 suggest using a bulk material whose dielectric constant equals 1.43. Although customized materials are available commercially or could be synthesized using effective medium theory [51, 58] their fabrication cost remains very expensive.

To keep the antenna manufacturing as simple as possible, a cylindrical lens made from Teflon ($\varepsilon_{r,1}$ = 2.1) has been synthesized. The final dimensions of the antenna prototype are the following: $r_{lens} = a_1 = 2.82$ cm, $L = 2$ cm, $W_z = 1$ cm, $W_y = 2.4$ cm, and $d = 0.65$ cm. The minimum height h_{min} of the lens, guarantying that the infinite and finite lenses have nearly the same radiation characteristics, is equal to 13.6 cm (according Eq. (2.26)).

To keep some margin with respect to this threshold value, a 16 cm long lens was manufactured. As the optimized dimensions of the feed horn are not standard, a specific horn was fabricated in aluminum using electrical discharge machining techniques. Both the lens antenna and the feed horn were fabricated and measured at IETR, University of Rennes [68].

The radiation patterns measured at 30 GHz in H- and E-planes are represented in Figs. 2.12a, b, respectively. They are in excellent agreement with those predicted by G1DMULT-based program and CST Microwave Studio®, confirming thereby the relevance of the empirical relation (2.26). Additional measurements have been performed from 26 to 40 GHz (Fig. 2.13). They show that the lens patterns are very stable over the whole frequency band proving that the design is wideband.

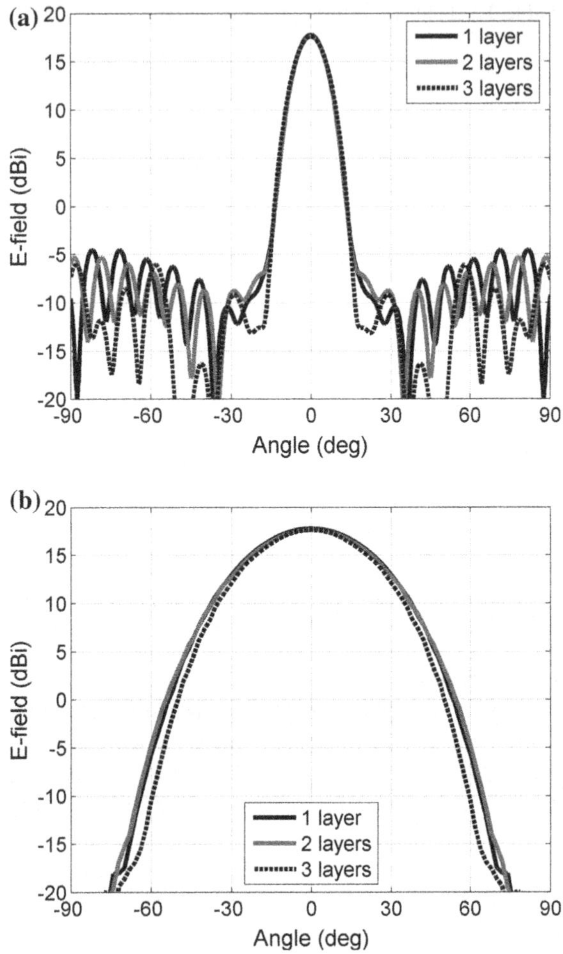

Fig. 2.11 Radiation patterns of the best designs at 30 GHz. The patterns have been computed using G1DMULT-based program. **a** H-plane. **b** E-plane. The corresponding lens geometries are detailed in Table 2.2

2.4.4 Size Reduction

Cylindrical lenses whose height is defined according to Eq. (2.26) produce nearly the same fan-beam patterns as lenses of infinite height. Nevertheless, applying this empirical criterion leads to large and bulky lens configurations. To overcome this limitation, the lens geometry has been optimized to minimize the influence of both lens bases, in particular to reduce the substantial ripples that are generated in E-plane for lenses of small height (Fig. 2.10b). These ripples originate from longitudinal resonant modes that contribute significantly to radiation in E-plane. This phenomenon is clearly highlighted on the amplitude near-field map represented in Fig. 2.14.

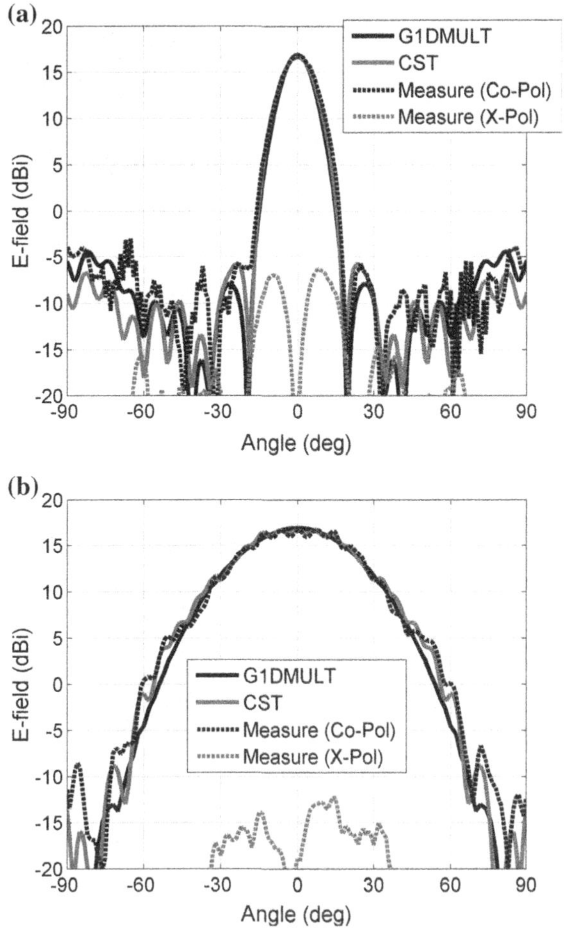

Fig. 2.12 Synthesized lens made from Teflon: radiation patterns measured and computed at 30 GHz. **a** H-plane. **b** E-plane

In order to reduce as much as possible the effects of lens truncation, two conical shapes are added to the cylinder bases (Fig. 2.15). As the resulting lens shape is rotationally symmetric, the ability to launch multiple beams and scan them around the lens axis is maintained. The height and permittivity of both cones are labeled h_c and $\varepsilon_{r,c}$, respectively. Their permittivity must be higher than the one of the cylindrical part of the lens to prevent possible total reflection that can happen when electromagnetic waves travels from a denser to a less dense medium.

Two cones made in Rexolite ($\varepsilon_{r,c} = 2.53$) have been designed to improve the radiation characteristics of the small lens represented in Fig. 2.15a. The height of the circular-cylindrical part is 4.8 cm. The lens is illuminated by the same horn as in Sect. 2.4.3. The cone height has been optimized using CST Microwave Studio® to minimize the ripple levels in E-plane ($h_c = 2$ cm). As a result the total antenna height equals 8.8 cm.

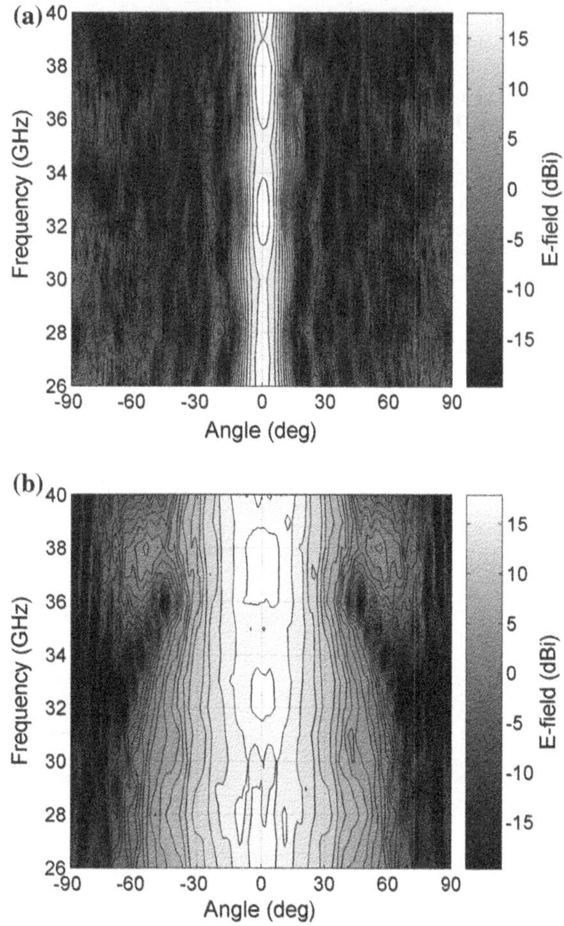

Fig. 2.13 Synthesized lens made in Teflon: radiation patterns measured in Ka-band. **a** H-plane. **b** E-plane. These patterns are quite frequency independent

The radiation patterns computed at 30 GHz are represented in Fig. 2.16, and they are compared to those of two pure cylindrical lenses: the first one has the same height (8.8 cm) as the lens with conical caps, and the height of the second one (16 cm) fulfills relation (2.26). The three lenses have nearly the same patterns in H-plane (Fig. 2.16a). However, by introducing two conical caps at the bases of small circular-cylindrical lenses, the ripple level in E-plane can greatly be reduced (Fig. 2.16b). This technique also provides a faster roll-off, compared to pure cylindrical lenses. The influence of cones on the amplitude near-field distribution is highlighted in Fig. 2.14b. It can be seen that the waves can propagate into the denser cone (without total reflection) and finally in the surrounding air as they approach the Rexolite-Air boundary at smaller incident angles.

The lens with conical caps has been fabricated and measured at University of Rennes [68]. Both cones have been assembled on the cylinder bases using a thin film of dielectric glue. The distance *d* separating the feed horn and the lens surface

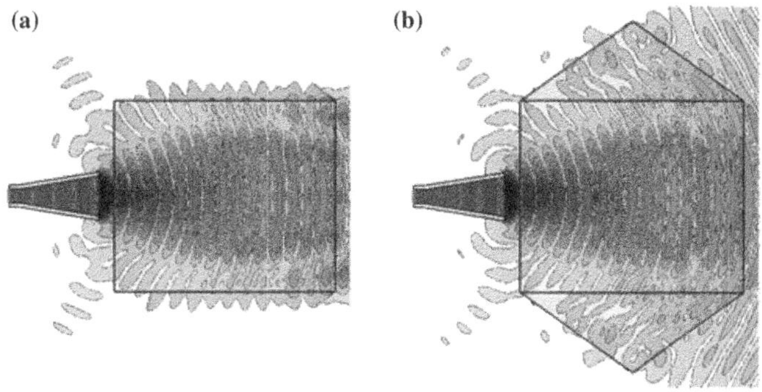

Fig. 2.14 Amplitude near-field maps (arbitrary units). **a** Original design. The lens parameters are the same as for the experimental model (Sect. 2.4.3), except the lens height that is roughly three times smaller than h_{min} (h = 4.8 cm). **b** Proposed design. The lens is the same as in (**a**) with addition of two cones. The cone parameters are the following: h_c = 2 cm, $\varepsilon_{r,c}$ = 2.53

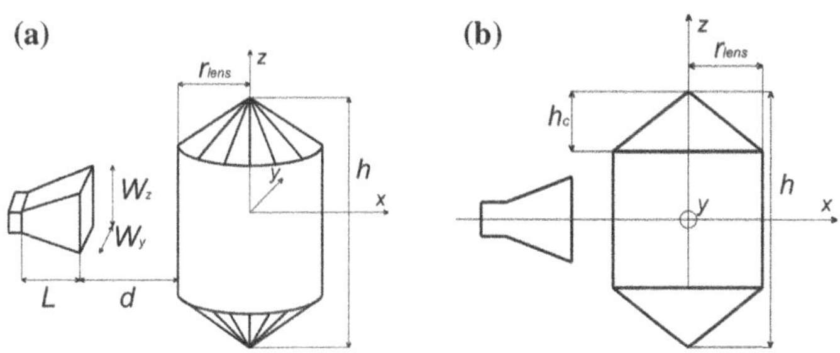

Fig. 2.15 Reduced-height circular-cylindrical dielectric lens with conical caps. The lens is illuminated by a pyramidal feed horn linearly polarized along z-axis. **a** 3-D view. **b** Cross-section view

is equal to 0.4 cm. The radiation patterns measured in both principal planes at 30 GHz are represented in Fig. 2.17. The agreement between numerical and experimental results is excellent, especially in E-plane where the ripple level in the main beam is smaller than 1.5 dB. The cross-polarization level is smaller than –21 dB and the measured antenna gain equals 17.66 dBi.

The height reduced design is also wideband as can be seen in Fig. 2.18. Measurements performed from 26 to 40 GHz show that the radiation patterns are really frequency independent over Ka-band with an expected increase of the gain as operating frequency increases. It should be noted that height reduced design shows even higher insensitivity of the radiation pattern to change in frequency compared to the original circular-cylindrical design.

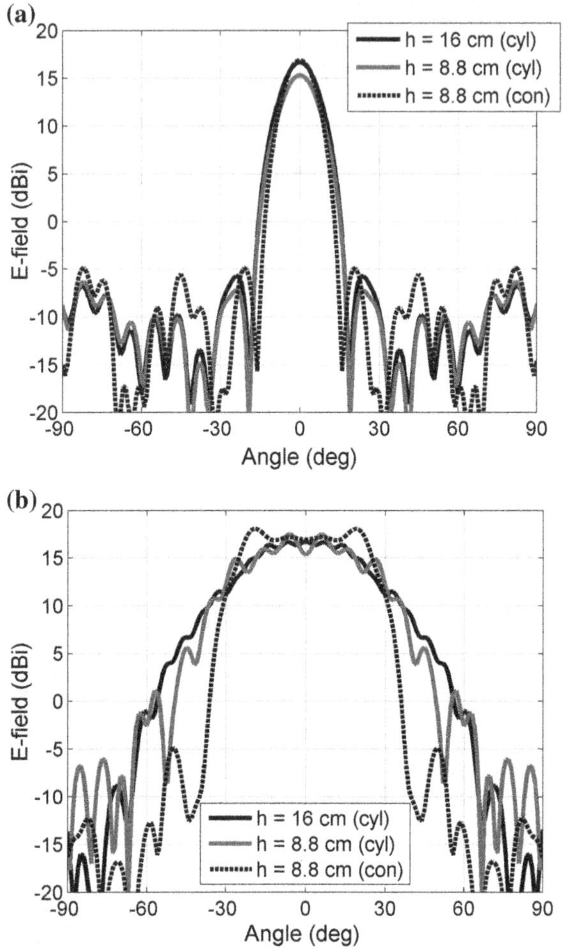

Fig. 2.16 Radiation patterns (computed at 30 GHz) of the reduced-size Teflon lens (h = 8.8 cm, $\varepsilon_{r,1}$ = 2.1) with conical caps (h_c = 2 cm, $\varepsilon_{r,c}$ = 2.53). Comparison with two pure cylindrical Teflon lenses of variable height (h = 8.8 cm or 16 cm). **a** H-plane. **b** E-plane. All lenses have the same radius ($r_{lens} = a_1$ = 2.82 cm) and are illuminated by the same horn as in Sect. 2.4.3

2.5 Concluding Remarks

A fast and reliable synthesis method is presented for designing multilayered spherical and circular-cylindrical dielectric lens antennas. In order to make the analysis program fast, thus enabling efficient merging with the global optimization routine, the multi-shell lens structure is analytically taken into account using Green's functions approach. The proposed analysis approach can successfully analyze lens structures with arbitrary number of layers and arbitrary type of excitation antenna. The latter is obtained by representing the measured or calculated far-field radiation pattern in terms of spherical harmonics in local coordinate system.

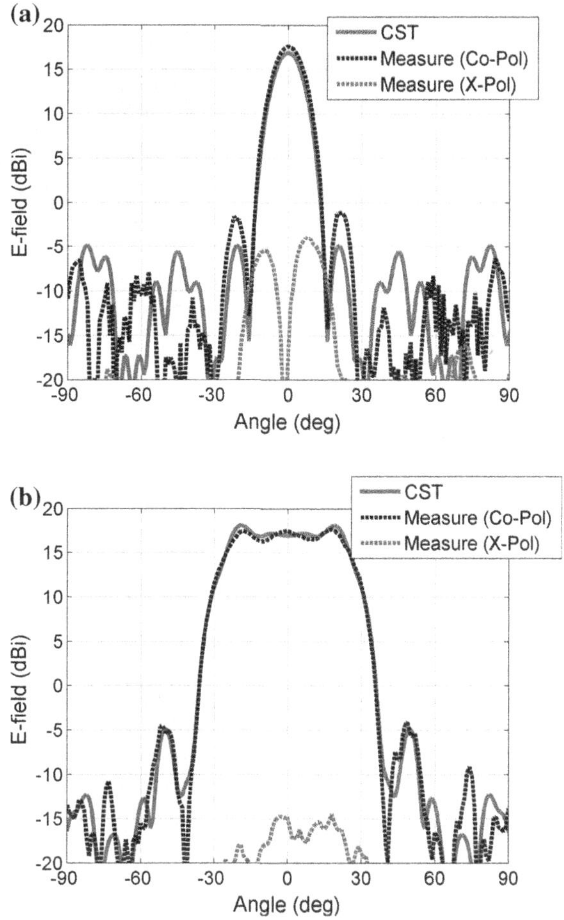

Fig. 2.17 Reduced-height Teflon lens with conical caps from Rexolite: radiation patterns measured and computed at 30 GHz. **a** H-plane. **b** E-plane

In the cylindrical case, the analysis approach assumes that the cylindrical structure is infinitely long, so we assess the range of validity of the analysis method. An empirical formula to estimate the minimum cylindrical lens height that provides similar far-field to the lens of infinite height is derived. For the cases where the lens size should further be reduced, a new height-reduction technique is proposed. It consists in adding conical caps to both bases of the cylinder. This technique allows a significant reduction of the ripple level in the elevation pattern of small cylindrical lenses, leading to high-quality fan-beams. Theoretical predictions are successfully verified by fabricating and characterizing various lens prototypes.

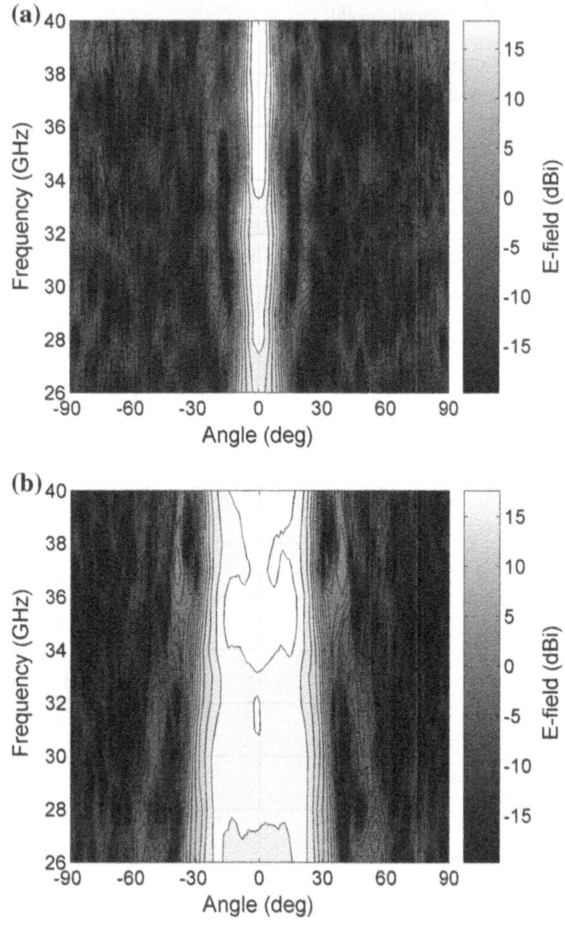

Fig. 2.18 Reduced-height Teflon lens with conical caps made in Rexolite: radiation patterns measured in Ka-band. **a** H-plane. **b** E-plane. The radiation pattern is quite similar at all frequencies, with expected gain increase as frequency increases (reproduced with permission from Ref. [68])

References

1. R.K. Luneburg, *Mathematical Theory of Optics* (Brown University Press, Providence, RI, 1944)
2. E. Braun, Radiation characteristics of the spherical Luneberg lens. IRE Trans. Antennas Propag. **4**(2), 132–138 (1956)
3. S.P. Morgan, Generalizations of spherically symmetric lenses. IRE Trans. Antennas Propag. **7**(4), 342–345 (1959)
4. T.L. Ap Rhys, The design of radially symmetric lenses. IEEE Trans. Antennas Propag. **18**(5), 497–506 (1970)
5. G. Godi, R. Sauleau, D. Thouroude, Performance of reduced size substrate lens antennas for millimeter-wave communications. IEEE Trans. Antennas Propag. **53**(4), 1278–1286 (2005)
6. A.D. Greenwood, J.-M. Jin, Finite-element analysis of complex axisymmetric radiating structures. IEEE Trans. Antennas Propag. **47**(8), 1260–1266 (1999)
7. J.A. Stratton, *Electromagnetic Theory* (McGraw Hill, New York, 1941)
8. R.F. Harrington, *Time Harmonic Electromagnetic Fields* (McGraw Hill, New York, 1961)

9. J.J. Mikulski, E.L. Murphy, The computation of electromagnetic scattering from concentric spherical structures. IEEE Trans. Antennas Propag. **11**, 169–177 (1963)
10. H. Mieras, Radiation pattern computation of a spherical lens using Mie series. IEEE Trans. Antennas Propag. **30**(6), 1221–1224 (1982)
11. S.S. Vinogradov, E.D. Vinogradova, P.D. Smith, Accurate modelling of a scanning Luneberg lens antenna Mie series approach, in *Proceedings of the International Conference on Electromagnetics in Advanced Applications, Turin, Italy, Sep. 1999* (1999), pp. 277–280
12. J.R. Sanford, Spherically Stratified Microwave Lenses, Ph.D. Thesis No. 1065, *École polytechnique fédérale de Lausanne (EPFL)*, 1992
13. J.R. Sanford, Scattering by spherically stratified microwave lens antennas. IEEE Trans. Antennas Propag. **42**, 690–698 (1994)
14. B. Fuchs, L. Le Coq, O. Lafond, S. Rondineau, M. Himdi, Design optimization of multishell Luneburg lenses. IEEE Trans. Antennas Propag. **55**, 283–289 (2007)
15. B. Fuchs, S. Palud, L. Le Coq, O. Lafond, M. Himdi, S. Rondineau, Scattering of spherically and hemispherically stratified lenses fed by any real source. IEEE Trans. Antennas Propag. **56**, 450–460 (2008)
16. W.C. Chew, *Waves and Fields in Inhomogeneous Media* (Van Nostrand Reinhold, New York, 1990)
17. H. Mosallaei, Y. Rahmat-Samii, Non-uniform Luneburg and 2-shell lens antennas: radiation characteristics and design optimization. IEEE Trans. Antennas Propag. **49**, 60–69 (2001)
18. T. Komljenovic, Z. Sipus, J.-P. Daniel, Scanning vehicular lens antennas for satellite communications, in *Proceedings of EuCAP'2010, Barcelona, Spain* (2010), pp. 1–4
19. Z. Sipus, D. Bojanjac, T. Komljenovic, Electromagnetic modeling of spherically stratified lenses illuminated by arbitrary sources. IEEE Trans. Antennas Propag. **63**(4), 1837–1843 (2015)
20. S. Rondineau, Modélisation de lentilles sphériques à gradient d'indice et sources conformes associées, Ph.D. dissertation, *University of Rennes 1, Rennes, France, Dec. 2002*
21. S. Rondineau, A.I. Nosich, J.-P. Daniel, M. Himdi, S.S. Vinogradov, MAR analysis of a spherical-circular printed antenna with finite ground excited by an axially symmetric probe. IEEE Trans. Antennas Propag. **52**(5), 1270–1280 (2004)
22. Z. Sipus, N. Burum, S. Skokic, P.-S. Kildal, Analysis of spherical arrays of microstrip antennas using moment method in spectral domain. IEE Proc. Microw. Antennas Propag. **153**, 533–543 (2006)
23. Z. Sipus, S. Skokic, M. Bosiljevac, N. Burum, Study of mutual coupling between circular stacked-patch antennas on a sphere. IEEE Trans. Antennas Propag. **56**, 1834–1844 (2008)
24. Z. Sipus, P.-S. Kildal, R. Leijon, M. Johansson, An algorithm for calculating Green's functions for planar, circular cylindrical and spherical multilayer substrates. Appl. Comput. Electromagn. Soc. J. **13**, 243–254 (1998)
25. L.B. Felsen, N. Marcuvitz, *Radiation and Scattering of Waves* (Prentice-Hall Inc., New Jersey, 1973)
26. W.Y. Tam, K.M. Luk, Resonances in spherical-circular microstrip structures of cylindrical-rectangular and wraparound microstrip antennas. IEEE Trans. Microw. Theory Tech. **39**, 700–704 (1991)
27. B. Fuchs, O. Lafond, S. Palud, L. Le Coq, M. Himdi, M.C. Buck, S. Rondineau, Comparative design and analysis of Luneburg and Half Maxwell fish-eye lens antennas. IEEE Trans. Antennas Propag. **56**, 3058–3062 (2008)
28. B. Schoenlinner, X. Wu, J.P. Ebling, G.V. Eleftheriades, G.M. Rebeiz, Wide-scan spherical-lens antennas for automotive radars. IEEE Trans. Microw. Theory Tech. **50**(9), 2166–2175 (2002)
29. J.R. Sanford, Analysis of spherical radar cross-section enhancers. IEEE Trans. Microw. Theory Tech. **43**(6), 1400–1403 (1995)
30. C.S. Lee, M.E. Rayner, *Luneberg lens antenna with multiple gimbaled RF feeds,* U.S.A. patent No. US 6266029 B1

31. J. Thornton, Wide-scanning multi-layer hemisphere lens antenna for Ka band. IEE Proc. Microw. Antennas Propag. **153**(6), 573–578 (2006)
32. A.J. Parfitt, J.S. Kot, G.L. James, The Luneburg lens as a radio telescope element, in *Proceedings of the IEEE Antennas and Propagation Society International Symposium, Salt Lake City* (2000), pp. 170–173
33. P.-S. Kildal, *Foundations of Antennas—A Unified Approach* (Studentlitteratur AB, 2000)
34. S. Stein, Addition theorems for spherical wave functions. Q. Appl. Math. **19**, 15–24 (1961)
35. O.R. Cruzan, Translational addition theorems for spherical vector wave functions. Quart. Appl. Math. **20**, 33–40 (1962)
36. D.L. Sengupta, T.M. Smith, R.W. Larson, Radiation characteristics of spherical array of circularly polarized elements. IEEE Trans. Antennas Propag. **16**, 2–7 (1968)
37. J.R. Sanford, Z. Sipus, Sidelobe reduction with array fed spherical lenses, in *Proceedings of the IEEE Antennas and Propagation Society International Symposium, Newport Beach* (1995), pp. 670–673
38. J.H. Bruning, Y.T. Lo, Multiple scattering of EM waves by spheres. Part I—Multiple expansion and ray optical solutions. IEEE Trans. Antennas Propag. **19**, 378–390 (1971)
39. CST Microwave Studio™, *Computer Simulation Technology AG*, www.cst.com, Darmstadt, Germany (2013)
40. G.D.M. Peeler, H.P. Coleman, Microwave stepped-index Luneburg lenses. IRE Trans. Antennas Propag. **6**, 202–207 (1958)
41. L.C. Gunderson, J.F. Kauffman, A high temperature Luneburg lens. IEEE Proc. **56**, 883–884 (1968)
42. R.E. Webster, Radiation patterns of a spherical luneburg lens with simple feeds. IRE Trans Antennas Propag. **6**, 301–302 (1968)
43. J. Robinson, Y. Rahmat-Samii, Particle swarm optimization in electromagnetics. IEEE Trans. Antennas Propag. **52**(2), 397–407 (2004)
44. D.W. Boeringer, D.H. Werner, Particle swarm optimization versus genetic algorithms for phased array synthesis. IEEE Trans. Antennas Propag. **52**(3), 771–778 (2004)
45. A.S. Gutman, Modified Luneberg lens. J. Appl. Phys. **25**, 855–859 (1954)
46. D.K. Cheng, Modified Luneberg lens for defocused source. IRE Trans. Antennas Propag. **8**, 110–111 (1960)
47. Y. Rahmat-Sammii, Reflector antennas, in *Antenna Engineering Handbook*, ed. by J.L. Volakis, Chap. 15 (McGraw-Hill, 2007)
48. V.S. Bulygin, T.M. Benson, Y.V. Gandel, A.I. Nosich, Full-wave analysis and optimization of a TARA-like shield-assisted paraboloidal reflector antenna using a Nystrom-type method. IEEE Trans. Antennas Propag. **61**(10), 4981–4989 (2013)
49. D. Bojanjac, A.I. Nosich, Z. Sipus, Design of spherical lens antennas using complex Huygens element, in *Proceedings of EuCAP'2014, The Hague, The Netherlands* (2014), pp. 1–3
50. G.P. Robinson, Three-dimensional microwave lens, in *Tele-Tech & Electronic Industry, Nov. 1954* (1954), p. 73
51. S. Rondineau, M. Himdi, J. Sorieux, A sliced spherical Lüneburg lens. IEEE Antennas Wirel. Propag. Lett. **2**, 163–166 (2003)
52. H.F. Ma, B.G. Cai, T.X. Zhang, Y. Yang, W.X. Jiang, T.J. Cui, Three-dimensional gradient-index materials and their applications in microwave lens antennas. IEEE Trans. Antennas Propag. **61**, 2561–2569 (2013)
53. M. Liang, W.-R. Ng, K. Chang, K. Gbele, M.E. Gehm, H. Xin, A 3-D Luneburg lens antenna fabricated by polymer jetting rapid prototyping. IEEE Trans. Antennas Propag. **62**, 1799–1807 (2014)
54. M.A. Mitchell, J.R. Sandford, Luneburg lens revival, in *Electronics&Wireless World, May 1989* (1989), pp. 456–458
55. W.R. Free, F.L. Cain, C.E. Ryan Jr., C.P. Burns, E.M. Turner, High-power constant-index lens antennas. IEEE Trans Antennas Propag. **22**, 582–584 (1974)
56. G.D.M. Peeler, D.H. Archer, A two-dimensional microwave Luneburg lens. IRE Trans. Antennas Propag. **1**, 12–23 (1953)

57. X. Wu, J. Laurin, Fan-beam millimeter-wave antenna design based on the cylindrical luneberg lens. IEEE Trans. Antennas Propag. **55**, 2147–2156 (2007)
58. K. Sato, H. Ujiie, A plate Luneberg lens with the permittivity distribution controlled by hole density. Electr. Commun. Jpn. **85**(9), 1–12 (2002)
59. C. Pfeiffer, A. Grbic, A printed, broadband Luneburg lens antenna. IEEE Trans. Antennas Propag. **58**, 3055–3059 (2010)
60. S. Maci, G. Minatti, M. Casaletti, M. Bosiljevac, Metasurfing: addressing waves on impenetrable metasurfaces. IEEE Antennas Wirel. Propag. Lett. **10**, 1499–1502 (2011)
61. M. Bosiljevac, M. Casaletti, F. Caminita, Z. Sipus, S. Maci, Non-uniform metasurface luneburg lens antenna design. IEEE Trans. Antennas Propag. **60**, 4065–4073 (2012)
62. O. Quevedo-Teruel, M. Ebrahimpouri, M.N.M. Kehn, Ultrawideband metasurface lenses based on off-shifted opposite layers. IEEE Antennas Wirel. Propag. Lett. **15**, 484–487 (2016)
63. L.C. Gunderson, An electromagnetic analysis of a cylindrical homogeneous lens. IEEE Trans. Antennas Propagat. **20**, 476–479 (1972)
64. G. Dubost, Flat linear radiating array applied on a cylindrical lens, in *MELECON'85, Madrid, Spain*, vol. 3 (1985), pp. 215–218
65. P. Wenig, R. Weigel, Analysis of a microstrip patch array fed cylindric lens antenna for 77 GHz automotive radar, in *IEEE AP-S International Symposium Digest, San Diego (CA), 5–11 July 2008*
66. T.T. Wu, R.W.P. King, H.-M. Shen, Circular cylindrical lens as a line-source electromagnetic-missile launcher. IEEE Trans. Antennas Propag. **37**(1), 39–44 (1989)
67. A.V. Boriskin, A.I. Nosich, Whispering-gallery and Luneburg-lens effects in a beam-fed circularly layered dielectric cylinder. IEEE Trans. Antennas Propag. **50**(9), 1245–1249 (2002)
68. T. Komljenovic, R. Sauleau, Z. Sipus, L. Le Coq, Layered circular-cylindrical dielectric lens antennas—synthesis and height reduction technique. IEEE Trans. Antennas Propag. **58**(5), 1783–1788 (2010)
69. R. Sauleau, B. Barès, A complete procedure for the design and optimization of arbitrarily-shaped integrated lens antennas. IEEE Trans. Antennas Propag. **54**(4), 1122–1133 (2006)

Chapter 3
Advanced Feeds for mm-Wave Antenna Systems

Jorge Teniente, Juan Carlos Iriarte, Iñigo Ederra and Ramon Gonzalo

Abstract Millimeter-wave antenna systems have traditionally required high performance feeds in order to fulfill its stringent requirements. Therefore, this goal has been achieved by corrugated horns. However, in the last years new applications mainly in the communication systems have driven the use of other types of antenna feed with slightly reduced performance but simpler manufacturing at mm-wave and submm-wave frequencies with improvements in cost reduction. These advanced profiles are usually based in smooth waveguide advanced profiles. Besides, the chapter describes different feed configurations based on metamaterial structures and several examples of metamaterial based or inspired antennas are considered. Then, this chapter covers the different alternatives currently used for mm-wave antenna feed: corrugated horns, spline horns, and metamaterial horns. At the end of each section, it includes some research successful results.

3.1 Introduction

Millimeter-wave antenna systems have traditionally required high performance feeds in order to fulfill its stringent requirements. Therefore, this goal has been achieved by corrugated horns. However, in the last years new applications mainly in the communication area have driven the use of other types of antenna feed with

J. Teniente (✉) · J.C. Iriarte · I. Ederra · R. Gonzalo
Electric and Electronic Engineering Department, Public University of Navarra,
Campus de Arrosadía, 31006 Pamplona, Navarra, Spain
e-mail: jorge.teniente@unavarra.es
URL: http://antenas.unavarra.es

J.C. Iriarte
e-mail: jcarlos.iriarte@unavarra.es

I. Ederra
e-mail: inigo.ederra@unavarra.es

R. Gonzalo
e-mail: ramon@unavarra.es

A. Boriskin and R. Sauleau (eds.), *Aperture Antennas for Millimeter and Sub-Millimeter Wave Applications*, Signals and Communication Technology,
DOI 10.1007/978-3-319-62773-1_3

slightly reduced performance but with simpler manufacturing and reduced cost. The simplicity of manufacturing is especially critical as frequency increases and in fact corrugated horns are extremely difficult and expensive to manufacture at submm-wave frequencies.

This chapter will cover the different alternatives currently used for mm-wave and submm-wave antenna feeds: corrugated horns, smooth-walled horns where multi-flare angle horns and splined horns will be introduced as alternatives to corrugated horns at mm-wave and submm-wave frequencies. Finally, the chapter will finish with the perspectives and actual research results regarding metamaterial-based feeds toward future applications.

3.2 Advanced Corrugated Horn Antennas

This section describes the actual corrugated feed horn technologies employed at mm-wave frequencies. It begins explaining the principles of the electrical behavior of the electrical fields inside a corrugated waveguide. The hybrid mode basis as a tool to help in the analysis of corrugated horns and the relation between waveguide modes and free space gaussian modes are explained to understand the behavior of the different types of corrugated horn antennas.

In fact, corrugated feed horns are difficult to manufacture as frequency increases and usually are not used above W-band because their manufacture results quite difficult. Corrugated horn fabrication above W-band usually needs using expensive electroforming techniques or even stacked rings if the weight is not an important parameter.

3.2.1 Introduction to Corrugated Horn Antennas

Nowadays, corrugated horns have become the preferred choice of feed antennas for use in high restrictive applications. This is because of their superior radiation performance, their high co-polar radiation pattern symmetry, and their low cross-polarization. Usually they are the preferred choice when a superior radiation pattern is needed for a specific application and the main market is spaceborne communications and science missions in fact.

The operation principle of corrugated horns can be physically explained by considering the way in which the corrugated wall affects the field distribution inside a corrugated waveguide (see Fig. 3.1). It can be demonstrated that the corrugations change the fields traveling through the waveguide to produce the desirable radiating properties of axial beam symmetry, low sidelobes and low cross-polarization [1].

A linear electric field for low cross-polar level will be desirable but it cannot be obtained with smooth waveguides that only support pure transverse electric (*TE*) or

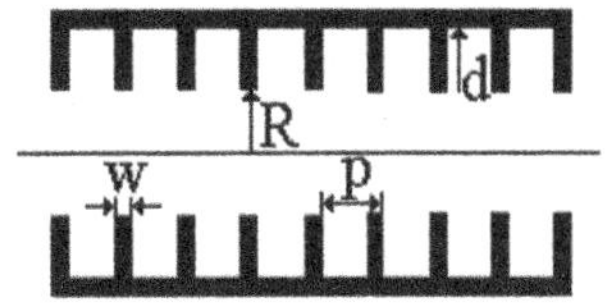

Fig. 3.1 Corrugated waveguide

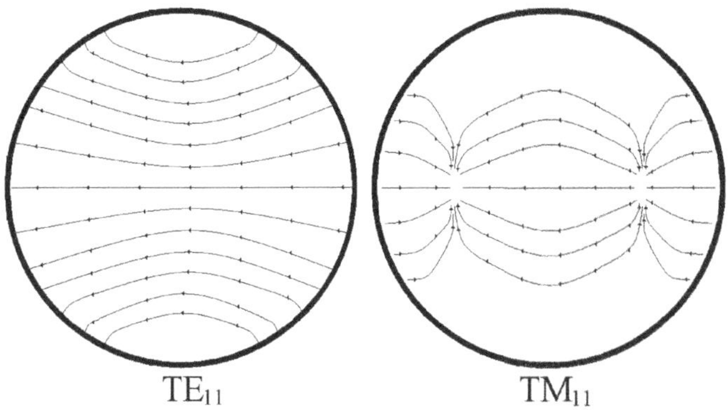

Fig. 3.2 TE_{11} and TM_{11} aperture electric fields

a pure transverse-magnetic (*TM*) modes. These modes have the aperture electric field lines curved (see Fig. 3.2). Therefore, a multimode horn should be designed. In [2], a special horn design to obtain an appropriate mode mixture by the addition of TE_{11} and TM_{11} modes in a particular proportion and phase was presented, but its bandwidth was very narrow.

It is well known that to define properly the field inside a circular waveguide the most known basis used is the *TE* and *TM* mode family which are a direct solution of the wave equation inside a smooth circular waveguide [3]. But if the waveguide is corrugated it could be also useful to define the field inside the waveguide by the family of hybrid modes *HE* and *EH* (those modes which does not present pure transverse components along the waveguide). So, in fact we can choose to define the field inside the corrugated horn antennas in terms of *TE* and *TM* modes or in terms of *HE* and *EH* modes.

Theoretically, the hybrid modes HE_{1n} present at the aperture of a circular waveguide perfectly linear electric field lines, (see Fig. 3.3). It is interesting to aid in the knowledge of corrugated horn antennas to deepen in the understanding of hybrid modes of a corrugated waveguide.

As a corrugated horn is fed via a circular waveguide that propagates the fundamental mode of a circular waveguide (TE_{11}) and considering that a corrugated horn is composed of symmetrical radius variations of such circular waveguide at its throat, the only hybrid modes to consider in a corrugated waveguide or horn antenna are those with $m = 1$, HE_{1n} and EH_{1n}.

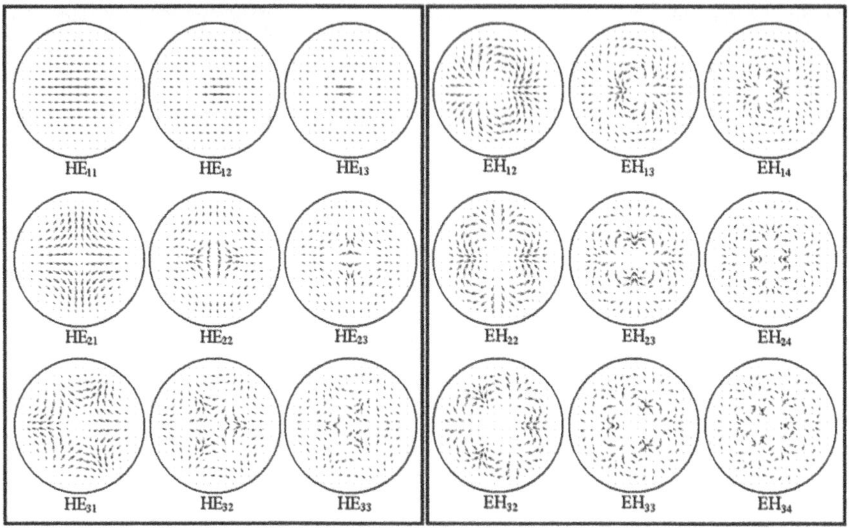

Fig. 3.3 HE_{mn} and EH_{mn} aperture electric fields

Table 3.1 HE_{11} mode decomposition in terms of TE_{1n} and TM_{1n} modes

TE modes	$TE_{11}\begin{cases}84.496\%\\0^\circ\end{cases}$	$TE_{12}\begin{cases}0.082\%\\180^\circ\end{cases}$	$TE_{13}\begin{cases}3.58\cdot10^{-3}\%\\180^\circ\end{cases}$	$TE_{14}\begin{cases}4.94\cdot10^{-4}\%\\180^\circ\end{cases}$
TM modes	$TM_{11}\begin{cases}14.606\%\\0^\circ\end{cases}$	$TM_{12}\begin{cases}0.613\%\\0^\circ\end{cases}$	$TM_{13}\begin{cases}0.121\%\\0^\circ\end{cases}$	$TM_{14}\begin{cases}0.039\%\\0^\circ\end{cases}$

Then, the HE_{11} mode is the fundamental mode of a corrugated waveguide and presents linear electric field at the aperture, (see Fig. 3.3). The mode can be expressed as a combination of TE and TM modes, as a generating basis that they are. Commonly, in the bibliography [1], the HE_{11} mode is supposed to be a combination of *85%* of TE_{11} and *15%* of TM_{11} with the adequate phase shift between them. But in fact, this mode mixture is not perfect, (*99.19%* efficient with HE_{11} mode), the perfect mode mixture in terms of smooth waveguide modes can be seen in Table 3.1.

It is also important to know the radiation properties of hybrid modes to understand the behavior of corrugated horn antennas. In Fig. 3.4, the radiation diagrams of the first HE_{1n} and EH_{1n} modes are represented. In that figure we can see, for example, the effect on the cross-polar component if a EH_{1n} mode is excited inside a corrugated horn. Or also from the same figure, the excellent radiation properties of the HE_{11} with nearly null cross-polar component and with a radiation diagram with quite low sidelobes can be observed.

The HE_{11} fundamental corrugated waveguide hybrid mode is then an excellent mode for radiation purposes. For a common corrugated horn with oversized

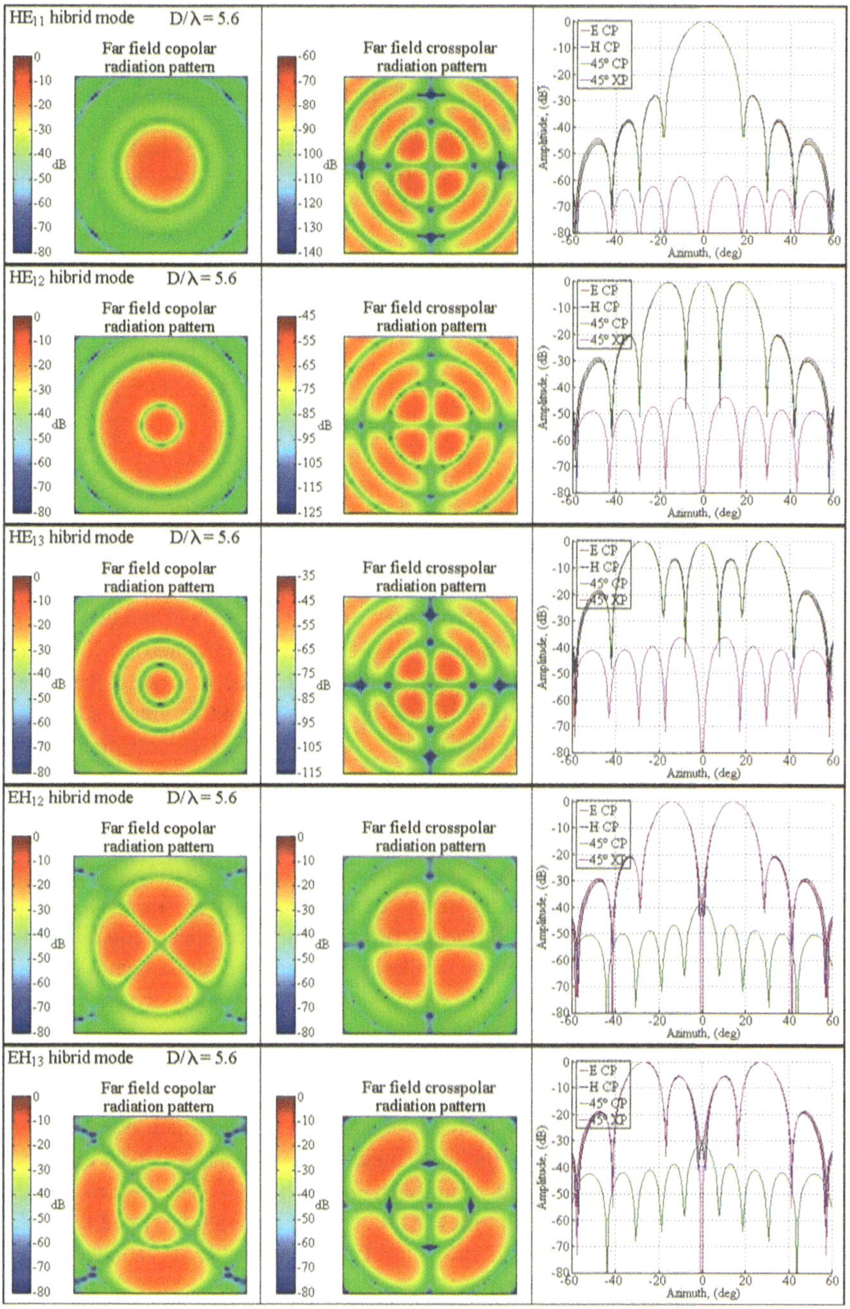

Fig. 3.4 Radiation properties of several HE_{1n} and EH_{1n} modes

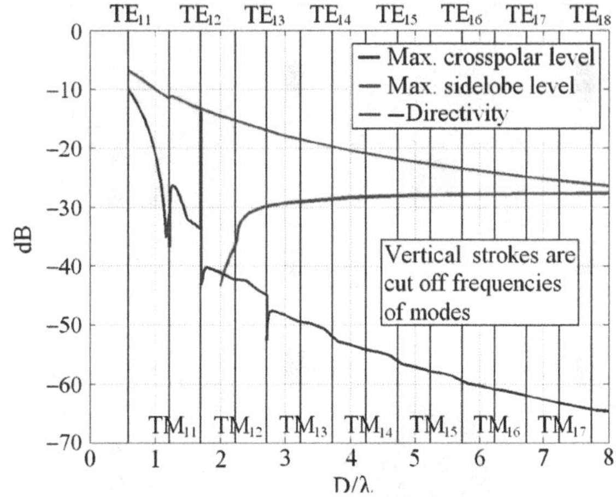

Fig. 3.5 Radiation properties of the HE_{11} hybrid mode against aperture diameter (D)

aperture, the radiated cross-polarized level will be very low, the aperture illumination efficiency will be high and the sidelobe level will be also quite low, (see Fig. 3.5).

As a conclusion of the radiation properties of the HE_{11} hybrid mode, it can be said that this mode is an excellent mode to be excited at the aperture of a corrugated waveguide, and it offers low cross-polar levels and quite low sidelobe levels. However, if very low sidelobes are required for a high restrictive antenna, and it is necessary to maintain low cross-polar levels, we would need several higher order HE_{1n} modes to be present at the aperture of the horn with appropriate amplitude and phase shift between them. This is in fact the purpose of advanced corrugated feed horn designs via optimization of the profiles and the corrugation parameters.

However, it is well known that one of the best ways to define a free space radiation from a horn antenna is by means of the paraxial free space modes, the gaussian modes, which are a solution of the paraxial free space equation. It is important also to remark the radiation similarity between the fundamental gaussian mode of a certain radiation pattern and the HE_{11} mode at a certain diameter aperture. In fact, the fundamental gaussian beam mode can be decomposed in terms of smooth waveguide modes (TE_{mn} and TM_{mn}) and also in terms of corrugated waveguide hybrid modes (HE_{mn} and EH_{mn}) at a certain horn aperture radius (R) with respect to the beamwaist (w_0) of the gaussian beam at such horn aperture.

From Fig. 3.6, we can see that the fundamental gaussian mode can be expressed completely as a combination of TE_{1n} and TM_{1n} smooth waveguide modes and also as a combination of HE_{1n} hybrid modes. At this point, it should be noted that HE_{11} mode has been always known as a gaussian-like mode because its radiated field is nearly a pure gaussian, in fact it is up to *98.1%* efficient with a fundamental gaussian beam of $R/w_0 = 1.554$ ($w_0/R = 0.6435$) (see Fig. 3.6b). To obtain at the aperture of a corrugated horn antenna a high efficient fundamental gaussian beam mode (this implies lower sidelobes), more hybrid modes must be present at the horn

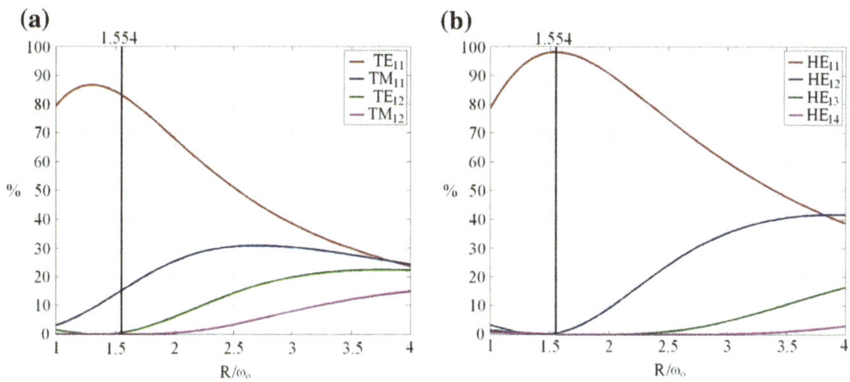

Fig. 3.6 **a** Fundamental gaussian mode decomposition in terms of TE and TM **b** Fundamental gaussian mode decomposition in terms of HE and EH

aperture as it has been introduced. This aspect will be covered in the following lines regarding advanced corrugated feed horn design.

3.2.2 Radially Corrugated Feed Horn Antennas

The design of circular horn antennas was based, for a long time, on the control of the waveguide mode mixture to excite an HE_{11} hybrid mode. It is well known from the previous part that this hybrid mode can be made up of approximately a combination of *85%* TE_{11} and *15%* TM_{11} smooth circular waveguide modes with an appropriate relative phasing between them. The starting field distribution is usually the TE_{11} mode of the circular waveguide under monomode operation, and by means of a proper step or taper in the horn radius, the right amount of TM_{11} (amplitude and phase) was excited (Potter-type horns) [1, 4, 5]. To get this mixture with nice radiating features, two main parameters had to be considered: the output diameter and the horn length. Since the coupling coefficient between waveguide modes is directly related to the waveguide slope change, for a given output radius that fixes the desired beamwidth, the change in horn length allows the designer to select the appropriate phasing in the *85%* of TE_{11} and *15%* of TM_{11} mode mixture obtaining the appropriate sidelobe and cross-polarization minimum levels. This type of horn antennas has been extensively used in the past and is known as Potter-type horns, [4]. Its drawback is the reduced bandwidth a design of this type could cover.

Another technique is based on corrugated circular waveguides and takes profit of the fact that this mode mixture corresponds to the fundamental mode of a circular corrugated waveguide, the HE_{11} mode. This technique reported in [1, 5–7] involves a gradual matching of the smooth circular guide to another corrugated one wherein the corrugation depth is smoothly tapered from $\lambda/2$ *to* $\lambda/4$. These two outlined techniques are combined in the so-called radially corrugated horn antennas with a matching device at their input port. In principle, corrugated horn antennas present a wider frequency response than Potter-type horns. Their design parameters are basically: corrugation parameters (period, duty cycle, depth, shape, etc.); length and profile of the $\lambda/2$-*to*-$\lambda/4$ impedance matching transformer; and the horn geometry in order to optimize the global performance of the horn. In the past, many of the applications involving high performance horn antennas have been equipped with radially corrugated horn antennas. Radially corrugated horn antennas are one of the best possibilities to accomplish very high radiation pattern requirements and they have been extensively used at mm-wave radiation applications with the only drawback of their difficulty of manufacture as the frequency increases.

The most common radially corrugated horns are the ones where the tapering of the horn is a constant slope. In this type of horns, the intention is to generate the HE_{11} fundamental hybrid mode inside the corrugated waveguide and to guide it smoothly to a certain aperture diameter. This type of horns avoided the coupling to any other hybrid mode because the designer thought that the rest of hybrid modes would ruin the radiation pattern. But every horn designer knows nowadays that this is not completely true since the addition of HE_{1n} modes does not affect cross-polar level but can improve the main beam radiation pattern lowering the sidelobes and helping to improve the Gaussian efficiency in the final radiation pattern, EH_{1n} hybrid modes must be avoided in any case since they add a lot of cross-polar radiation (see Fig. 3.3). Radially profiled corrugated feed horn antennas that improve the radiation pattern by means of optimization of their profile will be presented in the next section. In this section, as a simpler way to understand corrugated horns, the design of the ones that have a constant slope tapering is being covered.

To design a radially corrugated horn antenna for a specific application, the first thing we must define are the corrugation parameters. Such parameters that are frequency related can be chosen according to the corrugated waveguide (see Fig. 3.1) as follows:

- Input radius, R, must be of enough size to allow TE_{11} circular waveguide mode to be above cutoff at the lowest usable frequency and as small as possible to be in monomode operation at the highest cutoff frequency. Sometimes, this second condition cannot be met and the TM_{01} mode could be present at the input radius. This aspect is usually defined by the reflection coefficient requirement we must achieve since bigger input radius facilitates this parameter.
- The corrugation depth, d, should be around $\lambda/4$ with λ de free space wavelength at the central frequency (if there are several bands use the central frequency between although such frequency is not considered). This is a parameter to be optimized for every corrugation, so begin with a number around this $\lambda/4$ mentioned.
- The corrugation period, p, should be around $\lambda/3$ with λ the lowest frequency. Use a value rounded to the smallest integer in mm or use at least only one

decimal value. The corrugation period is constant along all the corrugated horn, if its dimension is optimized it complicates a lot the fabrication.
- The corrugation tooth width, *w*, should be around in between *p/2* to *p/5* with *p* the corrugation period. Use a value rounded to the nearest integer in mm or use at least only one decimal value. The thinner the corrugation tooth width, the lower the weight of the corrugated horn, but too thin values complicate a lot the fabrication, do not use values lower than 0.4 mm for this parameter if you plan to manufacture the mm-wave corrugated horn with a lathe. The corrugation tooth width is also constant along all the corrugated horn, if its dimension is optimized it complicates a lot the fabrication.

Once the corrugation parameters have been decided, the corrugated horn profile must be prepared. This preparation can be divided in to two parts, throat region and flare region:

- The throat region controls mainly the reflection coefficient result. To obtain a nice reflection coefficient for a corrugated horn antenna, the most common technique employed is to adapt the transition between smooth circular waveguide to corrugated waveguide. This is usually made via an impedance transformer with consists usually in a taper between $\lambda/2$ and $\lambda/4$ corrugation depths in the first 4 to 10 corrugations, (see Fig. 3.7). As a beginning point, use a linear taper in the first five corrugations and optimize this tapering for reflection coefficient improvement. The designer must consider that the first corrugation depth cannot be bigger than the input radius because the horn could not be manufactured in one single piece via a lathe. This aspect is in fact the main drawback of this type of corrugated horns.
- The flare region, (see Fig. 3.7), controls the radiation parameters. The designer should use as a starting point, the diameter given in Fig. 3.8 for a certain directivity. The flare angle selected must be the minimum possible considering the sidelobe level or the spillover power the application can assume. The

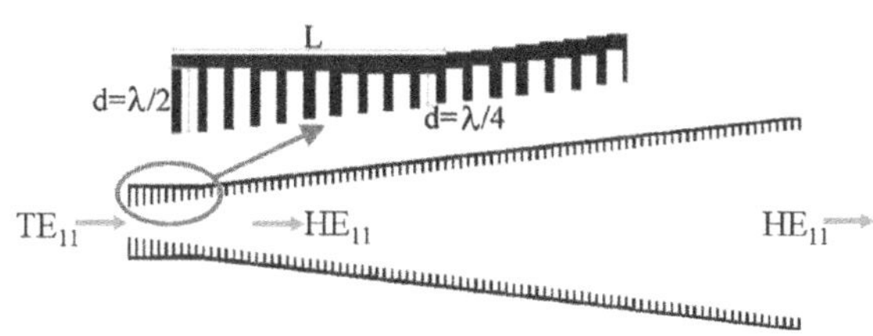

Fig. 3.7 Detail of a $\lambda/2$ to $\lambda/4$ impedance transformer at the throat of a corrugated horn antenna

designer must consider that a low flare angle means longer corrugated horn and more difficult to manufacture, especially as frequency increases.

As a practical example of how to design a linear taper corrugated feed horn, let's say that we want to design a 22 dB directivity corrugated horn antenna with sidelobes lower than –25 dB and as low return loss and cross-polar level as possible in the most compact profile. Analyzing carefully Fig. 3.8 and by means of a simple optimization via a mode matching software code [8, 9], the result leads to an antenna of 6.6° taper profile with the first corrugation depth of $0.52{\cdot}\lambda$ decreasing linearly with an impedance transformer of $1.9{\cdot}\lambda$ long to a corrugation depth for the rest of the antenna of $0.25{\cdot}\lambda$, see Fig. 3.9. The corrugation parameters have been selected as $p = \lambda/5$ and $w = p/3$. The resultant length of the profile is $17.8{\cdot}\lambda$ and presents a diameter of $4.86{\cdot}\lambda$. It is a quite long antenna for 22 dB directivity. This length cannot be shortened if we must have the sidelobe level at –25 dB or less by

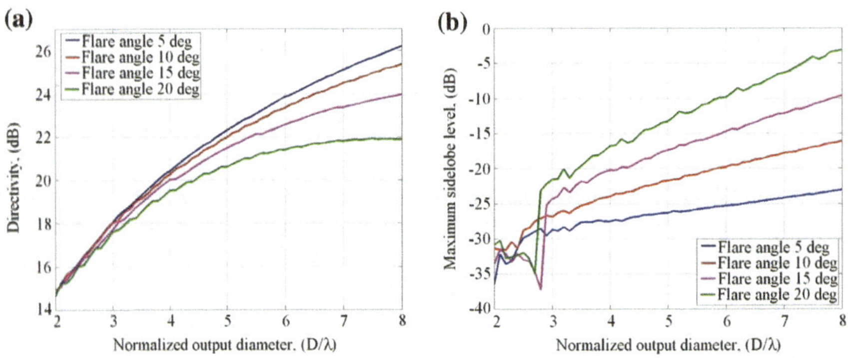

Fig. 3.8 **a** Directivity design curves for linear taper corrugated feed horns **b** Max. sidelobe level design curves for linear taper corrugated feed horns

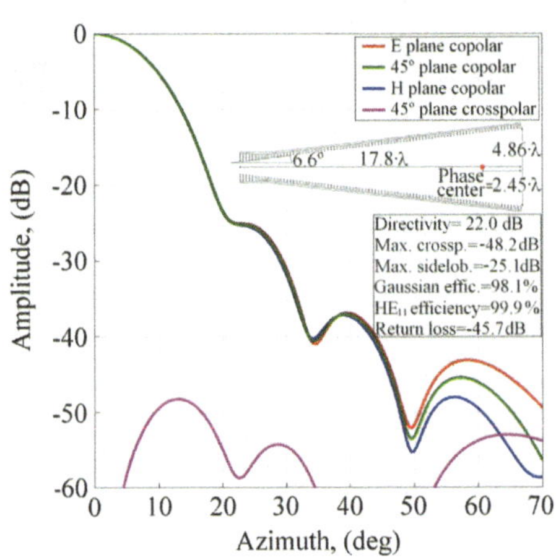

Fig. 3.9 Linear taper corrugated horn antenna design for 22 dB directivity and −25 dB maximum sidelobe level

any method if we maintain the linear taper corrugated profile, but in the next section we will learn that this can be made by means of an optimization of the profile.

3.2.3 Profiled Radially Corrugated Horn Antennas

In the last section, the linear taper corrugated feed horns have been covered, but there are not much parameters to improve their performance. To overcome the optimization limitations of the linear profiled corrugated horn antennas, during the 90s, several profiles by means of different formulas appeared in the scientific literature. The reason of such formulas was the availability of mode matching techniques (MM) for the design of corrugated horns. However, since the computational speed was not too much, the designers used several profile formulas to reduce the number of unknowns in the profiled corrugated feed horn design and then increase optimization speed.

A certain number of formulas appeared to solve this problem. In fact, every research group in corrugated horn antennas had their own preferences [10], such formulas were based in square roots, exponentials, sine-squared, polynomials, gaussian, etc., see Fig. 3.10.

One of the formulas used to profile radially corrugated horn antennas was the gaussian beam expansion formula that is in fact a form of a square root, [11]. This type of formula aroused to implement a perfect match between corrugated waveguide modes (mostly HE_{11} mode or similar mode mixtures) and the fundamental free space modes (fundamental gaussian mode, Ψ_{00}). By using this formula, the matching between the waveguide and the free space was almost perfect, being the most "natural" way to match the two media.

As a practical example of how to design a profiled corrugated feed horn with the profile defined via a gaussian beam expansion formula, let's say that we want to

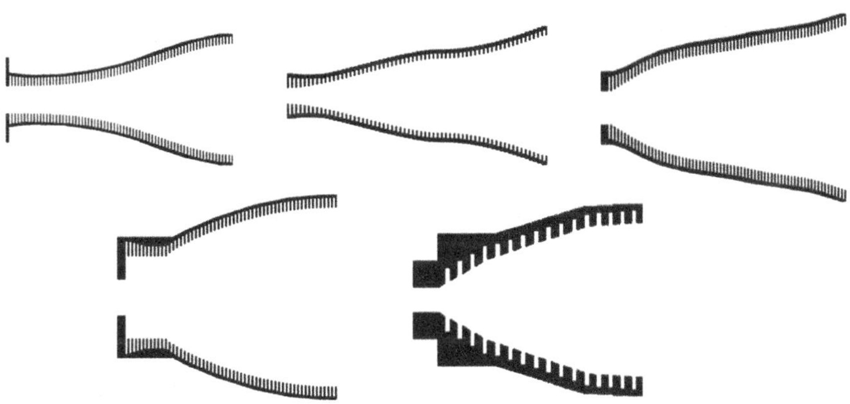

Fig. 3.10 Examples of profiled corrugated horn antenna designs

design again a 22 dB directivity corrugated horn antenna but now we are asking for sidelobes lower than –35 dB and as low return loss and cross-polar level as possible in the most compact profile. We use for the profiled radially corrugated feed horn definition two parts, the first part presents three sections: The first section at the beginning presents a quick and smooth change in the slope at the throat, (less than 10% of the total first part length). The second section has a linear taper profile to get quickly to the needed diameter (more than 65% of the total first part length). The third section (less than 25% of the total length) is a smoothed end to allow connectivity to the second part corrugated feed horn profile with a gaussian beam expansion formula, see Fig. 3.11.

Analyzing carefully Fig. 3.11, the result leads to an antenna of $11.07 \cdot \lambda$ length and $5.64 \cdot \lambda$ aperture diameter. The reduction in length, compared to the linear taper profile, is 38%. It must be also considered the improvement in radiation pattern with 10 dB lower sidelobe level. In fact the bigger aperture diameter is caused by such requirement of lower sidelobe level, the same radiation pattern would lead to even shorter solution with the same aperture diameter.

Nowadays, the availability of optimized mode matching software packages [8, 9] and the increased speed of computers make possible the optimization of a corrugated feed horn without the necessity to use a formula that defines the profile. In fact, usually the corrugated feed horn designers use a formula to define just the initial profile and after that, all corrugation radiuses (inner and outer) are optimized individually, the final appearance of the optimized radially corrugated feed horn is radically different from any formula and in fact the designer can adapt such optimization to generate the smallest size possible profiled radially corrugated horn and at the same time comply with all of the stringent requirements usually demanded for this type of feed horns.

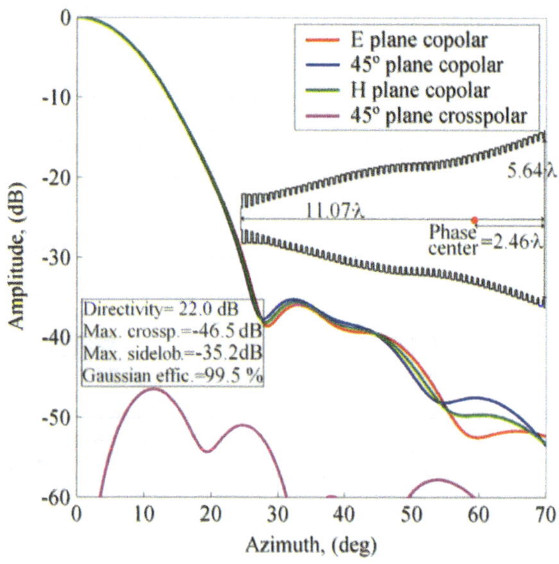

Fig. 3.11 Profiled radially corrugated feed horn designed and optimized with a combination of two sections defined by formulas for 22 dB directivity and −35 dB sidelobe level

3.2.4 Corrugated Horn Antennas that Combine Axial and Radial Corrugations

One of the main drawbacks for manufacture of radially corrugated feed horns is the impedance transformer at their throat. This part is difficult to manufacture in a piece by means of a lathe since the farther corrugations are also the deepest and indeed the ones where the fabrication tolerance is worse. This can be solved making such horn in several parts or by making the throat by means of stacked irises but in this section an elegant solution to this problem is given. This solution not only solves the common manufacture techniques; besides it even results in shortening the total profile by a significant amount.

The solution we are referring to, is a horn antenna that combines horizontal corrugations (known also as axial corrugations or chokes) for the throat region and vertical corrugations (known also as radial corrugations) for the flare region [12]. The design guidelines for this type of corrugated horn antennas were fully explained in [13].

This type of horn antenna achieves significant improvements on the following four parameters, when compared to a normal radially corrugated horn, maintaining the same radiation performance:

1. Shorter horn axial length
2. Improved return loss over a wide bandwidth
3. Smaller design computation complexity
4. Reduction of the manufacturing complexity (avoids the deep first radial corrugations)

Again, as a practical example of how to design a corrugated horn antenna that combines axial and radial corrugations, let's say that we want also to design a 22 dB directivity corrugated horn antenna but now we are asking for sidelobes lower than –40 dB and as low return loss and cross-polar level as possible in the most compact profile. We use for the throat region an axially corrugated part that presents a linear taper and six axial corrugations. For the flare region, we select a profile defined by a gaussian beam expansion formula, see Fig. 3.12. The solution leads to an antenna of only $7{\cdot}\lambda$ length and $6.3{\cdot}\lambda$ aperture diameter. The reduction in length compared to the optimized radially corrugated profile is a 37% and compared to the linear taper radially corrugated profile the reduction is a 61% and it should also be considered the improved radiation pattern with −40 dB sidelobe level, in fact again the bigger aperture diameter is caused by such requirement of lower sidelobe level, the same radiation pattern would lead to even shorter solution with the same aperture diameter as the previous solutions.

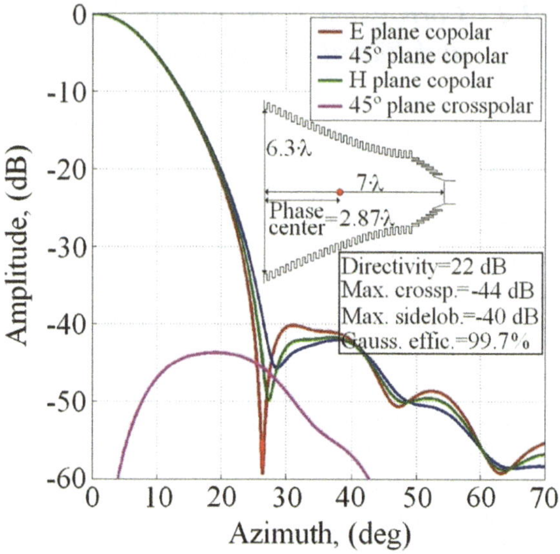

Fig. 3.12 Complete corrugated GPHA with a choked corrugated waveguide input for 22 dB directivity and −40 dB sidelobe level

3.2.5 *Examples*

In this section, two examples of advanced corrugated feed horn antennas are being presented. Both have been manufactured at mm-wave frequencies with two different manufacturing techniques and tested obtaining the expected performance simulations predicted.

3.2.5.1 Axial and Radial Corrugated Feed Horn Antenna Design for a mm-wave 80–100 GHz Body Scanner

An 80−100 GHz corrugated horn design was optimized to act as feed of a rotating reflector for a body scanner. The design was implemented by means of axial and radial corrugations and the main requirement was a very low spillover to achieve as pure as possible image from the body emission at mm-wave frequencies.

The design parameters were a FWHM (Full-Width Half-Maximum) of 14°, gaussian beam decay radiation pattern till 26 ° in azimuth (view angle from the subreflector) and sidelobes from 26° in azimuth lower than −30 dB. The maximum measured cross-polar level should be below −25 dB and the measured reflection coefficient should be lower than −20 dB.

The feed horn was designed with these specifications resulting in a total length of 46.3 mm and an output diameter of 36 mm, (see Fig. 3.13). The manufacture was made by means of iris rings of 0.7 mm and 0.2 mm sequentially stacked to form the radially corrugated part. The axially corrugated part was manufactured with a high-precision milling machine.

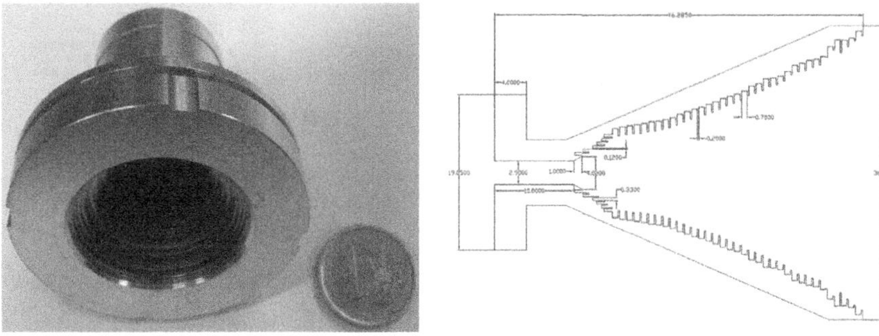

Fig. 3.13 Manufactured axial and radial corrugated feed horn antenna design for a mm-wave 80–100 GHz body scanner

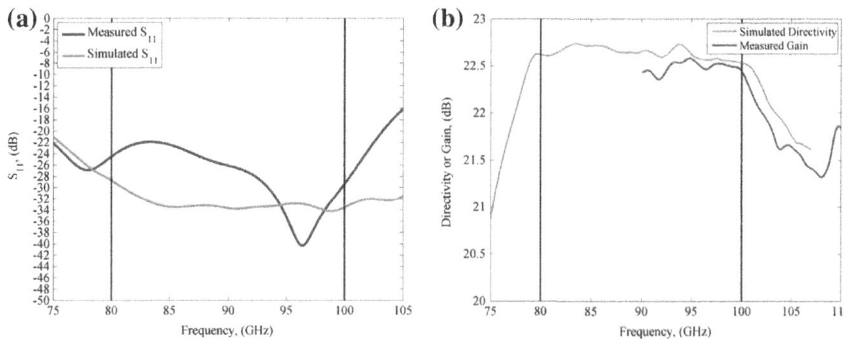

Fig. 3.14 Simulated and measured reflection coefficient **a** and gain **b** for the axial and radial corrugated feed horn antenna design for a mm-wave 80–100 GHz body scanner

Simulated and measured reflection coefficient can be seen in Fig. 3.14a. The measured reflection coefficient is lower than −22 dB in the whole band. Regarding the gain, see Fig. 3.14b, losses lower than 0.2 dB were measured as well. Measured radiation patterns can be checked in Fig. 3.15, the measured cross-polar level is always below −28 dB for the whole band. The measured main beam radiation pattern decay is as expected from simulations, approximately of 14° FWHM.

3.2.5.2 Radially Profiled Corrugated Feed Horns for ESA Funded MARSCHALS Airborne System

MARSCHALS was a three-channel limb sounder for the European Space Agency (ESA) that flew on an aircraft. The designer of the mission was especially concerned about the sidelobes generated by the feed horns or the rest of the optics. The aim of the MARSCHALS project is the limb observation, and therefore any power introduced via a sidelobe pointing the earth should be consider as noise decreasing

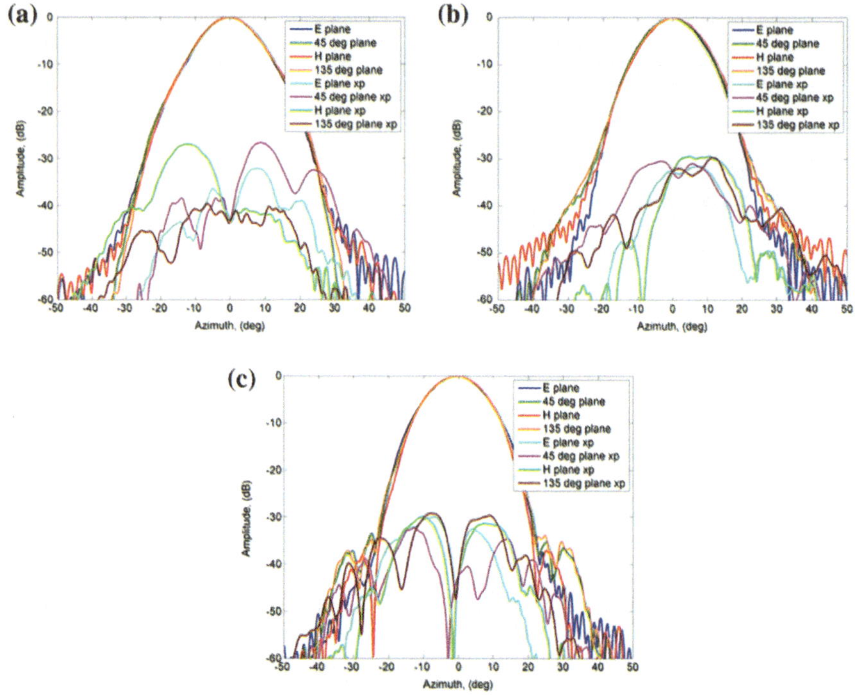

Fig. 3.15 Measured radiation patterns at 80 GHz (**a**), 90 GHz (**b**) and 100 GHz (**c**) for the axial and radial corrugated feed horn antenna design for a mm-wave 80–100 GHz body scanner

the performance of the whole system. The sidelobe requirement was *–35* dB in both feed horn designs.

The requirements for these corrugated horn antennas (called antenna in band C and antenna in band D) are as follows:

Other parameters common to both antennas are as follows:

- Sidelobes: It was required sidelobes to be less than *−35* dB (with no "shoulders" on the main lobe). Obviously, the lower the better.
- Peak cross-polar: below *−35* dB (again, the lower the better)
- Feed waveguide: Full height rectangular waveguide (*0.762* × *0.381* mm).
- Horn Length (flange to aperture): Maximum *40* mm.

The manufacture of the complete corrugated feed horn was made by means of high-precision electroforming including the rectangular to circular transition to the rectangular feed waveguide of *0.762* × *0.381* mm dimension. To reduce the manufacture complexity, the corrugation tooth width should be exactly half of the corrugation period and constant along the whole antenna.

The bandwidth requirements were not too tight, *2.8%* for the C-band antenna and 1.9% for the D-band antenna. The resultant directivity, by means of translation

Table 3.2 MARSCHALS feed horns requirements

Band	Frequency range (GHz)	Center frequency (GHz)	Beamwaist radius (mm)
C	*316.5–325.5*	*321.00*	*2.078*
D	*342.2–348.8*	*345.50*	*2.003*

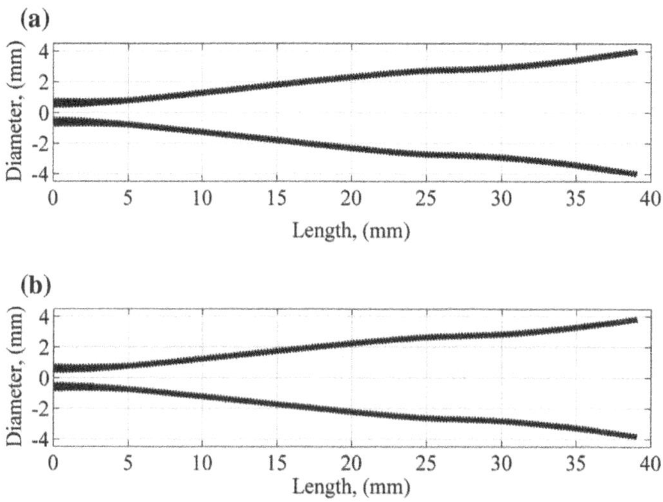

Fig. 3.16 **a** MARSCHALS feed horn proposed profile for C-band **b** MARSCHALS feed horn proposed profile for D-band

of the beamwaist radius values of Table 3.2, was quite high; *2.078* mm beamwaist radius at *321* GHz in C-band antenna means an illumination at *16*° of *–35* dB and therefore a directivity of *26 dB*. As well, the *2.003* mm beamwaist radius at *345.5 GHz* in D-band antenna means an illumination at *15.5*° of *–35* dB and therefore a directivity of *26.3* dB.

The transition from rectangular to circular waveguide was designed ending in a circular diameter of *0.762* mm. This diameter was used as the input waveguide diameter for both antennas. After the optimization of both profiles, these were the results:

- The total length of the C-band antenna is *39.1* mm ($41.9 \cdot \lambda$) and the output diameter $8.29 \cdot \lambda$ (*7.74* mm). A picture of the antenna is shown in Fig. 3.16a.
- The total length of the D-band antenna is *39.1* mm ($45.1 \cdot \lambda$) and the output diameter $8.53 \cdot \lambda$ (*7.398* mm). A picture of the antenna is shown in Fig. 3.16b.

The rectangular to circular transition design was very simple and it was a direct cut of *15*° angle of the circular input waveguide of each antenna in a *0.711* mm length to result in the 0.762×0.381 mm input rectangular waveguide (see Fig. 3.17).

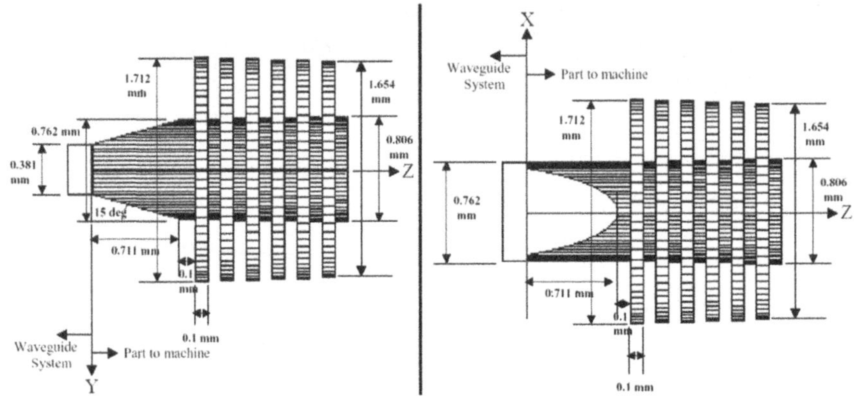

Fig. 3.17 MARSCHALS feed horns throat region with rectangular to circular transition

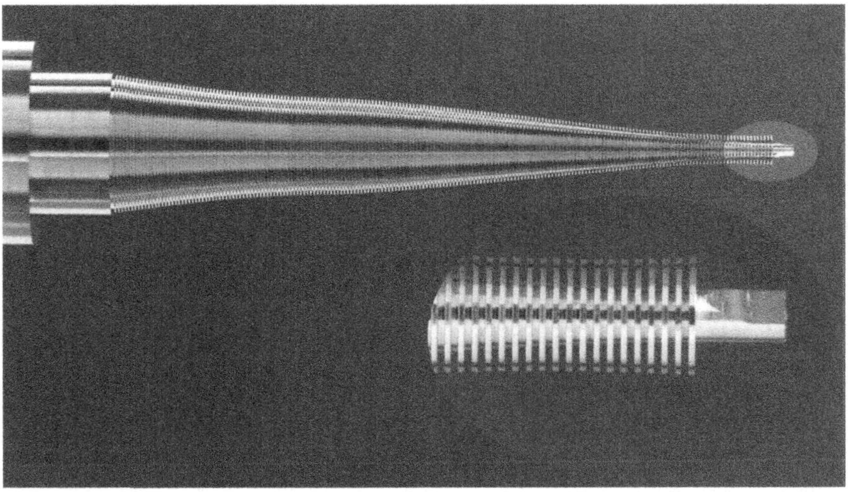

Fig. 3.18 Mandrel of one of the MARSCHALS feed horns

The antenna was manufactured using an electroforming technique. Electroforming is a metal forming process that forms parts through electrodeposition or electroplating on a model, known in the industry as a mandrel. The mandrel was made by a milling machine and a picture of it can be seen in Fig. 3.18. Metallic mandrels are pretreated chemically to allow subsequent separation of the finished electroform. The outer surface of the mandrel forms the inner surface of the form. A thick layer of electroplating copper is applied until the plate itself is strong enough to be self-supporting. The mandrel is dissolved away after forming. The surface of the finished part that was in intimate contact with the mandrel is rendered in fine detail with respect to the original, and is not subject to the shrinkage that

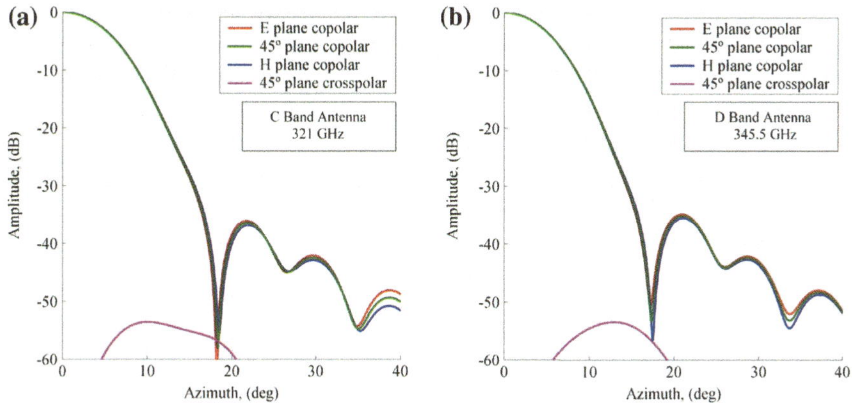

Fig. 3.19 Simulated far-field radiation pattern of both MARSCHALS feed horns

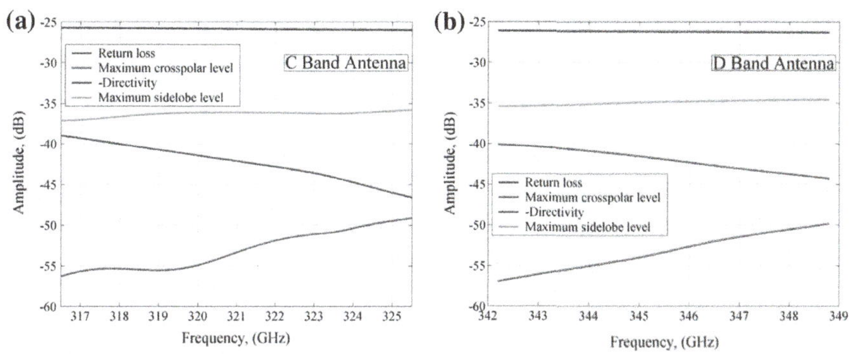

Fig. 3.20 Simulated far-field behavior of both MARSCHALS feed horns

would normally be experienced in a foundry cast metal object, or the tool marks of a milled part. It can be observed in such Fig. 3.18, the extremely high-precision manufacturing of the mandrel remembering that the corrugation tooth width is only *0.1* mm.

Simulated results for both antennas at central frequency can be found in Figs. 3.19 and 3.20. Both antennas meet all the specifications widely.

3.3 Smooth-Walled Feed Horns

This section describes the technology concerning the smooth-walled feed horn technology for mm-wave and submm-wave frequencies. It begins with a short introduction to smooth-walled feed horn profiles and continues with the main techniques to design this type of feed horns, multi-flare angle and spline profiled.

This type of horns presents worse radiation characteristics compared to corrugated profiled ones and are much longer. However, these are much simpler and cheaper to manufacture at high frequencies, so many times are the preferred choice. This problem regarding manufacture of corrugated horns at frequencies above 100 GHz becomes particularly acute when many tens, hundreds, or even thousands of horns are required for large format focal plane array receivers. Therefore, there has been much recent interest in the use of easy-to-fabricate smooth-walled horns [14] at these shorter wavelengths.

3.3.1 Introduction to Smooth-Walled Feed Horn Profiles

Typically, smooth-walled horns either have step or flare angle discontinuities, or smoothly varying interior profiles. The shapes of the horn interiors are optimized numerically by mode matching techniques [8, 9] to generate a balance of higher order waveguide modes which leads to a uniform aperture illumination and hence far-field patterns with high beam circularity, low sidelobe levels and low cross-polarization.

The simplest form of a smooth-walled horn is a Potter-type horn [15], which has only one step discontinuity near the throat followed by a conical flaring section, (see Fig. 3.21). The step discontinuity excites the TM_{11} mode at about 15% of the total incident power of the incoming TE_{11} mode. Both fields propagate along the conical flaring section until they arrive in phase at the horn aperture.

Figure 3.22 shows the electric field of the TE_{11} and TM_{11} modes, and it can be seen clearly that the addition of both in the correct proportion and in phase produces a linearly polarized electric field at the aperture of the horn. The combination of 85% of TE_{11} and 15% of TM_{11} leads in fact to an HE_{11} hybrid waveguide mode, (see Table 3.1), being the mode mixture that generates automatically a linear taper corrugated horn. As it was indicated in the section regarding corrugated horns, this mode combination generates a highly uniform field that produces a radiation pattern with low sidelobes level and low cross-polarization approximating to the performance of a corrugated feed horn. But the main disadvantage of the Potter horn is that it has a narrow operating bandwidth (5–8%).

However, the performance of a Potter-type horn can be substantially improved by increasing the number of discontinuities of the horn. These types of horns are known nowadays as multi-flare angle feed horns [16] and are being covered in the

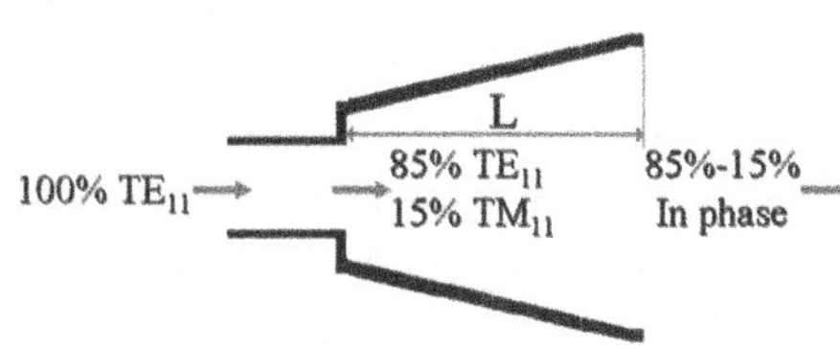

Fig. 3.21 Principle of operation of a Potter-type horn

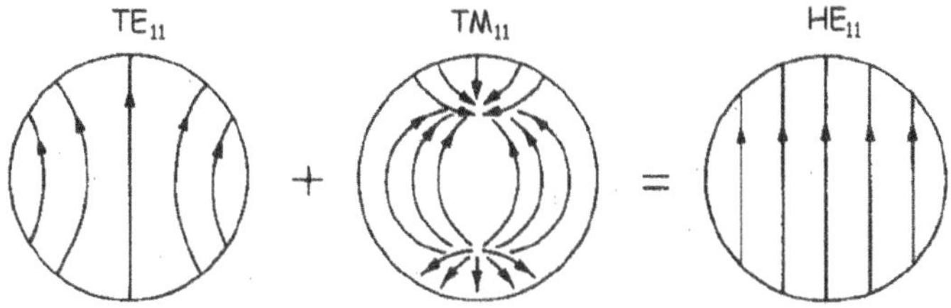

Fig. 3.22 Combination of modes pursued in a Potter-type horn

Fig. 3.23 Principle of operation of a Turrin-type horn

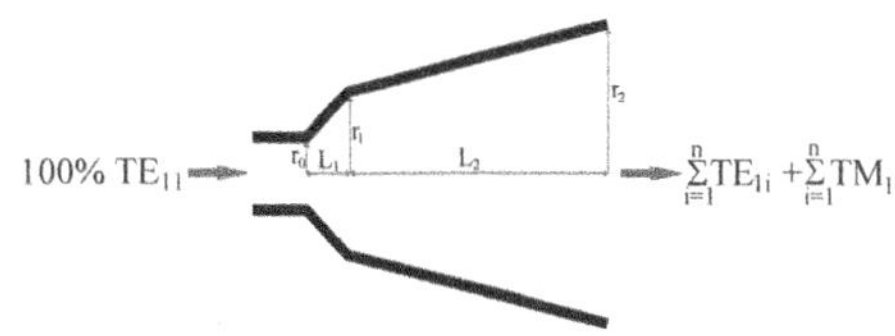

following section. On the other hand, there is another method of optimizing a smooth-walled feed horn profile with even better results, this method consists in the optimization of the coefficients of a spline curve to meet the specific radiation parameters needed for a certain application [14], these types of horns will be covered in the next to the following section.

3.3.2 Smooth-Walled Multi-flare Angle Feed Horns

The Potter-type horn antenna exhibits nice radiation properties, but in a reduced bandwidth. To improve the performance, the designer increases the number of discontinuities of the horn. With some more parameters to optimize, the designer can generate a carefully chosen combination of the higher order modes which could widen the operating bandwidth of the horn and maintain the nice radiation properties.

The simplest multi-flare angle horn is the Turrin-type horn [17], see Fig. 3.23. This type of horn starts with a short conical section with a cone angle and then is followed by another conical section of lower angle.

After the Turrin-type horn that only has two different angles, (see Fig. 3.23), the designer can add as many discontinuities in the throat region of the horn as he needs, these types of horns are then known as multi-flare angle horns, (see Fig. 3.24). The optimized depth of these discontinuities (R_0, R_1, R_2, R_3…), the horn length and flare angles (L_1, L_2, L_3…), can be predicted using modal matching in conjunction with optimization algorithms to derive the desired radiation properties in a certain bandwidth.

Step or flare angle discontinuities near the throat of the horn will, unavoidably, excite other higher modes (TE_{1n} and TM_{1n}) in addition to the desired TM_{11} mode.

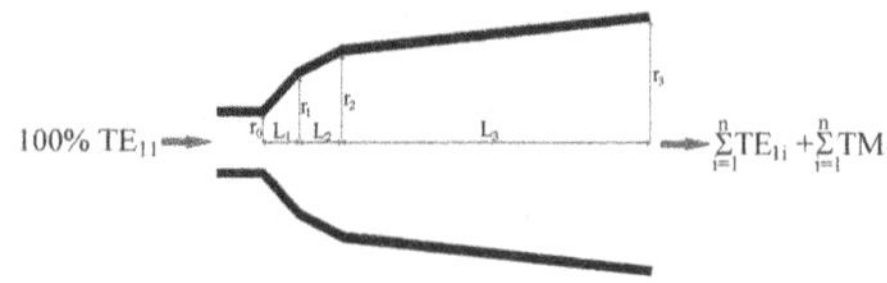

Fig. 3.24 Multi-flare angle smooth waveguide feed horn with three steps

Fig. 3.25 Direct metallic laser sintering manufactured 2 × 2 antenna array of multi-flared angle smooth-walled feed horns for a communications system working at 330 GHz (courtesy of ANTERAL S.L.)

These higher order modes will affect the aperture field in a complicated, frequency dependent way, making the design of multi-flare angle horns that give good performance over a finite bandwidth more difficult. Fortunately, the effect of these modes on the far-field pattern of a particular horn can be predicted very accurately using the numerical modal matching techniques. Since such modal matching techniques can be used to calculate the far-field patterns, it is possible also to use this technique in conjunction with suitable optimization algorithms to determine the optimum horn profiles for good performance over a particular bandwidth [7, 8].

These types of horns are very suitable for manufacture at mm-wave and submm-wave frequencies since they can be made very quickly and cheaply by repeated drilling with a properly shaped electrode as machine tool, [18] or even by additive manufacturing techniques with a certain postprocessing of the manufactured result. In Fig. 3.25, a set of four feed multi-flare angle smooth-walled feed horns for a communication system at 330 GHz is presented. Such submm-wave horn array was manufactured by means of an additive manufacturing technique called direct metallic laser sintering, the array needed a postprocessing with properly shaped electrode to be fully functional.

3.3.3 Smooth-Walled Spline Profiled Feed Horns

The smooth-walled profiles defined in the previous section are formed by steps, but it could be interesting to have the possibility to design a smooth-walled profile whose taper is smoothly opening without any discontinuity. To fill this gap, in 2004, the researchers from CSIRO in Australia presented their contribution, and they called it the smooth-walled spline profile [14]. This profile was developed as a substitute of corrugated horns for mm-wave and submm-wave frequencies since the manufacturing complexity is much reduced. Their performance is not as good as a corrugated horn but it is quite acceptable considering a bandwidth around 30% can be easily achieved for cross-polar levels below −30 dB. In fact, in many cases (as feed for multibeam reflector antennas) smooth-walled spline profiled feed horns have replaced corrugated horns for onboard satellite payload [19].

The spline profile is a numeric function that is piecewise defined by polynomial functions and which possesses a high degree of smoothness at the places where the polynomial pieces connect, see Fig. 3.26. In fact, the term spline was adopted from the name of a flexible strip of metal commonly used by drafters to assist in drawing curved lines.

To design a smooth-walled spline profile feed horn antenna for a specific application, we must define first the initial dimensions of the horn we want to develop. We need as an initial approach three parameters: input radius, output aperture radius and total length. Once they are defined, we should decide how many optimization points we need. Many optimization points will allow a better result, but the optimization time will be longer; a few optimization points (in the Fig. 3.26 only five variables are used) allow a quicker optimization time, but perhaps they are not enough to reach the requirements needed.

If the designer needs to achieve a better result, he must increase the number of optimization radiuses, in fact in [20], the authors increase the number of optimizable radiuses till 20 unknowns improved the result significantly.

As an example, a K/Ka-band smooth-walled spline profile feed horn for a communications satellite is presented, see Fig. 3.27.

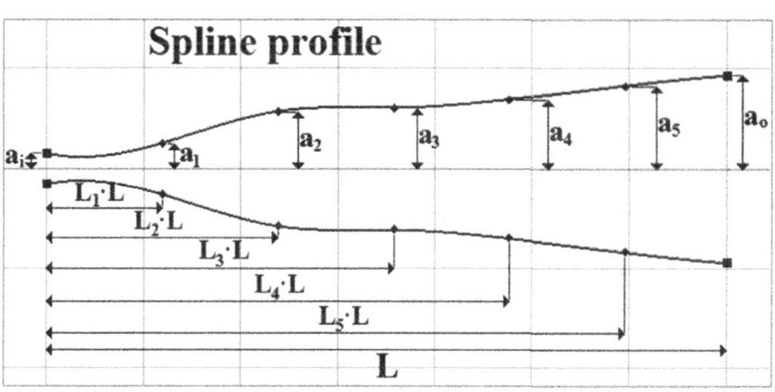

Fig. 3.26 Spline profile feed horn geometry

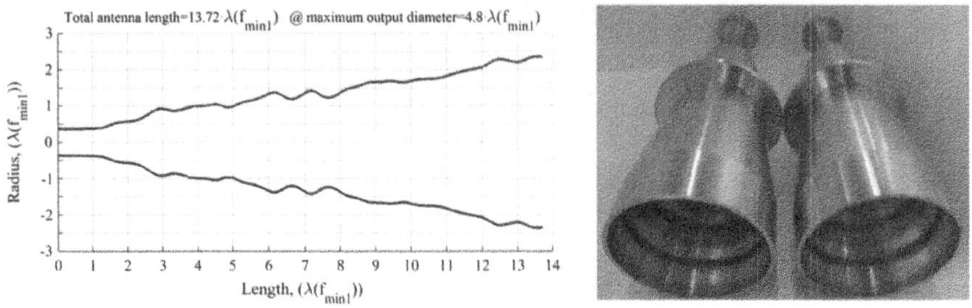

Fig. 3.27 Optimized smooth-walled spline profile feed horn for onboard a communications satellite (courtesy of ANTERAL S.L.)

Such a feed horn has more than a 42.3% bandwidth and the partial bandwidths were 3.5% and 2.4%. Therefore, it is a wideband design and its requirements are extremely severe for a smooth-walled horn, since a return loss value better than 40 dB and a cross-polar level below −26 dB were required for both frequency bands, and they are quite apart in frequency. In addition, it is very short for a spline profile.

The design employs 40 radiuses to define the profile (see Fig. 3.27). The outcome is that of a rather curved inside for smooth-walled horn, but the results (see Figs. 3.28 and 3.29) show that the requirements are met.

The resulting radiation patterns can be seen in Fig. 3.29. This horn antenna was manufactured and is successfully operating onboard a geostationary satellite, where it is part of a feed horn array for spot beam communications.

3.4 Metamaterial Based Feeds

This section describes different feed configurations based on metamaterial structures. The field of metamaterials is very broad and several examples of metamaterial based or inspired antennas are considered along this book. In this case, metamaterial antennas which resemble or create horn-like antennas are treated. First the metallic walls of the horns will be substituted by a tailored surface so that different properties are achieved. These antennas are based on the so-called hard and soft surfaces, which can be considered as special cases of metasurfaces. Finally, horn antennas will be constructed based on 3D electromagnetic bandgap (EBG) structures. Their fundamentals and several examples are given in the following sections.

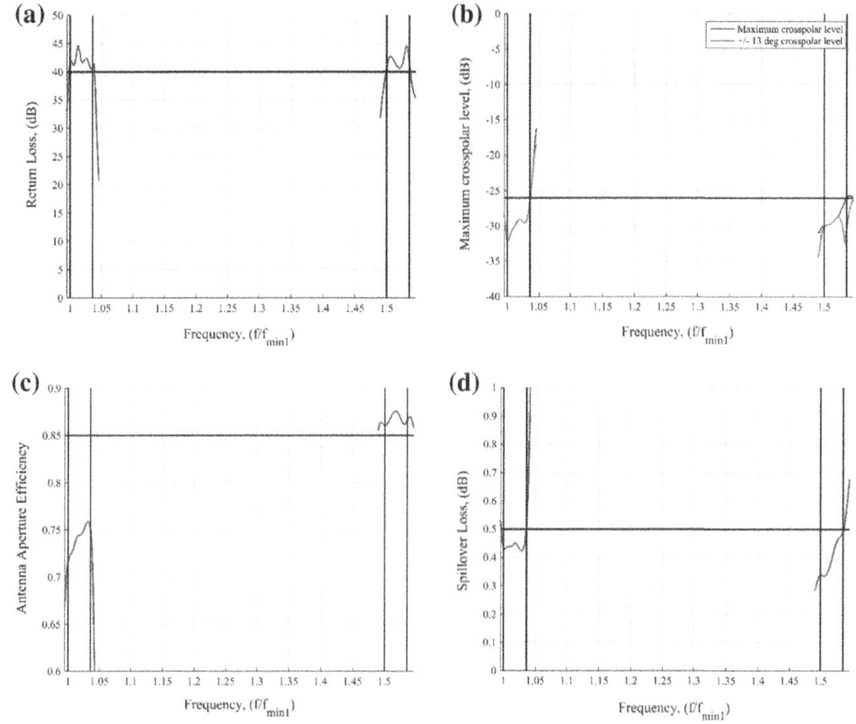

Fig. 3.28 Results of the optimized smooth-walled spline profile: **a** Return loss **b** Cross-polar level **c** Aperture efficiency **d** Spillover loss above 13°

3.4.1 Soft and Hard Horns

Corrugated horn antennas can be considered as a particular case of the more general concept of hybrid mode horns [21]. These antennas support linear polarized modes provided that an anisotropic boundary condition exists in the horn walls. The condition to be fulfilled by the boundary impedances is called balanced hybrid condition:

$$Z_{TE}Z_{TM} = \eta_0^2,$$

where Z_{TE} and Z_{TM} are the TE and TM boundary impedances and η_0 is the free space wave impedance.

Two families of antennas satisfy this condition, leading to the so-called hard and soft horns [21]. In the soft horn case, the boundary conditions correspond to:

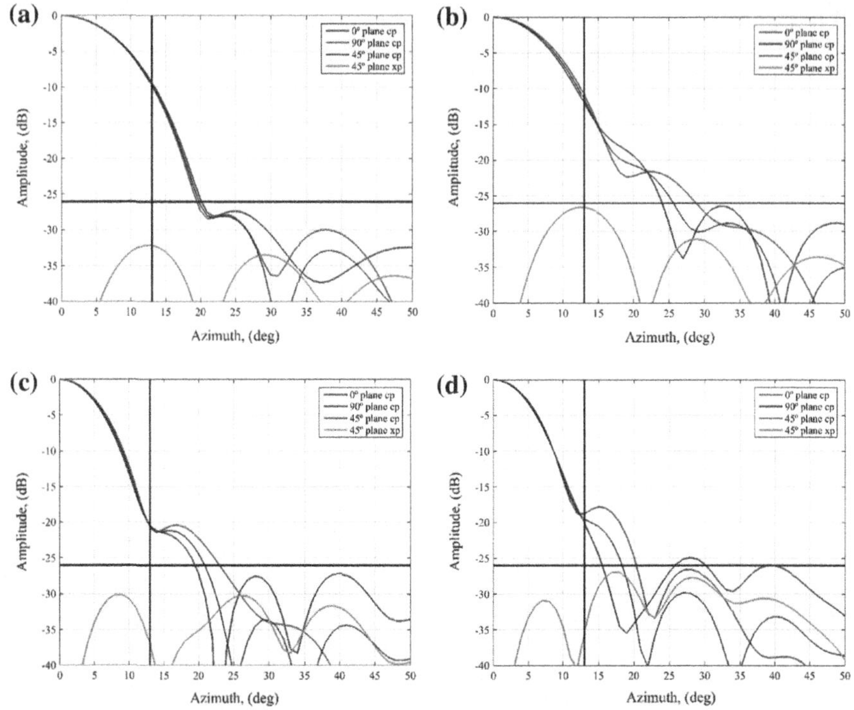

Fig. 3.29 Far-field radiation patterns of the optimized smooth-walled spline profile: **a** At f_{min1} **b** At $1.04 \cdot f_{min1}$ **c** At $1.5 \cdot f_{min1}$ **d** At $1.54 \cdot f_{min1}$

$$Z^{TE} = Z_x = \frac{E_x}{H_z} = 0$$
$$Z^{TM} = Z_z = \frac{E_z}{H_x} = \infty$$

whereas in the hard horn they are:

$$Z^{TE} = Z_x = \frac{E_x}{H_z} = \infty$$
$$Z^{TM} = Z_z = \frac{E_z}{H_x} = 0$$

Soft horns support tapered aperture distributions which lead to very low side lobe radiation patterns. Conversely, hard horns provide uniform aperture distributions which correspond to very high aperture efficiencies.

These hybrid mode horns have been implemented with different techniques, which are reviewed in [22]. These include longitudinal and transversal corrugated horns, dielcore horns, strip loaded and metamaterial or metasurface wall horns.

This last case constitutes the most novel implementation and, thanks to the flexibility of metamaterials, a very promising alternative to the conventional implementations. The following sections describe different implementations of these concepts.

3.4.2 Metamaterial Horns

Metamaterial-based horn antennas, as schematically shown in Fig. 3.30, can be employed for alternative implementations of the dielcore horn designs. Since metamaterials allow obtaining values of the dielectric constant lower than 1, both hard and soft surface designs can be constructed based on the dual dielcore horn. In both cases the core dielectric can be taken as air, which requires the outer dielectric to have dielectric constant lower than 1, which can be achieved by a metamaterial. This has additional advantages, since removing the central core allows reducing mass and losses and improves reflections in the interface between the air-filled feeding waveguide and the dielectric loaded horn antenna.

The concept behind these antennas was first demonstrated by [22]. Simulation results showed that the performance of these antennas complied with the expectations. However, the results were based on theoretical materials with ideal parameters.

The challenge of finding a metamaterial realization which complied with the required parameters was sorted out by [22]. The developed metamaterial is based on a wire grid with wires in the three coordinate axis directions, so that the structural elements of the metamaterial affect all the field components and full control of the surface impedance can be achieved. The first experimental demonstration of a horn antenna based on this metamaterial wall was carried out by the same group [23]. A photograph of the manufactured antenna is shown in Fig. 3.31a. The achieved pattern was symmetric, with low sidelobes along the full super-extended C-band. A comparison between the E-plane pattern of this antenna and a trifurcated and standard pyramidal horn is presented in Fig. 3.31b. Lower sidelobes than in the conventional counterparts have been achieved by means of the metamaterial surface.

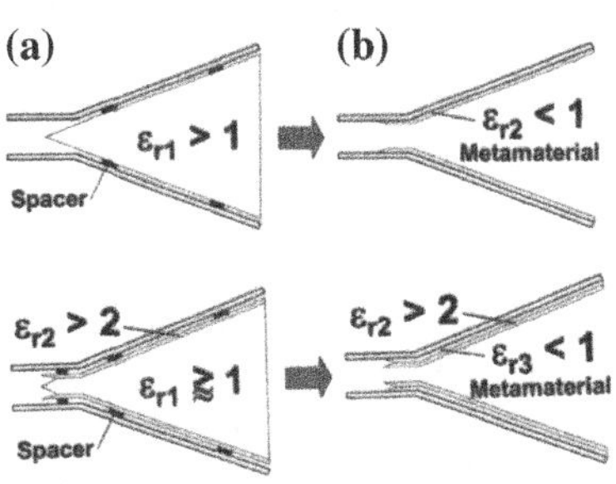

Fig. 3.30 Schematic of hard and soft antenna implementations using metamaterials as alternatives to the dual dielcore horns

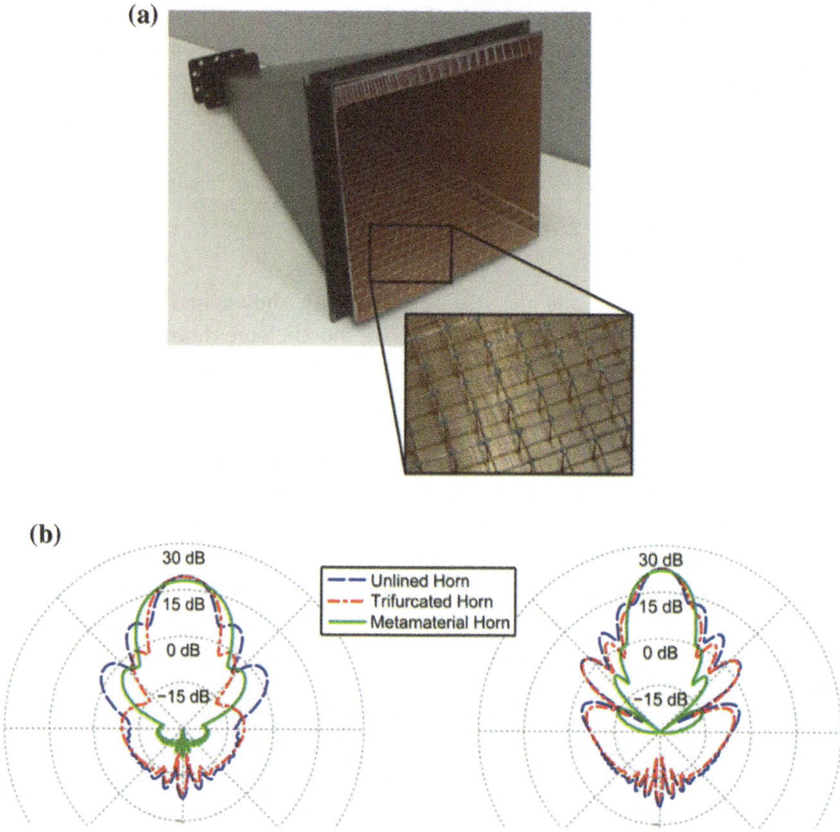

Fig. 3.31 **a** Photograph of metamaterial horn prototype, with inset showing the wire grid used to create the metamaterial surface. **b** Comparison of the radiation pattern on the metamaterial horn, a standard pyramidal horn and trifurcated horn [23]

In order to simplify the antenna manufacturing a planar metasurface was presented in [24], where a metasurface implemented on a printed circuit board (PCB) was used to create a soft horn antenna. Photographs of the prototype are shown in Fig. 3.32. The achieved performance shows sidelobes at the −30 dB level, in agreement with the soft nature of the horn. These sidelobes are nearly 20 dB lower than those obtained with a similar smooth wall pyramidal horn and are maintained in the whole Ku-band.

An example of the measured patterns is presented in Fig. 3.32b. Very symmetrical patterns are achieved leading to low cross-polarization levels (around 30 dB).

One problem associated with these antennas is the need of a matching section to reduce reflections at the interface between the smooth wall and the metamaterial sections of the horn. In this case, a dielectric section with an optimized profile is

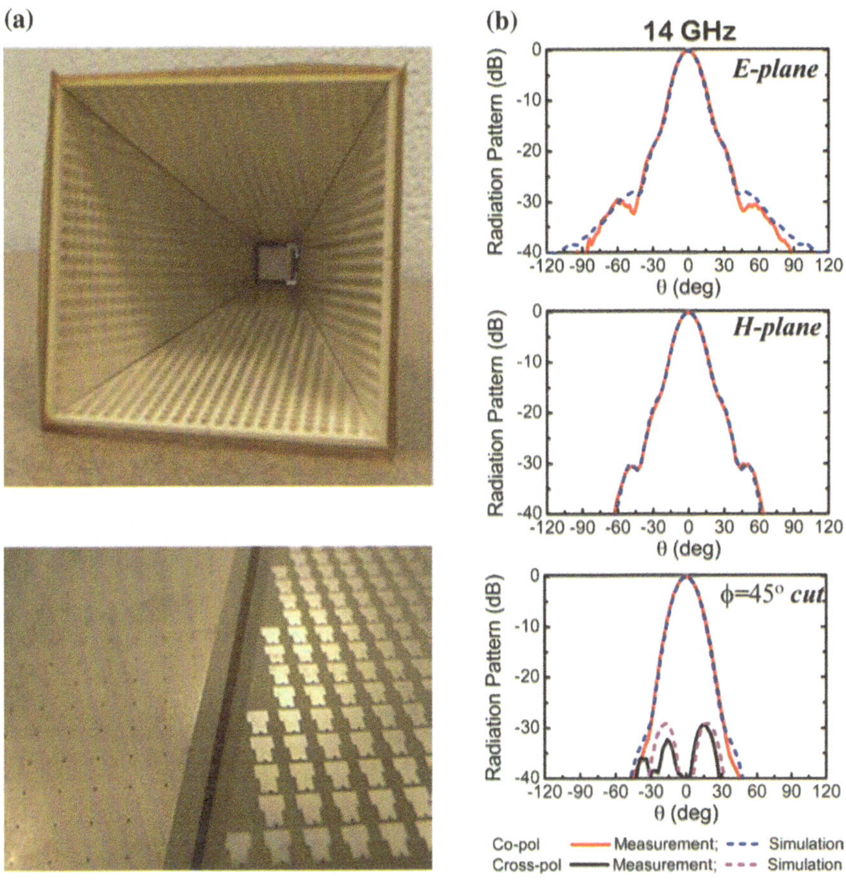

Fig. 3.32 **a** Photographs of the fabricated metahorn antenna and the metasurface liners on the PCB. **b** Measured and simulated E-plane, H-plane, and $\phi = 45°$ plane cut co- and cross-polarized radiation patterns of the metahorn at 14 GHz [24]

used for this purpose. This requires a brute force optimization which can be avoided if a tapered transition is implemented in the metamaterial wall liner. This approach has been followed in [25].

Finally another implementation of this concept was used in [25], applied to a conical horn antenna. The metasurface was based on a mushroom structure, see Fig. 3.33.

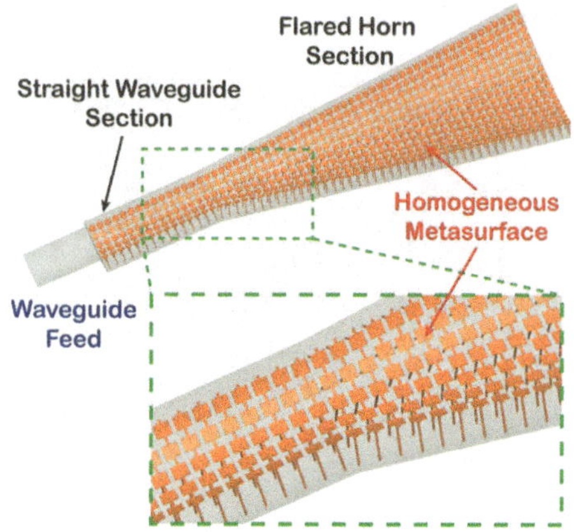

Fig. 3.33 Schematic of the conical metamaterial horn antenna proposed in [25]

3.4.3 EBG Horn Antennas

Electromagnetic band gap materials can be considered as a type of metamaterials. They are 3D periodic arrangements of elements, usually dielectric, which, by properly choosing their shape and spatial position present a frequency range where no propagation of electromagnetic radiation is allowed [26]. This frequency range is called the bandgap and its use has been proposed to reduce the effect of surface waves in planar antennas [27]. This effect can be obtained with planar configurations, usually based on printed board technologies or with 3D structures, where the material must be tailored in three dimensions.

In this last case, if defects are created in the otherwise perfect periodic structure, cavities and waveguides can be created. These can be considered as linear defects and their performance is similar to that of metallic waveguides. Therefore, the same techniques that are used in order to create horn antennas can also be applied in this case.

Horn-like structures have been proposed based on 2D and 3D EBG structures [28]. However, we will focus on those horn configurations based on 3D EBG structures. Among them, the most used EBG for these experiments is the woodpile or layer-by-layer structure [29, 30], due to its simplicity and relatively easy manufacturing up to the submillimeter wave range [31, 32]. As a matter of fact, it is in these high frequency ranges where this type of configurations can be of interest thanks to their low losses due to the absence of metals (provided the EBG structure is fabricated with sufficient precision).

3.4.4 Sectoral Horn Antennas

The simplest way to create a horn antenna based on a woodpile structure is by altering the bars of one of its layers. Removing one bar will create a waveguide and by tilting those bars that form the waveguide walls a sectoral antenna is created. Given the polarization of the mode in the woodpile feeding waveguide, this antenna corresponds to a H-plane sectoral horn. The radiation properties of the antenna depend on the flare angle and the antenna length, as in standard metallic sectoral horns.

These configurations were first studied by Weily et al. [33]. An example of the radiation pattern achieved can be seen in Fig. 3.34b. It is worth noting the non-symmetric E-Plane cut, due to the asymmetry of the woodpile in the stacking direction. Moreover, given the adiabatic nature of the transition, good impedance matching was achieved between the EBG waveguide section and the horn. In [34], the same group proposed a transition between rectangular and EBG waveguides with good performance. This configuration was employed to create an array of such antennas with which scanning performance was demonstrated. The radiation pattern of the 16-element array for different scanning angles is shown in Fig. 3.34c.

3.4.5 Evanescently Fed EBG Horn Antenna Arrays

Another way of controlling the radiation E-plane is by arraying two sectoral EBG antennas. They can be simultaneously fed by means of the so-called evanescent coupling method [35]. In this configuration, an EBG waveguide is used to couple the two sectoral horns by properly adjusting their relative position, as shown in Fig. 3.35. This way, narrower E-plane cut can be obtained while maintaining the same H-plane of the individual sectoral antennas, as shown in Fig. 3.36, where the simulated and measured antenna radiation patterns obtained with this antenna are compared.

3.4.6 Pyramidal Horn Antennas

As mentioned above, the most natural sectoral horn antenna which can be created taken as base the woodpile structure is a sectoral H-horn antenna, where the electric field is parallel to the woodpile stacking direction. Thus, the E-plane dimensions of such horn antennas are determined by the thickness of the woodpile bars, since

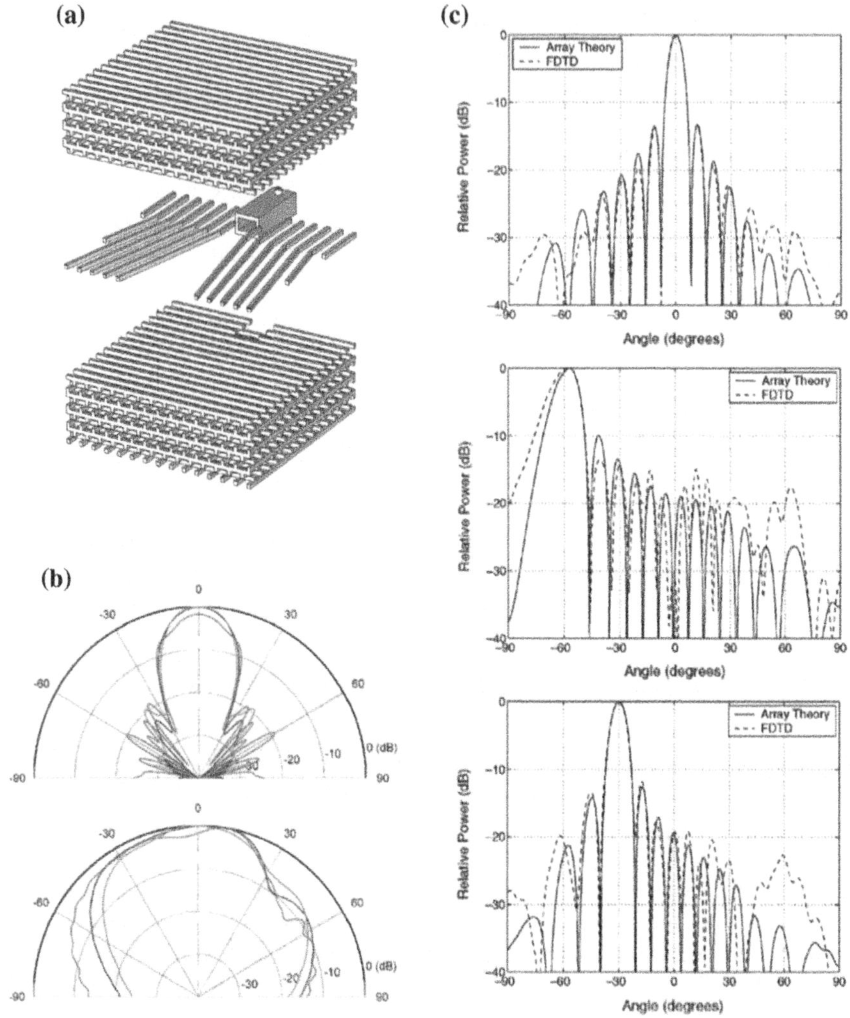

Fig. 3.34 **a** Perspective view of the woodpile EBG sectoral horn antenna [34]. **b** Radiation pattern **c** Array theory and FDTD computed E-plane radiation patterns for the 16-element linear array for scan angles of: **a** 0°, **b** 60°, **c** 30°

these horns are created within a single woodpile layer. Such a narrow and predetermined aperture leads to a broad beam in the E-plane, which is nonsymmetrical due to the absence of mirror symmetry in the stacking direction of the woodpile structure.

Reducing the E-plane beamwidth would require creating a pyramidal horn antenna. The difficulty of creating a pyramidal horn based on a defect-containing woodpile structure lies in the complexity of the embedding medium itself.

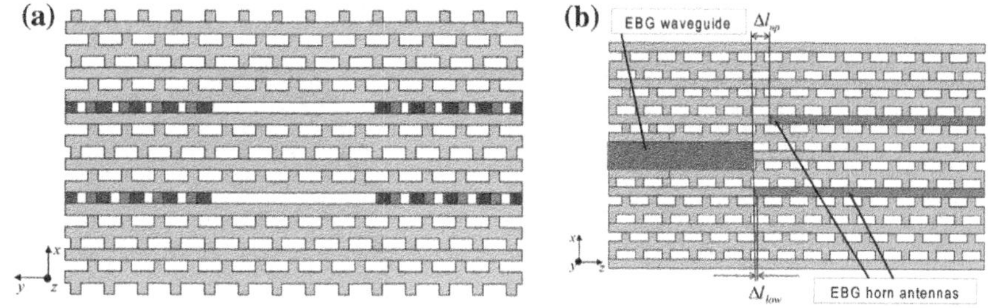

Fig. 3.35 Front and side view of the evanescently fed double-EBG horn antenna. Dark areas represent the feeding EBG waveguide region and the EBG horn antenna layers [35]

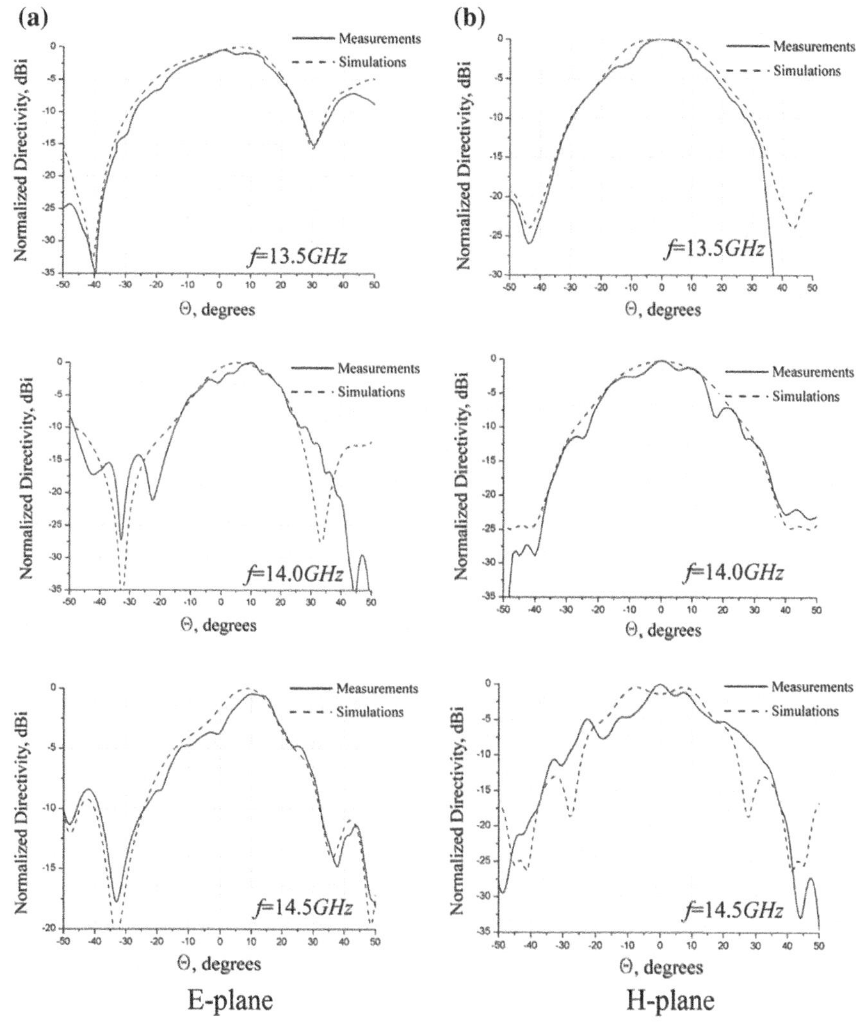

Fig. 3.36 Comparison between simulation and measurements [35]

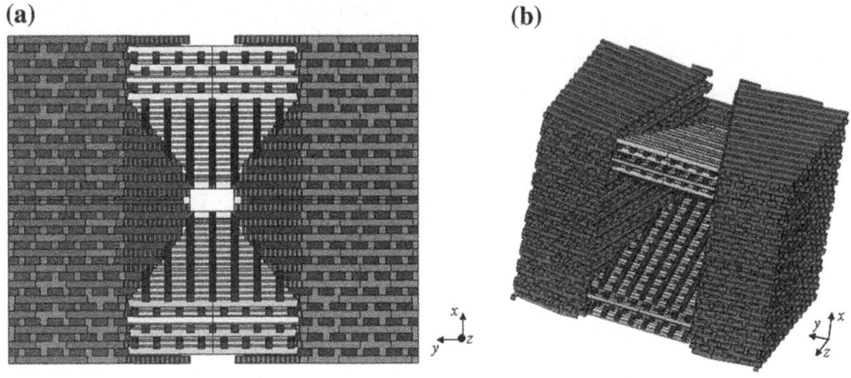

Fig. 3.37 Woodpile based symmetrical EH-horn antenna (a—front view, b—3D view) [36]

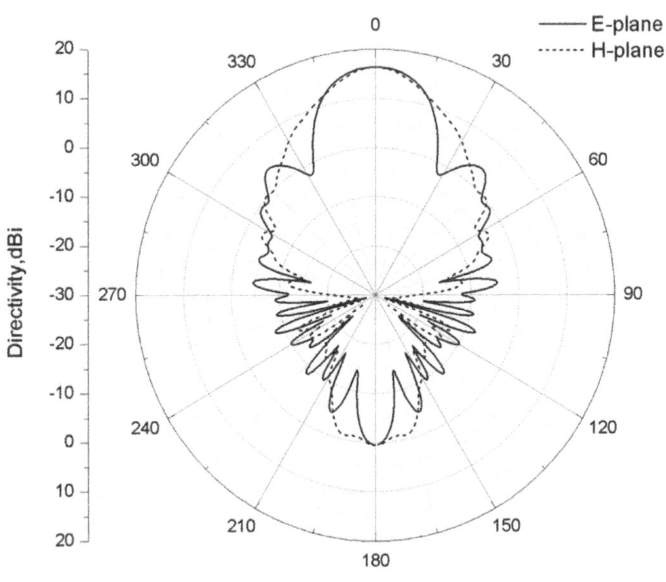

Fig. 3.38 Radiation pattern of the EBG pyramidal horn in Fig. 3.37 [36]

As shown in [36], simple solutions based on cutting several woodpile layers lead to resonant structures. A solution to his problem can be obtained if the woodpile layers are tilted as shown in Fig. 3.37.

This EBG pyramidal horn allows simultaneously controlling the E- and H-planes of the radiation patterns. An example of the pattern that can be obtained is presented in Fig. 3.38. Its main drawback comes from the manufacturing difficulties, since the layer-by-layer manufacturing approach followed for the woodpile is not valid anymore.

References

1. A.D. Olver, P.J.B. Clarricoats, A.A. Kishk, L. Shafai, Microwave horns and feeds, in *IEE Electromagnetic waves series 39* (The Institution of Electrical Engineers, 1994)
2. P.D. Potter, A new horn antenna with suppressed sidelobes and equal beamwidths. Microw. J. **6**, 71–78 (1963)
3. S. Ramo, J.R. Whinnery, T. van Duzer, in *Fields and Waves in Communication Electronics* (Wiley, 1994)
4. P. D. Potter, A new horn antenna with suppressed sidelobes and equal beamwidths. Microw. J. **VI**, 71–78 (1963)
5. P.J.B. Clarricoats, A.D. Olver, Corrugated horns for microwave antennas, in *IEE Electromagnetics Waves Series 18*, Chap. 5, 6 and 7 (Peter Peregrinus, 1984)
6. G.L. James, Analysis and design of TE_{11} to HE_{11} corrugated cylindrical waveguide mode converters, in *IEEE Transactions on Microwave Theory and Techniques*, vol. MTT-29, pp. 1059–1066, 1981
7. B.M. Thomas, G.L. James, K.J. Greene, Design of high-performance wideband corrugated horns for Cassegrain antennas, *IEEE Transactions on Antennas and Propagation*, vol. AP-34, pp. 750–757, 1986
8. Mician µWave Wizard Software. http://www.mician.com
9. Ticra's CHAMP Software. http://www.ticra.com/products/software/champ
10. C. Granet, G.L. James, Design of corrugated horns: a primer. IEEE Antennas Propag. Mag. **47**(2), 76–84 (2005)
11. R. Gonzalo, J. Teniente, C. del Río, Improved radiation pattern performance of horn antennas with gaussian profiled shapes. IEEE Trans. Antennas Propag. **50**(11), 1505–1513 (2002)
12. J. Teniente, R. Gonzalo, C. del-Río, Choked gaussian antenna: extremely low sidelobe compact antenna design, in *IEEE Antennas and Wireless Propagation Letters*, vol. 1, pp. 200–202, 2002
13. J. Teniente, A. Martínez, B. Larumbe, A. Ibañez y R. Gonzalo, Design guidelines of horn antennas that combine horizontal and vertical corrugations for satellite communications, in *IEEE Transactions on Antennas and Propagation*, vol. 63, No. 4, pp. 1314–1323, 2015
14. C. Granet, G.L. James, R. Bolton, G. Moorey, A smooth-walled spline-profile horn as an alternative to the corrugated horn for wide band millimeter-wave applications. IEEE Trans. Antennas Propag. **52**(3), 848–854 (2004)
15. P.D. Potter, A new horn antenna with suppressed sidelobes and equal beamwidths. Microw. J. **6**, 71–78 (1963)
16. J. Leech, B.K. Tan, G. Yassin, Smooth walled feed horns for mm and submm radio astronomy, in *6th UK, Europe, China Millimeter Waves and THz Technology Workshop (UCMMT)*, 2013
17. R.H. Turrin, Dual mode small-aperture antennas. IEEE Trans. Antennas Propag. **15**(2), 307–308 (1967)
18. J. Leech, B.K. Tan, G. Yassin, P. Kittara, S. Wangsuya, J. Treuttel, M. Henry, M.L. Oldfield, P.G. Huggard, Multiple flare-angle horn feeds for sub-mm astronomy and cosmic microwave background experiments. Astron. Astrophys. **532** (2011)
19. J. Teniente, D. Valcázar, B. Larumbe, A. Martínez, A. Ibañez, R. Gonzalo, Feed horn antennas for data downlink and uplink spaceborne communications, in *10th European Conference on Antennas and Propagation, EuCAP*, 2016
20. L. Zeng, C.L. Bennett, D.T. Chuss, E.J. Wollack, A low cross-polarization smooth-walled horn with improved bandwith, IEEE Trans. Antennas Propag. **58**(4) (2010)
21. E. Lier, P.S. Kildal, Soft and hard horn antennas. IEEE Trans. Antennas Propag. **36**(8), 1152–1157 (1988)
22. E. Lier, Review of soft and hard horn antennas, including metamaterial-based hybrid-mode horns. IEEE Antennas Propag. Mag. **52**(2), 31–39 (2010)

23. C.P. Scarborough, Q. Wu, D.H. Werner, E. Lier, R.K. Shaw, B.G. Martin, Demonstration of an octave-bandwidth negligible-loss metamaterial horn antenna for satellite applications. IEEE Trans. Antennas Propag. **61**(3), 1081–1088 (2013)
24. Q. Wu, C.P. Scarborough, B.G. Martin, R.K. Shaw, D.H. Werner, E. Lier, X. Wang, A Ku-Band dual polarization hybrid-mode horn antenna enabled by printed-circuit-board metasurfaces. IEEE Trans. Antennas Propag. **61**(3), 1089–1098 (2013)
25. Q. Wu, C.P. Scarborough, D.H. Werner, E. Lier, R.K. Shaw, Inhomogeneous metasurfaces with engineered dispersion for broadband hybrid-mode horn antennas. IEEE Trans. Antennas Propag. **61**(10), 4947–4956 (2013)
26. J.D. Joannopoulos, R.D. Meade, J.N. Winn, *Photonic Crystals* (Princeton University Press, 1995)
27. P. de Maagt, R. Gonzalo, J.C. Vardaxoglou, J.M. Baracco, Photonic bandgap antennas and components for microwave and (sub)millimetre wave applications. IEEE Trans. Antennas Propag. **51**(10), 2667–2677 (2003)
28. A.R. Weily, K.P. Esselle, B.C. Sanders, Photonic crystal horn and array antennas. Phys. Rev. E. **68**, 016609-1-016609-6 (2003)
29. K.M. Ho, C.T. Chan, C.M. Soukoulis, R. Biswas, M. Sigalas, Photonic band gaps in three dimensions: New layer-by-layer periodic structures. Solid State Commun. **89**(6), 413–416 (1994)
30. H.S. Sözüer, J. Dowling, Photonic band calculations for woodpile structures. J. Mod. Opt. **41** (2), 231–239 (1994)
31. R. Gonzalo, B. Martínez, C.M. Mann, H. Pellemans, P. Haring-Bolivar, P. de Maagt, A low-cost fabrication technique for symmetrical and asymmetrical layer-by-layer photonic crystals at submillimeter-wave frequencies. IEEE Trans. Microw. Theory Tech. **50**(10), 2384–2392 (2002)
32. B. Martínez, I. Ederra, R. Gonzalo, B. Alderman, L. Azcona, P.G. Huggard, B. de Hon, A. Hussain, S.R. Andrews, L. Marchand, P. de Maagt, Manufacturing tolerance analysis, fabrication, and characterization of 3-D submillimeter-wave electromagnetic-bandgap crystals. IEEE Trans. Microw. Theory Tech. **55**(4), 672–681 (2007)
33. A.R. Weily, K.P. Esselle, B.C. Sanders, Layer-by-layer photonic crystal horn antenna. Phys. Rev. E **70**, 4–037602 (2004)
34. A.R. Weily, K.P. Esselle, T.S. Bird, B.C. Sanders, Linear array of woodpile EBG sectoral horn antennas. IEEE Trans. Antennas Propag. **54**(8), 2263–2274 (2006)
35. I. Khromova, I. Ederra, J. Teniente, R. Gonzalo, K.P. Esselle, Evanescently-fed electromagnetic band gap horn antennas and arrays. IEEE Trans. Antennas Propag. **60**(6), 2635–2644 (2012)
36. I. Khromova, I. Ederra, R. Gonzalo, B.P. de Hon, Symmetrical pyramidal horn antennas based on EBG structures. PIER B **29**, 1–22 (2011)

Chapter 4
Flat Corrugated and Bull's-Eye Antennas

Miguel Beruete, Unai Beaskoetxea and Tahsin Akalin

Abstract This chapter is focused on a family of antennas recently proposed, whose common denominator is that they consist in a flat metallic plane with a central narrow aperture surrounded by corrugations. This new family of antennas has occupied a very important place in the development of communication technologies and solutions for several communication applications since, compared to higher volume structures, as horn or parabolic antennas, they present equal or even higher radiation characteristics. An introduction, which also serves as an historical overview, is presented in the first place, emphasizing the relation with both extraordinary transmission structures demonstrated initially at optical wavelengths as well as with leaky-wave antennas, developed mainly in microwaves. Afterwards, the physical mechanism for radiation and the main guidelines for the design of these antennas are discussed. Next, an overview of some of the most appealing designs and results related with this technology is presented, putting special emphasis in terahertz-band applications. Finally, tips on foreseen future trends are summarized to conclude the chapter.

M. Beruete (✉) · U. Beaskoetxea
Antennas Group-TERALAB, Universidad Pública de Navarra, 31006 Pamplona, Spain
e-mail: miguel.beruete@unavarra.es

U. Beaskoetxea
e-mail: unai.beaskoetxea@unavarra.es

T. Akalin
Institute of Electronics, Microelectronics and Nanotechnology (IEMN), Lille University, Lille, France
e-mail: Tahsin.Akalin@iemn.univ-lille1.fr

A. Boriskin and R. Sauleau (eds.), *Aperture Antennas for Millimeter and Sub-Millimeter Wave Applications*, Signals and Communication Technology,
DOI 10.1007/978-3-319-62773-1_4

4.1 Introduction: Extraordinary Transmission Resonance and Evolution Towards Metallic Corrugated Antennas

Daily experience teaches us that the amount of light able to penetrate an aperture perforated on a metallic plane is proportional to the size of the aperture. More technically, the power coupled through the hole is limited by diffraction and, for sufficiently small apertures, diffraction rings will appear at the output, a fact that is customarily exploited to demonstrate the wave nature of light in typical lab experiments. Quantifying the fraction of power that goes through a hole perforated in an infinite metal film is a classical problem that was solved by Bethe and Bouwkamp some time ago [1, 2]. Using standard diffraction theory, they established that this power is proportional to $(r/\lambda)^4$, where r is the radius of the aperture and λ is the operation wavelength. The previous expression confirms our intuition putting in evidence that for wavelengths much larger than the hole diameter only a small fraction of the incoming power can traverse the aperture.

The theory of Bethe and Bouwkamp provided a satisfactory solution and remained unquestioned for more than 30 years. However, in 1998 the scenario changed radically when Ebbesen et al. [3] reported an experiment where high transmittance peaks appeared in an aluminium plate perforated with a periodic matrix of sublambda holes at wavelengths as large as ten times the diameter of the circular apertures. This result came as a surprise, since the holes were operating in cutoff and still were able to produce a high transmittance, in apparent contradiction with Bethe-Bouwkamp's theory.[1] For this reason, this phenomenon was called Extraordinary Optical Transmission (EOT). Ebbesen's experiment was done in the near infrared band where metals admit a Drude model and have a negative real part of permittivity, as corresponds to an overdense plasma. This negative permittivity property allows the coupling of surface plasmon polaritons (SPPs), which are surface waves sustained at the metal-dielectric (also known as metal-insulator, MI) interface [4]. SPPs are bound modes and hence are not directly accessible by an external incident wave. The extra momentum of the tangent wave vector necessary to excite SPPs can be achieved by an external prism or by using a periodic structure, as was the case in the experiment by Ebbesen et al.. Thus, the explanation proposed to elucidate the high transmittance was the coupling of light with SPPs excited through the periodic matrix of holes.

Ebbesen's result ignited the spark of modern SPP research, leading to the so-called "surface-plasmon resurrection" [5] (excellent reviews can be found in [6, 7]). A model able to embody the physics involved in the extraordinary transmission of light through an array of subwavelength holes in a metal film was presented by Martín-Moreno et al. [8]. With this analysis, it was established that the phenomenon

[1]Strictly speaking, Ebbesen's results do not contradict Bethe-Bouwkamp's theory. Recall that the latter is valid for an isolated hole in an infinite metal film, rather different from the hole matrix studied by Ebbesen.

depended crucially on the periodic structure. Even more, a minimal model able to catch the main underlying physics of the phenomenon was presented.

Following the path initiated by Ebbesen, other structures capable of enhancing the transmission through small apertures were investigated. In [9], one-dimensional metallic gratings with very narrow slits periodically distributed over a dielectric substrate were studied for impinging light beams with polarization perpendicular to the aperture. The transmission resonances obtained were related to the excitation of two types of electromagnetic modes: (i) the coupling between the SPPs in the two surfaces of the grating, when the impinging wavelength was approximately equal to the grating period; and (ii) waveguide modes in the slits when the wavelength was much larger than the period. It was pointed out that those effects appeared for any range of frequencies, provided the width of the slits was small in comparison to the period and the frequency of the incident wave was lower than the plasma frequency of the metal.

Further studies were carried on in [10], analyzing structures formed only by holes and by holes and periodically inserted dimples. It was observed that the zero-order transmission could be improved at certain wavelengths for the holes and dimples structure, although the rest of the spectrum remained almost equal. This proved that the light coupling depended mainly on the period of the corrugations rather than on their shape. Given that the dimples do not emit but do contribute to the peaks, the authors suggested that the surface modulation caused by the dimples helped to couple the incident photons to SPPs, whereas the holes also contributed to the opposite operation coupling again the surface plasmons to photons. Taking this idea to the limit, they demonstrated that a single sub-wavelength aperture surrounded by a dimple array had a transmittance enhancement proportional to λ^4/r^2 compared to a hole in a flat plane and that the transmission peak could be tuned in frequency by just changing the period of the dimples. In [11], it was showed that the use of metamaterial shaped apertures rather than typical geometries (ellipsoidal or quadrilateral holes), i.e., Babinet metamaterials [12, 13], led to a reinforcement of the EOT effect. More specifically, a higher transmittance was observed at lower frequencies, while the transmission efficiency at higher frequencies remained almost still.

The natural step forward was to test other periodic structures flanking a central aperture. In [14], two different structures were analyzed: a small hole surrounded by concentric annular grooves, a structure called Bull's-Eye and a narrow slit surrounded by straight grooves (respective schemes are shown in Fig. 4.1a and b). In either case, an enhancement of transmission was demonstrated both when the grooves where carved on the input and output faces. It was found that grooves on the input face contribute to a more efficient light harvesting from the incident wave, which was then directed towards the central aperture and radiated almost isotropically, with a clear diffractive pattern. On the other hand, with grooves on the output face, light harvesting at the input did not change compared to a single aperture on a flat metallic plane, but radiated a directive beam at the output with its maximum pointing perpendicularly to the surface. It was also confirmed that when the output observation angle was shifted away from the normal direction, the maximum

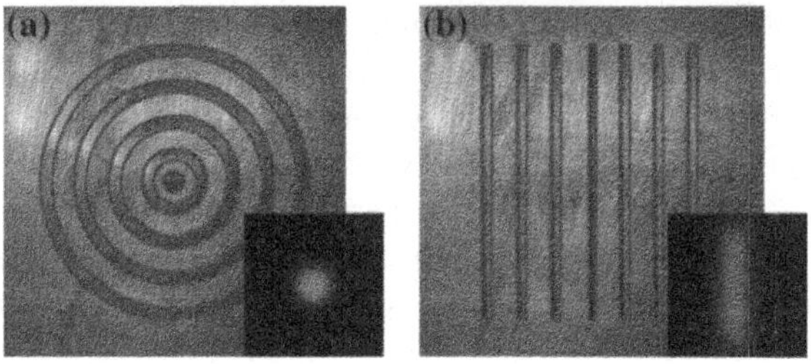

Fig. 4.1 Schematic of **a** Bull's-Eye with circular aperture and **b** slit aperture with parallel grooves. *Insets* observed optical response of the corresponding sample illuminated from the back at its wavelength of maximum transmission. Artwork inspired in the results of Lezec et al.

transmission intensity dropped and the peak splitted into two peaks, moving each one towards lower and higher wavelengths respectively.

As a theoretical complement, in [15] the beaming mechanism for this structure was described in detail, analyzing the angular distribution of light at the exit side. It was shown that the beaming intensity saturated for a certain number of grooves. It was also concluded that the emission process basically consists of two steps: first, the output side of the slit emits a primary beam directly into free space and into the grooves; then, these grooves reradiate light to vacuum and into other corrugations. It was demonstrated that the grooves' depth plays an important role in the beaming, as it shifts the wavelength at which the maximum occurs, as well as the achievable maximum beam intensity. These cavity modes—controlled by the grooves' dimensions—and the coupling between indentations—maximum at wavelengths commensurable with the period—lead to the surface resonances responsible for the observed beaming properties. These results were experimentally confirmed afterwards in [16]. It was proved that the maximum boost of transmission takes place when three radiation mechanisms coincide at the same frequency: the previously mentioned groove cavity mode excitation, the slit waveguide mode and a third one corresponding to the in-phase groove reemission, dependent on the period of the groove array.

An important fact disregarded up to now in this brief historical introduction, is that the theoretical analyses derived to explain extraordinary transmission through hole arrays or through small apertures on corrugated planes showed high transmittance peaks even assuming a perfect electric conductor instead of a Drude model for the metallic plane. This suggested that the phenomenon was far more general and that it could be reproduced in other frequency regimes where genuine SPPs do not exist, like microwaves, millimeter-waves or even the terahertz range in which metals admit a high conductivity model. The first demonstration of EOT through hole arrays at millimeter-waves was published in 2004 [17] showing definitely that the phenomenon was universal and it could appear in any band of the electromagnetic spectrum, regardless of the exact metal model employed. The key to get a high transmittance peak is to design a coarse hole matrix, ensuring that the cutoff frequency of the apertures falls above the onset of the first diffraction order. If this condition is fulfilled and the metal plate is thin enough so as to avoid a strong decay of the power transmitted through the holes, the EOT peak will appear at a frequency slightly below the onset of the first diffraction order. An experimental study showing the importance of the number of holes in the EOT resonance was reported soon after [18] putting into evidence that the strength of the transmission is basically controlled by the number of periods in the array. A theoretical analysis of the finite structure effects was presented in [19]. Further numerical and experimental studies in [19, 20] proved that the enhancement and bandwidth also depended on the hole diameter.

Regarding corrugated structures, enhanced transmission was demonstrated for a single slit surrounded by two narrow and deep corrugations at both sides of the slab [21]. Using a scaled version of the single slit surrounded by multiple corrugations presented in [14], enhanced transmission and strong beaming was demonstrated at microwaves [22]. Later on, a thorough experimental study showing the enhancement with input and output corrugations was reported at millimeter-waves [23]. In [24, 25], the microwave enhanced transmission and its angle-dependence was experimentally studied for a subwavelength circular aperture surrounded by annular concentric grooves. All these experiments are the precursors of the flat metallic antennas that we will consider in the next section.

Finally, it has to be remarked that the corrugated structures have been presented here as an evolution of extraordinary transmission, which is in the roots of the antennas discussed in the chapter. However, an alternative point of view is to employ the leaky-wave (LW) formalism to explain the behavior of these structures. Recently, in [26], a thorough study of the beam-forming physical mechanism of flat horns was presented. There, the regimes of the surface waves and the leaky waves are clearly distinguished and the behavior of corrugated flat horn antennas is meticulously analyzed and depicted by means of representations of near- and far-fields at each regime. A LW analysis will be presented later in this chapter, for the sake of completeness.

4.2 A New Antenna Family: Small Apertures on Corrugated Metallic Planes

4.2.1 First Generation: Antennas at Microwaves

Up to now, all the structures discussed are similar to frequency selective surfaces or spatial filters, in the sense that they act on an incident wave and produce a response at the output. However, the beaming property of corrugated structures makes them interesting candidates for high-gain antennas with the advantage of having a much reduced volume and weight than conventional solutions. The capability of presenting beaming characteristics comparable to horn or parabolic antennas, among others, has made them a field of intensive research for more than a decade. Indeed, the results obtained in [22, 23] served as a basis for a new family of antennas based on a small aperture on a corrugated metallic plane. These antennas differ from the previous structures in that power is injected by a waveguide attached to the back of a metallic plane, and is coupled to the output through a resonant slot instead of a narrow slit.

In [27] the first member of this family was presented: a metallic Bull's-Eye antenna characterized by a central radiating slot surrounded by annular corrugations and optimized to a wavelength approximately equal to the period of the grooves, see Fig. 4.2a. A relatively high gain beam of approximately 21 dB was measured at the operation frequency of 16.5 GHz, as shown in the gain curve of Fig. 4.2b. For the comparison, an EMCO-3115 horn antenna was used (black curve). This antenna, from now on called Test Antenna 1, is a 24.4 cm × 15.9 cm double-ridged waveguide horn which operates in the band from 750 MHz to 18 GHz. For all the band of interest (10–18 GHz) it presents an almost constant gain of more than 12 dB. This corresponds to an aperture efficiency $e_a = 6.3\%$. Comparing Fig. 4.2c and d it is evident that the antenna has different radiation patterns in the principal planes. In the E-plane a narrow main lobe pointing at broadside with well-defined nulls can be identified, whereas in the H-plane the radiation pattern is smoother. This is a general characteristic of this kind of antennas and can be explained by attending to the current distribution shown in Fig. 4.2e, calculated with the commercial simulator CST Microwave Studio. The current density is higher in vertical direction, which corresponds to E-plane, than in the horizontal direction, which corresponds to the H-plane. In the E-plane (Fig. 4.2c) the antenna presents an angular width of 6° and less than −15 dB side lobe level. In the H-plane, by contrast, it presents a wider beamwidth at −3 dB of 12.3°, (Fig. 4.2d). It has relatively low cross-polarization levels for E- and H-plane (30 dB and 25 dB, respectively), as shown in Fig. 4.2c and d respectively. In Fig. 4.2f, a three-dimensional plot of the radiation pattern has been included for the sake of completeness.

This design opened the path to other schemes such as the antenna with 6 straight grooves on each side depicted in Fig. 4.3a and published in [28]. The experimental results showed a gain of 18 dB ($e_a = 6.5\%$), 6 dB higher than the Test Antenna 1, see Fig. 4.3b. In this case, due to the absence of revolution symmetry, the antenna presented a wider beam at −3 dB of 12° in the E-plane and 30° in the H-plane. Besides, the prototype showed lower side lobes for a range of about ±30° at the E-plane, see Fig. 4.3c. As it was pointed out before, the current density is distributed preferentially along the vertical (E-field) direction, which is the reason for the larger beamwidth observed in the H-plane, Fig. 4.3d.

With the aim of reducing the size of the antenna and hence improve the aperture efficiency, a miniaturized 1-D planar antenna operating at 17 GHz and consisting of

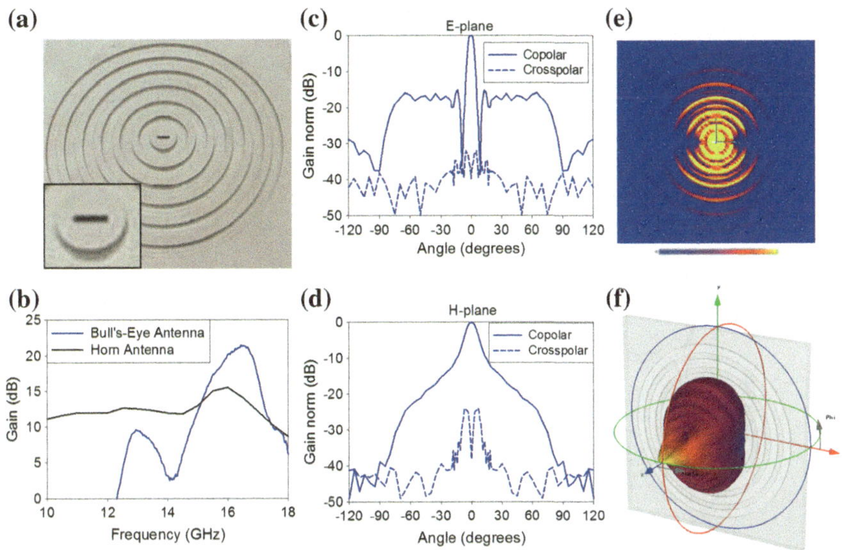

Fig. 4.2 **a** Picture of the metallic Bull's-Eye antenna. **b** Measured gain versus frequency for prototype (*blue curve*) and test horn antenna (*black curve*). Measured co- (*solid line*) and cross-polar (*dashed line*) E- and H-planes, (**c**) and (**d**), respectively. **e** Surface current distribution on the Bull's-Eye antenna surface. **f** Three-dimensional plot of the radiation pattern. Reprinted with permission from IEEE

a resonant slot and two straight corrugations, was also experimentally and numerically studied in [29], see picture in Fig. 4.4a. A relatively high gain of near 10 dB (e_a = 17.2%) was measured, a value slightly lower than that of the Test Antenna 1, see Fig. 4.4b. The radiation pattern showed a –3 dB angular width at of 34° in the E-plane and 60° in the H-plane, Fig. 4.4c and d respectively. Attending to the characteristics of the prototype it can be described as a weakly directive antenna with broadside gain enhancement of 4.3 dB compared to the non-corrugated metallic plane. Nevertheless, it must be taken into account that the lower gain and wide beaming is balanced out by the reduced physical dimensions compared to those of the horn antenna. It must be emphasized that the behavior of this antenna is more similar to a parasite-element radiating structure than to a LW antenna, where a periodic grating is needed. This explains the observed relatively low gain, compared to the previous antenna with 6 grooves on each side.

A compact antenna design was discussed in [30]. By means of the use of the slot's transversal resonance, instead of the longitudinal resonance, the antenna thickness can be reduced. This was applied to the antenna shown in the bottom antenna of Fig. 4.5a. The thickness can be further reduced by filling the corrugations with a dielectric material so that the grooves' depth can be reduced. In the antenna shown on the top of Fig. 4.5a the corrugations were filled with glass-fiber

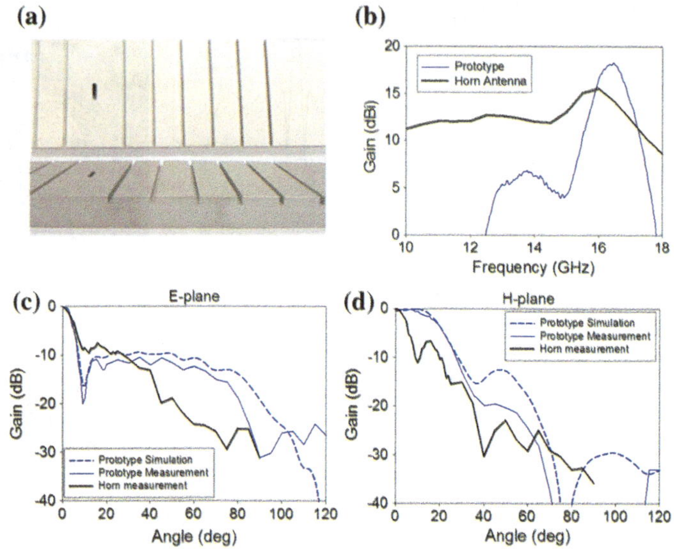

Fig. 4.3 **a** Pictures of the metallic low profile 12 corrugations antenna. **b** Measured gain versus frequency for the antenna prototype (*blue curve*) and test horn antenna (*black curve*). E-plane (**c**) and H-plane (**d**) radiation pattern for the simulated prototype (*dashed blue*), measured prototype (*solid blue*) and measured test horn antenna (*solid black*)

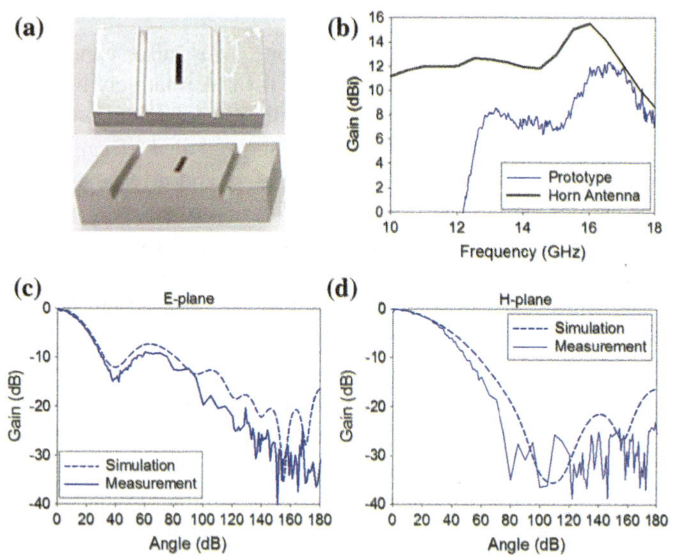

Fig. 4.4 **a** Pictures of the metallic miniaturized antenna. **b** Measured gain versus frequency for the antenna prototype (*blue curve*) and test horn antenna (*black curve*). **c** E-plane and **d** H-plane radiation patterns for simulated prototype (*dashed blue curve*) and measured prototype (*solid blue curve*)

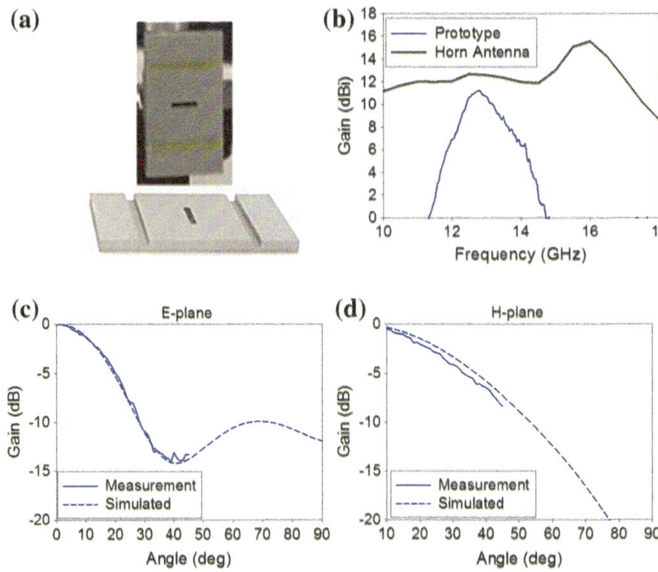

Fig. 4.5 **a** Picture of the metallic miniaturized flat antennas: dielectric filled grooves (*top*) and air-filled grooves (*bottom*). **b** Measured gain versus frequency for the dielectric filled grooves prototype (*blue curve*) and test horn antenna (*black curve*). **c** E-plane and **d** H-plane radiation patterns for the simulated dielectric filled grooves prototype (*dashed blue curve*) and measured prototype (*solid blue curve*)

(with permittivity $\varepsilon_r = 4$) to reduce the grooves' depth by half. By applying these modifications, a reduction by a factor of over 3 of the thickness compared with the previous prototype was achieved. According to the experimental results of Fig. 4.5b, there is a clear improvement in the radiation pattern with a gain of 11.3 dB ($e_a = 39\%$), more than 5 dB larger than a flat metallic plane and barely 1 dB less than the Test Antenna 1. Figure 4.5c and d show the E- and H-plane radiation diagrams, respectively. The measured –3 dB angular width for E-plane is 30° and for H-plane is 52°.

A dual-band low-profile antenna was reported in [28], taking advantage of the double $\lambda/2$ resonance of the slot located at $f_1 = 13$ GHz and $f_2 = 16.5$ GHz. The transversal resonance, which corresponds to $f_1 = 13$ GHz is fixed by the slot width, whereas the longitudinal resonance, given at $f_2 = 16.5$ GHz, is fixed by the slot's depth. In this case, two sets of corrugations were disposed at the output surface, each one separated from the slot a distance equal to nearly the wavelengths fixed by the resonances, see Fig. 4.6a. The patterning of the surface led to a gain near 10 dB at both frequencies of design (aperture efficiency over 23%), Fig. 4.6b, nearly the same values obtained for the Test Antenna 1 at f_1 and f_2. The measured beamwidth at –3 dB was 20° and 30° in the E-plane (Fig. 4.6c and d), and 50° and 40° in the H-plane (Fig. 4.6e and f), at f_1 and f_2, respectively.

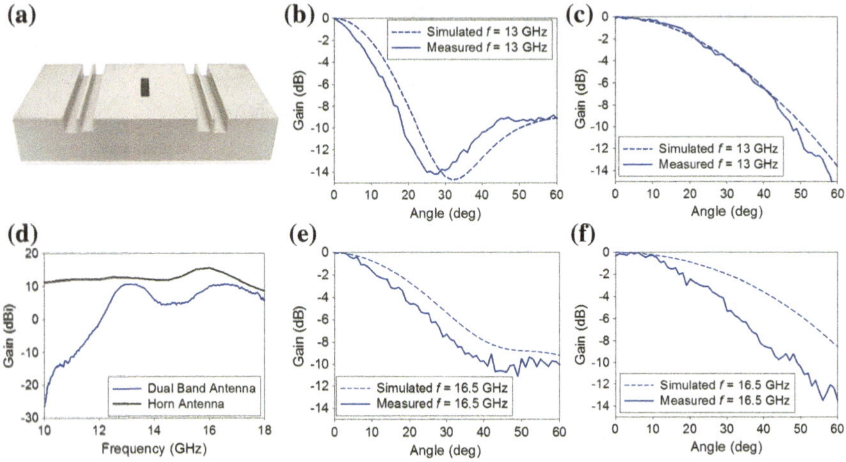

Fig. 4.6 **a** Picture of the metallic dual band flat antenna. **b** Measured gain versus frequency for the dual band prototype (*blue curve*) and test horn antenna (*black curve*). Simulated (*dashed curves*) and measured (*solid curve*) E-plane at **c** f_1 = 13 GHz and **d** f_2 = 16.5 GHz. Idem for H-plane at **e** f_1 = 13 GHz and **f** f_2 = 16.5 GHz

These antennas are included in the first generation of waveguide fed corrugated antennas developed in the Public University of Navarre (UPNA), being a landmark and design line for further coming structures. Figure 4.7 gathers all these antennas, characterized by their low profile and operating at low frequencies in the microwave range, and makes a comparison of the observed gain for each of them with a standard Horn antenna.

4.2.2 *Towards Terahertz Frequencies*

One-dimensional (1-D, i.e., slit + grooves) and two-dimensional (2-D, i.e., Bull's-Eye) antennas based on a radiating slot surrounded by corrugations were designed and experimentally measured in the terahertz range in [31]. First, the feasibility to develop antennas and the enhancement in the transmission at this range of frequencies was proved, with the prototype shown in Fig. 4.8a. Two structures with 10 corrugations working at f = 0.56 THz were designed and analyzed, differing only in the corrugation shape: triangular and rectangular. Simulations showed a gain enhancement compared to a flat plane of 10 dB for both antennas with a narrow beamwidth of $\theta_{-3\mathrm{dB}}$ = 5.8° and 6.8° and side lobe levels of 8.6 and 9.7 dB in the E-plane for the triangular and square grooves, respectively. Due to fabrication restraints, only the antenna with triangular grooves was fabricated and experimentally characterized. As expected, the structure showed large transmission enhancement at the working frequency, fixed by the period of the

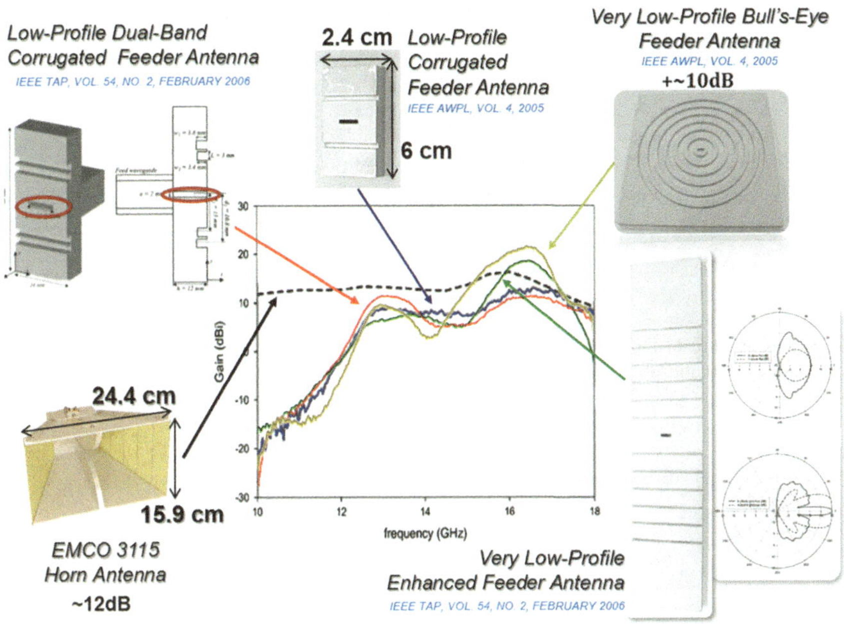

Fig. 4.7 Compendium of the first generation of low profile leaky wave antennas operating at microwave frequencies

corrugations and the transversal resonance of the slot. A secondary peak at 0.8 THz and with much lower amplitude was noticed in the spectrum, corresponding to the longitudinal slot resonance. However, it was clear that the corrugations, optimized for the design frequency of 0.56 THz, had little effect at 0.8 THz, Fig. 4.8b. In the same article a Bull's-Eye antenna was also experimentally analyzed, see Fig. 4.9a. In this case, the time-domain waveforms for different sets of concentric annular corrugations with variable depths were recorded, see Fig. 4.9b. It was first considered the electric field at the input face of the aperture, arising from the directly incident THz beam, which evanescently tunneled through the subwavelength aperture. Then, from these measurements, it was clear that the beaming mechanism could be explained as the sum of the directly transmitted and time-delayed-groove-contributions at the output face, Fig. 4.9c.

Even though there is a vast research on LW antennas, the THz range is still a relatively unexplored field concerning to leaky corrugated structures, with few designs as the one presented above or the one in [32] where a 560 μm × 280 μm terahertz low-profile pyramidal horn antenna was loaded with parallel straight V-grooves of different depth, see a picture in Fig. 4.10a and b. The corrugated structure presented at 0.49 THz a narrower beam, (Fig. 4.10c), enhancing its gain more than 3 dB compared to the non-loaded horn (15.6 dB).

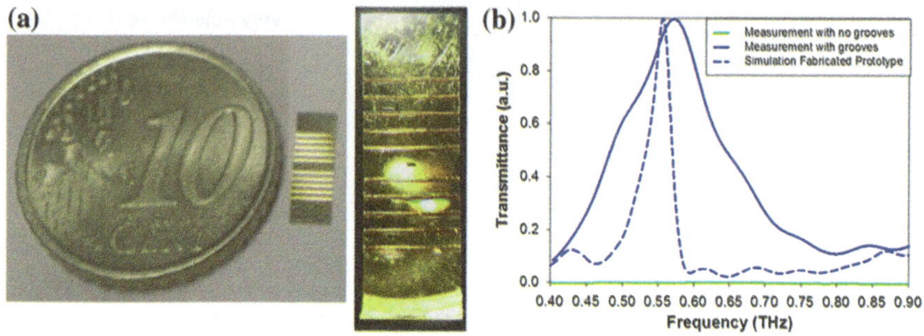

Fig. 4.8 **a** Picture of the metallic low profile antenna operating at 0.56 THz. **b** THz Bull's-Eye antenna. **c** Schematic of output beaming forming

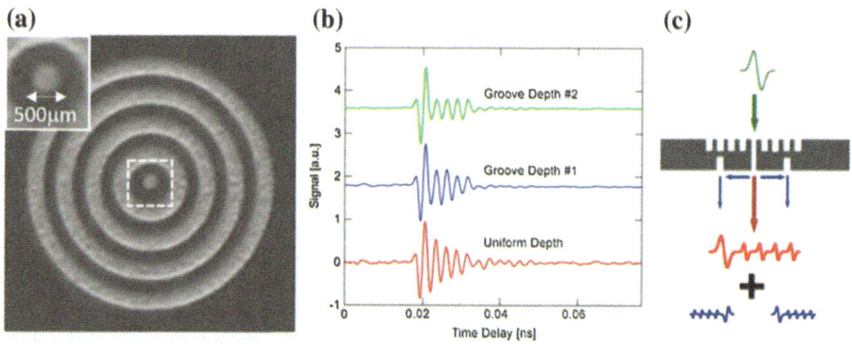

Fig. 4.9 **a** Picture of the metallic low profile antenna operating at 0.56 THz. **b** THz Bull's-Eye antenna. **c** Schematic of output beaming forming

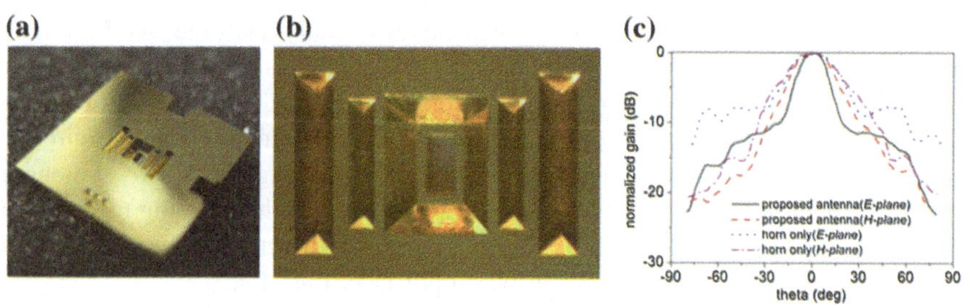

Fig. 4.10 **a** Micromachined V-grooves antenna. **b** Close view of manufactured antenna. **c** Normalized E- and H-plane for patterned antenna (*black* and *blue dotted curves*) and horn antenna (*red* and *pink dashed curves*). Reprinted with permission from John Wiley & Sons, Ltd

A typical drawback of quantum cascade lasers is their large beam divergence due to the small emission aperture, Fig. 4.11c. In [33–36] this problem was overcome by integrating periodic subwavelength corrugations onto the output facet of semiconductor lasers, Fig. 4.11a. With this solution, the beam divergence could be highly reduced, Fig. 4.11d. This structure, named laser collimator or laser beam

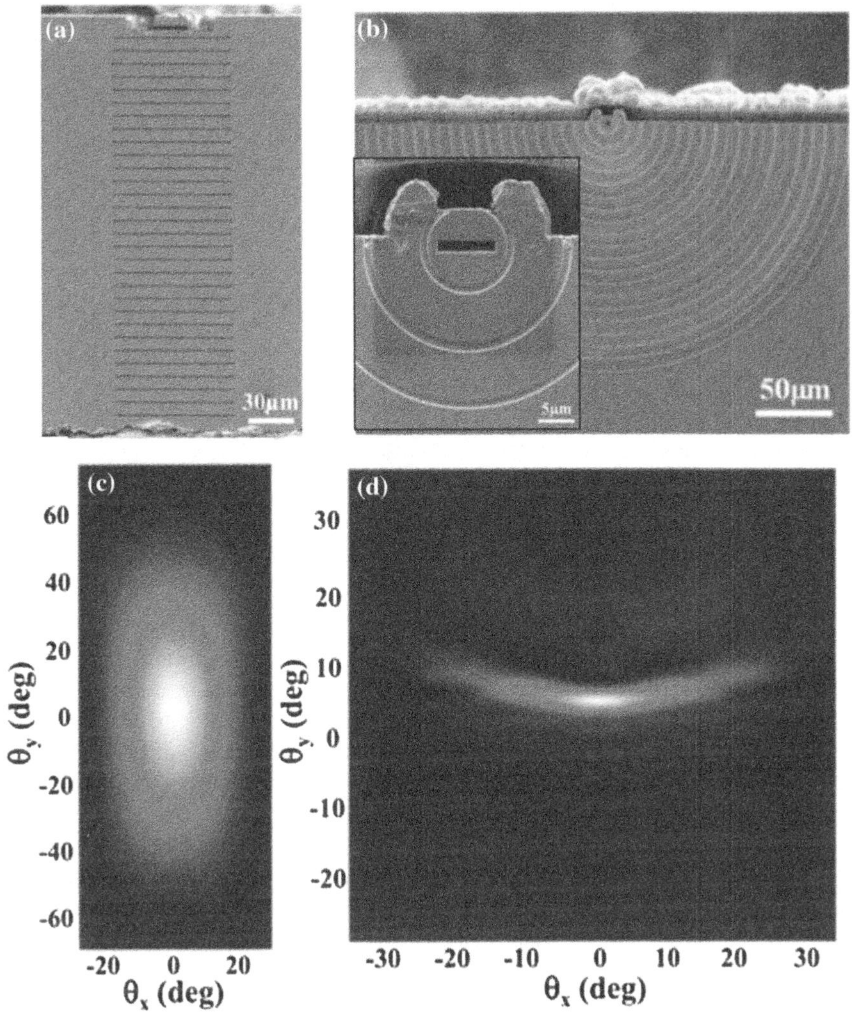

Fig. 4.11 **a** Detail of a 1D beam collimator. **b** 2D (Bull's-Eye) beam collimator (*Inset* detail of the structure). **c** Measured 2D far-field intensity distributions of the original unpatterned device. **d** Measured 2D far-field intensity distributions of the device patterned with a 1D slit-grating structure. Reprinted with permission from John Wiley & Sons, Ltd

shaper, achieves the reduction of divergence from $\sim 180°$ to $\sim 10°$, an over 10 dB directivity improvement and an increase factor of about six for the power collection efficiency. It was also proved that using a concentric ring grating rather than a 1D grating, Fig. 4.11b, the beam divergence was reduced not only for the laser polarization direction, but also for the parallel one. As it happens with the previously mentioned antennas, the collimator design can be scaled to work from the visible to the far-IR regimes.

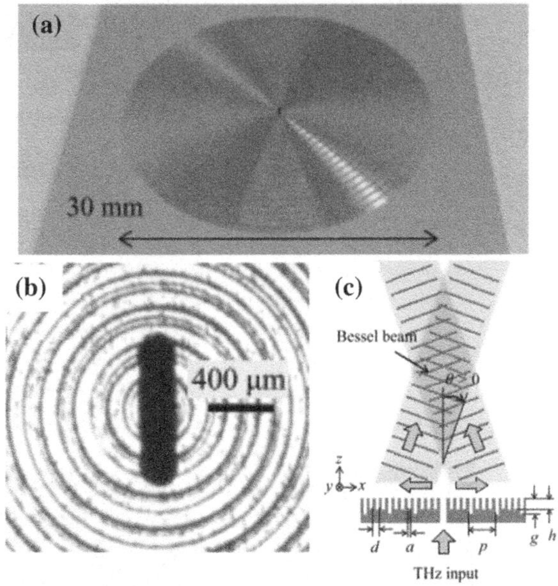

Fig. 4.12 **a** Photograph of THz Bessel beamformer. **b** Detail of the central slot in the beamformer. **c** Schematic of the Bessel beam generation mechanism. Reprinted with permission from AIP Publishing LLC

In [37] it was shown the usefulness of Bull's-Eye corrugated structures as Bessel beam generators embeddable in solid-state terahertz sources, in contrast to conventional axicon lenses approach typically used for Bessel beam generation which are usually bulky systems. In that work, a Bull's-Eye operating at 0.29 THz was fabricated, Fig. 4.12a and b, and characterized through terahertz time-domain spectrometry, observing a non-diffractive behavior within a distance greater than 20λ. In Fig. 4.12c it is shown how the beam generation mechanism is identical to that observed for the antennas discussed above. A much more exhaustive review of the Bessel beam sources as well as the physical mechanism involved in the beam generation is carried out in Chap. 8.

4.2.3 Latest Developments

Nearly a decade after the first Bull's-Eye paper, an antenna of this type was proposed as a good solution to embed in CubeSat devices (miniaturized satellites for space research composed of multiple small cubic units) taking advantage of the antenna's low profile [38]. In this publication, the effect of shifting the center of each of the 7 annular corrugations was studied, observing a variation of the radiated beam direction. For the concentric corrugations case, the antennas displayed a gain of 19.6 dB, whereas it was reduced as the center was shifted and the radiated beam gradually pointed away from broadside, although the beamwidth barely varied. We can see how, once again, this kind of antennas surpasses in the field of space

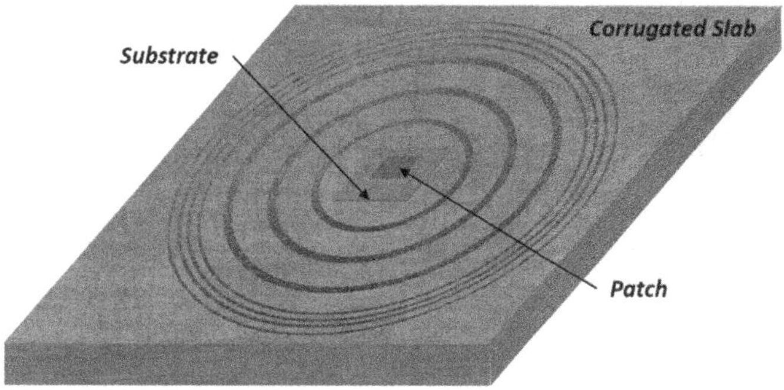

Fig. 4.13 Schematic of patch over a corrugation loaded substrate

communications (among others) classical parabolic antennas, providing high gain, unconventional beaming properties and lightweight, which results of high interest in this area.

Another example of the application of a Bull's-Eye geometry in a metallic plane is that reported in [39], in which a patch over a substrate was placed on top of a metallic annular corrugated structure, increasing strongly the directivity, as shown schematically in Fig. 4.13. An extra annular soft-surface configuration placed after the Bull's-Eye served to reduce the backward radiation, preventing the diffraction of the surface wave on the edge of the structure. This diffraction produces a degradation of the antenna pattern and a decrease of the antenna efficiency. This is due to the nearly infinite surface impedance observed in the soft surface and the null value of all components of the electric field in any polarization direction. This antenna presented a gain of 19.45 dB, more than 12 dB larger gain than the non-patterned ground + patch, and a narrow beamwidth of 7.9°.

In [40] a Bull's-Eye antenna with sinusoidal corrugations working at millimeter waves was presented, see picture in Fig. 4.14a. In this case, the aim of the study was the comparison between a sinusoidal pattern and the previously analyzed rectangular groove antennas. Similar results were obtained for a high number of periods, with the added value of a soft surface with no corners in the case of the sinusoidal profile, which could reduce multipactor effects, interesting for applications where medium/high power handling is needed. The manufactured 20 period antenna presented a gain of 28.9 dB (Fig. 4.14b) with a –24 dB side lobe level and a very narrow 1.2° beam in the E-plane as shown in Fig. 4.14c. The H-plane is shown in Fig. 4.14d for completeness.

In [41–44] two Bull's-Eye antennas in planar technology were thoroughly analyzed. These antennas consisted of a series of annular concentric metallic strip gratings placed over a grounded dielectric slab and fed by a surface wave launcher (SWL) (scheme shown in Fig. 4.15) or an aperture coupled slot (ACS). There, the

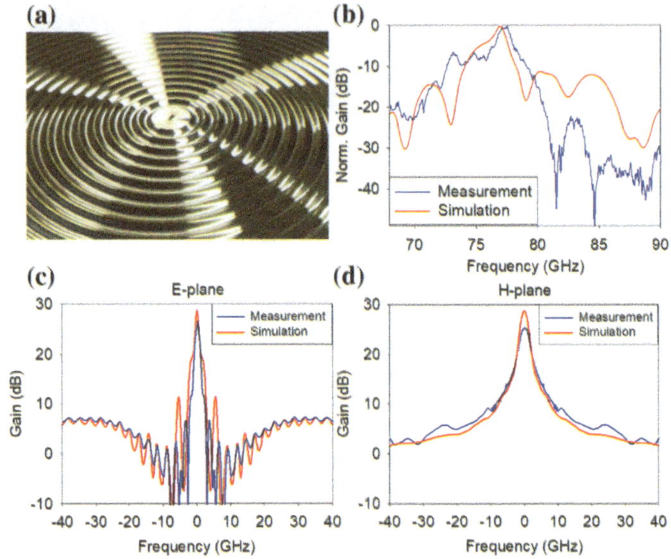

Fig. 4.14 **a** Picture of the metallic sinusoidal grooves Bull's-Eye antenna. **b** Simulated (*red curve*) and measured (*blue curve*) gain versus frequency. Simulated (*red*) and measured (*blue*) E- and H-planes, (**c**) and (**d**) respectively

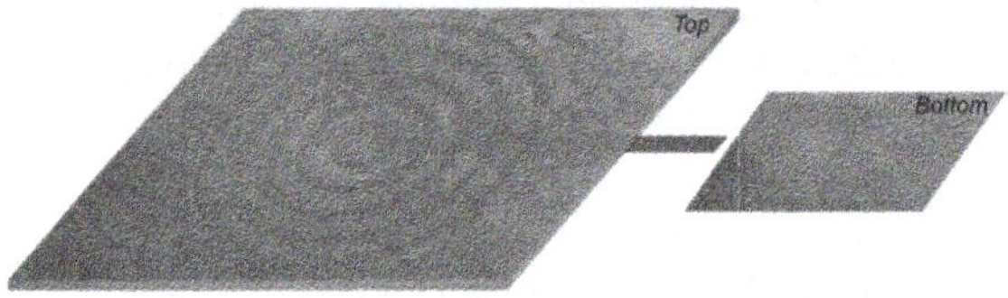

Fig. 4.15 Schematic of a metallic strip grating over a grounded dielectric slab fed by a SWL antenna

performance of different geometries as well as suitable optimizations were proposed. For the Bull's-Eye fed by SWL, a beam of near 15 dB and approximately 10° beamwidth was measured at 19.8 GHz.

Closer in time, an off-axis beaming 3D-printed antenna operating at 96 GHz was proposed [45]. This structure was manufactured by means of SLA (stereolithography) and then metal coated by a several step chemical and plating process, Fig. 4.16a. The antenna presented 7 annular off-center corrugations which permitted a tilted beaming pointing towards 16.5° and a 17 dB peak gain, with −10 dB side lobe level. Figure 4.16b displays the measured gain vs. frequency for E-plane, where the tilted beaming at 96 GHz is observed. This tilted beaming is possible due to the in-phase radiation of upper and lower halves of the structure at the same angle. This structure is a clear example of where the interest is veering to: corrugated leaky wave antennas were initially proposed as an alternative solution to classical high-volume antennas as horns, since they offered better radiation

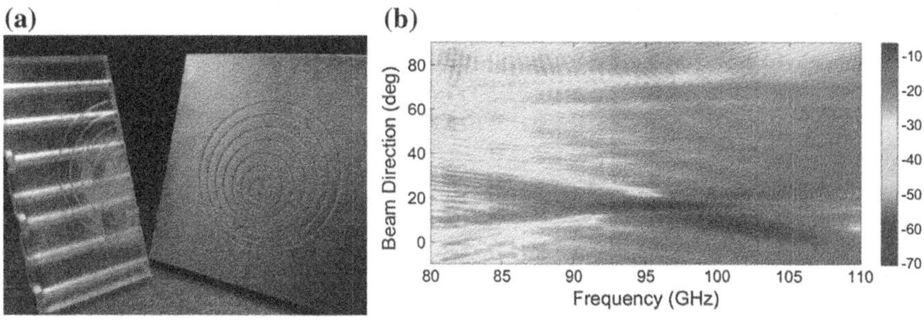

Fig. 4.16 **a** 3D-printed off-axis beaming Bull's-Eye antenna. **b** Measured realized gain versus frequency in dB for E-plane

characteristics and, at the same time, presented lower profiles and lighter weights. The following natural step is to reduce the weight of this new family of antennas. Evidently, the metallization of plastic structures (fabrication method that also facilitates manufacturing of complex structures) is a very interesting solution which, sooner or later, will surely become a hot research topic. This result demonstrates that this type of antennas can be produced employing additive manufacturing which is nowadays envisioned as the fabrication technology of the future and a major actor of the fourth industrial revolution.

The following table sums up some of the radiation characteristics at E-plane for all the above mentioned antennas (Table 4.1).

A quick glance at the table enables us to verify that the best performance in terms of gain is obtained, among the presented structures, for the sinusoidal corrugations antenna. Nevertheless, note that in this case, it is also the structure with the highest number of periods (N = 20 periods) which leads us to a low aperture efficiency antenna, typical of corrugated leaky wave antennas. Meanwhile, the Patch + Bull's-Eye antenna, while showing high gain and relatively low side lobe level, displays a wider beam compared to that presented by the sinusoidally modulated antenna. Thus, it is easy to understand that there is no ideal design which matches any requirement, and it is highly important to carry out a preliminary study so as to choose the design that correctly fits our needs.

4.3 Classical Leaky-Wave Antennas (LWA) and the Problem of Broadside Radiation

As mentioned in the Introduction, corrugated structures can be analyzed using the LW formalism [46]. Indeed Jackson et al. argued in [47], that the theories based on extraordinary transmission could be complemented with the LW theory to explain the mechanism of narrow beaming in corrugated structures [47, 48]. In short, they argued that when a metallic surface is loaded with a periodic array of grooves with periodicity larger than the wavelength and the operating frequency is raised enough,

Table 4.1 Flat leaky wave antennas overview table

Antenna	Frequency (GHz)	Gain (dB)	Side lobe level (dB)	Beamwidth (°)	Physical aperture (normalized to λ^2)
Low profile Bull's-Eye [27]	16.5	21	–15	6	158
12 straight grooves [28]	16.5	18	–10	12	48
Miniaturized [29]	17	10	–8	34	4.62
Dielectric loaded grooves [30]	13	11.3	–9	30	4.62
Dual-band [28]	13/16.5	10.4/9	–9/–9	20/30	4.62
Wedge corrugations [31]	560	15.88	–8.6	5.8	94
Micromachined V-grooves [32]	490	15.6	~–12	18.5	223
CubeSat Bull's-Eye [38]	60	19.6	–14	~10	400
Patch + Bull's-Eye [39]	12.64	19.45	–15	7.9	17.75
Sinusoidal corrugated Bull's-Eye [40]	77	28.9	–24	1.2	1370
Surface wave launcher Bull's-Eye [42]	19.8	~15	~15	~10	126
Tilted beaming 3D-printed antenna [45]	96	17	–10	3.5	5

a leaky plasmon mode is achieved, which radiates as it propagates along the periodic structure [46, 49].

A leaky wave antenna (LWA) is a member of the traveling-wave antennas family [50]. It is a waveguiding structure with a mechanism able to leak power along its length [51]. Due to this leakage, the waveguide presents a complex propagation wavenumber consisting in a phase constant β, which controls the beaming angle, and a leakage constant α. A large or small α corresponds to a large or low leakage affecting the size of the effective aperture of the antenna, which results in a wide or narrow beamwidth, respectively. This kind of antennas present a LW that attenuates in the propagating direction ($\alpha > 0$) and an amplitude that either increases transversely away from the guiding surface ($\alpha_t < 0$), forward leaky wave, or decreases ($\alpha_t > 0$), backward leaky wave. For this reason, forward LWs have been called improper waves and backward LWs proper waves. Anyway, in the forward case, as the LW only exists in a portion of space in the direction normal to the structure, the amplitude increases up to a point and then decreases [52]. Furthermore, due to the fact that the phase constant β varies with frequency, the beam direction can be changed sweeping the frequency, as discussed in [53, 54].

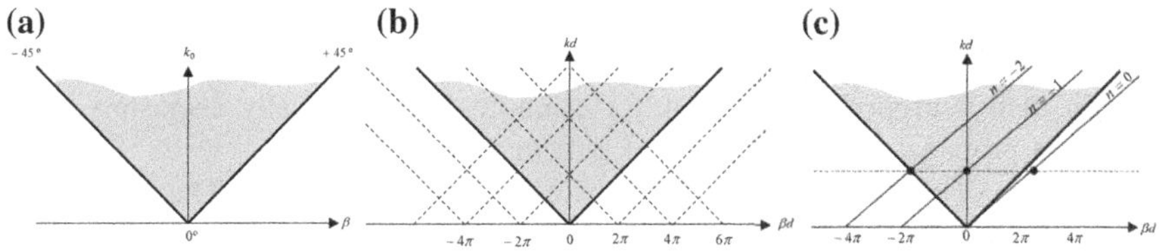

Fig. 4.17 **a** Dispersion diagram for non-corrugated metallic surface. **b** Dispersion diagram for periodically corrugated metallic surface. **c** Fast modes within the radiation region of the dispersion diagram for a corrugated metallic surface

The width of the emerging beam is also controllable by the configuration of the antenna. Thus, for a given beam direction θ and knowing L, the angular width is: $\Delta\theta = \lambda_0/(L\cos\theta)$. The easiest and earliest example of this kind of structures is a simple rectangular waveguide with a continuous slit cut along its side, presented by W.W. Hansen in 1940 [55].

Depending on the geometry of the guiding structure two kinds of LWAs are distinguished: uniform and periodic LWAs. Although the literature surrounding LWAs is extensive, only the structures that serve to the purpose of this chapter will be reviewed here. Thus, more emphasis will be made on periodic antennas:

- Uniform LWAs: in these type of antennas, no periodic modulation is present along the length of the guiding structure. In uniform structures the dominant mode, the space harmonic $n = 0$, is a fast wave that radiates whenever the structure is open. This kind of antennas can also be catalogued in two groups: 1-D and 2-D LWAs.
- Periodic LWAs: are loaded with a periodic structure along the guiding structure's length. In comparison with uniform antennas, periodic LWAs possess a dominant slow-wave mode. The phase velocity of this mode is greater than the velocity of light. This implies that even though the structure is open, it does not radiate. This is why it needs a periodic array to radiate. Although these antennas can scan from backward endfire into part of the forward quadrant, there is a narrow region around broadside where radiation is not allowed due to an open stop band. This issue will be explained thoroughly further in this section. Earliest structures of this kind of antennas consisted in the periodically modulation of a dielectric waveguide like the one presented by King [56]. A typical example is that analyzed in [57–60], which basically consists in a dielectric rectangular rod loaded with periodically distributed metal strips. In another geometry, the metallic strips were replaced by grooves [61].

It is easy to see the resemblance between the periodic LWAs and the antennas previously discussed, in particular [22, 23, 28–31, 40, 62, 63]. All of these antennas consisted of a central radiating slot perforated on a metallic slab surrounded by periodically distributed corrugations with periodicity d. For a non-corrugated flat slab, the dispersion curve is like that displayed in Fig. 4.17a. As all these antennas are made of good electric conductors, we can approximate $\beta_0 \approx k$, where β_0 is the phase constant of the $n = 0$ space harmonic (or fundamental mode), where n is the index of the space harmonic, and $k = 2\pi/\lambda$ is the free-space wavenumber. When a metallic slab is properly modulated, for example, by periodically disposing corrugations, a propagating surface wave is generated.

Thus, the dispersion diagram shows two straight lines at ±45°. The corrugation patterning permits the coupling of the power emanating from the slot to a TM_z mode surface wave that propagates away from the aperture. The presence of the grating produces an infinite number of space harmonics (also called Floquet modes), Fig. 4.17b, that depending on the frequency, will be fast or slow. All these space harmonics are related to each other by:

$$\beta_n d = \beta_0 d + 2\pi n \tag{4.1}$$

where β_n is the phase constant of the n-th space harmonic and β_0 is the phase constant of the $n = 0$ space harmonic, corresponding to the travelling wave supported by a non-corrugated flat metallic plane (valid only for weak modulations).

As stated in [54, 64], the beam direction of LWAs can be swept varying the frequency, due to the fact that the phase constant changes as frequency is swept. In general, and when the surface modulation is weak, it is possible to guess the beam direction θ_n of the n-th radiated space harmonic following the next equation:

$$\sin(\theta_n) = \frac{\beta_n}{k} = \frac{\beta_0}{k} + \frac{2\pi n}{kd} \tag{4.2}$$

where $k = 2\pi/\lambda$ the free-space wavenumber. If the wave propagates along a metal-air interface we can approximate $\lambda_0 \approx k$. Furthermore, for a practical antenna, it is desired only the $n = -1$ space harmonic to radiate inside the radiation region, Fig. 4.17c. The radiation region is the area of the dispersion diagram for which $kd > \beta d$, i.e., the modes possess a phase velocity larger than the light. So, recalling that $\beta_0 \approx k = 2\pi/\lambda$ and the relation between β_0 and β_n, for the mode $n = -1$, Eq. (4.2) becomes:

$$\sin(\theta_{-1}) \approx 1 - \frac{\lambda}{d} \tag{4.3}$$

Thus, for a beam pointing at broadside, $\theta_{-1} = 0°$, d must be chosen equal to λ. Nevertheless, as discussed in [49], pure broadside radiation is not possible. Radiation at $\theta_{-1} = 0°$ would imply $\beta_{-1} = 0$, which would also imply that the leaky mode becomes a standing wave, as attenuation constant drops to zero, resulting in an open stop band in which radiated power sharply decreases [49, 64].

Already in [65], it was attempted to overcome the stop-band inconvenient present in most periodic LWAs. It was foreseen that for a structure like those in [66, 67], basically a dielectric rod with a single metallic strip grating, the stop-band could be highly reduced if a double-strip was used per unit cell. Later on, methods

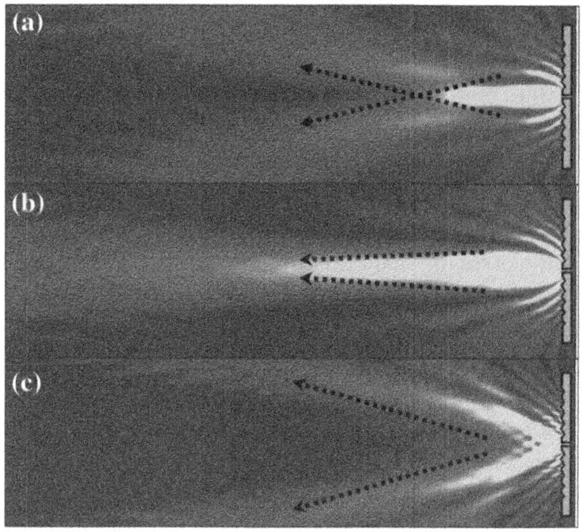

Fig. 4.18 Radiated beams for **a** frequencies $f < f_0$, **b** frequencies $f \sim f_0$ and **c** frequencies $f > f_0$

to optimize broadside radiation in uniform [68] and periodic [69] LWAs were discussed.

As explained in [49], the open stop band effect in periodic LWAs is inevitable. But in the antennas analyzed in the previous section, this limitation was circumvented by choosing a proper β_{-1} so that the beams corresponding to each half of the structure point towards a scan angle slightly diverted from $\theta_{-1} = 0°$. The maximum of radiation appears just before or immediately after the null point, depending on the wavelength-period relation. With regard to the radiation diagram, at the optimized frequency, opposed beams will be so close that they will appear to merge as a single broadside beam. For a certain range around that frequency, the emitted beam will appear as a less directive wider beam and will ultimately be detached into two different beams, corresponding to each half.

This characteristic was experimentally studied in [40] and in more detail in [70], where the behavior of the radiated beams in the vicinity of the operation frequency f_0 was analyzed. For frequencies $f < f_0$, the leaky-mode corresponded to a backward-wave. Given the symmetry of the antenna, the upper and lower halves radiated in the opposite half: the upper-half in the lower-half plane direction and vice versa, Fig. 4.18a. At $f \sim f_0$, an apparently single beam was obtained, Fig. 4.18b. Then, for $f > f_0$, a forward leaky wave is supported by each half and the beams split again in two, radiating each half of the structure in its own half, Fig. 4.18c. Both beams diverged as frequency was shifted towards higher frequencies.

4.4 Design Tips and Guidelines

In this section it is intended to offer a few tips and basic guidelines to design flat leaky wave corrugated antennas. Two types of geometries will be considered: straight parallel and annular corrugations. For both cases, the possibility of designing a tilted beam radiating structure will be shown. An intuitive procedure to design a leaky-wave flat corrugated antenna was presented in [27–30, 40]. Here, we summarize the main tips and guidelines that must be followed for the design of this type of antennas.

The first goal that must be accomplished is ensuring a good matching between the input waveguide and the antenna. This is usually done by designing a central resonant slot, with a relatively high-quality factor. Depending on the needs, the resonant aperture can be designed to resonate at one or two frequencies. The transversal resonance is governed by the slot width, $s_x \sim \lambda_0/2$, whereas the longitudinal resonance is given by the slot depth (or metal plate thickness) $s_z \sim \lambda_0/2$, although in this case the approximation is rougher since the resonance is due to an open cavity, Fig. 4.19a. In either case, the height of the slot must be relatively small $s_y \ll \lambda_0$. This ensures a high-quality factor and hence a good matching which is of interest in this type of antennas. Although a high-quality factor implies a reduced bandwidth, in this case this is not a strong limitation, since the periodic structure automatically imposes a narrow bandwidth for broadside radiation. For dual band operation, the chosen frequencies must be relatively far apart. In fact, as stated in [71], when both s_x and s_z tend to $\lambda_0/2$, longitudinal and transversal resonances might be affected and differ from their isolated behaviour. This is due to the necessary fulfillment of Foster's Theorem, which forces a zero to exist in between two poles (resonances) in the frequency response of passive structures, such as slots (identifiable as an open Fabry-Perot cavity) [72]. This means that there cannot be two

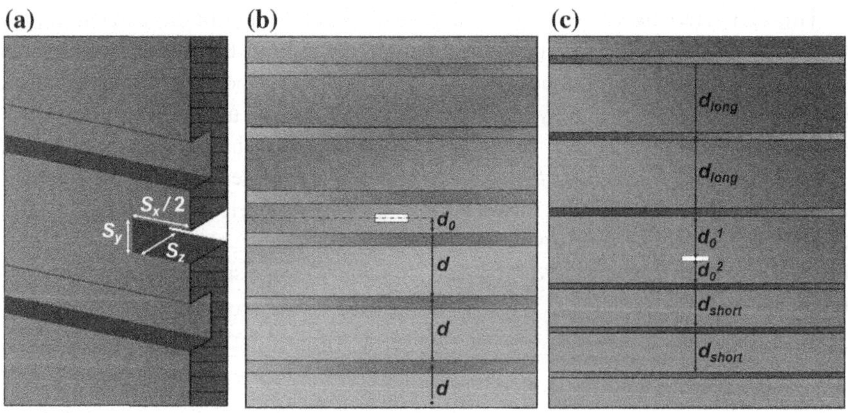

Fig. 4.19 **a** Detail of slot. **b** Front view detail of straight parallel corrugations. **c** Front view detail of asymmetric straight parallel corrugations

identical resonances at the same frequency in the same cavity. Thus, in order to design an antenna with dual frequency operation, s_z must be slightly larger than λ_0.

Designing the grating is probably the most complex issue and some considerations must be taken into account. For a given periodicity d, there is a certain frequency range for which only the $n = -1$ mode radiates and a narrower range for which the beam points close to broadside. Thus, from (4.3) it is easily obtained that for broadside radiation $d \approx \lambda_0$, recalling that beaming at exactly $\theta_{-1} = 0°$ is not possible. Regarding the optimal number of periods, it was pointed out in [28] that for approximately $n = 12$ grooves (6 at each side of the slot) a "gain saturation" point is reached. So, $n = 12$ can be chosen as a first guess to obtain a large directivity with a reasonable side length.

Thus, on the basis that the maximum directivity is obtained at broadside radiation for periods $d \approx \lambda_0$, the metallic plate should have a side length of at least $L_y = (n + 2)\lambda_0$ in the direction of the currents, i.e., the E-plane. As a first guess, a width of the metallic plate of $L_x = 7s_x$ (with slot width $s_x \sim \lambda_0/2$) has been used in previous works, although the final width is fixed further on, as a part of the optimization of the structure. The slab thickness L_z usually coincides with the slot depth, s_z.

For a broadside beaming structure, the first groove must be placed at an offset distance d_0 from the central slot. The rest of the grooves are periodically distributed following the first one with period d, Fig. 4.19b. The offset distance d_0 for the first pair of opposite corrugations must be such that the intense reactive field near the slot discontinuity does not affect the excited LW. In addition, for broadside radiation this distance must be such that all the corrugations radiate in-phase. A convenient way to optimize the offset distance is to place the first pair of grooves almost touching the slot and then sweep the separation with a numerical simulator, monitoring the return losses response, until matching is obtained.

For off-axis beaming, i.e., beam pointing in a direction other than broadside, the groove distribution is a little bit trickier, as corrugations must be asymmetrically distributed at both sides of the slot, Fig. 4.19c. This implies that different offset distances must be separately optimized for both halves. In this case, period d_{long} for one half of the structure for a beam pointing to $\theta_{-1} \neq 0°$ is obtained using Eq. (4.3). The period d_{short} for the other half must be calculated for a beam pointing towards the opposite direction, i.e.,$-\theta_{-1}$. Thus, from Eq. (4.3) we extract two equations:

$$\begin{aligned} d_{long} &\approx \frac{\lambda}{(1 - \sin(\theta_{-1})} \\ d_{short} &\approx \frac{\lambda}{(1 + \sin(\theta_{-1})} \end{aligned} \tag{4.4}$$

where d_{long} corresponds to the half structure which supports a forward leaky-wave and d_{short} to the part which supports a backward leaky-wave. For these asymmetric structures, the number of grooves is not equal and an optimization can help to

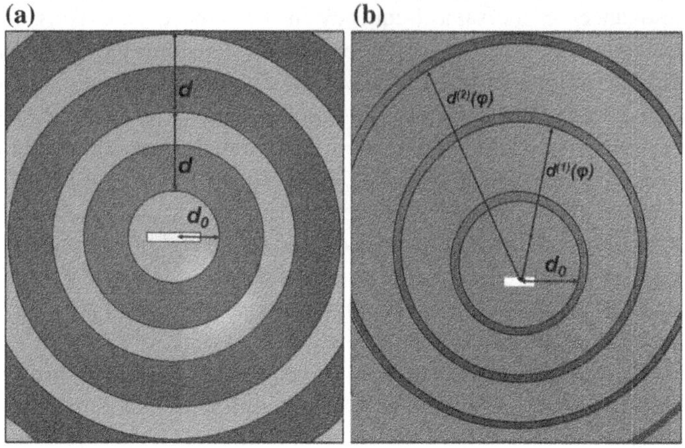

Fig. 4.20 **a** Front view detail of annular concentric corrugation. **b** Front view detail of off-center annular corrugations

design the antenna with little effort. Normally, a larger number of grooves is needed for the half-side which presents a lower period.

However, a better performance can be obtained for both broadside and off-axis beaming antennas if annular corrugations are employed, rather than parallel straight corrugations, Fig. 4.20a. This is due to the in-phase interaction between the cylindrical wavefront launched by the slot and the annular grooves. For the $\theta_{-1} \approx 0°$ case, concentric annular corrugations surrounding the central slot must be placed with a separation of *d*, obtained with (4.3). Period and offset values obtained for the straight corrugations case can be taken in this configuration as a seed for a new optimization.

If $\theta_{-1} \neq 0°$ is desired, the equation governing elliptical corrugations, taking the slot as the origin of coordinates, is as follows:

$$d^{(n)}(\varphi) = \frac{d_0 + n\lambda}{1 - \sin\theta_{-1}\sin\varphi} \tag{4.5}$$

where $d^{(n)}(\varphi)$ is the distance from the slot to the *n*-th groove at angle φ, contained on the plane of the surface, d_0 is the distance from the slot to the first groove at $\varphi = 0°$ (*x* axis) *and* θ_{-1} is the elevation angle at which the beam points, Fig. 4.20b.

It has been previously shown in Sect. 4.2 that the grooves' design can vary depending on the needs and the available resources. Here, for simplicity, reference will only be made to the rectangular geometry. Normally, the depth of the grooves is less than $\lambda/4$ and the width is very small compared to the wavelength at the operation frequency. An adequate way to obtain these values, is to carry out an optimization where $\lambda/4$ and $\lambda/8$ are taken as seeds for the depth and the width, respectively. It must be taken into account that, for the off-axis beaming case, grooves' dimensions differ at both halves. Thus, the optimization of the whole structure for this case becomes more complex than for the broadside (symmetric) one, as more parameters must be included in the calculation. Anyway it was recently shown, in the optical regime, that it was possible to harvest even more light by means of a corrugated structure if wider grooves were used [73]. We have recently applied this concept for a miniaturized antenna working at millimeter-waves with a high aperture efficiency of 32% [74].

If any possible pitfall should be pointed out, it would be the need of an optimization routine for each modification introduced in the structure. For example, the sole introduction of an extra corrugation in an optimized N-period 1D structure, would require a review of the corrugations' parameters, so as to recover the optimum throughput, as well as the resizing of the slot due to a possible frequency-shifting of the cavity resonance.

Regardless of the potential need of several optimizations to obtain an efficient structure, following these few tips it is possible to design a basic corrugated leaky wave antenna whose performance can be fine-tuned by successive numerical optimizations. It is not necessary to remark the scalability of this kind of structures, as its presence in different frequency regimes has already been shown in Sect. 4.2. However, it is noteworthy that the design of the corrugated surfaces at any frequency, from microwaves to the optical regime, follows the same tips above mentioned and only the need of different sources and the proper metallic modelling at each frequency must be separately addressed.

4.5 Discussion and Future Trends

Leaky wave antennas came to light several decades ago. Despite being a classical subject, novel antennas appear from time to time, based on different topologies and coming up with new designs which merge design characteristics specific to other antennas, achieving unparalleled throughputs taking into account their low profile. Among them, corrugated antennas stand out in a privileged position due to their interesting radiation performance, showing higher gains and narrower beams compared to those emitted by standard horn antennas. This has led to the development of several geometries and different capabilities, which is evidenced with the wide range of antennas shown in Sect. 4.2. The large diversity of these antennas and the possibility of scaling them in frequency, allow their use in a large variety of applications, overcoming the challenge of having antennas which presented good radiation characteristics and reduced volume and weight (essential for certain applications, as space communications).

Although corrugated LWAs are also known for presenting a relatively low aperture efficiency, they compensate this issue by a large gain comparable or even higher than that presented in larger volume antennas, such as horn antennas. Anyway, latest designs and research on LWAs are focused in attaining lower side lobe levels and exploiting in a better way the surface of the corrugated structure. It is also of high interest the development of antennas which achieve pure radiation at broadside, and several solutions have been proposed using different technologies.

Modern 3D printing technologies can also be applied to corrugated antennas, giving rise to lighter structures. In that respect, in the recent literature are found purely plastic structures with a thin metallic coating, whose configuration allows the radiation of beams at desired angles. This off-axis beaming had been previously achieved [75–78], but the possibility of fast manufacturing, low price, light weight and easiness of its design are added values that stress the interest of 3D-printed corrugated antennas.

Finally, the interest of this type of structures goes beyond simple corrugated antennas oriented towards communication applications. In recent times, similar structures been proposed in the field of renewable energies, as heat harvesting devices for the photosensitive cells arranged in photovoltaic panels [79, 80].

Micro lenses play a major role in the nowadays thriving sector of the miniaturized optics, with applications in fiber coupling, optical switching, laser collimation, imaging… The joint use of lenses and corrugated structures may result in sensors with increased responsivity and noise decrease [81–83].

Corrugated structures have also found application in imaging in the visible and infrared spectrum. Gratings have been incorporated to multicolor sensing devices or colour filters, replacing typically used filters and glass prisms, which were used to disperse light, and taking advantage of their capability of selective conversion between free-space waves and spatially confined modes [84].

4.6 Conclusions

In this chapter the high interest of corrugated antennas has been shown through the analysis of a vast number of prototypes and applications over a large range of the electromagnetic spectrum. It has been shown the evolution of these antennas since the extraordinary optical transmission was replicated in the microwave and millimeter ranges until the latest designs at THz frequencies.

A brief analysis of the leaky mechanism has been presented, the problem that these structures encounter when broadside beaming is needed and the existing solutions to bypass it, as the pointing of opposed radiating beams to a direction close to broadside. Furthermore, a basic guideline to design broadside and off-axis beaming corrugated antennas has been proposed for both 1D and 2D geometries.

Applications which go far beyond the field of the communications have been introduced, as efficient frequency selective devices found in the range of optics or heat harvesting devices in the field of renewable energies, becoming manifest the large number of possibilities which are on the fringes of the antennas field.

As the drive for miniaturization increases, more efficient and power saving systems will be demanded in fields like optical computing, imaging, sensing and, without a doubt, telecommunication. Thus, as long as the manufacturing and handling processes allow it, patterned surfaces will surely find application in the upcoming technological developments. We have already demonstrated that these antennas are compatible with modern additive manufacturing techniques even at relatively high frequencies. These facts allows us to foresee a brilliant future for planar corrugated antennas all along the electromagnetic spectrum.

References

1. H.A. Bethe, Theory of diffraction by small holes. Phys. Rev. **66**, 163–182 (1944). doi:10.1103/PhysRev.66.163
2. C.J. Bouwkamp, Diffraction theory. Rep. Prog. Phys. **17**, 35–100 (1954). doi:10.1088/0034-4885/17/1/302
3. T.W. Ebbesen, H.J. Lezec, H.F. Ghaemi, T. Thio, P.A. Wolff, Extraordinary optical transmission through sub-wavelength hole arrays. Nature **391**, 667–669 (1998)
4. W.L. Barnes, A. Dereux, T.W. Ebbesen, Surface plasmon subwavelength optics. Nature **424**, 824–830 (2003). doi:10.1038/nature01937
5. Surface plasmon resurrection, Nat. Photon. **6**, 707 (2012)
6. F.J. García de Abajo, Colloquium: light scattering by particle and hole arrays. Rev. Mod. Phys. **79**, 1267–1290 (2007). doi:10.1103/RevModPhys.79.1267
7. F.J. Garcia-Vidal, L. Martin-Moreno, T.W. Ebbesen, L. Kuipers, Light passing through subwavelength apertures. Rev. Mod. Phys. **82**, 729–787 (2010). doi:10.1103/RevModPhys.82.729
8. L. Martín-Moreno, F.J. García-Vidal, H.J. Lezec, K.M. Pellerin, T. Thio, J.B. Pendry, T.W. Ebbesen, Theory of extraordinary optical transmission through subwavelength hole arrays. Phys. Rev. Lett. **86**, 1114–1117 (2001). doi:10.1103/PhysRevLett.86.1114
9. J.A. Porto, F.J. Garcia-Vidal, J.B. Pendry, Transmission resonances on metallic gratings with very narrow slits. Phys. Rev. Lett. **83**, 2845–2848 (1999). doi:10.1103/PhysRevLett.83.2845
10. D.E. Grupp, H.J. Lezec, T. Thio, T.W. Ebbesen, Beyond the Bethe limit: tunable enhanced light transmission through a single sub-wavelength aperture. Adv. Mater. **11**, 860–862 (1999). doi:10.1002/(SICI)1521-4095(199907)11:10<860::AID-ADMA860>3.0.CO;2-V
11. W.C. Chen, N.I. Landy, K. Kempa, W.J. Padilla, A subwavelength extraordinary-optical-transmission channel in babinet metamaterials. Adv. Opt. Mater. **1**, 195–195 (2013). doi:10.1002/adom.201370017
12. F. Falcone, T. Lopetegi, M.A.G. Laso, J.D. Baena, J. Bonache, M. Beruete, R. Marqués, F. Martín, M. Sorolla, Babinet principle applied to the design of metasurfaces and metamaterials. Phys. Rev. Lett. **93**, 197401 (2004). doi:10.1103/PhysRevLett.93.197401
13. T. Zentgraf, T.P. Meyrath, A. Seidel, S. Kaiser, H. Giessen, C. Rockstuhl, F. Lederer, Babinet's principle for optical frequency metamaterials and nanoantennas. Phys. Rev. B Condens. Matter Mater. Phys. **76**, 4–7 (2007). doi:10.1103/PhysRevB.76.033407
14. H.J. Lezec, A. Degiron, E. Devaux, R.A. Linke, L. Martin-Moreno, F.J. Garcia-Vidal, T.W. Ebbesen, Beaming light from a subwavelength aperture. Science **297**, 820–822 (2002). doi:10.1126/science.1071895
15. L. Martín-Moreno, F.J. García-Vidal, H.J. Lezec, A. Degiron, T.W. Ebbesen, Theory of highly directional emission from a single subwavelength aperture surrounded by surface corrugations. Phys. Rev. Lett. **90**, 167401 (2003). doi:10.1103/PhysRevLett.90.167401
16. F.J. García-Vidal, H.J. Lezec, T.W. Ebbesen, L. Martín-Moreno, Multiple paths to enhance optical transmission through a single subwavelength slit. Phys. Rev. Lett. **90**, 213901 (2003). doi:10.1103/PhysRevLett.90.213901
17. M. Beruete, M. Sorolla, I. Campillo, J.S. Dolado, L. Martín-Moreno, J. Bravo-Abad, F. J. García-Vidal, Enhanced millimeter-wave transmission through subwavelength hole arrays. Opt. Lett. **29**, 2500–2502 (2004). doi:10.1364/OL.29.002500

18. M. Beruete, M. Sorolla, I. Campillo, J.S. Dolado, Increase of the transmission in cut-off metallic hole arrays. IEEE Microw. Wirel. Compon. Lett. **15**, 116–118 (2005). doi:10.1109/LMWC.2004.842852
19. M. Beruete, M. Sorolla, I. Campillo, J.S. Dolado, L. Martín-Moreno, J. Bravo-Abad, F. J. García-Vidal, Enhanced millimeter wave transmission through quasioptical subwavelength perforated plates. IEEE Trans. Antennas Propag. **53**, 1897–1903 (2005). doi:10.1109/TAP.2005.848689
20. J. Bravo-Abad, A. Degiron, F. Przybilla, C. Genet, F.J. García-Vidal, L. Martín-Moreno, T. W. Ebbesen, How light emerges from an illuminated array of subwavelength holes. Nat. Phys. **2**, 120–123 (2006). doi:10.1038/nphys213
21. A.P. Hibbins, J.R. Sambles, C.R. Lawrence, Gratingless enhanced microwave transmission through a subwavelength aperture in a thick metal plate. Appl. Phys. Lett. **81**, 4661 (2002). doi:10.1063/1.1527704
22. M. Beruete, I. Campillo, J.S. Dolado, J.E. Rodríguez-Seco, E. Perea, M. Sorolla, Enhanced microwave transmission and beaming using a subwavelength slot in corrugated plate. IEEE Antennas Wirel. Propag. Lett. **3**, 328–331 (2004). doi:10.1109/LAWP.2004.839461
23. M. Beruete, M. Sorolla, I. Campillo, J.S. Dolado, Subwavelength slotted corrugated plate with enhanced quasioptical millimeter wave transmission. IEEE Microw. Wirel. Compon. Lett. **15**, 286–288 (2005). doi:10.1109/LMWC.2005.845753
24. M.J. Lockyear, A.P. Hibbins, J.R. Sambles, C.R. Lawrence, Surface-topography-induced enhanced transmission and directivity of microwave radiation through a subwavelength circular metal aperture. Appl. Phys. Lett. **84**, 2040–2042 (2004). doi:10.1063/1.1688001
25. M.J. Lockyear, A.P. Hibbins, J.R. Sambles, C.R. Lawrence, Enhanced microwave transmission through a single subwavelength aperture surrounded by concentric grooves. J. Opt. A Pure Appl. Opt. **7**, S152–S158 (2005). doi:10.1088/1464-4258/7/2/020
26. S. Steshenko, M. Zhadobov, R. Sauleau, A.A. Kirilenko, A.V. Boriskin, Beam-forming capabilities of waveguide feeds assisted by corrugated flanges. IEEE Trans. Antennas Propag. **63**, 5548–5560 (2015). doi:10.1109/TAP.2015.2487990
27. M. Beruete, I. Campillo, J.S. Dolado, J.E. Rodríguez-Seco, E. Perea, F. Falcone, M. Sorolla, Very low-profile "bull's eye" feeder antenna. IEEE Antennas Wirel. Propag. Lett. **4**, 365–368 (2005). doi:10.1109/LAWP.2005.851104
28. M. Beruete, I. Campillo, J.S. Dolado, E. Perea, F. Falcone, M. Sorolla, Dual-band low-profile corrugated feeder antenna. IEEE Trans. Antennas Propag. **54**, 340–350 (2006). doi:10.1109/TAP.2005.863380
29. M. Beruete, I. Campillo, J.S. Dolado, J.E. Rodríguez-Seco, E. Perea, F. Falcone, M. Sorolla, Low-profile corrugated feeder antenna. IEEE Antennas Wirel. Propag. Lett. **4**, 378–380 (2005). doi:10.1109/LAWP.2005.857297
30. M. Beruete, I. Campillo, J.S. Dolado, J.E. Rodríguez-Seco, E. Perea, F. Falcone, M. Sorolla, Very low profile and dielectric loaded feeder antenna. IEEE Antennas Wirel. Propag. Lett. **6**, 544–548 (2007). doi:10.1109/LAWP.2007.909969
31. M. Beruete, U. Beaskoetxea, M. Zehar, A. Agrawal, S. Liu, K. Blary, A. Chahadih, X.-L. Han, M. Navarro-Cía, D. Etayo, A. Nahata, T. Akalin, M. Sorolla, Terahertz corrugated and bull's-eye antennas. IEEE Trans. Terahertz Sci. Technol. **3**, 740–747 (2013). doi:10.1109/TTHZ.2013.2287096
32. H. Lu, X. Lv, Z. Gao, Y. Liu, Experimental radiation characteristics of micromachined terahertz low-profile corrugated horn antenna. Microw. Opt. Technol. Lett. **57**, 364–367 (2015). doi:10.1002/mop.28849
33. N. Yu, J. Fan, Q.J.Q.J. Wang, C. Pflügl, L. Diehl, T. Edamura, M. Yamanishi, H. Kan, F. Capasso, Small-divergence semiconductor lasers by plasmonic collimation. Nat. Photonics **2**, 564–570 (2008). doi:10.1038/nphoton.2008.152
34. N. Yu, Q.J. Wang, M.A. Kats, J.A. Fan, S.P. Khanna, L. Li, A.G. Davies, E.H. Linfield, F. Capasso, Designer spoof surface plasmon structures collimate terahertz laser beams. Nat. Mater. **9**, 730–735 (2010). doi:10.1038/nmat2822

35. N. Yu, R. Blanchard, J. Fan, Q.J. Wang, C. Pflügl, L. Diehl, T. Edamura, S. Furuta, M. Yamanishi, H. Kan, F. Capasso, Plasmonics for laser beam shaping. IEEE Trans. Nanotechnol. **9**, 11–29 (2010). doi:10.1109/TNANO.2009.2029099
36. N. Yu, Q. Wang, F. Capasso, Beam engineering of quantum cascade lasers. Laser Photon. Rev. **6**, 24–46 (2012). doi:10.1002/lpor.201100019
37. Y. Monnai, D. Jahn, W. Withayachumnankul, M. Koch, H. Shinoda, Terahertz plasmonic Bessel beamformer. Appl. Phys. Lett. 21101 (2015). doi:10.1063/1.4905445
38. C.J. Vourch, T.D. Drysdale, V-band "bull"s eye' antenna for cubeSat applications. IEEE Antennas Wirel. Propag. Lett. **13**, 1092–1095 (2014). doi:10.1109/LAWP.2014.2327852
39. C. Huang, Z. Zhao, X. Luo, Application of "bull's eye" corrugated grooves integrated with artificially soft surfaces structure in the patch antenna to improve radiation performance. Microw. Opt. Technol. Lett. **51**, 1676–1679 (2009). doi:10.1002/mop.24443
40. U. Beaskoetxea, V. Pacheco-Peña, B. Orazbayev, T. Akalin, S. Maci, M. Navarro-Cía, M. Beruete, 77 GHz high gain bull's-eye antenna with sinusoidal profile. IEEE Antennas Wirel. Propag. Lett. **14**, 205–208 (2015). doi:10.1109/LAWP.2014.2360215
41. P. Baccarelli, P. Burghignoli, G. Lovat, S. Paulotto, A novel printed leaky-wave "bull-eye" antenna with suppressed surface-wave excitation, in *IEEE Antennas and Propagation Society Symposium, 2004*. IEEE (2004), pp. 1078–1081. doi:10.1109/APS.2004.1329861
42. S.K. Podilchak, S. Paulotto, P. Burghignoli, Y.M.M. Antar, P. Baccarelli, A.P. Freundorfer, G. Lovat, A printed "bull-eye" leaky-wave antenna fed by a non-directive surface-wave launcher, in *2nd European Wireless Technology Conference, EuWIT* (2009), pp. 81–83
43. S.K. Podilchak, P. Baccarelli, P. Burghignoli, A.P. Freundorfer, Y.M.M. Antar, Analysis and design of annular microstrip-based planar periodic leaky-wave antennas. IEEE Trans. Antennas Propag. **62**, 2978–2991 (2014)
44. S.K. Podilchak, Y.M.M. Antar, A. Freundorfer, P. Baccarelli, P. Burghignoli, S. Paulotto, G. Lovat, Planar antenna for continuous beam scanning and broadside radiation by selective surface wave suppression. Electron. Lett. **46**, 613 (2010). doi:10.1049/el.2010.0074
45. U. Beaskoetxea, S. Maci, M. Navarro-Cía, M. Beruete, 3D-printed 96 GHz bull's-eye antenna with off-axis beaming. IEEE Trans. Antennas Propag. **65**, 17–25 (2017). doi:10.1109/TAP.2016.2628322
46. A.A. Oliner, A. Hessel, Guided waves on sinusoidally-modulated reactance surfaces. IRE Trans. Antennas Propag. **7**, 201–208 (1959). doi:10.1109/TAP.1959.1144771
47. D.R. Jackson, T. Zhao, J.T. Williams, A.A. Oliner, Leaky surface-plasmon theory for dramatically enhanced transmission through a subwavelength aperture, Part II: Leaky-wave antenna model, in *IEEE Antennas Propagation Society International Symposium*, Columbus, OH (2003), pp. 1095–1098. doi:10.1109/APS.2003.1219426
48. A.A. Oliner, D.R. Jackson, Leaky surface-plasmon theory for dramatically enhanced transmission through a subwavelength aperture, Part I : Basic features, in *Proceedings of IEEE AP-S Symp. Radio Science Meeting*, Columbus, OH (2003), pp. 1095–1098. doi:10.1109/APS.2003.1219425
49. D.R. Jackson, A.A. Oliner, T. Zhao, J.T.T. Williams, Beaming of light at broadside through a subwavelength hole: leaky wave model and open stopband effect. Radio Sci. **40**, 1–12 (2005). doi:10.1029/2004RS003226
50. C.H. Walter, *Traveling Wave Antennas* (McGraw Hill, 1965)
51. R.C. Johnson, H. Jasik, *Antenna Engineering Handbook* (McGraw Hill, 2007)
52. A. Ishimaru, *Electromagnetic Wave Propagation, Radiation and Scattering* (Prentice Hall, Englewood Cliffs, 1991)
53. L.O. Goldstone, A.A. Oliner, Leaky-wave antennas I: rectangular waveguides. IRE Trans. Antennas Propag. **7** (1959). doi:10.1109/TAP.1959.1144702
54. F. Monticone, A. Alù, Leaky-wave theory, techniques, and applications: from microwaves to visible frequencies. Proc. IEEE **103**, 793–821 (2015). doi:10.1109/JPROC.2015.2399419
55. W.W. Hansen, *Radiating Electromagnetic Waveguide* (1946)
56. D.D. King, Dielectric image line. J. Appl. Phys. **23**, 699–700 (1952). doi:10.1063/1.1702285

57. M. Chen, B. Houshmand, T. Itoh, FDTD analysis of a metal-strip-loaded dielectric leaky-wave antenna. IEEE Trans. Antennas Propag. **45**, 1294–1301 (1997). doi:10.1109/8.611250
58. S. Kobayashi, R. Lampe, R. Mittra, S. Ray, Dielectric rod leaky-wave antennas for millimeter-wave applications. IEEE Trans. Antennas Propag. **29**, 822–824 (1981). doi:10.1109/TAP.1981.1142669
59. K.L. Klohn, R.E. Horn, H. Jacobs, E. Freibergs, Silicon waveguide frequency scanning linear array antenna. IEEE Trans. Microw. Theory Tech. **26**, 764–773 (1978). doi:10.1109/TMTT.1978.1129484
60. R.E. Horn, H. Jacobs, K.L. Klohn, E. Freigbergs, Single-frequency electronic-modulated analog line scanning using a dielectric antenna. IEEE Trans. Microw. Theory Tech. **30**, 816–820 (1982). doi:10.1109/TMTT.1982.1131144
61. F.K. Schwering, S.-T. Peng, Design of dielectric grating antennas for millimeter-wave applications. IEEE Trans. Microw. Theory Tech. **31**, 199–209 (1983). doi:10.1109/TMTT.1983.1131458
62. M. Beruete, I. Campillo, J.S. Dolado, E. Perea, F. Falcone, M. Sorolla, Very low-profile "bull's-eye" feeder antenna. IEEE Antennas Wirel. Propag. Lett. **4**, 365–368 (2005). doi:10.1109/LAWP.2005.851104
63. M. Beruete, I. Campillo, J.S. Dolado, J.E. Rodríguez-Seco, E. Perea, M. Sorolla, Enhanced microwave transmission using a subwavelength slot in corrugated plate, in *IEEE International Symposium* on *Antennas* and *Propagation*, vol. 3A (2005), pp. 14–17. doi:10.1109/APS.2005.1552161
64. A.A. Oliner, D.R. Jackson, Leaky-wave antennas, in *Antenna Engineering Handbook*, ed. by J.L. Volakis (Mc Graw-Hill, New York, 2007), pp. 11-1-56
65. M. Guglielmi, D.R. Jackson, Broadside radiation from periodic leaky-wave antennas. IEEE Trans. Antennas Propag. **41**, 31–37 (1993). doi:10.1109/8.210112
66. T. Rozzi, R. De Leo, A. Morini, Analysis of the "microstrip-loaded inset dielectric waveguide", in *IEEE MTT-S International Microwave Symposium Digest* (1989), pp. 923–926. doi:10.1109/MWSYM.1989.38873
67. T. Rozzi, L. Ma, Scattering by dipoles in inset dielectric guide and application to millimetric leaky wave antennas, in *17th European Microwave Conference 1987*, pp. 543–548. doi:10.1109/EUMA.1987.333662
68. G. Lovat, P. Burghignoli, D.R. Jackson, Fundamental properties and optimization of broadside radiation from uniform leaky-wave antennas. IEEE Trans. Antennas Propag. **54**, 1442–1452 (2006). doi:10.1109/TAP.2006.874350
69. P. Burghignoli, G. Lovat, D.R. Jackson, Analysis and optimization of leaky-wave radiation at broadside from a class of 1-D periodic structures. IEEE Trans. Antennas Propag. **54**, 2593–2604 (2006). doi:10.1109/TAP.2006.880725
70. U. Beaskoetxea, M. Navarro-Cía, M. Beruete, Broadband frequency and angular response of a sinusoidal bull's eye antenna. J. Phys. D Appl. Phys. **49**, 265103 (2016). doi:10.1088/0022-3727/49/26/265103
71. M. Beruete, P. Rodríguez-Ulibarri, V. Pacheco-Peña, M. Navarro-Cía, A.E. Serebryannikov, Frozen mode from hybridized extraordinary transmission and Fabry-Perot resonances. Phys. Rev. B. **87**, 205128 (2013). doi:10.1103/PhysRevB.87.205128
72. N. Engheta, R.W. Ziolkowski, *Metamaterials: Physics and Engineering Explorations* (Wiley, 2006)
73. D.Y. Na, K.-Y. Jung, Y.B. Park, Transmission through an annular aperture surrounded with corrugations in a PEC plane. IEEE Antennas Wirel. Propag. Lett. **14**, 179–182 (2015). doi:10.1109/LAWP.2014.2360130
74. U. Beaskoetxea, M. Beruete, High aperture efficiency wide corrugations bull's-eye antenna working at 60 GHz. IEEE Trans. Antennas Propag. **65**(6), 3226–3230 (2017). doi:10.1109/TAP.2017.2696423

75. M. Nannetti, F. Caminita, S. Maci, Leaky-wave based interpretation of the radiation from holographic surfaces, in *IEEE Antennas and Propagation Society, AP-S International Symposium (Digest)* (2007), pp. 5813–5816. doi:10.1109/APS.2007.4396873
76. S. Cakmakyapan, A.E. Serebryannikov, H. Caglayan, E. Ozbay, Spoof-plasmon relevant one-way collimation and multiplexing at beaming from a slit in metallic grating. Opt. Express **20**, 26636–26648 (2012). doi:10.1364/OE.20.026636
77. B.H. Fong, J.S. Colburn, J.J. Ottusch, J.L. Visher, D.F. Sievenpiper, Scalar and tensor holographic artificial impedance surfaces. IEEE Trans. Antennas Propag. **58**, 3212–3221 (2010). doi:10.1109/TAP.2010.2055812
78. G. Minatti, M. Faenzi, E. Martini, F. Caminita, P. De Vita, D. Gonzalez-Ovejero, M. Sabbadini, S. Maci, Modulated metasurface antennas for space: synthesis, analysis and realizations. IEEE Trans. Antennas Propag. **63**, 1288–1300 (2015). doi:10.1109/TAP.2014.2377718
79. J.J. Greffet, R. Carminati, K. Joulain, J.P. Mulet, M. Stéphane, Y. Chen, Coherent emission of light by thermal sources. Lett. Nat. **416**, 61–64 (2002). doi:10.1103/PhysRevB.69.155412
80. S. Han, Theory of thermal emission from periodic structures. Phys. Rev. B. **80**, 1–10 (2009). doi:10.1103/PhysRevB.80.155108
81. S. Krishna, S.D. Gunapala, S.V. Bandara, C. Hill, D.Z. Ting, Quantum dot based infrared focal plane arrays. Proc. IEEE **95**, 1838–1852 (2007). doi:10.1109/JPROC.2007.900969
82. J.A. Shackleford, R. Grote, M. Currie, J.E. Spanier, B. Nabet, Integrated plasmonic lens photodetector. Appl. Phys. Lett. **94**, 8–11 (2009). doi:10.1063/1.3086898
83. A. Harrer, B. Schwarz, R. Gansch, P. Reininger, H. Detz, T. Zederbauer, A.M. Andrews, W. Schrenk, G. Strasser, Plasmonic lens enhanced mid-infrared quantum cascade detector. Appl. Phys. Lett. **105** (2014). doi:10.1063/1.4901043
84. T. Xu, Y.-K. Wu, X. Luo, L.J. Guo, Plasmonic nanoresonators for high-resolution colour filtering and spectral imaging. Nat. Commun. **1**, 59 (2010). doi:10.1038/ncomms1058

Chapter 5
Reflectarrays

Sean Victor Hum

Abstract High-gain antennas are a crucial component of microwave, millimetre, and sub-millimetre wave communication and sensing systems. Many systems, such as satellite communication links, point-to-point terrestrial links, deep-space communication links, RADAR systems, and remote-sensing systems depend on high-gain antennas that can be realized in low-profile form at low cost. Combining the best features of reflector antennas and antenna arrays, reflectarrays are promising planar antenna solution for these applications, providing good efficiency and a high degree of flexibility in synthesizing arbitrary antenna patterns. Furthermore, they can be engineered to provide reconfigurable beam-forming capabilities using a variety of approaches. This chapter describes the fundamental operating and design principles of reflectarrays, both in fixed and reconfigurable form. Several practical design examples are included to help readers build confidence with designing this type of space-fed array. Important developments in the field, particularly in the active research area of reconfigurable reflectarrays, are summarized and important future challenges enumerated. Finally, the state of the art in reflectarray applications is reviewed, illustrating the promising future of this emergent antenna architecture.

5.1 Reflectarray Fundamentals

5.1.1 Motivation

High-gain antennas are a crucial component of microwave, millimetre, and sub-millimetre wave communication and sensing systems. These antennas are paramount in many systems including satellite communication links, point-to-point terrestrial links, deep-space communication links, RADAR (including synthetic

S.V. Hum (✉)
Edward S. Rogers Sr. Department of Electrical & Computer Engineering,
University of Toronto, 10 King's College Road, Toronto, ON M5S 3G4, Canada
e-mail: sean.hum@utoronto.ca

A. Boriskin and R. Sauleau (eds.), *Aperture Antennas for Millimeter and Sub-Millimeter Wave Applications*, Signals and Communication Technology,
DOI 10.1007/978-3-319-62773-1_5

aperture RADAR) systems, and remote-sensing systems. Generally, these antennas either take the form of aperture antennas such as reflectors and lenses, or antenna arrays, depending on system requirements. On the one hand, aperture antennas offer a straightforward solution for producing high-gain patterns. However, they tend to be bulky and heavy, which can be a major drawback in spaceborne platforms where launch weight and volume are crucial driving factors. In situations where beam-steering is needed, traditional aperture antennas require mechanical scanning to implement this capability. On the other hand, antenna arrays offer a flexible antenna platform from which high-gain antenna patterns, including potentially adaptive ones, can be realized. Phased arrays, in particular, provide adaptive beam-scanning and beam-shaping capabilities. However, the main drawbacks of arrays are the high complexity and potential losses in the beam-forming network. The latter are particularly pronounced at millimetre and sub-millimetre wave frequencies. In the case of phased arrays, there is also a substantial hardware footprint as each array element (or sub-array in some cases) requires a dedicated RF signal chain to handle transceiver functions. This leads to very high implementation costs in phased arrays since this hardware needs to be duplicated at the element or sub-array level.

The issues facing both apertures and arrays have spurred research into alternative antenna platforms. Most significantly, *reflectarray* antennas have emerged as promising candidates for realizing high-gain, high-performance antenna platforms. As the name suggests, the reflectarray combines features of both reflector antennas and antenna arrays. A typical microstrip reflectarray is shown in Fig. 5.1a. In its simplest form, a reflectarray can be thought of as an array of electromagnetic scatterers illuminated by a feed antenna, much like a traditional reflector antenna. The array and scatterers are usually planar. In order to emulate the behaviour of a curved reflector, each scatterer must be configured to produce a phase shift to

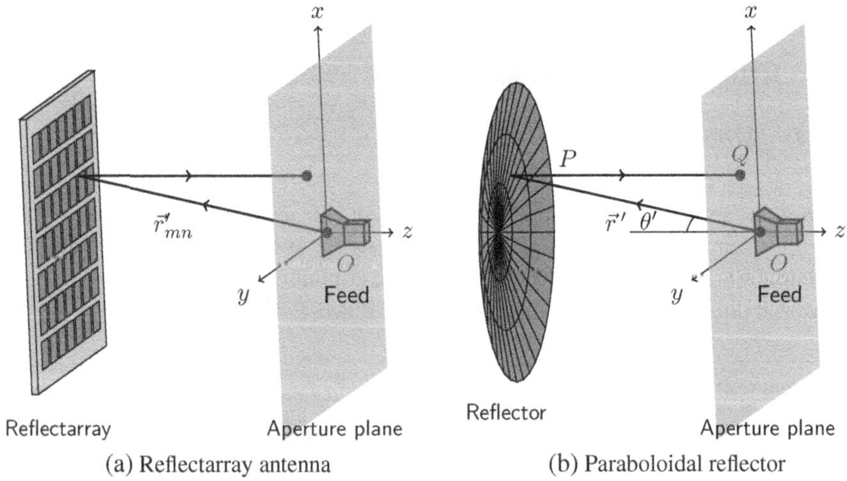

(a) Reflectarray antenna (b) Paraboloidal reflector

Fig. 5.1 Comparison of a reflectarray and a reflector antenna compensate for the path lengths and subsequent phase delays introduced between the feed antenna and reflector surface. As such, a reflectarray can be thought of as an array of spatial phase shifters that electrically deform the surface to resemble the electromagnetic response of a physical reflector, over a given frequency range. A myriad of techniques for realizing these phase shifters have emerged over the past few decades, and have inspired research along many fronts. While most research focuses on realizing fixed-beam reflectarrays, increasingly research has focused on reconfigurable reflectarrays capable of tuning the response of the scatterers to provide adaptive beam-steering and beam-forming capabilities.

5.1.2 Brief History of Development

Flat reflectors have been long sought after as replacements for bulkier curved reflectors. The first reflectarray implementation was reported by Berry et al. in 1963 [1]. The unit cells consisted of short-circuited rectangular waveguide sections, whose lengths were chosen appropriately to realize the correct delays to collimate radiation from the primary feed. However, the concept did not capture widespread interest, likely because the waveguide array itself was bulky, similar to the shaped-reflector it aimed to replace. The advent of microstrip antenna technology was a game-changer for reflectarrays, since it allowed them to be realized in true low-profile form. Advances in microstrip antenna technology revived interest in reflectarrays in the late 1980s/early 1990s [2, 3], sparking considerable advances in the design of various planar units.

Early reflectarray elements were based on resonators, based on the fact that pronounced shifts in the phase of the scattered field can be produced in the vicinity of the resonant frequency of the scatterer. Such designs employed printed dipoles/patches slots, rings, and other interesting shapes to realize the resonator. Equally popular in early designs was using delay lines in the form of stubs connected to microstrip patches, to achieve the required phase shifts. Physical (sequential) rotation of the elements could be used to realize phase shifting of circularly polarized waves. These phase-shifting techniques are described more detail in Sect. 5.2.2. Basically, by manipulating the physical characteristics of the elements, the desired phase profiles for beam collimation, shaped beams, and other radiation patterns could be realized by treating the reflectarray as a type of antenna array with only control of the phase of the radiated fields from each element. Early reflectarrays tended to follow standard rules for arrays to avoid grating lobes, usually employing half-wavelength spacing between elements. Much of the early effort in reflectarrays throughout the 1990s was motivated by their potential in satellite communications.

In the 2000s, important advances were made to expand the functionality and utility of reflectarrays. One important advance was that researchers recognized that the bandwidth of reflectarrays needed to be significantly improved if they were to realistically compete with reflectors. This was initially done by coupling multiple resonators together, either to support dual/multi-band operation or to increase the bandwidth about a single operating frequency. However, researchers soon saw that in order to achieve good performance in terms of phase control and polarization purity over large bandwidths, the reflectarray had to be seen as less as a discrete array of scatterers and more as an inhomogeneous surface impedance, which initiated the development of reflectarrays based on sub-wavelength (versus half-wavelength) unit cells. Many modern wideband implementations of reflectarrays employ sub-wavelength unit cells and are analyzed in terms of this framework.

A second major advance was the realization of reconfigurable functionality from reflectarrays. By adding electronic tuning capability to the unit cell, the phase profile of the reflectarray can be dynamically manipulated, allowing for functionality such as adaptive beam-scanning, on-the-fly pattern synthesis, and even advanced features such as frequency reconfigurability. Advances in both microwave components as well as tunable microwave materials have spurred research and development in a great variety of beam-forming platforms that includes not only reflectarrays, but related space-fed arrays such as transmitarrays, as discussed in Chap. 6. We will consider developments in this area in more detail in Sect. 5.4.

5.1.3 Basic Operating Principle

The operating principle of a reflectarray can be rationalized from both the perspective of a reflector antenna and an antenna array. In their most basic form, reflectarrays emulate curved reflectors to synthesize pencil beams. As such, consider a basic axially symmetric reflector shown in Fig. 5.1b that produces a pencil beam along the reflector (z) axis. The reflector achieves collimation of radiation from a feed placed at the origin O by ensuring that for each ray emanating from the feed and subsequently reflected from the reflector, that the total path length traversed by the ray is constant. Referring to Fig. 5.1b, for each point P considered on the reflector surface,

$$\overline{OP} + \overline{PQ} = r' + r' \cos\theta' = \text{constant}, \tag{5.1}$$

where the point Q is located in an imaginary aperture perpendicular to the reflector axis, in front of the reflector. A standard paraboloidal reflector results if the constant in Eq. (5.1) is set to be $2f$, where f is the focal length of the reflector. Then

$$r' = f \sec^2\left(\frac{\theta'}{2}\right), \quad \theta' \leq \theta_0, \tag{5.2}$$

where θ_0 is the subtended half-angle of the reflector. Incidentally, from a ray optics perspective, this reflector shape also produces a local angle of reflection such that all rays emanating from the feed point O are parallel to the reflector axis.

The path length relation given by Eq. (5.1) can also be stated in terms of electrical path length. If the free space wavenumber is denoted as k_0, then $-k_0(\overline{OP}+\overline{PQ})=\text{constant}$, where a negative sign has been shown explicitly to indicate phase delay. The reflectarray in Fig. 5.1a can be configured to achieve the same behaviour as a reflector antenna by making all the electrical path lengths at one frequency or a range of frequencies constant for each element on the reflectarray surface. Each element of the reflectarray produces a phase shift ψ_{mn} between the incident and reflected waves. Furthermore, the second term of the equation is not needed since the aperture plane can be defined to be coincident with the reflectarray surface. If we consider a reflectarray in which the reflectarray elements have been configured in a rectangular lattice and use a standard row–column indexing scheme as shown in Fig. 5.1a, then the design equation for reflectarrays can be written as

$$-k_0 r'_{mn} + \psi_{mn} = \text{constant}, \tag{5.3}$$

where the indices m and n are used to index each element. ψ_{mn} is typically constrained to fall within an interval $[0, 2\pi]$, and thus a factor of $2\pi N, N \in \mathbb{Z}$ is typically added to the constant so that $\psi_{mn} \in [0, 2\pi]$. It is worth noting that the remaining constant in the equation is a phase offset that can be used as a degree of freedom in determining where the phase wraps in the resulting phase mask of the reflectarray are produced.

At this point, the choice of ψ_{mn} produces a constant phase across the aperture plane after the incident fields have been reflected from the reflectarray surface. From an array perspective, it is as if the element excitations are uniform in phase. Continuing with this interpretation, it is straightforward to see that the beam can be steered off the reflector axis to a direction given by a unit vector $\hat{r}_0$ by superimposing a phase gradient on top of the existing phases,

$$-k_0(r'_{mn} - \vec{R}_{mn} \cdot \hat{r}_0) + \psi_{mn} = \text{constant} + 2\pi N, \tag{5.4}$$

where $\vec{R}_{mn}$ is a position vector of the mnth element within the plane of the reflectarray. This equation is the standard pencil beam-forming equation for reflectarrays. This antenna array perspective offers a simple way to see how the phase shifts are derived. However, a corresponding reflector perspective is also possible: just as a curved reflector employs Snell's law to achieve beam collimation, the same is also true of reflectarrays. However, a more generalized version of Snell's Law is required [4].

Consider a cross section of a 1D reflectarray surface shown in Fig. 5.2. The element spacing is Δx. Two rays are shown that are associated with an incoming plane wave incident upon the reflectarray. A phase shift $\psi(x_A)$ is applied by a reflectarray element at point A, and a phase shift $\psi(x_B)$ by the corresponding element at point B. If the angle of incidence is $-\theta_i$, the phase gradient between points A and B determines the angle of reflection produced by the reflectarray, θ_r.

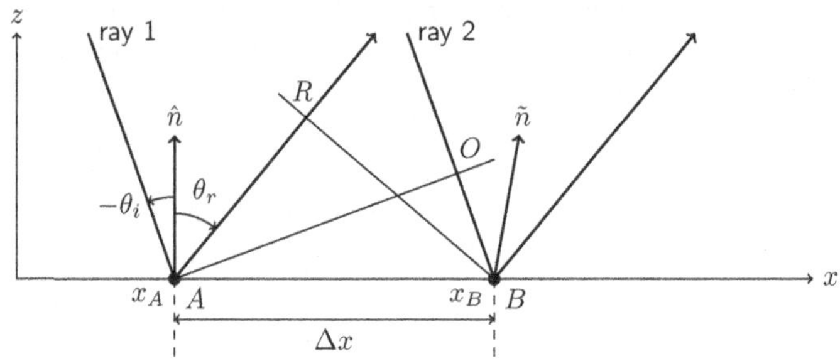

Fig. 5.2 Generalized Snell's law

The generalized Snell's law can be found by observing that the phase shift incurred along path $\overline{AR}$ plus that from $\psi(x_A)$ must be equal to that incurred along path $\overline{OB}$ plus $\psi(x_B)$,

$$\psi(x_A) - k_0\Delta x \sin\theta_r = \psi(x_B) + k_0\Delta x \sin\theta_i. \tag{5.5}$$

Rearranging this equation and taking the limit as $\Delta x \to 0$,

$$\lim_{\Delta x \to 0} \sin\theta_r + \sin\theta_i = -\frac{1}{k_0}\frac{\partial\psi(x)}{\partial x}. \tag{5.6}$$

Clearly, we can see that Snell's law for normal media ($\theta_r = -\theta_i$) does not apply here because of the phase gradient imposed by the reflectarray surface. In fact, the effect of this spatial phase shift is to produce an effective normal $\tilde{n}$ that is not coincident with the normal to the physical reflectarray surface, about which the incident and reflected rays appear to obey Snell's law. Hence, reflectarrays can be seen as synthesizing a set of effective normals such that the emerging rays from the surface are parallel to the desired direction of propagation. For a two-dimensional reflectarray,

$$\sin\theta_r \cos\theta_r + \sin\theta_i \cos\theta_i = -\frac{1}{k_0}\frac{\partial\psi(x,y)}{\partial x} \tag{5.7a}$$

$$\sin\theta_r \sin\theta_r + \sin\theta_i \sin\theta_i = -\frac{1}{k_0}\frac{\partial\psi(x,y)}{\partial y} \tag{5.7b}$$

With the basic calculation of the required reflectarray element phases concluded, we turn our attention to the realization of the required phase shifts by the reflectarray elements. This forms the core part of the reflectarray design process: devising and implementing a suitable reflectarray element or unit cell for implementing the phase-shifting operation needed on the reflector surface. Literally, hundreds of different designs have been proposed over the years, including both fixed and reconfigurable variations. A survey of different element types will be presented in more detail in Sect. 5.2.2. For illustration purposes, here, we explain the operation of the reflectarray using a simple reflectarray element, a square microstrip patch.

The element geometry is shown in Fig. 5.3a. The element consists of a conductive patch of dimensions $l \times w$ printed on a dielectric substrate. The substrate has a thickness h and a dielectric constant ϵ_r, and is backed by a conducting ground plane. Patches in the reflectarray are situated in a periodic arrangement corresponding to a rectangular unit cell of dimensions $a \times b$. For simplicity, we consider a simple square patch $(w = l)$ situated within a square unit cell $(a = b)$.

The scattering characteristics of the patch are determined by analyzing the patch in a periodic environment. This is accomplished using Floquet analysis and is presented in more detail in Sect. 5.2.1. The patch is illuminated with a plane wave, for example, at normal incidence, and the scattered field is analyzed in terms of the resulting reflection coefficient Γ. Typically, with low loss conductors and dielectrics, the magnitude of the reflection coefficient is close to unity, and the phase is of most interest. An example of the phase shifts produced by cells of different lengths l is plotted in Fig. 5.3b. It can be seen that for a given patch length l, an (inverted) S-shaped phase curve is produced, resulting from the resonant behaviour of the patch. This results in a phase curve that nearly spans a 2π radian phase range, and it is this pronounced phase shift that allows patch-like scatterers to act as spatial phase shifters. As the patch length increases, the resonant frequency decreases, resulting in the multitude of curves shown in the plot. At the design frequency f_0, it is possible to evaluate the phases as a function of patch length, resulting in the phase curve plotted in Fig. 5.4.

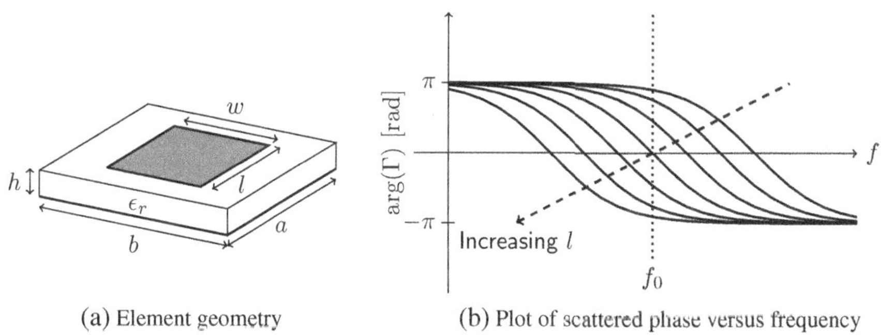

Fig. 5.3 Microstrip patch reflectarray element and its scattering characteristics

Fig. 5.4 Plot of scattered phase versus patch length

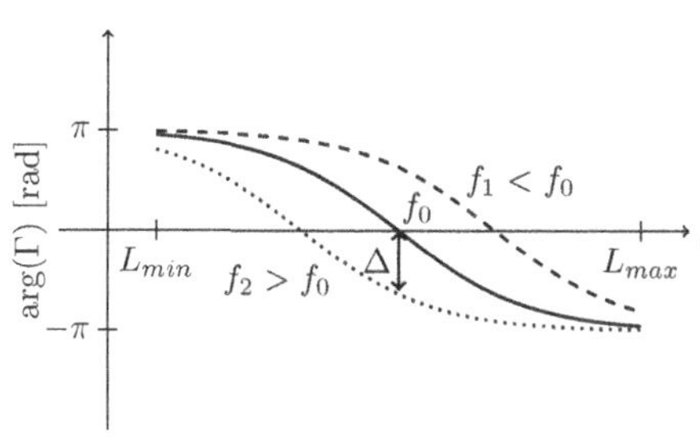

This plot is a design curve for a reflectarray unit cell. It informs the designer which patch length l must be chosen to yield a given phase shift. In general, the realizable patch lengths are limited ($l \in [L_{min}, L_{max}]$), given that the patch must fit within the confines of the periodic cell ($L_{max} < a$). This, coupled with the fact that a single-pole resonator like a patch cannot fully span a 2π scattered phase range even if its length was unconstrained, means that there will a range of phase shifts that are not achievable with a cell of this type. Generally, the region is small and leads to only small phase errors across the aperture and subsequent changes in the radiation pattern from the reflectarray. A more significant issue is the fact that the required phases (in the achievable range) can only be realized at the design frequency f_0. As seen in the plot, at frequencies in the vicinity of f_0, a different S-curve is produced, leading to an error Δ between the design phase at f_0 and the actual phase at the frequency of interest (e.g. f_1 in the plot). A common rule of thumb is that the bandwidth of the reflectarray is constrained to frequencies where the phase error is $\Delta \leq \pm\pi/8$. Beyond this range, phase errors become pronounced enough to significantly impact the ability of the reflectarray to form a pencil beam in the required direction with the required gain. As a result, single-pole reflectarrays such as those based on the unit cell here can only achieve very narrow bandwidths of operation, typically on the order of a few percent.

Fundamentally, the bandwidth problem stems from the fact that reflectarray element must not only compensate for the phase shift in reflectors but also correct for the difference in delay between the flat reflector and curved-reflector surrogate. Hence, fundamentally, true time delay (TTD) reflectarray elements are required to achieve wideband operation. In the case of a broadside pencil beam, for example, the ideal reflection phase from the element is $\psi_{mn} = -k_0 r'_{mn}$, which is a linear function of frequency. Such a phase response cannot be realized by a tuned-resonator design with a finite number of resonators, and necessitates other design types which will be discussed in Sect. 5.2.2.

Reflectarrays can realize a diverse range of radiation patterns in addition to traditional pencil beams, owing to the flexibility in manipulating the phase of the aperture field. A wide variety of patterns have been demonstrated, including contoured beams, cosecant beams, multiple beams, and many others. While the detailed presentation of synthesis methods is beyond the scope of this chapter, we briefly overview a few techniques here. One simple technique is to use a shaped-reflector as a surrogate for a reflectarray and determine the phase shifts using ray-tracing [5]. A number of phase-only synthesis techniques have also been proposed that are applicable to reflectarrays, such as the intersection approach [6], which can be subsequently implemented using alternating projections [7], and the matrix projection method [8]. There is also a class of methods that have been successfully applied to reflectarray synthesis problems that use evolutionary algorithms to determine the element phase distributions. These include particle swarm optimization (PSO) [9], genetic algorithms [10], hybrid techniques, and others.

5.1.4 Reflectarray Variations

The majority of reflectarray designs are single-reflector designs as shown in Fig. 5.1a whereby a moderate-directivity antenna such as a horn antenna is used as the feed. Such reflectarrays consist of a single reflecting surface which can be fed from a point centred in front of the reflector (a centre- or prime focus-fed reflectarray), or at a point offset from the centre point (offset-fed reflectarray). Offset reflectors are useful in situations where one wishes to minimize the impact of feed blockage, particularly when the feed size is a significant fraction of the reflectarray size. Offset reflectarray geometries can be related to their curved reflector counterparts [11].

Another reflectarray architecture uses multiple reflectors, usually two, to accomplish beam-forming as shown in Fig. 5.5a. Most such reflectarrays are designed to emulate classical two-reflector designs such as Cassegrain or Gregorian reflectors. This is done to reduce the distance between the feed antenna and reflecting surface. In such cases, the sub-reflectarray phase distribution is chosen so that it produces an effective hyperboloidal shape (for Cassegrain reflectors) or an ellipsoidal shape (for Gregorian reflectors), while the main reflectarray maintains a paraboloidal phase distribution. The corresponding design equations for the sub-reflectarray are easily found [12].

Finally, a special type of dual-reflector antenna is known as the folded reflectarray, which is illustrated in Fig. 5.5b. Similar to a folded reflector, a polarizing grid is placed in front of the reflector, and is used to reflect radiation from the feed back on to the reflectarray, and in some cases, to manipulate the amplitude distribution on the reflectarray. As such, the main purpose of such designs is to reduce the physical size of the reflector system while maintaining the equivalent focal length of a single reflector. In reflectarray form, the main reflector must accomplish a polarization-twisting operation so that the scattered fields, which have been collimated by the reflectarray, can pass through the polarizing grid. The details of how this can be accomplished will be discussed in Sect. 5.2.2.

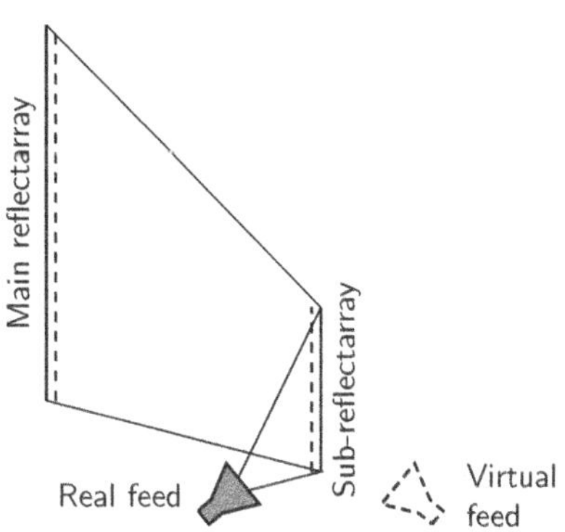

(a) Dual reflector

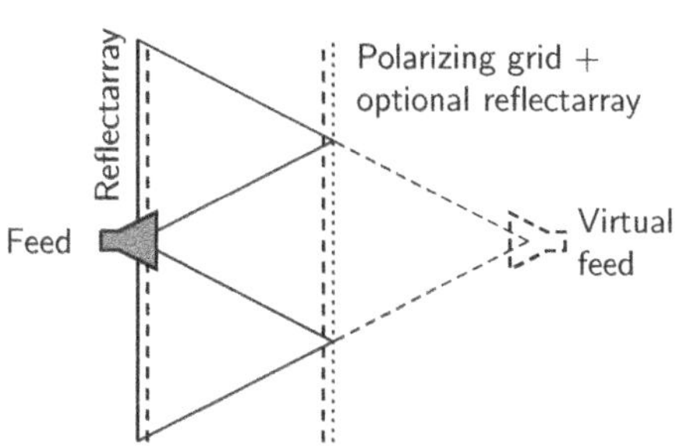

(b) Folded reflector

Fig. 5.5 Reflectarray architectures

5.2 Reflectarray Design at the Unit Cell Level

5.2.1 Floquet Analysis

Reflectarray unit cell analysis revolves around treating the unit cells as if they were in an infinite array, or periodic, environment. Reflectarrays are not strictly periodic structures, because the cell geometry changes from cell to cell in order to achieve feed collimation and beam-forming. Nevertheless, the changes from cell to cell are generally small enough that locally, each cell can be treated as if it were surrounded by identical unit cells. The use of the so-called *local periodic assumption* greatly simplifies the analysis of reflectarray unit cells while fairly accurately capturing the mutual coupling effects with adjacent unit cells. Approaches for accounting for perturbations in nearby cells in order to yield more accurate phase curves have been proposed, but employing the periodic assumption generally yields acceptable results.

Before discussing reflectarray unit cell analysis further, a brief review of two-dimensional periodic structures is in order. Consider a planar surface current that flows in a plane parallel to the *xy*-plane. This current is defined spatially using a function $f(x, y)$ which is definable over one period of the unit cell, $-a/2 \leq x \leq a/2$ and $-b/2 \leq y \leq b/2$. The current density is repeated periodically in both the *x*- and *y*-directions. For example, the *x*-component of the current density can be written as

$$J_x(x,y) = \sum_{m=-\infty}^{\infty} \sum_{m=-\infty}^{\infty} f(x - ma, y - nb). \tag{5.8}$$

Without loss of generality, we can impose a linear phase gradient on the current distribution, since linear phase gradients are also periodic, albeit generally with a different spatial period than that of the unit cell itself. Let the scalar current distribution be

$$J_x(x,y) = \sum_{m=-\infty}^{\infty} \sum_{m=-\infty}^{\infty} f(x - ma, y - nb) \exp(-jk_{x0}ma - jk_{y0}nb), \tag{5.9}$$

where k_{x0} and k_{y0} denote the phase gradients in the *x*- and *y*-directions between two successive unit cells. Selecting the phase shifts to be

$$k_{x0} = k_0 \sin\theta_0 \cos\phi_0 \tag{5.10a}$$

$$k_{y0} = k_0 \sin\theta_0 \sin\phi_0 \tag{5.10b}$$

establishes a beam from the infinite array pointing in the direction (θ_0, ϕ_0). The two-dimensional Fourier transform of this current distribution is

$$\tilde{J}_x(k_x, k_y) = \frac{4\pi^2}{ab}\tilde{f}(k_x, k_y) \sum_{m=-\infty}^{\infty} \sum_{m=-\infty}^{\infty} \delta(k_x - k_{xmn})\delta(k_y - k_{ymn}), \tag{5.11}$$

where $\tilde{f}$ denotes the two-dimensional Fourier transform of $f(x, y)$, $\delta(\cdot)$ is the Dirac delta function, and

$$k_{xmn} = k_{x0} + \frac{2m\pi}{a} \tag{5.12a}$$

$$k_{ymn} = k_{y0} + \frac{2n\pi}{b}. \tag{5.12b}$$

It can be seen from Eq. (5.11) that the current distribution in the spectral domain is defined at discrete spectral points. If we transform back to the spatial domain, then

$$J_x(x, y) = \frac{4\pi^2}{ab} \sum_{m=-\infty}^{\infty} \sum_{m=-\infty}^{\infty} \tilde{f}(k_{xmn}, k_{ymn}) \exp(-jk_{xmn}x - jk_{ymn}y) \tag{5.13}$$

which is known as the *Floquet series* of $J_x(x, y)$. It can be seen that each discrete spectral point corresponds to a corresponding plane wave weighted by a factor of $\tilde{f}(k_{xmn}, k_{ymn})$, which is known as a *Floquet mode*. If

$$k_{xmn}^2 + k_{ymn}^2 \leq k_0^2 \tag{5.14}$$

then the Floquet mode is a propagating plane wave whose direction in the spherical coordinate system is described according to

$$k_{xmn} = k_0 \sin\theta_{mn} \cos\phi_{mn} \tag{5.15a}$$

$$k_{ymn} = k_0 \sin\theta_{mn} \sin\phi_{mn} \tag{5.15b}$$

$$k_{zmn} = k_0 \cos\theta_{mn}. \tag{5.15c}$$

If Eq. (5.14) is not satisfied, then the Floquet mode is evanescent. Knowledge of propagating Floquet modes is important in identifying the ways in which a periodic structure can radiate, since large unit cell sizes can produce *grating lobes* which are associated with higher order propagating Floquet modes. For reflectarray designs, we are principally interested in the behaviour of the fundamental propagating Floquet modes.

Our analysis of a reflectarray unit cell begins by defining the scatterer to be within a periodic environment as shown in Fig. 5.6b. The scatterer is shown as a free-standing planar structure in the $z = 0$ plane that supports symmetric scattering on both sides of $z = 0$. The scatterer can be thought of as being placed with a periodic waveguide whose walls correspond to periodic boundary conditions

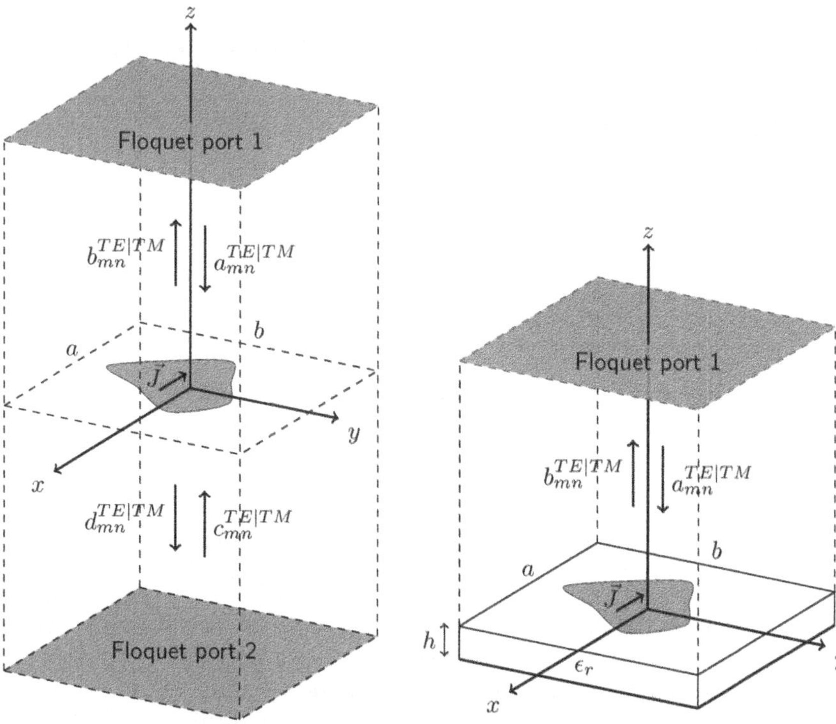

(a) Two-port periodic structure (b) One-port periodic structure

Fig. 5.6 Periodic unit cell and excitation

(PBCs) and whose axis is along z-axis. A Floquet port is placed at both ends of the waveguide to introduce incident waves towards the structure. Each mode represents an incident wave propagating at an angle (θ_0, ϕ_0) towards the scatterer, establishing incident wavenumbers according to Eq. (5.10a). Structuring the analysis of the unit cell in this way, the goal is to derive the generalized scattering matrix (GSM) relating the incident and scattered waves associated for all the Floquet modes, which is written as

$$\begin{bmatrix} \mathbf{b}^{\mathrm{TE}} \\ \mathbf{b}^{\mathrm{TM}} \\ \mathbf{d}^{\mathrm{TE}} \\ \mathbf{d}^{\mathrm{TM}} \end{bmatrix} = \begin{bmatrix} \mathbf{S}_{11}^{\mathrm{TE,TE}} & \mathbf{S}_{11}^{\mathrm{TE,TM}} & \mathbf{S}_{12}^{\mathrm{TE,TE}} & \mathbf{S}_{12}^{\mathrm{TE,TM}} \\ \mathbf{S}_{11}^{\mathrm{TM,TE}} & \mathbf{S}_{11}^{\mathrm{TM,TM}} & \mathbf{S}_{12}^{\mathrm{TM,TE}} & \mathbf{S}_{12}^{\mathrm{TM,TM}} \\ \mathbf{S}_{21}^{\mathrm{TE,TE}} & \mathbf{S}_{21}^{\mathrm{TE,TM}} & \mathbf{S}_{22}^{\mathrm{TE,TE}} & \mathbf{S}_{22}^{\mathrm{TE,TM}} \\ \mathbf{S}_{21}^{\mathrm{TM,TE}} & \mathbf{S}_{21}^{\mathrm{TM,TM}} & \mathbf{S}_{22}^{\mathrm{TM,TE}} & \mathbf{S}_{22}^{\mathrm{TM,TM}} \end{bmatrix} \begin{bmatrix} \mathbf{a}^{\mathrm{TE}} \\ \mathbf{a}^{\mathrm{TM}} \\ \mathbf{c}^{\mathrm{TE}} \\ \mathbf{c}^{\mathrm{TM}} \end{bmatrix}, \tag{5.16}$$

where $\mathbf{a}_{\mathbf{mn}}^{\mathbf{TE|TM}}$ and $\mathbf{b}_{\mathbf{mn}}^{\mathbf{TE|TM}}$ are vectors containing the amplitudes of the incident and reflected waves for each Floquet mode from port 1, respectively, and $\mathbf{c}_{\mathbf{mn}}^{\mathbf{TE|TM}}$ and $\mathbf{d}_{\mathbf{mn}}^{\mathbf{TE|TM}}$ are the corresponding amplitudes for port 2. Generally, unit cells are chosen to be $\lambda_0/2$ or smaller to avoid grating lobes (where $\lambda_0/2$ is the free space wavelength), so only two fundamental Floquet modes can be excited and propagate: the TE_{00} mode and the TM_{00} mode.

Next, we consider the reflectarray case shown in Fig. 5.6b, whereby a ground plane has been introduced at $z = -h$ and a substrate with dielectric constant ε_r occupies the space between $z = -h$ and $z = 0$. In general, the scatterer could also be a multi-layer structure with multiple dielectrics forming stratified layers below $z = 0$. In either case, the waves c_{mn} and d_{mn} are constrained by the reflection coefficient seen looking in the $z < 0$ direction at $z = 0$. The reflectarray unit cell can be thought of as a one-port device with

$$\begin{bmatrix} \mathbf{b}^{\mathrm{TE}} \\ \mathbf{b}^{\mathrm{TM}} \end{bmatrix} = \begin{bmatrix} \mathbf{S}_{11}^{\mathrm{TE,TE}} & \mathbf{S}_{11}^{\mathrm{TE,TM}} \\ \mathbf{S}_{11}^{\mathrm{TM,TE}} & \mathbf{S}_{11}^{\mathrm{TM,TM}} \end{bmatrix} \begin{bmatrix} \mathbf{a}^{\mathrm{TE}} \\ \mathbf{a}^{\mathrm{TM}} \end{bmatrix} = \mathbf{S}_{11} \begin{bmatrix} \mathbf{a}^{\mathrm{TE}} \\ \mathbf{a}^{\mathrm{TM}} \end{bmatrix}. \tag{5.17}$$

The GSM of the reflectarray unit cell is usually first evaluated for normal incidence ($\theta_0 = 0$). In the consideration of unit cell designs, it is also important to evaluate the GSM at oblique angles, which is readily facilitated using the GSM by changing θ_0 and ϕ_0. This leads to a GSM that is angle-dependent, and subsequently S-curves that are a function of the angle of incidence. Generally, there is a region of angles of incidence over which the GSM is moderately stable, allowing the S-curve at normal incidence to be used for first-cut design purposes. However, rigorous designs consider all possible angles of incidence for a particular reflectarray and feed geometry, and employing individual S-curves to account for the various different angles. This is more time-consuming, but leads to more accurate results.

In the literature, most numerical formulations for determining the GSM employ the method of moments (MoM) in the spectral domain [13, 14]. These formulations solve the EFIE on the metallic surfaces of the unit cell to yield the surface current densities, which can, in turn, be used to calculate the GSM. Modern commercial computer-aided design (CAD) tools generally support periodic structure analysis within a Floquet waveguide, and support the computation of the periodic structure GSMs. Numerical techniques are powerful and accurate, but at the early design stages, analytical models can also be useful. A presentation of a potential analytical analysis technique is deferred until reconfigurable reflectarrays are introduced in Sect. 5.4.

5.2.2 *Unit Cell Design Types and Methods*

Over the past few decades, many different reflectarray unit cell designs have been proposed for meeting a multitude of design requirements. To add to this mix, there has recently been significant activity in the area of reconfigurable reflectarray unit cells, which allow the scattered field phases of the reflectarray to be dynamically manipulated. To explain the operation of these unit cells, we begin with fixed unit cell designs and divide them according to the basic principle by which the phase shifting action is realized. In Sect. 5.4, we will see how the design methods translate to reconfigurable reflectarrays elements.

5.2.2.1 Tuned-Resonator Unit Cells

Most of the first planar reflectarray designs exploited the fact that a planar resonator, such as a microstrip patch or dipole, resonates when the electrical length of the resonator corresponds to one-half of a wavelength (higher order resonances are possible, but cell sizes are usually limited to be around a half-wavelength to avoid the onset of grating lobes). About the resonance point, there are dramatic changes in the phase of the scattered field of the unit cell as either frequency or the resonant dimensions of the cell are changed. This is the basic principle of the variable-size patch element discussed in Sect. 5.1.3. Exploiting the resonant behaviour of the scatterer, phase ranges of close to 360° can be realized using basic shapes such as dipoles, patches, loops, dog-bone shapes, and many others. A few of these shapes are shown in Fig. 5.7a. The phase of the scattered field is manipulated by modifying one or more dimensions of the metallic scatterers. Generally, the slope of the phase curve versus scatterer dimension is influenced by the overall shape of the scatterer. Sharp phase slopes are usually avoided, since they increase the sensitivity of the phase to the manufacturing tolerances of the scatterer. More importantly, sharp S-curves also translate to sharp S-curves as a function of frequency, whose non-linear shape has a significant impact on the phase bandwidth of the unit cell.

The bandwidth of the unit cell can be significantly improved by linearizing the phase response of the cell. This is accomplished by using a design with multiple scatterers with closely spaced, coupled resonances. This has the effect of making the phase curve more linear in a region, while simultaneously extending the phase range to make the cell emulate its true time delay counterpart. Conversely, resonators with substantially different resonant frequencies can be integrated together to realize multi-band unit cell designs.

Unit cell designs based on the tuned-resonator approach can be further classified into how they manipulate the incoming polarization of the wave. First, unit cells designed using this approach is generally linearly polarized. Most unit cells are

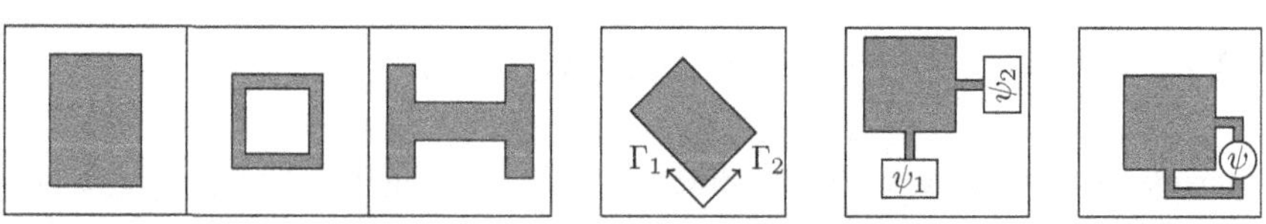

(a) Various tuned-resonator LP cells (b) Resonator twist cell (c) Guided-wave LP cell (d) Guided-wave twist cell

Fig. 5.7 Reflectarray unit cell design methods

co-polarized, meaning that the reflected field is the same polarization as the incident field. It is possible to devise unit cell designs that phase orthogonally polarized waves independently, realizing a *dual-polarized* unit cell design. For example, the rectangular patch element shown in Fig. 5.7a can control the reflection phase of two polarizations: the vertical polarization is phase-shifted by modifying the length of the patch, and the horizontal polarization by the width of the patch. Finally, there are some unit cells that twist the incoming polarization. Such cells are referred to as *cross-polarized*. Usually, the need to twist the polarization is needed for specific purposes, such as implementing a so-called "trans-reflector" for folded reflectarrays discussed in Sect. 5.1.4. This requires that a dual-polarized reflectarray element that is designed to achieve specific phase shifts to the two orthogonal polarizations. In the case of a linearly polarized reflectarray element, if the incident wave is polarized at a 45° angle with respect to the reflectarray element axis, then the incoming electric field vector can be decomposed into two orthogonal components, as shown in Fig. 5.7b. These two components are then acted upon by two different reflection coefficients Γ_1 and Γ_2 controlled by the length and width of the patch. The dimensions of the element can be chosen to produce a relative phase shift of 180° between the two components in order to twist the polarization of the incoming signal by 90°. Meanwhile, the choice of dimensions is also chosen to manipulate overall phase shift applied commonly to both components, which can be tailored for collimating the feed radiation similar to a single-polarization reflectarray.

5.2.2.2 Guided-Wave Unit Cells

A guided-wave unit cell design transitions the space-wave impinging upon the reflectarray element to a guided-wave. The phase shifting operation is implemented in the guided-wave portion of the element, using traditional guided-wave phase shifters. The cell is shown in Fig. 5.7c, where a transmission line is coupled to each polarization of the patch and connected to corresponding reflective phase shifters producing phase shifts ψ_1 and ψ_2 for each polarization (only one phase shifter and line are needed for a single-polarization cell). In the case of fixed reflectarray elements, the phase shifter is usually an open-circuited stub connected to a patch element. By varying the length of the stub, the phase shift is equal to $-2\beta\ell$, where ℓ is the length of the stub and β is the phase constant along the line.

This technique can be seen as part of a class of true time delay (TTD) reflectarray elements that attempt to reproduce the required *delay* profile required of the reflectarray, rather than simply the phase. This is important in the design of broadband reflectarrays. However, it is worth pointing out that there are limits to the delays that can be realized (since there are limits to how long the stub can be made given the confines of the unit cell) and the fact that coupling between the patch and stub tends to still be narrowband. Furthermore, the introduction of stubs can interfere with the scattering response of the cell, particularly the cross-polarization level, necessitating the need for symmetric reflectarray layouts to reduce cross-polarization.

Polarization-twisting elements can also be realized using the approach of manipulating the two reflection coefficients associated with each polarization of the cell in a dual-polarization version. However, an easier technique is to simply cross-connect the two polarization ports of the antenna as shown in Fig. 5.7d. A suitable phase shifter or delay line can be introduced in the path to change the phase of the outgoing wave.

5.2.2.3 Circularly Polarized, Sequential Rotation Unit Cells

Circular polarization (CP) reflectarray elements fall into a special category because they are easily implemented using the so-called *sequential rotation technique* [15]. Instead of changing the size of the element to change the associated reflection coefficients, in the sequential rotation technique the physical rotation angle α of the element is used to control the phase shift, as illustrated in Fig. 5.8. Here a split ring resonator is used to illustrate a generic cell which can realize different responses in the principal directions, while maintaining symmetry about a local coordinate system.

Mathematically, the operation of the cell can be explained using the diagram and coordinate system illustrated in Fig. 5.8. An incident wave, nominally travelling in the $-z$ direction, impinges upon the cell at normal incidence. This wave may be composed of both left-hand and right-hand circularly polarized waves in general, and is described according to

$$\vec{E}^i = a^{rh}(\hat{x}+j\hat{y})e^{jk_0z} + a^{lh}(\hat{x}-j\hat{y})e^{jk_0z}, \tag{5.18}$$

where $a^{rh|lh}$ denote the amplitudes of each circularly polarized component. The cell is rotated physically by an angle α as shown. To determine the reflected wave, the wave must be decomposed into its x- and y-components. Locally, the unit cell is described using a GSM $\mathbf{S}'$ aligned with the $x' - y'$ coordinate system shown in Fig. 5.8, which the incident wave can be translated into using a rotation matrix. It can then be shown that the reflected wave is given by [16]

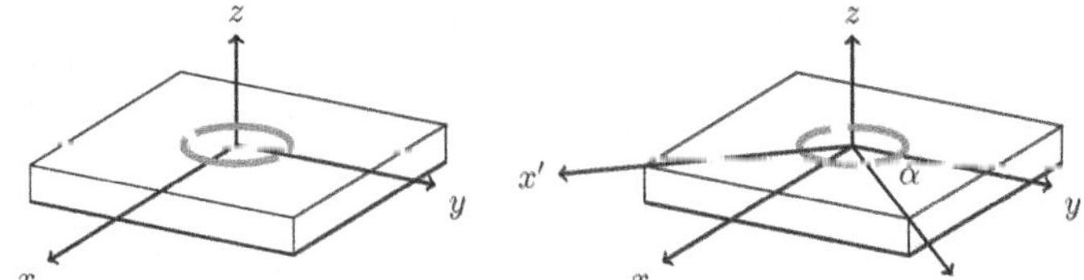

Fig. 5.8 Example geometry and coordinate system for CP reflectarray unit cell

$$\vec{E}^r = (\Gamma_c^{rh} a^{rh} + \Gamma_x^{rh} a^{lh})(\hat{x} - j\hat{y})e^{-jk_0 z} + (\Gamma_c^{lh} a^{lh} + \Gamma_x^{lh} a^{rh})(\hat{x} + j\hat{y})e^{-jk_0 z}, \quad (5.19)$$

where $\Gamma_c^{lh|rh}$ and $\Gamma_x^{lh|rh}$ refer to the co- and cross-polarized components of the corresponding polarization components, respectively, and

$$\Gamma_c^{rh} = \left[\frac{1}{2}(S'_{11} - S'_{22}) + jS'_{12}\right] e^{j2\alpha} \quad (5.20a)$$

$$\Gamma_c^{lh} = \left[\frac{1}{2}(S'_{11} - S'_{22}) - jS'_{12}\right] e^{-j2\alpha} \quad (5.20b)$$

$$\Gamma_x^{lh} = \Gamma_x^{rh} = \frac{1}{2}(S'_{11} + S'_{22}). \quad (5.20c)$$

An ideal CP cell is based on a unit cell that has perfect isolation between the TE and TM modes $\left(|S'_{12}| = |S'_{21}| = 0\right)$, and maximum co-polarization reflection for each mode with a 180° phase shift between them $\left(|S'_{11}| = |S'_{22}| = 1, |\angle S'_{11} - \angle S'_{22}| = \pi\right)$. With this phase shift, polarization sense of the reflected component wave is the same as the incident wave. For example, for a purely right-hand circularly polarized (RHCP) wave, these conditions lead to $\Gamma_c^{rh} = e^{j2\alpha}$ and $\Gamma_x^{rh} = 0$. We see that the phase shift introduced to the RHCP wave is proportional to 2α, twice the physical rotation angle of the cell. In fact, according to Eq. (5.20b), an incident LHCP wave will undergo a similar phase shifting process, so it makes no difference whether the cell is used for RHCP or LHCP systems. We also see in both cases that the cross-polarization level is established purely by S'_{11} and S'_{22}, and that for minimum cross-polarization level their magnitudes should be equal while the phase difference between them should be 180°.

The sequential rotation technique is an easy and straightforward way to implement phase shifting in CP reflectarrays. Once a suitable polarization-twisting cell has been defined, it is a simple matter to rotate the elements to achieve the required phase shift. A number of cells have been proposed for this purpose, including cells based on stubs, rings, ring-slot resonators, and others. Generally, only single-polarization cells have been proposed, since achieving dual-polarization CP designs is quite challenging. Reconfigurable variants will be considered in Sect. 5.4.

5.3 Reflectarray Analysis at the Array Level

5.3.1 Pattern Calculation

Once a suitable reflectarray element has been designed and characterized, the next step is to assemble the unit cells and feed system into the desired reflectarray geometry, and analyze the performance of the resulting system. The simplest way to predict reflectarray patterns is using antenna array analysis, which employs the principle of superposition. Knowing the individual excitation amplitudes and phases of the scatterers composing the reflectarray, the radiated fields from the reflectarray can be easily found via an array factor calculation. As such, the following assumptions are made:

- Mutual coupling between elements, beyond that predicted using the periodic analysis (infinite array assumption), is not considered;
- The element factor is usually considered to be a uniformly illuminated rectangular aperture whose dimensions correspond to that of the unit cell $a \times b$; and
- The array itself does not interact with the feeding system, i.e. there is no near-field coupling between the array and the feed system.

Consider the reflectarray shown in Fig. 5.1a, which has the feed located at the origin. To calculate the radiated fields from the reflectarray, the electric field incident upon each element must be determined. For real (non-analytical) feeds, this can be determined from full-wave simulations or measured experimentally. At the early design stages of a reflectarray, an analytical feed model is usually easier to work with. In this case, the feed is assumed to be a source of spherical waves weighted spatially by the radiation pattern of the feed. Mathematically, the incident field upon the mnth reflectarray element can then be expressed as

$$\vec{E}^i_{mn} = \hat{e}^i_{mn}\left[\eta_0 \frac{P_f}{2\pi} G_f(\theta'_{mn}, \phi'_{mn})\right]^{1/2} \frac{e^{-jkr'_{mn}}}{r'_{mn}}, \tag{5.21}$$

where η_0 is the intrinsic impedance of free space, P_f is the power applied to the feed, $G_f(\theta', \phi')$ is the gain pattern of the feed and $(\theta'_{mn}, \phi'_{mn})$ and r'_{mn} are the angles and distance from the phase centre of the feed to the mnth reflectarray element. The unit vector $\hat{e}^i_{mn}$ depends on the polarization of the feed $\hat{p}$, and is equal to

$$\hat{e}^i_{mn} = \frac{(\hat{r}'_{mn} \times \hat{p}) \times \hat{r}'_{mn}}{\left|(\hat{r}'_{mn} \times \hat{p}) \times \hat{r}'_{mn}\right|}. \tag{5.22}$$

Locally, at each reflectarray unit cell along the surface of the reflectarray, the incident electric field is treated as a plane wave arriving at an angle of incidence $(\theta'_{mn}, \phi'_{mn})$. As such, the incident field consists of only two fundamental propagating Floquet modes: the TE_{00} mode and the TM_{00} mode, which are designated a^{TE}_{mn} and a^{TM}_{mn}. These can be determined solely from the tangential components of the incident field at each cell. The GSMin Eq. (5.17) is then used to compute the reflected electric field amplitudes b^{TE}_{mn} and b^{TM}_{mn}. Since the tangential components are oriented in the xy-plane, we can write the GSM relation as

$$\begin{bmatrix} b^x_{mn} \\ b^y_{mn} \end{bmatrix} = \begin{bmatrix} \Gamma^{mn}_{xx}(\theta'_{mn}, \phi'_{mn}) & \Gamma^{mn}_{yx}(\theta'_{mn}, \phi'_{mn}) \\ \Gamma^{mn}_{xy}(\theta'_{mn}, \phi'_{mn}) & \Gamma^{mn}_{yy}(\theta'_{mn}, \phi'_{mn}) \end{bmatrix} \begin{bmatrix} a^x_{mn} \\ a^y_{mn} \end{bmatrix}. \tag{5.23}$$

This process can also be used to treat feeds not located at the origin (e.g. in the case of offset reflectors), by defining a local feed coordinate system, expressing the radiated fields from the feed in that coordinate system, and transforming these fields into the incident fields on the reflectarray using a transformation matrix.

The far-field radiation pattern can now be easily computed using array theory. Each reflectarray element can be treated as an aperture of dimensions $a \times b$, which is considered to be uniformly illuminated by the reflected electric fields for each polarization. The element factor is found through asymptotic evaluation of the radiated fields from a rectangular aperture, which is assumed to be embedded in an infinite ground plane. Essentially, the equivalent currents are evaluated over the aperture, and the radiated (reflected) electric field from the mnth element is given by

$$\vec{E}^r_{mn} \approx jF\frac{k_0 e^{-jk_0 r}}{2\pi r}\left[\hat{\theta}\left(b^x_{mn}\cos\phi + b^y_{mn}\sin\phi\right) + \hat{\phi}\cos\,\theta\left(-b^x_{mn}\sin\phi + b^t_{mn}\cos\phi\right)\right]$$
$$\exp[jk_0(x_{mn}\sin\,\theta\,\cos\phi + y_{mn}\sin\,\theta\,\sin\phi], \tag{5.24}$$

where

$$F = ab\,\mathrm{sinc}\left(\frac{k_x a}{2}\right)\mathrm{sinc}\left(\frac{k_y b}{2}\right)\exp(-jk_0 z'\cos\,\theta). \tag{5.25}$$

Here, $\mathrm{sinc}(x) = \sin(x)/x$ and z' is the location of the reflectarray. The reflected electric amplitudes b^x_{mn} and b^y_{mn} act as the weights for the array factor, so that the total radiated field can be written as

$$\vec{E}^r = jF\frac{k_0 e^{-jk_0 r}}{2\pi r}\sum_m\sum_n\left[\hat{\theta}\left(b^x_{mn}\cos\phi + b^y_{mn}\sin\phi\right) + \hat{\phi}\cos\,\theta\left(-b^x_{mn}\sin\phi + b^t_{mn}\cos\phi\right)\right]$$
$$\exp[jk_0(x_{mn}\sin\,\theta\,\cos\phi + y_{mn}\sin\,\theta\,\sin\phi], \tag{5.26}$$

A rectangular lattice has been assumed for the reflectarray geometry, and hence row–column indexing has been employed in the summations. It is worth noting that each component of $\vec{E}^r$ can be computed quite quickly using the inverse 2D discrete Fourier transform [17]. This can be a major advantage when evaluating the radiated fields from large reflectarrays, since the Fast Fourier Transform can be employed to greatly accelerate the calculations.

The array factor method for computing the radiated fields from the reflectarray is quite accurate. Since the GSM captures the co- and cross-polarization behaviour of the constituent elements of the reflectarray, most of the dominant contributors to pattern features are included. Diffraction from the edges of the reflectarray is not accounted for. Interestingly, despite the assumption of local periodicity among the reflectarray unit cells, this simplified analysis does a very good job in predicting the pattern when comparing to measurements or full-wave simulations of the entire reflectarray. More accurate formulations have been proposed, whereby the spectral domain Green's function of the array elements is considered and used to form a continuous spectrum of equivalent currents over the entire reflectarray aperture [18]. This mitigates the problem of having discontinuities in the equivalent currents between unit cells, which leads to more accurate results, especially in the case of highly aperiodic reflectarrays.

5.3.2 Gain and Aperture Efficiency

The peak gain G of a reflectarray is related to the aperture efficiency η_{ap} via

$$G = \eta_{ap} \frac{4\pi}{\lambda_0^2} A_p \tag{5.27}$$

where A_p is the physical area of the reflectarray. The aperture efficiency is determined from

$$\eta_{ap} = \eta_{spillover} \eta_{taper} \eta_{ohmic} \eta_{diel} \eta_{phase} \eta_{blockage} \eta_{specular}, \tag{5.28}$$

where only the dominant contributors to efficiency have been listed. Each term is described as follows.

The spillover efficiency $\eta_{spillover}$ is determined from the ratio of the power collected from the reflectarray to that emitted from the feed. This is easily approximated by treating reflectarray unit cells as ideal rectangular collectors, so that

$$\eta_{spillover} = \frac{\sum_m \sum_n P_{mn}^{inc} ab}{\int_0^{2\pi} \int_0^{\pi} U_f(\theta', \phi') \sin \theta' d\theta' d\phi'}, \tag{5.29}$$

where P_{mn}^{inc} is the incident power density at the mnth cell and $U_f(\theta, \phi)$ is the radiation intensity from the feed. A straightforward way for calculating the taper efficiency η_{taper} is to employ the array factor approximation of the radiated field 5.26) for computing the radiated fields with and without the feed-induced taper in place. The taper efficiency can then be found from the ratio

$$\eta_{taper} = \frac{|E^r(\theta_0, \phi_0)|^2_{\text{actual}}}{|E^r(\theta_0, \phi_0)|^2_{\text{incident field uniform}}}, \tag{5.30}$$

where (θ_0, ϕ_0) refers to the beam pointing direction and E^r refers to the desired co-polarized field component. The reflected fields from the unit cells in this calculation should not include ohmic or dielectric losses from the element which would further modify the taper levels. Instead, the losses are removed and accounted separately in the efficiency terms η_{ohmic} and $\eta_{dielectric}$ separately, considering the analysis of the reflectarray with the appropriate loss mechanisms in place. Ohmic loss here considers both conductor losses and thermal losses within potential reconfiguration mechanisms (e.g. the parasitic resistance of varactor diodes).

Phase errors over the aperture contribute to a phase error efficiency term η_{phase} in (5.35). These phase errors arise from imperfect phasing of the elements, resulting from limited phase range of the elements, the discrete number of phase states available from the library of elements (or finite number of bits available in the case of digitally controlled reconfigurable elements), and phase errors arising in general from imperfections, frequency dispersion, etc. The phase error efficiency can also be determined using array factor techniques using

$$\eta_{phase} = \frac{|E^r(\theta_0, \phi_0)|^2_{\text{actual}}}{|E^r(\theta_0, \phi_0)|^2_{\text{ideal phases}}}. \tag{5.31}$$

where the numerator is evaluated with the actual phases employed in the reflectarray, and the denominator is evaluated assuming perfect (ideal) phases without errors.

The final two terms in (5.35) are the feed blockage and specular reflection losses, respectively. Feed blockage results from a shadow cast by the feed when it is placed in front of the reflector surface. It is usually mitigated by using an offset feed configuration. It can only be predicted using full-wave simulations of the entire reflectarray and feed system. Specular reflection losses result from imbalances between the reflectarray element and ground plane currents, particularly away from the operating frequency of the reflectarray. This imbalance gives rise to stray reflections in the specular direction creating a beam known as a *feed image lobe* [19]. Similar to feed blockage, full-wave analysis of the entire reflectarray is needed to predict the exact loss to specular reflections. Generally specular reflection losses do not significantly reduce the gain of the reflectarray, but they can increase the stray radiation in other directions if not controlled carefully.

5.4 Reconfigurable Reflectarrays

The addition of reconfigurable capabilities to reflectarrays opens the possibility to dynamic beam-steering and shaping, allowing for on-the-fly reconfiguration of the antenna pattern. There are many scenarios that can benefit from adaptive, high-gain patterns including satellite systems (especially satcom-on-the-move systems), RADAR, point-to-point microwave links, and remote-sensing (e.g. synthetic aperture RADAR). Beam-forming capabilities will also be foremost in next-generation (5G) communications systems, which employ higher frequencies and hence have stronger needs for high-gain antennas. With the associated narrower antenna beams, adaptive beam-forming becomes a natural requirement.

Reflectarrays and other space-fed arrays offer compelling advantages over other antennas in this arena. Unlike reflector antennas, which must be mechanically scanned and offer very limited possibilities for adaptive re-shaping of the beam, the aperture fields can be modified in electronically tunable reflectarrays by modifying the reflectarray phase distribution. While phased arrays offer the same capability, this is achieved at a much higher cost due to the replication of transmit/receive modules for each antenna (or sub-array, at least) in the array. Additionally, the transmission line feed networks associated with phased arrays can become significantly lossy at high frequencies, especially in the mm-wave range. Beam-forming networks also become very complex in the structure for very large phased arrays. Spatially fed arrays remove the need to lay out complex feed networks and mitigate much of the loss associated with transmission line feeds, replacing it partly with taper and spillover losses typical of reflector antennas, which are easier to control.

Reconfigurable reflectarrays can be placed into two classes in terms of methods used to realize beam reconfiguration. The first method uses a scanned feed and fixed (non-reconfigurable) reflectarray surface. The feed is moved to different locations in front of the reflectarray. This scans the beam off the reflector axis, and is commonly used in parabolic reflectors and their reflectarray counterparts [20]. Parabolic, spherical, and multi-focal phase distributions can be used on reflectarray aperture to manage the scanning range of the beam. Hence, the reconfiguration is still mechanical, albeit in a different sense than a reflector which is mechanically rotated to point the beam. Some degree of "electronic" beam-steering can be achieved by electronically switching between multiple feeds located at required focal points to produce the corresponding beams, allowing for multi-beam capabilities [9]. The second method is to use a stationary feed, but a reflectarray surface that can be electronically manipulated to change the phase distribution and corresponding aperture fields [21]. This method offers the most flexibility in terms of the patterns that can be realized from the antenna, as well as the potential speed at which the aperture can be reconfigured, since large macroscopic movement of parts is not required. This method is the focus of this section.

5.4.1 Enabling Technologies

At the heart of a reconfigurable reflectarrays is a unit cell design with electronically tunable characteristics. These characteristics in turn depend on microwave and mm-wave technologies that allow for electronic reconfiguration of the cell in some way [21]. A summary of a wide range of enabling technologies is tabulated in Table 5.1. An ideal reflectarray obviously should be low cost, consume little or no power, exhibit low loss, be fast to reconfigure, and offer low implementation complexity. Additionally, generally digital control is preferred to analogue control

Table 5.1 Enabling technologies for reconfigurable reflectarrays

Type	Technology	Maturity	Digital/analogue control	Complexity	Cost	Loss (μ-wave / mm-wave)	Bias power consumption	Linearity	Configuration time
Lumped-elements	PIN diodes	High	D	Low	Low	H/H	High	Low	Fast
	Varactor diodes	High	A	Low	Low	H/H	Low	Low	Fast
	RF-MEMS	Mod.	D	Low	Mod.	L/M	Low	High	Mod.
Hybrid	Ferroelectric films	Mod.	A	Mod.	Mod.	M/L	Low	Mod.	Fast
Tunable materials	Liquid crystal	Low	A	High	High	H/L	Low	Low	Mod.
	Graphene	Low	A	High	High	H/L	Low	Low	Fast
	Photoconductive	Low	A	High	Mod.	H/H	High	Low	Fast
Mechanical	Fluidic	Low	A	High	Mod.	M/H	Low	High	Slow
	Micro-motors	Low	A	High	Mod.	L/–	Low	High	Slow

schemes, and in certain applications, mature technologies are preferable (e.g. space-qualified devices for satellite communications). However, as we will see, as one migrates to the millimetre wave frequencies, we will see that more of these ideals must be sacrificed.

Each technology is broadly classified according to its type in the first column. The first technology type, lumped-element, refers to electronic components that can be introduced into the reflectarray unit cell at discrete, spatially localized points. Radio frequency (RF) switches are the most obvious example, which can be realized both in semiconductor form (such as from PIN diodes and FET switches), and in the form of RF microelectromechanical systems (MEMS) switches. Tunable reactances are also commonly used, usually as capacitances which can be realized from semiconductors (varactor diodes), and also in MEMS form, where both tunable capacitors and tunable inductors have been realized. Semiconductor-based technologies exhibit significant levels of loss which reduce the reflection efficiency of reflectarray unit cells. They also exhibit nonlinear behaviour at large RF signal levels, which can lead to signal distortion problems when used in transmitters. For these reasons, MEMS devices have grown in popularity, especially MEMS switches, since they generally mitigate these issues.

At the other end of the spectrum are tunable materials, which instead provide tunable constitutive parameters (permittivity, permeability, or conductivity) in the underlying material platform order to realize *distributed* tuning of the reconfigurable unit cell. An excellent example of these is liquid crystal (LC) materials. These materials have a tunable permittivity tensor which is dependent on a DC or low-frequency AC signal biasing field applied to the material. Any structure, such as patch, using LC as its substrate will have a resonant frequency (and corresponding scattered field phase) dependent on the properties of the substrate permittivity. Graphene is an emerging technology, which in the terahertz frequency range exhibits a surface impedance which can be tuned by manipulating the doping parameters of the graphene as well an applied electric field (the so-called graphene “field effect”). Photoconductive materials, such as semiconductors, can exhibit sensitivity to an applied light source to change the carrier concentration (and hence conductivity) of a predefined region. This region can range from being localized to a very small area (realizing a photoconductive switch, which is more of a lumped-element type of reconfiguration) to the entire element itself (or absence thereof, when the light is off).

Somewhere between these two extremes lie hybrid technologies, which are neither distributed nor spatially localized as in the case of lumped-elements. Devices based on ferroelectric films fall into this category, since the ferroelectric films are a type of tunable material, but the corresponding devices are often realized in component form.

Finally, mechanical solutions tend to fall into their own unique category, though they largely achieve distributed type of tuning. For example, we have seen in Sect. 5.2.2 that physical rotation of CP elements can be used to directly manipulate the phase shift of the unit cell. Mounting such cells on micro-motors, for example, can be used to change the phase shifter. Micro-fluidic devices generally have been used to move highly conductive liquid “slugs” of galistan and other materials in the vicinity of resonant reflectarray unit cells to change the resonance characteristics and induce a phase shift.

5.4.2 *Realizing Phase-Tunable Reflectarray Unit Cells*

As in the case of fixed reflectarray cells, reconfigurable unit cells can be classified according to the same schemes used in Sect. 5.2.2. Examples of the application of these techniques are shown via examples in Fig. 5.9 and discussed in detail in the sections below. With each technique, varying degrees of phase control can be achieved depending on the underlying technology. For example, devices with a discrete number of states (e.g. those relying on switches) may only have limited phase resolution. While devices with as low as 1-bit (2-state) phase control have been successfully realized, designers must be mindful in the tradeoff that exists between the effective number of bits in the phase shifter, and the corresponding impact on the antenna pattern, with particular attention to the gain and sidelobe levels [22]. For example, phase quantization at the 1-bit level generally leads to a 3-4 dB reduction in gain. The loss is reduced to less than 1 dB if 2-bit control is used, and less than 0.2 dB for 3-bit control.

5.4.2.1 Tuned-Resonator Cells

Tuned-resonator cells can be implemented using the widest range of enabling technologies discussed in the previous section, since perturbing the resonant frequency of a cell can be realized by virtually any technique. Beginning with lumped components, tunable reactances can easily be integrated with resonators, and the

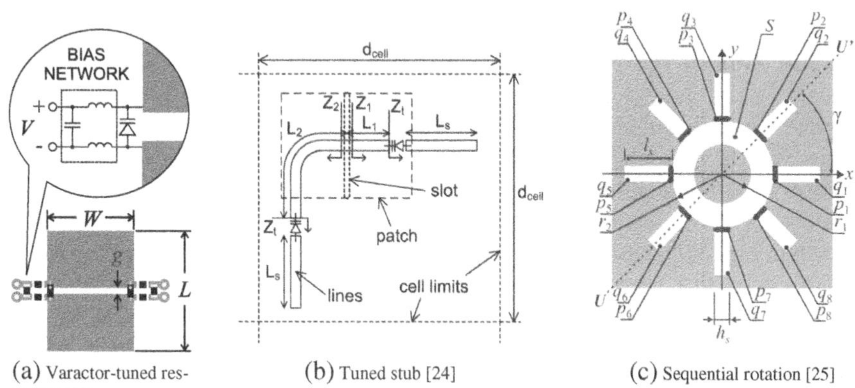

(a) Varactor-tuned resonator [23] (b) Tuned stub [24] (c) Sequential rotation [25]

Fig. 5.9 Reconfigurable reflectarray element examples

reactance electronically manipulated to produce the necessary changes in resonant frequency and corresponding phase shift in the scattered field. An example of this type of cell is illustrated conceptually in Fig. 5.9a, where a pair of varactor diodes is used to change the resonant frequency of a patch antenna. Through the proper coupling of the tunable reactances to the patch, very large phase ranges close to 360° can be achieved. Furthermore, MEMS-tunable reactances such as MEMS capacitors can also replace varactor diodes to decrease losses and improve linearity, while maintaining a high effective number of bits.

Tunable reactances integrated with reflectarray elements change the resonant frequency of the resonator by changing the effective length of the resonator. Dual-LP variations are also possible [26], and even photo-induced plasmas have been used in place of switches [27]. To mitigate the loss, linearity, and power consumption issues with these approaches, numerous examples of RF-MEMS switch-controlled reflectarray elements have emerged in previous years which coupled switches and resonators together to achieve phase control [28, 29]. In general, switch-based designs are often preferred, since they are readily compatible with control circuits and do not need continuously variable voltages to tune the phase.

Distributed tuning techniques also have the most impact on tuned-resonator cells, since small changes in constitutive parameters can be translated to large changes in the scattered phase with proper designs of the unit cell. Tunable dielectrics have the most potential in this regard. For example, liquid crystal materials employing LC molecules in the nematic state possess an anisotropic permittivity tensor which can be tuned between $\epsilon_{\parallel} - \epsilon_{\perp}$. Placing a resonator such as a patch [30–32] on such a material can lead to pronounced changes in the phase as the magnitude of the DC or AC biasing field is varied. Similarly, ferroelectric films also produce changes in the permittivity tensor in the presence of an applied biasing field and can also be used as substrate layers to produce phase shifters and phase shifting reflectarray unit cells [33].

5.4.2.2 Guided-Wave Cells

Guided-wave reconfigurable unit cells work very similarly to their fixed counterparts. Instead of coupling to a delay line, the scatterer is coupled to a tunable guided-wave phase shifter. The phase shifter must be a one-port reflection-type phase shifter, so very often tunable delay lines are used. One such as in the example shown in Fig. 5.9b, where a rectangular patch is aperture-coupled to a continuously tunable delay line. Switched-delay lines are also used to produce a more wideband response from the reflectarray. PIN diode-based versions have been realized [34, 35]. Minimizing the loss of the delay lines often means that MEMS switches are preferable for this purpose, and several experimental demonstrations have been produced [36, 37]. Since reconfigurable capabilities add cost to each unit cell, a useful strategy that can be employed is to share the same switched-delay line between a group (or sub-array) of elements.

Most guided-wave reconfigurable reflectarray elements are co-polarized. However, cross-polarized (polarization-twisting) elements open up the possibility of using the two-port microwave and mm-wave circuits to interconnect the two polarization ports, as shown in Fig. 5.7d. An important function that can be added to reflectarrays in this configuration is amplification. This function is traditionally difficult to add to co-polarized arrays, because of two major challenges: (1) the need for a one-port reflection-mode amplifier, which is difficult to design and make stable; and (2) a fundamental requirement of isolation between the incident and reflected wave ports. Cross-polarized designs are more amenable to the introduction of amplification since two-port amplifiers are readily available, and the cross-polarization discrimination of the element can be made very high. As a result, a number of interesting and truly "active" designs (i.e. those with a reflection coefficient magnitude larger than 1) have emerged: those based on amplifiers alone (fixed-beam reflectarrays) [38, 39] and those combining amplification and phase shifting [40, 41]. Such designs offer advantages in transmitters where the spatial combining feature of reflectarrays mitigates losses normally present in transmission line manifolds. This advantage is more compelling at mm- and sub-mm wave frequencies, where the need for power combining is stronger and the losses in these networks are most pronounced.

5.4.2.3 Sequential Rotation Cells

Finally, reconfigurable variations on the sequential rotation unit cell introduced in Sect. 5.2.2 are a straightforward extension of the idea. The cell can be physically rotated using an actuator such as a micro-motor. Novel MEMS micro-motors have been realized for this purpose [42]. However, a slightly easier approach is to electromagnetically rotate the element through the use of RF switches, which is the approach illustrated in Fig. 5.9c. Here, a series of switches are placed along the circumference of a slot, and a pair of switches diametrically opposite one another are actuated to produce the rotation effect shown earlier in Fig. 5.8. This concept has been successfully deployed in a number diode-based [25] and MEMS-based demonstrators [16, 43].

A recent innovation in this area is to make dual-polarization CP reflectarray cells. That is, the reflectarray cell produces a unique phase shift for the left-handed component and another unique phase shift for the right-handed component. Such reflectarrays can be useful in applications utilizing circular polarization, such as satellite systems. This can be achieved by stacking two circular polarization selective surfaces (CPSSs) on top of each other: one layer which is transparent to one polarization type (e.g. RHCP) while being reflective to the other, and a second is having the reverse behaviour. The first layer is structured to reflect and manipulate the phase of the LHCP signal through the principle of sequential rotation, while an RHCP signal incident upon this layer passes through unaffected to the second layer. The second layer also modifies the reflection phase of the RHCP signal through sequential rotation; it then passes unaffected through the first layer. This concept has been successfully demonstrated using a unique combination of CPSSs and can potentially be realized in a switched, reconfigurable form [44].

5.4.3 Modelling Considerations

There are several frameworks that assist in the modelling of reconfigurable reflectarray elements. Both apply to the case where lumped-elements such as switches, tunable reactances, and other components are introduced in the unit cell. The first is the concept of a mixed-mode generalized scattering matrix, which assists in predicting the overall scattering behaviour of the unit cell without resorting to exhaustive full simulations with the lumped components in different states. The second is an equivalent circuit model that can be used to predict the scattering behaviour of simple unit cells even before the numerical simulation stage. Both are described next.

5.4.3.1 Mixed-Mode Generalized Scattering Matrix

The GSM representation of reflectarray unit cells can be extended to the case where lumped-element components are introduced in the reflectarray. For example, the capacitively loaded cell shown in Fig. 5.9a employs two varactor diodes as tunable capacitances. A straightforward way to determine the S-curve for such an element is to introduce a reactance in the electromagnetic description of the unit cell (e.g. in the impedance matrix for MoM formulations), and sweep the capacitance over a series of simulations run to trace out the S-curve. While this process is correct, it unnecessarily lengthens the simulation time of structure, since the underlying scatterer is linear and hence obeys network theory. Furthermore, when the scatterer involves a large number of reconfigurable elements, the number of configurations to simulate quickly can become unmanageable. For example, a cell equipped with N switches has 2^N unique states.

A much more computationally efficient approach is to extract a mixed-mode GSM from the unit cell. This approach is illustrated in Fig. 5.10a, which considers a unit cell which has L lumped-elements introduced in its structure. Here, we assume that only the fundamental Floquet modes are of interest, and hence we only have two Floquet ports at the input to the periodic waveguide. A series of L discrete ports $3, 4, \ldots k, \ldots N$, $N = L + 2$ are introduced at the locations of the lumped-element component loading points.

Using this description, we then describe the overall scattering response of the unit cell in terms of the two spatial ports and L lumped ports using

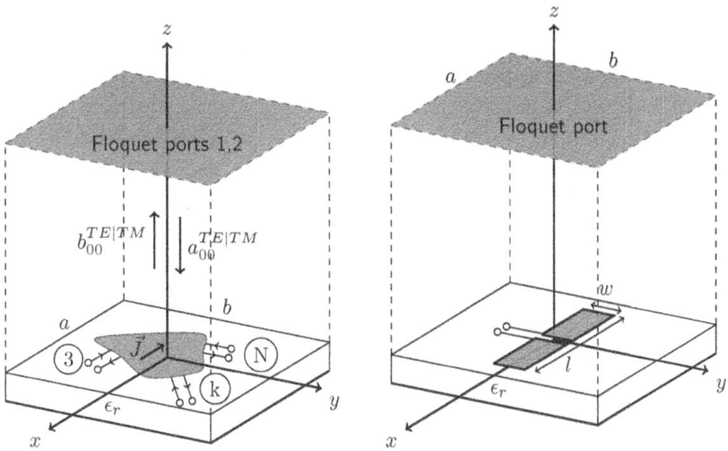

(a) Mixed-mode GSM of a reflectarray unit cell (b) Dipole in driven mode

Fig. 5.10 Scattering scenarios for modelling

$$\begin{bmatrix} \mathbf{b}_{12} \\ \mathbf{b}_{3N} \end{bmatrix} = \begin{bmatrix} \mathbf{S}_A & \mathbf{S}_B \\ \mathbf{S}_C & \mathbf{S}_D \end{bmatrix} \begin{bmatrix} \mathbf{a}_{12} \\ \mathbf{a}_{3N} \end{bmatrix}, \tag{5.32}$$

where we see we now have a mix of Floquet ports $(1,2)$ and lumped-element ports $(3,4,\ldots k,\ldots N)$ associated with the incident and reflected waves. Each port has its own associated port impedance Z_{0k}. For the Floquet ports, these impedances are the associated TE and TM wave impedances associated with ports 1 and 2, while for the lumped ports, an arbitrary reference impedance can be employed (usually an impedance of 50 Ω is used.) Note that the incident and reflected waves are normalized according to the conventional power wave definitions, i.e. a_k and b_k are normalized by the square root of the port impedance associated with the kth port.

When actual components are introduced to load each port, the associated reflection coefficients Γ_k are determined for each state the lumped components can be in. A reflection coefficient vector is then defined,

$$\Gamma = \text{diag}(\Gamma_3, \Gamma_4, \ldots, \Gamma_k, \ldots, \Gamma_L) \tag{5.33}$$

which relates $\mathbf{a}_{3N}$ and $\mathbf{b}_{3N}$ through

$$\mathbf{a}_{3N} = \Gamma \mathbf{b}_{3N}. \tag{5.34}$$

Imposing this constraint on Eq. (5.32) and solving yields

$$\mathbf{b}_{12} = \underbrace{\left[\mathbf{S}_A + \mathbf{S}_B \Gamma (\mathbf{I} - \mathbf{S}_D \Gamma)^{-1} \mathbf{S}_C\right]}_{\mathbf{S}_M} \mathbf{a}_{12}, \tag{5.35}$$

where **I** is the identity matrix and $\mathbf{S}_M$ is the resulting 2×2 GSM resulting from the loading of the reflectarray unit cell in a given state of lumped-element configurations Γ. This approach to modelling is very powerful, because it only requires the determination of the initial mixed-mode scattering matrix shown in Eq. (5.32) by electromagnetic simulations or other means. Once this matrix is known, $\mathbf{S}_M$ can be evaluated algebraically using Eq. (5.35). Usually, Γ can be easily computed by analytical means or using circuit simulators in the case of complicated loading networks (e.g. amplifiers). This technique opens the possibility up of optimizing the unit cell when many load combinations are possible, for example, using genetic algorithms or particle swarm optimization.

5.4.3.2 Equivalent Circuit Modelling of Reconfigurable Reflectarray Elements

For tuned-resonator types of reconfigurable reflectarrays, a further tool one can use in determining a "first-cut" reflectarray design is an equivalent circuit model (ECM) for the element. Such tools have an advantage of being faster to evaluate during the initial design stages, when a design is trying to determine the optimal size and loading of a potential reflectarray element. Here, we employ a Floquet-based framework for deriving an equivalent circuit model that is versatile and easily to evaluate [45]. In particular, unlike many other models proposed in the literature, this model does not require full-wave simulations to determine component values, and can treat a variety of reconfigurable element cases.

The ECM is derived by first considering the behaviour of a dipole scatterer in the driven mode, in an infinite array environment. An example of a dipole oriented along the x-direction of dimensions $l \times w$ is shown in Fig. 5.10b. The dipole is initially treated in the driven mode where an excitation is placed across the terminals shown. Infinite dipole arrays have been considered at length in the literature. The dominant current component on the dipole is expanded in terms of a Floquet series as

$$\vec{I}(x,y) = \hat{x} \sum_m \sum_n \tilde{f}(k_{xmn}, k_{ymn}) e^{-jk_{xmn}x - jk_{ymn}y}, \tag{5.36}$$

where $\tilde{f}$ is the Fourier transform of the x-directed current distribution $f(x,y)$ on the dipole. In general, the current distribution on the patch is assumed to be separable such that $f(x,y) = g(x)h(y)$. Using a common formulation, the active input impedance seen looking into the dipole for a particular scan angle (θ, ϕ) can subsequently be derived as [46]

$$Z^{FL}(k_{x0}, k_{y0}) = \frac{l^2}{ab}\frac{4}{\pi^2} \sum_m \sum_n \left[\frac{k_{ymn}^2}{y_{mn}^{TE}} + \frac{k_{xmn}^2}{y_{mn}^{TM}} \right] \frac{G_{mn}^2 H_{mn}^2}{k_0^2 - k_{zmn}^{+2}}, \tag{5.37}$$

where

$$k_{zmn}^{+} = \sqrt{k_0^2 - k_{xmn}^2 - k_{ymn}^2} \tag{5.38}$$

is the z-component of the wavenumber in the $z > 0$ region, and G_{mn} and H_{mn} are the Fourier transforms of the current distribution in the x- and y-directions evaluated at the corresponding wavenumbers k_{xmn} and k_{ymn}, respectively. Meanwhile, the quantities y_{mn}^{TE} and y_{mn}^{TM} represent the equivalent admittance experienced by a TE $_{zmn}$ and TM $_{zmn}$ current source at the air-dielectric interface, respectively. This admittance is the sum of the modal admittance in the $z > 0$ region, $Y_{mn}^{TE+|TM+}$, and the input admittance of the grounded dielectric slab,

$$jB_{mn}^{TE|TM} = -jY_{mn}^{TE-|TM-}\cot(k_{zmn}^{-}h). \tag{5.39}$$

$Y_{mn}^{TE-|TM-}$ are the TE $_{zmn}|$ TM $_{zmn}$ model admittances for the $z<0$ region. Hence,

$$y_{mn}^{TE|TM} = Y_{mn}^{TE+|TM+} + jB_{mn}^{TE|TM}. \tag{5.40}$$

The modal admittances are given by

$$Y_{mn}^{TE+} = \frac{k_{zmn}^{+}}{\omega\mu_0}, \; Y_{mn}^{TE-} = \frac{k_{zmn}^{-}}{\omega\mu_0} \tag{5.41a}$$

$$Y_{mn}^{TM+} = \frac{\omega\,\epsilon_0}{k_{zmn}^{+}}, \; Y_{mn}^{TM-} = \frac{\omega\,\epsilon_0\,\epsilon_r}{k_{zmn}^{-}}. \tag{5.41b}$$

Equation (5.44) can be reorganized to explicitly show the contributions from the TE and TM modes,

$$Z^{FL}(k_{x0}, k_{y0}) = \sum_m \sum_n T_{mn}^{TE}\frac{1}{y_{mn}^{TE}} + \sum_m \sum_n T_{mn}^{TM}\frac{1}{y_{mn}^{TM}}, \tag{5.42}$$

where

$$T_{mn}^{TE} = \frac{l^2}{ab}\frac{4}{\pi^2}\frac{k_{ymn}^2}{k_0^2 - k_{zmn}^{+2}}G_{mn}^2 H_{mn}^2 \tag{5.43}$$

and

$$T_{mn}^{TM} = \frac{l^2}{ab}\frac{4}{\pi^2}\frac{k_{xmn}^2}{k_0^2 - k_{zmn}^{+2}}G_{mn}^2 H_{mn}^2. \tag{5.44}$$

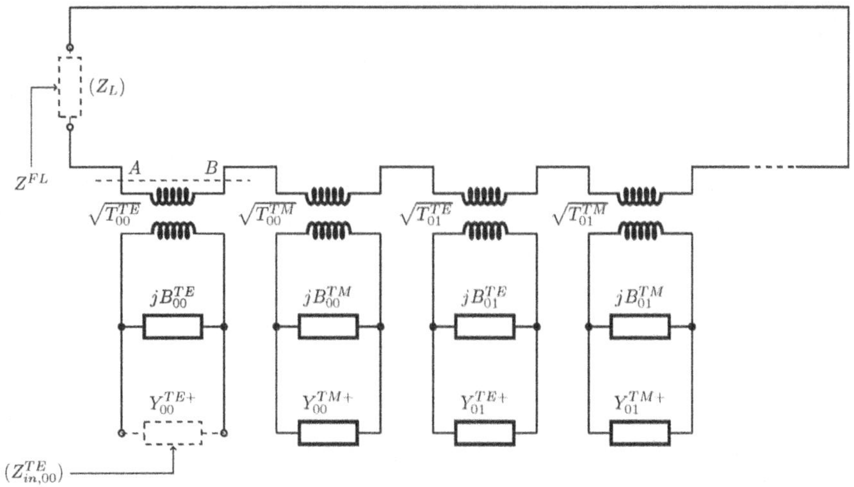

Fig. 5.11 Equivalent circuit for a dipole an infinite array environment. In the driven mode (Fig. 5.10b), Z_L is absent, Y_{00}^{TE+} is present, and Z_{FL} is measured. In the scattering mode, Z_L is present, Y_{00}^{TE+} is absent, and $Z_{in,00}^{TE}$ is measured

A close examination of (5.37) reveals that the input impedance Z^{FL} to the dipole is an infinite series summation of impedances $1/y_{mn}^{TE|TM}$ scaled by transformers with turns ratios $\sqrt{T_{mn}^{TE/TM}}$. This suggests the equivalent circuit shown in Fig. 5.11, where Z_L is absent, Y_{00}^{TE+} is present, and Z_{FL} is measured.

This circuit can easily be turned around to treat the scattering scenario whereby the dipole in Fig. 5.10 is excited externally by a Floquet mode and a load places across the dipole terminals. For example, consider excitation of the TE_{00} mode. In Fig. 5.11, Z_L is now introduced, Y_{00}^{TE+} is removed, and $Z_{in,00}^{TE}$ is measured as shown. The input impedance seen looking upward across the interface $\overline{AB}$ is

$$Z_{00}^{TE'} = Z_L + \sum_{m\neq 0}\sum_{n\neq 0} T_{mn}^{TE}\frac{1}{y_{mn}^{TE}} + \sum_{m}\sum_{n} T_{mn}^{TM}\frac{1}{y_{mn}^{TM}}. \tag{5.45}$$

Then the input impedance of the desired TE_{00} Floquet mode can be calculated as

$$Z_{in,00}^{TE} = \left[\left(\frac{Z_{00}^{TE'}}{T_{00}^{TE}}\right)^{-1} - jY_{00}^{TE-}\cot\left(k_{z00}^{-}h\right)\right]^{-1}. \tag{5.46}$$

Knowing the input impedance of the TE_{00} mode, the associated reflection coefficient and corresponding S-curve can be found.

The ECM is straightforward to evaluate. To populate the ECM, the transformer turns ratios $T_{mn}^{TE|TM}$ must be found, which in turn relies on a reasonably accurate description of the current distribution on the scatterer when illuminated by the co-polarized field. For the case of a rectangular patch scatterer in which width w in Fig. 5.10b becomes comparable to the cell width b, the patch current distribution is specified as follows. First, in the x-direction, the current distribution is described by a half-wave sinusoid along a lossless TEM transmission line with effective phase constant β_{eff}, characteristic impedance Z_c and fringing capacitance C_f associated with common patch antennas. These are calculated using the standard quasi-static formulae for microstrip lines. The capacitance C_f loads each end of the transmission line. This allows the current distribution in the x-direction to be found as

$$g(x) = I_i \cosh(j\beta_{eff}x) - \frac{V_i}{Z_c}\sinh(j\beta_{eff}x), \quad 0 \leq |x| \leq L/2, \tag{5.47}$$

where V_i is an arbitrary applied voltage to the circuit and I_i is the resulting input current at the input to the transmission line. In the y-direction, the patch currents are non-uniform with a singularity near the edges of the patch. In fact, the current distribution, which is determined by the tangential magnetic fields along the patch, is the dual of the electric field distribution that occurs in the gaps of the complementary structure, a metal strip grating. Therefore, for a TE-polarized incident wave, the surface current density on the patch is equal to

$$h(y) = \frac{Ce^{-j\beta y}}{\sqrt{(W/2)^2 - y^2}} \tag{5.48}$$

where C is a constant. G_{mn} and H_{mn} is readily found by numerically taking the Fourier transforms of Eqs. (5.47) and (5.48), after which the turns ratios can be found from Eqs. (5.43) and (5.44). The remainder of the components in the ECM of Fig. 5.11 are found analytically from Eqs. (5.41a) and (5.39).

5.5 Design Examples

In this section, we consider two design examples for reflectarrays designed at 10 GHz. The first is a fixed reflectarray based on varying the shape of a square patch, while the second is reconfigurable reflectarray, implemented using the tuned-capacitor method outlined in the previous section. In both cases the scatterer is single patch-like scatterer printed on a grounded dielectric substrate. The key parameters of the reflectarray design are summarized as follows. The reflectarray is realized on a 1-mm thick substrate with $\epsilon_r = 3.02$. It consists of 625 elements arranged in a 25 × 25 array with a cell period of 15 mm. The f/D ratio is 1, the reflectarray is linearly polarized, and the desired beam direction is $\theta = 30°, \phi = 0°$. For the examples considered here, conductor and dielectric losses are ignored.

5.5.1 Fixed Reflectarray

The first reflectarray example is a fixed-beam reflectarray realized using patch-based scatterers of varying size, corresponding to the tuned-resonator approach described in Sect. 5.2.2. For simplicity, the patch is kept square in size, though dipole-shaped and rectangular-shaped patches can also be used. The size of the patch is used to manipulate the phase of the fields co-polarized with a linearly polarized incident field.

To create the S-curve for the reflectarray, both analytical and numerical approaches can be employed. The equivalent circuit model described in Sect. 5.4.3. can be evaluated for a variety of patch dimensions to produce the desired S-curve. Since there is no lumped-element loading present, the load Z_L is simply set to 0 Ω (a short circuit), and the model evaluated for a variety of patch dimensions according to a parametric sweep. The ECM is evaluated using 60 TE and 60 TM spatial harmonics and results in the S-curve shown in Fig. 5.12. For comparison, a spectral domain method of moments code employing Rao-Wilton-Glisson basis functions is used to compute the GSM of the same cell, and the results are overlaid in the same plot. It can be seen that the results of both methods align well, and that the ECM can be used to accurately predict the response of the unit cell.

The analysis shows that the cell possesses approximately 330° of phase range, which is adequate for forming a beam. The missing 30° of phase range has only a very mild impact on the sidelobe levels. The moderate slope in middle of the phase curve is a good feature, since high slopes increase the sensitivity of the design to fabrication errors. The slope can be reduced at the expense of phase range by, for example, increasing the substrate height, and hence the designer is afforded some freedom in the design process.

Next, the full array is designed based on the reflectarray unit cell. The required phases to collimate the fields from the feed and produce the beam in the required direction are computed using Eq. (5.4). The required phases are then reconciled against the phase curve and mapped to the corresponding patch lengths required at

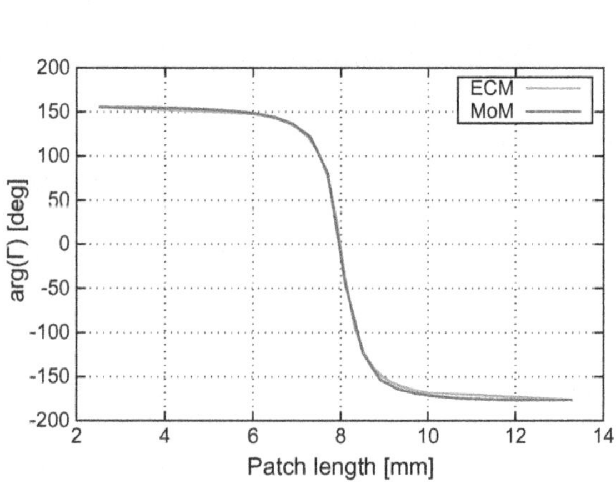

Fig. 5.12 Phase curve for the fixed reflectarray unit cell as predicted by the ECM and MoM techniques

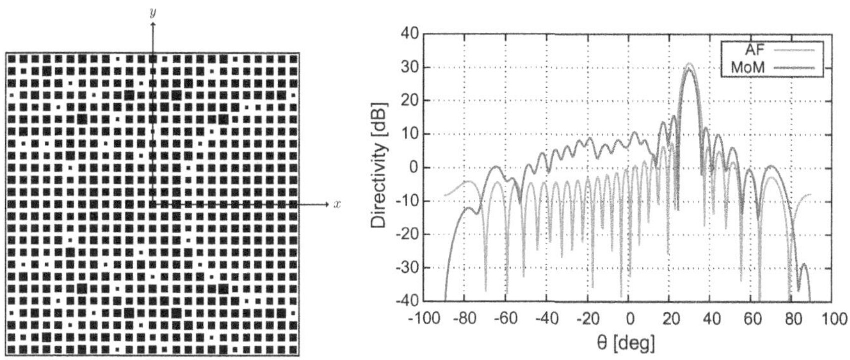

(a) Phase mask (b) Directivity

Fig. 5.13 Fixed reflectarray example

10 GHz via a lookup table, and a corresponding mask for the reflectarray is shown in Fig. 5.13a. Note that a variable in the mapping is the arbitrary phase offset on the right-hand side of Eq. (5.4), which can be used to determine where the phase-wraps occur in the reflectarray element distribution. Further, this can be used as a degree of freedom in the case where the reflectarray element has loss, in order to best distribute the lossy (lower amplitude) cells across the reflectarray to yield the best performance according to the required specifications. This tends to be more meaningful in when using reconfigurable reflectarray elements, which can have more loss compared to fixed elements in general.

Once the phase distribution across the reflectarray has been established, the feed is usually optimized to yield a desired taper level across the array. With the given f/D ratio and a 12.5-wavelength reflectarray diameter, a $\cos^q(\theta)$ feed with $q = 6$ is chosen to produce an approximate -12 dB edge taper in the principal planes of the reflectarray. The feed is chosen to be linearly polarized in the x-direction. Using Eq. (5.21), the incident field levels at the centre of each reflectarray element can be computed and used as inputs to the local reflected field calculations via the GSM in Eq. (5.17).

The radiated fields from the reflectarray are determined from Eq. (5.26), and the corresponding directivity from the array is computed. Loss terms such as spillover loss, surface wave losses, reflection efficiency, feed blockage, etc., have not been considered here, since a simplified model of the reflectarray and feed is being analyzed, but these calculations can be included in order to determine the overall gain of the reflectarray. The resulting directivity of the array is plotted in Fig. 5.13b, which clearly shows a pencil beam pointing in the $\theta = 30°, \phi = 0°$ direction.

To complete the analysis, the results are compared with a method of moments simulation of the same reflectarray, which more accurately accounts for mutual coupling effects between the elements. The feed is assumed to be completely decoupled from the reflectarray. The simulation employs an integral formulation involving the Green's function for a radiating element on a grounded dielectric slab,

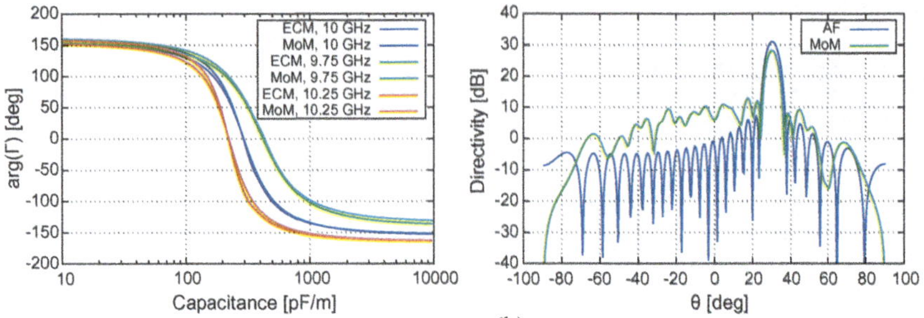

(a) Phase curve of reconfigurable unit cell as predicted by the ECM and MoM techniques

(b) Directivity pattern of the reconfigurable reflectarray as predicted by the AF and MoM techniques

Fig. 5.14 Reconfigurable reflectarray example

and as such, the finite nature of the substrate and ground plane in a real reflectarray is not accounted for. The simulation results are overlaid with the array factor results in Fig. 5.14b. We see that the AF technique is generally quite good at predicting the overall shape of the beam, though the more rigorous MoM analysis reveals stray scattering that increases the overall sidelobe levels (and correspondingly reduces the peak directivity). Overall, very good correlation is obtained between the two methods.

5.5.2 *Reconfigurable Reflectarray*

The second reflectarray example is based on a unit cell whose phase can be tuned dynamically using a tunable capacitance such as a varactor diode. Modelled after a successful reconfigurable reflectarray implementation and demonstration [47], the unit cell is based on the design shown in Fig. 5.9a whereby two halves of a patch are serially loaded by a variable capacitance. For simplicity in the analysis, the capacitance is assumed to uniformly load the gap between the two patch-halves, but in reality this can be replaced with a number of varactor diodes placed across the gap (nominally two). In this situation, the equivalent circuit model should be modified to account for the gap capacitance and current crowding inductance in the vicinity of the varactor diodes, by incorporating these parasitics into the diode model [47]. The parasitics intrinsic to the diode itself can also be included.

To determine the patch dimensions, the equivalent circuit model described in Sect. 5.4.3 can be tested for a variety of patch dimensions to produce the desired S-curve. For the example, the patch dimensions here are chosen to be 8.3 mm $\times$8.3 mm. The ECM is evaluated using the same number of spatial harmonics as before and results in the S-curve shown in Fig. 5.14a. For comparison, results from the spectral domain MoM codes are overlaid in the same plot. It can be seen that the results of both methods align well, and that the ECM can be used to accurately predict the response of the unit cell. Also shown are the corresponding S-curves at adjacent frequencies, to illustrate the frequency-dependent behaviour of the unit cell. We see that as the frequency is changed, the phase curves deviate from the nominal 10 GHz case, which introduces phase errors and eventual pattern and directivity degradation as the difference between the operating and actual frequencies becomes larger. However, for a narrowband reflectarray, the cell design is sufficient.

Using an analogous process as for the fixed reflectarray case, the beam is synthesized by mapping the required phase shifts as required by Eq. (5.4) to the capacitances at 10 GHz. In practice, the required capacitances can be further mapped to control voltages in the case that a varactor diode is used to implement the variable capacitance, using the corresponding $C(V)$ relation for the diode. Results of both the array factor calculation and the MoM analysis are shown in Fig. 5.14b, again illustrating good correlation between the two methods.

The practical design of reconfigurable reflectarrays involves an additional step whereby a suitable control network needs to be introduced for managing the reconfigurable components of the antenna. The realization of the bias network is highly implementation-specific, and while an exhaustive discussion is not possible here, some best practices can be deduced from studying the example here. For the diode-based cell employed in the example, a local biasing network needs to be introduced similar to the scheme shown in Fig. 5.9a. In this case, the bias network consists of a pair of RF chokes, in the form of self-resonant inductors, and an RF short composed of a self-resonant capacitor, through which a DC bias voltage is applied. The bias network serves to form both a high-impedance load to prevent RF currents on the patch from travelling to the bias wires, and a low-impedance load to short out such currents should they make it to the interface with the wires. This is not the only way to realize bias networks; distributed components such as stubs can also be used in place of lumped components. Furthermore, in the case of components with zero DC current draw (such as varactor diodes), resistors serve equally well in place of inductive RF chokes, and are less frequency-dependent.

The presence of the bias network can also introduce undesired scattering, especially if the feeding traces are long. Efforts can be undertaken to minimize this scattering. The use of a symmetric bias network, such as that shown in Fig. 5.9a, maintains low cross-polarization. In the case of reflectarrays, the bias network itself can often be routed to be behind the ground plane of the reflectarray, effectively shielding it from incoming radiation. However, if the bias network must co-exist near or on the same layer as the scatterers, resistively loading the lines has been shown to be effective, especially in transmitarrays where bias lines cannot be conveniently tucked behind ground planes. Resistive lines can be realized by depositing resistive material on the substrate and patterning it into the required bias lines, or by periodically loading metallic lines with resistors.

5.6 Practical Examples

To round out the design examples presented in Sect. 5.5, this section presents some samples of practical implementations taken from the literature. The samples illustrate each of the element design approaches discussed in Sect. 5.2.2. Emphasis is placed on full-scale experimental demonstrations of reconfigurable reflectarrays, since they exemplify the operating principles of reflectarrays while simultaneously illustrating useful beam-scanning capabilities.

The performance of reflectarray designs based on lumped-element tuning is strongly affected by packaging and parasitic issues at millimetre wave and sub-mm wave frequencies. As such, distributed tuning based on the tunable resonator approach mitigates many of these issues. As discussed in Sect. 5.4.2, liquid crystal materials possess a tunable bulk permittivity tensor which can be exploited to tune the resonant frequency of microstrip elements. An example of a recent F-band design is shown in Fig. 5.15 [48]. As illustrated in Fig. 5.15a, the unit cell employs a multi-resonant design based three unequal length parallel dipoles placed within the unit cell to extend the phase bandwidth of the unit cell around 100 GHz. The unit cells include a bias line to allow a low-frequency biasing field to be applied to the LC substrate beneath the dipoles, allowing the permittivity tensor to be manipulated. LC materials with $\Delta\epsilon = 4.6$. were employed in the design. A 52×54 element offset-fed reflectarray was fabricated with one-dimensional scanning capabilities. Some examples characteristics of the radiation patterns for different scan angles are shown in Fig. 5.15b. Overall, the work achieved a very wide scanning range and an aperture efficiency of 18.5%. Most of the inefficiency is attributable to absorption losses in the LC material, but considering the frequency of operation, the results are quite impressive.

The guided-wave approach has also proved useful in the millimetre wave frequency range. A 60 GHz design employing PIN diodes is shown in Fig. 5.16 [34]. As shown in Fig. 5.16a, the unit cell employs a stub-based phase shifter design, whereby the length of the stub is controlled by the state of the switch.

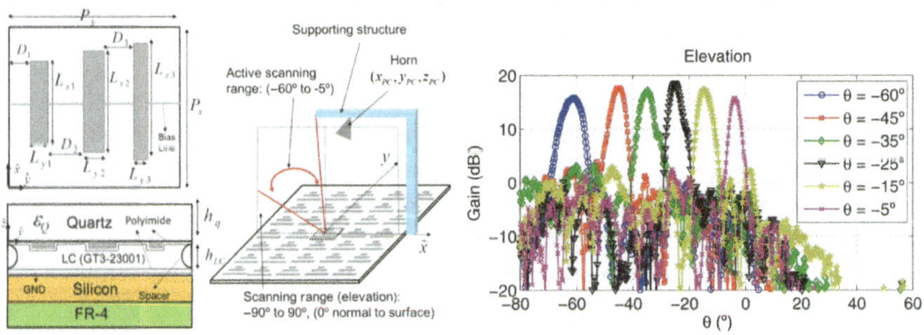

(a) Unit cell and array layout (b) Measured radiation pattern for elevation scanning

Fig. 5.15 Liquid-crystal based reflectarray operating at 100 GHz [48]

Biasing of the diode is accomplished by placing control circuits behind the reflectarray ground plane, as shown in Fig. 5.16b. This implements one-bit phase control, providing the reflectarray elements with 0° or 180° of phase shifting capability. A 160 × 160 array with independent control of each element was fabricated and characterized at 60.25 GHz. Typical beam patterns illustrating the beam-scanning capabilities of the array are shown in Fig. 5.16c, d. The achieved aperture efficiency of the array is about 9% (−10.5 dB). The bulk of the losses originates from phase quantization losses due 1-bit control ($\eta_{phase} \approx -4$ dB) and diode losses ($\eta_{ohmic} \approx -4$ dB). These underscore the balance designers try to achieve between complex, high-performance elements (with high phase resolution and low loss) and low-complexity elements which can be implemented at a much lower cost. Keeping costs down is especially important in large reconfigurable apertures such as the one in this work.

The final example considered employs a reconfigurable element utilizing the sequential rotation concept. CP designs are important for satellite applications, where an additional and imposing requirement is operation at two frequencies (transmit and receive). A dual-band K/Ka-band reflectarray design has been developed working at 24.4 and 35.5 GHz [16]. The reflectarray utilizes a unit cell whereby ring-slots of two different radii are interleaved to form the reflectarray surface, as shown in Fig. 5.17a. Each ring-slot contains 6 MEMS switches that can be actuated in pairs to achieve electronic rotation of the cell, allowing for 3 phase

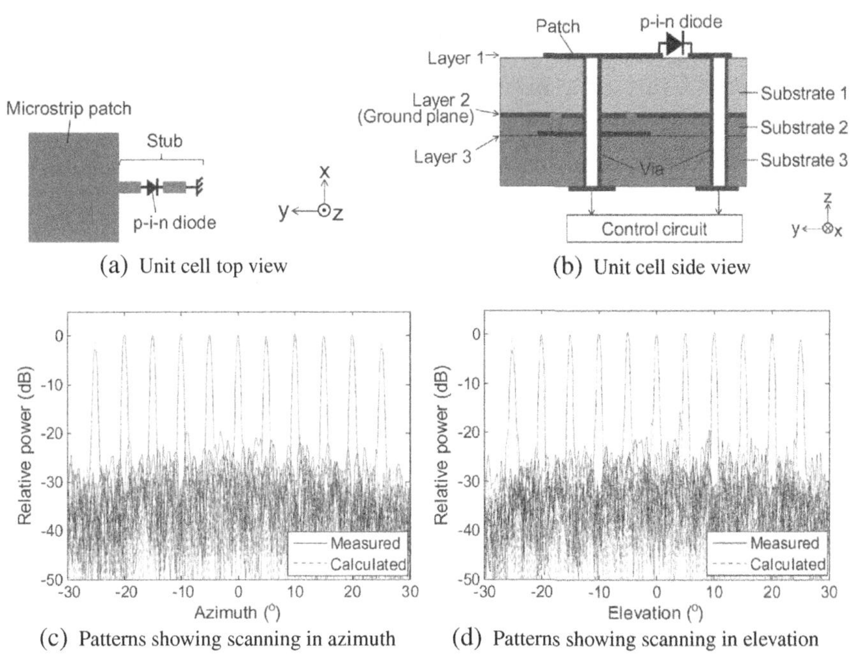

Fig. 5.16 Switched-stub reflectarray operating at 60 GHz [34]

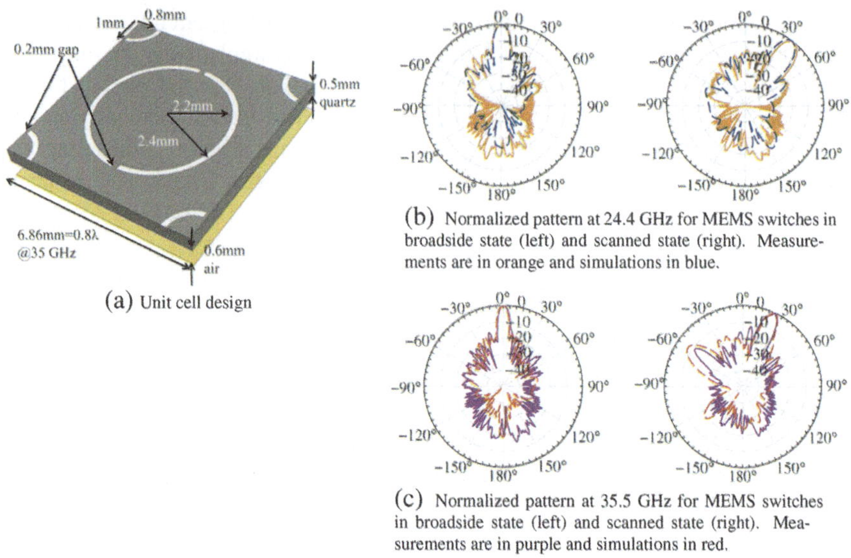

(a) Unit cell design

(b) Normalized pattern at 24.4 GHz for MEMS switches in broadside state (left) and scanned state (right). Measurements are in orange and simulations in blue.

(c) Normalized pattern at 35.5 GHz for MEMS switches in broadside state (left) and scanned state (right). Measurements are in purple and simulations in red.

Fig. 5.17 Dual-band reflectarray based on sequential rotation [16]

states per ring-slot. A prototype was realized with actual MEMS devices (albeit in a frozen state), and the achieved co-polarized radiation patterns are shown for each frequency in Fig. 5.17b, c. A grating lobe appears in the 35.5 GHz results since an interleaved design results in a fixed 6.86-mm cell period that is used for both frequency bands. This results in an electrical periodicity of $0.56\lambda_0$ at 24.4 GHz and $0.86\lambda_0$ at 35.5 GHz, and highlights the challenges of designing dual-band reflectarrays. Nevertheless, this is an acceptable compromise and this example illustrates the potential of MEMS devices in future generations of reflectarrays.

5.7 Applications

Reflectarrays are promising candidates as replacements for reflector antennas in a number of current and next-generation applications. As low-profile, high-gain antennas with potential adaptive beam-forming capabilities, they offer a number compelling advantages, especially at millimetre and sub-mm wave frequencies where array antennas become increasingly lossy and complex. Reflectarrays offer the best features of both reflector antennas and arrays, providing the gain and efficiency of reflectors while offering the flexibility and potentially adaptability of an antenna array. The economical nature of reflectarrays can also make them attractive in applications where there are high setup costs (e.g. as replacements for shaped reflectors, which require costly moulds) or a large amount of hardware at the

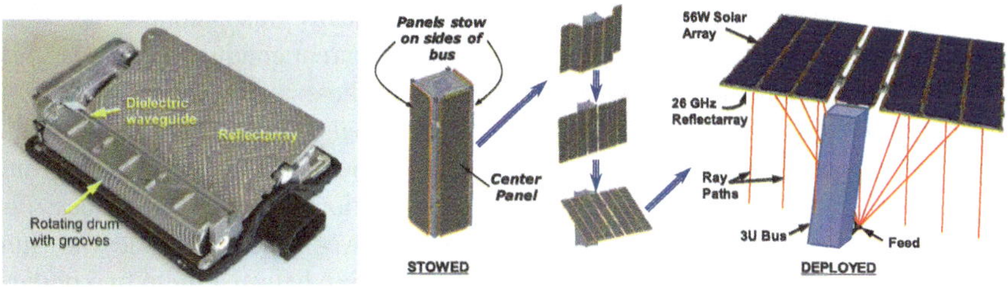

(a) Automotive RADAR unit [49]   (b) ISARA antenna [50]

Fig. 5.18 Reflectarray application examples

unit cell level (e.g. in phased arrays, where a large amount of RF hardware is duplicated at an element or sub-array level).

As high-gain antennas, reflectarrays are potential antenna candidates in RADAR applications. Perhaps one of the most successful applications of reflectarrays has been in automotive RADAR systems at 77 GHz, where the technology has been successfully commercialized into compact, mechanically steered RADAR units [49]. Folded reflectarrays offer further advantages here by allowing the radar unit to be reduced in size while maintaining a narrow-beam. An example of a commercial reflectarray RADAR is shown in Fig. 5.18a.

Reflectarrays have always held the promise of being replacements for reflector antennas in space applications. Indeed, microstrip reflectarrays had their humble beginnings as a research topic at the Jet Propulsion Laboratory in the 1990s where it was hoped that so-called "inflatable" reflectarrays could be deployed on satellites and other spacecraft to reduce launch volume and mass. Fast-forwarding to the present day, this dream has been fulfilled with the launch of the NASA ISARA (Integrated Solar Array and Reflectarray Antenna) cubesat in 2016, which was the first demonstration in space of a multi-faceted reflectarray on the back side of a solar array that deploys in space after launch [50]. This concept, shown in Fig. 5.18 (b), has become a central feature on NASA's upcoming MarCO (Mars Cube One) mission to Mars, which uses an X-band reflectarray as a bent-pipe telecom link to transmit entry, descent, and landing (EDL) data to Earth [51]. The MarCO reflectarray is based on the square patch element discussed in the example, and is a deployable design.

Many other features of reflectarrays have been highlighted as advantageous for space applications. Foremost of these is the ease at which the beam can be shaped, opening up the possibility of shaped coverage patterns for satellites [52, 53]. As mentioned earlier, the ability to use reflectarrays in such applications could be a potential cost-saving measure as the need for fabricating shaped-reflector moulds is eliminated. Reflectarrays have also been identified as promising candidates in a number of remote-sensing applications. For example, reflectarrays have been proposed as an interferometric instrument for measuring ocean height [54], as a replacement for cylindrical reflectors in the Titan cloud precipitation radar and altimeter [55], and for use as antennas in synthetic aperture RADAR (SAR) [56].

It is possible to realize reflectarrays that are optically transparent, either from materials that are optically transparent, such as transparent conducting oxides [57], or from elements that are only partially opaque over the extent of the unit cell [58]. It is possible to overlay such elements over solar panels on spacecraft to allow the large surface area to be used for both solar power collection and as a large aperture for communications. This enables a number of additional operational modes. For spacecraft missions at the L_2 Lagrangian point between the Earth and the Sun, where the spacecraft maintains the same relative position between these two bodies, pointing the solar panel and communications array in the same direction is a necessary requirement that such an antenna fulfills. In other satellite missions not requiring full-time communication with the ground station, the satellite's attitude control system could be used to point the solar array/reflectarray at the Earth only when it is in the range of the ground station, and pointed at the sun the rest of the time.

Terrestrial communication systems can also greatly benefit from reflectarray antennas. Reflectarrays have been proposed for local multi-point distribution (LMDS) communication systems at mm-wave frequencies, since they offer an easy way to shape the beam according to the requirements of the base station [59]. An interesting capability of reflectarrays is their ability to manipulate the propagation characteristics of wireless channels. For example, the frequency selective nature has been shown to improve the characteristics of non-line-of-sight channels when reflectarrays are introduced as passive reflectors in the environment [60]. A similar concept can be employed to control radio wave propagation across different bands within buildings [61].

Finally, as reflectarrays move into the upper-mm-wave and sub-mm wave ranges, possibilities emerge for imaging, particularly among reconfigurable reflectarrays. Large 1-bit electronically controlled reflectarrays have been proposed for imaging at 60 GHz [34], and recent advancements in liquid crystal-based devices could enable imaging and communications within the F-band [62].

5.8 Future Trends

The wide scope and variety of applications discussed in Sect. 5.7 illustrates the progress reflectarrays is making in industrial and scientific fields. Still, the field is germane for future research, especially in the mm-wave and sub-mm wave frequency range.

Bandwidth is a well-known limitation of reflectarrays, especially when they are compared to the curved reflectors they aim to replace, which are very wideband. Many reflectarray designs capitalize on resonances with the scatterer to achieve the pronounced changes in phase about an operating point. Even guided-wave designs, which can employ wideband phase-shifting devices, are ultimately constrained by the bandwidth of the antennas they couple to. As a result, elementary reflectarray designs are usually constrained to have bandwidths on the order of 10% or less.

Dual- and multi-band designs have been explored to increase the attractiveness of reflectarrays, particularly in satellite applications, which separate uplink and downlink frequencies. This can be achieved by integrating multiple resonators with significantly differing resonant frequencies together [63, 64]. Alternatively, the bandwidth of a single band can be increased by coupling resonators closely spaced in frequency, such as by stacking patches together [65] or printing dipoles in close proximity to each other in the same layer [32], which can be advantageous in simplifying the layout of reconfigurable reflectarrays based on tunable materials [62].

Guided-wave designs can have an advantage here provided the delay devices (e.g. delay lines) can be integrated within the confines of the unit cell, and the antenna portion engineered to be suitably wideband. However, planar wideband antennas are difficult to realize, and the introduction of large delay lines can cause secondary scattering effects. Recently, a class of TTD reflectarrays has emerged that address these issues. The first is reflectarray design where the elements are designed to emulate the response of a Bessel filter, which achieves as flat a group delay as possible in the passband. Using planar scatterers to realize the required inductive and capacitive elements of a Bessel filter, reflectarray designs with bandwidths in excess of 67% and fidelity factors close to 1 can be achieved [66]. Another approach employs miniaturized-element FSSs to approximate a TTD response by using multiple layers of sub-wavelength square patches, leading to a 40% fractional bandwidth. A key idea in both concepts is the use of sub-wavelength elements, which in turn realizes something analogous to an impedance surface (or stack of impedance surfaces), which has long been known to be linked to improving reflectarray bandwidth [67]. Translating the bandwidths achieved from these approaches to reconfigurable reflectarrays remains an open problem.

As frequencies increase to the mm-wave and especially sub-mm wave ranges, the behaviour of metals moves into the plasmonic regime, necessitating the change in the analysis of reflectarray elements at these frequencies [68–71]. In some cases, it is preferable to move to dielectrics for realizing reflectarray elements [72]. While the much shorter wavelengths entail significant novel practical considerations, the operating principle is essentially the same as discussed in this chapter, at microwave frequencies.

An equally important future research direction is the development of reconfigurable reflectarray technologies in the mm- and sub-mm wave ranges. For example, lumped-element loading as a reconfiguration mechanism, has only been demonstrated up to around 60 GHz for the case of PIN diodes, and 120 GHz in the case of guided-wave-based MEMS devices [73]. Beyond this frequency, tunable material platforms are required, since the electrical size and parasitics associated with component-based tuning become too difficult to deal with. Liquid crystal technology has been the most promising thus far, achieving practical operation in the F-band [62]. Still, loss remains an important issue here and efforts are underway to engineer LC materials with lower losses in the mm-wave range.

In the sub-mm wave range, graphene has shown significant potential as a tunable material platform for realizing reconfigurable reflectarrays [74, 75]. Employing the so-called graphene "field effect", the phase of a unit cell at 1.3 THz can in principle be controlled by manipulating the complex conductivity of a graphene sheet from a localized bias voltage. Such novel devices may open the doors to reconfigurable THz reflectarrays in the future.

References

1. D.G. Berry, R.G. Malech, W.A. Kennedy, The reflectarray antenna. IEEE Trans. Antennas Propag. **11**(6), 645–651 (1963)
2. R.E. Munson, H.A. Haddad, J.W. Hanlen, Microstrip reflectarray for satellite communication and radar cross-section enhancement or reduction, US Patent 4,684,952 (1987)
3. J. Huang, Microstrip reflectarray, in *1991 Antennas Propagation Society International Symposium Digests*, vol. 2 (1991), pp. 612–615
4. R. Leberer, W. Menzel, A dual planar reflectarray with synthesized phase and amplitude distribution. IEEE Trans. Antennas Propag. **53**(11), 3534–3539 (2005)
5. D.M. Pozar, S.D. Targonski, R. Pokuls, A shaped-beam microstrip patch reflectarray. IEEE Trans. Antennas Propag. **47**(7), 1167–1173 (1999)
6. O.M. Bucci, G. Franceschetti, G. Mazzarella, G. Panariello, Intersection approach to array pattern synthesis. IEE Proc. H Microw. Antennas Propag. **137**(6), 349–357 (1990)
7. J.A. Zornoza, J.A. Encinar, Efficient phase-only synthesis of contoured-beam patterns for very large reflectarrays. Int. J. RF Microw. Comput. Aided Eng. **14**(5), 415–423 (2004)
8. A.K. Bhattacharyya, Projection matrix method for shaped beam synthesis in phased arrays and reflectors. IEEE Trans. Antennas Propag. **55**(3), 675–683 (2007)
9. P. Nayeri, F. Yang, A.Z. Elsherbeni, Design of single-feed reflectarray antennas with asymmetric multiple beams using the particle swarm optimization method. IEEE Trans. Antennas Propag. **61**(9), 4598–4605 (2013)
10. M. Mussetta, P. Pirinoli, G. Dassano, R.E. Zich, M. Orefice, Experimental validation of a genetically optimized microstrip reflectarray, in *Antennas and Propagation Society International Symposium, 2004*, vol. 1. (IEEE, 2014), pp. 9–12
11. P. Nayeri, A.Z. Elsherbeni, and F. Yang, The analogy between offset configurations of parabolic reflectors and reflectarrays, in *2015 IEEE International Symposium on Antennas and Propagation USNC/URSI National Radio Science Meeting* (2015), pp. 2127–2128
12. E. Almajali, D. McNamara, J. Shaker, M.R. Chaharmir, Derivation and validation of the basic design equations for symmetric sub-reflectarrays. IEEE Trans. Antennas Propag. **60**(5), 2336–2346 (2012)
13. C. Wan, J.A. Encinar, Efficient computation of generalized scattering matrix for analyzing multilayered periodic structures. IEEE Trans. Antennas Propag. **43**(11), 1233–1242 (1995)
14. M. Zhou, E. Jorgensen, O.S. Kim, S.B. Sorensen, P. Meincke, O. Breinbjerg, Accurate and efficient analysis of printed reflectarrays with arbitrary elements using higher-order hierarchical Legendre basis functions. IEEE Antennas Wirel. Propag. Lett. **11**, 814–817 (2012)
15. H.R. Phelan, Spiraphase reflectarray for multitarget radar. Microw. J. **20**, 67 (1977)
16. C. Guclu, J. Perruisseau-Carrier, O. Civi, Proof of concept of a dual-band circularly-polarized RF MEMS beam-switching reflectarray. IEEE Trans. Antennas Propag. **60**(11), 5451–5455 (2012)
17. C. Tienda, M. Arrebola, J.A. Encinar, G. Toso, Analysis of a dual-reflect array antenna. IEE Proc. Microw. Antennas Propag. **5**(13), 1636–1645 (2011)
18. M. Zhou, S.B. Sorensen, E. Jorgensen, P. Meincke, O.S. Kim, O. Breinbjerg, An accurate technique for calculation of radiation from printed reflectarrays. IEEE Antennas Wirel. Propag. Lett. **10**, 1081–1084 (2011)

19. E. Almajali, D.A. McNamara, J. Shaker, M.R. Chaharmir, Feed image lobes in offset-fed reflectarrays: diagnosis and solution. IEEE Trans. Antennas Propag. **62**(1), 216–227 (2014)
20. S.R. Rengarajan, Scanning and defocusing characteristics of microstrip reflectarrays. IEEE Antennas Wirel. Propag. Lett. **9**, 163–166 (2010)
21. S.V. Hum, J. Perruisseau-Carrier, Reconfigurable reflectarrays and array lenses for dynamic antenna beam control: a review. IEEE Trans. Antennas Propag. **62**(1), 183–198 (2014)
22. B. Wu, A. Sutinjo, M.E. Potter, M. Okoniewski, On the selection of the number of bits to control a dynamic digital MEMS reflectarray. IEEE Antennas Wirel. Propag. Lett. **7**, 183–186 (2008)
23. S.V. Hum, M. Okoniewski, R.J. Davies, Realizing an electronically tunable reflectarray using varactor diode-tuned elements. IEEE Microw. Wirel. Compon. Lett. **15**(6), 422–424 (2005)
24. M. Riel, J. Laurin, Design of an electronically beam scanning reflectarray using aperture-coupled elements. IEEE Trans. Antennas Propag. **55**(5), 1260–1266 (2007)
25. J. Rodriguez-Zamudio, J.I. Martinez-Lopez, J. Rodriguez-Cuevas, A.E. Martynyuk, Reconfigurable reflectarrays based on optimized spiraphase-type elements. IEEE Trans. Antennas Propag. **60**(4), 1821–1830 (2012)
26. R. Pereira, R. Gillard, R. Sauleau, P. Potier, T. Dousset, X. Delestre, Dual linearly-polarized unit-cells with nearly 2-bit resolution for reflectarray applications in X-band. IEEE Trans. Antennas Propag. **60**(12), 6042–6048 (2012)
27. M.R. Chaharmir, J. Shaker, M. Cuhaci, A. Sebak, Novel photonically-controlled reflectarray antenna. IEEE Trans. Antennas Propag. **54**(4), 1134–1141 (2006)
28. J. Perruisseau-Carrier, A.K. Skriverviky, Monolithic MEMS-based reflectarray cell digitally reconfigurable over a 360° phase range. IEEE Antennas Wirel. Propag. Lett. **7**, 138–141 (2008)
29. H. Legay, Y. Cailloce, O. Vendier, G. Caille, J. Perruisseau-Carrier, M. Lathi, J.P. Polizzi, U. Oestermann, P. Pons, N. Raveu, Satellite antennas based on MEMS tunable reflectarrays, in *Proceedings of the 2007 European Conference on Antennas Propagation (EuCAP2007)* (2007)
30. A. Moessinger, R. Marin, S. Mueller, J. Freese, R. Jakoby, Electronically reconfigurable reflectarrays with nematic liquid crystals. Electron. Lett. **42**(16), 899–900 (2006)
31. W. Hu, R. Cahill, J. Encinar, R. Dickie, H. Gamble, V. Fusco, N. Grant, Design and measurement of reconfigurable millimeter wave reflectarray cells with nematic liquid crystal. IEEE Trans. Antennas Propag. **56**(10), 3112–3117 (2008)
32. G. Perez-Palomino, J.A. Encinar, M. Barba, E. Carrasco, Design and evaluation of multi-resonant unit cells based on liquid crystals for reconfigurable reflectarrays. IEE Proc. Microw. Antennas Propag. **6**(3), 348–354 (2012)
33. R.R. Romanofsky, Advances in scanning reflectarray antennas based on ferroelectric thin-film phase shifters for deep-space communications. Proc. IEEE **95**(10), 1968–1975 (2007)
34. H. Kamoda, T. Iwasaki, J. Tsumochi, T. Kuki, O. Hashimoto, 60-GHz electronically reconfigurable large reflectarray using single-bit phase shifters. IEEE Trans. Antennas Propag. **59**(7), 2524–2531 (2011)
35. E. Carrasco, M. Barba, J.A. Encinar, X-band reflectarray antenna with switching-beam using PIN diodes and gathered elements. IEEE Trans. Antennas Propag. **60**(12), 5700–5708 (2012)
36. O. Bayraktar, O.A. Civi, T. Akin, Beam switching reflectarray monolithically integrated with RF MEMS switches. IEEE Trans. Antennas Propag. **60**(2), 854–862 (2012)
37. E. Carrasco, M. Barba, B. Reig, C. Dieppedale, J.A. Encinar, Characterization of a reflectarray gathered element with electronic control using ohmic RF MEMS and patches aperture-coupled to a delay line. IEEE Trans. Antennas Propag. **60**(9), 4190–4201 (2012)
38. M.E. Bialkowski, A.W. Robinson, H.J. Song, Design, development, and testing of X-band amplifying reflectarrays. IEEE Trans. Antennas Propag. **50**(8), 1065–1076 (2002)
39. R.W. Clark, G.H. Huff, J.T. Bernhard, An integrated active microstrip reflectarray element with an internal amplifier. IEEE Trans. Antennas Propag. **51**(5), 993–999 (2003)

40. L. Cabria, J.Á. García, J. Gutiérrez-Ríos, A. Tazón, J. Vassal'lo, Active reflectors possible solutions based on reflectarrays and Fresnel reflectors. Int. J. Antennas Propag. Article ID 653952, 13 (2009)
41. K.K. Kishor, S.V. Hum, An amplifying reconfigurable reflectarray antenna. IEEE Trans. Antennas Propag. **60**(1), 197–205 (2012)
42. R.H. Phillion, M. Okoniewski, Improving the phase resolution of a micromotor-actuated phased reflectarray, in *2008 Microsystems and Nanoelectronics Research Conference (MNRC 2008)* (2008), pp. 169–172
43. H. Legay, B. Pinte, M. Charrier, A. Ziaei, E. Girard, R. Gillard, A steerable reflectarray antenna with MEMS controls, in *Proceedings of the 2003 IEEE International Symposium Phased Array System and Technology* (2003), pp. 494–499
44. S. Mener, R. Gillard, R. Sauleau, C. Cheymol, P. Potier, Design and characterization of a CPSS-based unit-cell for circularly polarized reflectarray applications. IEEE Trans. Antennas Propag. **61**(4), 2313–2318 (2013)
45. S.V. Hum, B. Du, Equivalent circuit modeling for reflectarrays using Floquet modal expansion, IEEE Trans. Antennas Propag., **65**(3), 1131–1140 (2017)
46. A.K. Bhattacharyya, *Phased Array Antennas: Floquet Analysis, Synthesis, BFNs and Active Array Systems* (Wiley-Interscience, 2006)
47. S.V. Hum, M. Okoniewski, R.J. Davies, Modeling and design of electronically tunable reflectarrays. IEEE Trans. Antennas Propag. **55**(8), 2200–2210 (2007)
48. G. Perez-Palomino, M. Barba, J.A. Encinar, R. Cahill, R. Dickie, P. Baine, M. Bain, Design and demonstration of an electronically scanned reflectarray antenna at 100 GHz using multiresonant cells based on liquid crystals. IEEE Trans. Antennas Propag. **63**(8), 3722–3727 (2015)
49. W. Menzel, A. Moebius, Antenna concepts for millimeter-wave automotive radar sensors. Proc. IEEE **100**(7), 2372–2379 (2012)
50. R.E. Hodges, M.J. Radway, A. Toorian, D.J. Hoppe, B. Shah, A.E. Kalman, ISARA—integrated solar array and reflectarray CubeSat deployable Ka-band antenna, in *2015 IEEE International Symposium on Antennas and Propagation USNC/URSI National Radio Science Meeting* (2015), pp. 2141–2142
51. R.E. Hodges, D.J. Hoppe, M.J. Radway, N.E. Chahat, Novel deployable reflectarray antennas for CubeSat communications, in *2015 IEEE MTT-S International Microwave Symposium* (2015)
52. J.A. Encinar, J.A. Zornoza, Three-layer printed reflectarrays for contoured beam space applications. IEEE Trans. Antennas Propag. **52**(5), 1138–1148 (2004)
53. M. Zhou, O. Borries, E. Jørgensen, Design and optimization of a single-layer planar transmit-receive contoured beam reflectarray with enhanced performance. IEEE Trans. Antennas Propag. **63**(4), 1247–1254 (2015)
54. R.E. Hodges, M. Zawadzki, A reflectarray antenna for use in interferometeric ocean height measurement, in *2005 IEEE Aerospace Conference* (2005), pp. 1131–1139
55. C. Han, S. Hsu, K. Chang, J. Huang, A Ku/Ka-dual band reflectarray to emulate a cylindrical reflector for titan cloud precipitation radar and altimeter, in *2007 IEEE Antennas and Propagation Society International Symposium* (2007), pp. 1445–1448
56. C. Tienda, M. Younis, P. Lopez-Dekker, P. Laskowski, Ka-band reflectarray antenna system for SAR applications, in *The 8th European Conference on Antennas and Propagation (EuCAP 2014)* (2014), pp. 1603–1606
57. C. Kocia, S.V. Hum, Design of an optically transparent reflectarray for solar applications using indium tin oxide. IEEE Trans. Antennas Propag. **64**(7), 2884-2893 (2016)
58. P. Dreyer, M. Morales-Masis, S. Nicolay, C. Ballif, J. Perruisseau-Carrier, Copper and transparent-conductor reflectarray elements on thin-film solar cell panels. IEEE Trans. Antennas Propag. **62**(7), 3813–3818 (2014)
59. J.A. Zornoza, R. Leberer, J.A. Encinar, W. Menzel, Folded multilayer microstrip reflectarray with shaped pattern. IEEE Trans. Antennas Propag. **54**(2), 510–518 (2006)

60. Q. Chen, Reflectarray development for improving NLOS radio channel, in *2013 Asia-Pacific Microwave Conference Proceedings (APMC)* (2013), pp. 654–656
61. L. Li, Q. Chen, Q. Yuan, K. Sawaya, T. Maruyama, T. Furuno, S. Uebayashi, Frequency selective reflectarray using crossed-dipole elements with square loops for wireless communication applications. IEEE Trans. Antennas Propag. **59**(1), 89–99 (2011)
62. G. Perez-Palomino, P. Baine, R. Dickie, M. Bain, J.A. Encinar, R. Cahill, M. Barba, G. Toso, Design and experimental validation of liquid crystal-based reconfigurable reflectarray elements with improved bandwidth in F-band. IEEE Trans. Antennas Propag. **61**(4), 1704–1713 (2013)
63. C. Han, C. Rodenbeck, J. Huang, K. Chang, A C/Ka dual frequency dual layer circularly polarized reflectarray antenna with microstrip ring elements. IEEE Trans. Antennas Propag. **52**(11), 2871–2876 (2004)
64. M.R. Chaharmir, J. Shaker, H. Legay, Dual-band Ka/X reflectarray with broadband loop elements. IET Microw. Antennas Propag. **4**(2), 225–231 (2010)
65. J.A. Encinar, J.A. Zornoza, Broadband design of three-layer printed reflectarrays. IEEE Trans. Antennas Propag. **51**(7), 1662–1664 (2003)
66. L. Liang, S.V. Hum, Design of an UWB reflectarray as an impedance surface using Bessel filters, IEEE Trans. Antennas Propag. **64**(10) 4242-4255 (2016)
67. D.M. Pozar, Wideband reflectarrays using artificial impedance surfaces. Electron. Lett. **43**(3), 148–149 (2007)
68. M. Farmahini-Farahani, H. Mosallaei, Birefringent reflectarray metasurface for beam engineering in infrared. Opt. Lett. **38**(4), 462–464 (2013)
69. T. Niu, W. Withayachumnankul, B.S.-Y. Ung, H. Menekse, M. Bhaskaran, S. Sriram, C. Fumeaux, Experimental demonstration of reflectarray antennas at terahertz frequencies. Opt. Express **21**(3), 2875–2889 (2013)
70. X. Chen, L. Huang, H. Mühlenbernd, G. Li, B. Bai, Q. Tan, G. Jin, C.-W. Qiu, S. Zhang, T. Zentgraf, Dual-polarity plasmonic metalens for visible light. Nat. Commun. **3**, 1198 (2012)
71. N. Yu, N. Yu, P. Genevet, M.A. Kats, F. Aieta, Light propagation with phase discontinuities: generalized laws of reflection and refraction. Science **334**(6054), 333–337 (2011)
72. L. Zou, W. Withayachumnankul, C.M. Shah, A. Mitchell, M. Bhaskaran, S. Sriram, C. Fumeaux, Dielectric resonator nanoantennas at visible frequencies. Opt. Express **21**(1), 1344–1352 (2013)
73. Z. Du, A. Tamminen, J. Ala-Laurinaho, J. Säily, P. Rantakari, A. Luukanen, A.V. Räisänen, Design and optimization of reconfigurable reflectarray element with MEMS phase shifter, in *2013 7th European Conference on Antennas and Propagation (EuCAP)* (2013), pp. 2422–2426
74. E. Carrasco, J. Perruisseau-Carrier, Reflectarray antenna at terahertz using graphene. IEEE Antennas Wirel. Propag. Lett. **12**, 253–256 (2013)
75. E. Carrasco, M. Tamagnone, J. Perruisseau-Carrier, Tunable graphene reflective cells for THz reflectarrays and generalized law of reflection. Appl. Phys. Lett. **102**(10), 104103 (2013)

Chapter 6
Transmitarray Antennas

Laurent Dussopt

Abstract Transmitarray antennas have attracted a lot of interest for the last 10 years to address the challenges of high-directivity reconfigurable antenna designs for emerging applications at millimeter-wave frequencies. This chapter presents the main principles of operation of transmitarray antennas, their main advantages and challenges, and typical envisioned applications. The main achievements and research orientations are summarized. Next, a theoretical model is presented enabling an accurate and efficient simulation and optimization; generic theoretical results are included to illustrate the main optimization trade-offs. In the final section, practical examples taken from the most recent state of the art are presented covering passive, beam-steering, and reconfigurable antennas in different frequency ranges as well as innovative illumination schemes to reduce the total height of these antennas. Finally, the conclusion highlights some important perspectives and research topics toward future developments.

6.1 Introduction

6.1.1 Principles and Development

For the last 10–15 years, transmitarray antennas have triggered a lot of interest and became popular to enable high-gain, high-efficiency, wideband, low-cost, and reconfigurable antennas for emerging millimeter-wave applications. They achieve a similar function as dielectric lenses, but are fabricated on planar technologies (e.g., Printed Circuit Boards) and materials, which makes them lightweight, conformable, and suitable for integrating active devices for reconfiguration. Transmitarray antennas are typically composed of one or several focal sources illuminating a planar arrangement of sub-wavelength phase-shifting unit cells designed to define a predetermined phase distribution across the aperture and therefore a predetermined

L. Dussopt (✉)
CEA, LETI, MINATEC Campus, 17 rue des martyrs, 38054 Grenoble, France
e-mail: laurent.dussopt@cea.fr

A. Boriskin and R. Sauleau (eds.), *Aperture Antennas for Millimeter and Sub-Millimeter Wave Applications*, Signals and Communication Technology,
DOI 10.1007/978-3-319-62773-1_6

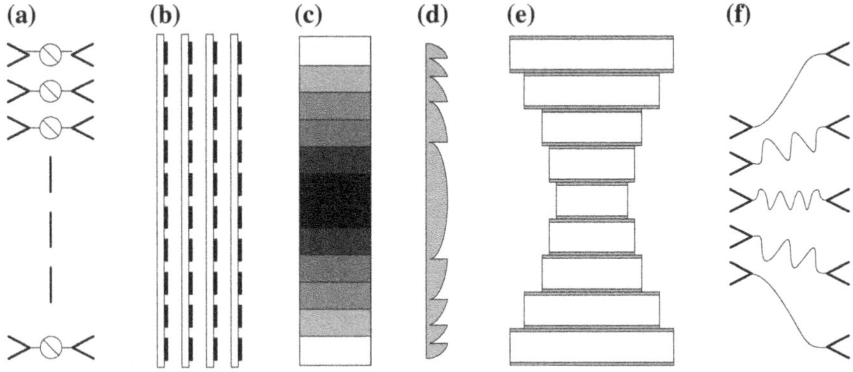

Fig. 6.1 Different structures of artificial lens antennas based on phase-shifters (**a**), multiple coupled layers (**b**), planar lens (**c**), Fresnel lens (**d**), metal waveguide lens (**e**), delay lines based on constrained lenses (**f**)

radiation characteristic in far-field or near-field. A wide variety of structures based on this principle have been investigated and named using different terminologies such as discrete lens, phase-shifting surface, planar lens, flat lens, array lens, or transmitarray (Fig. 6.1). In this document, the generic term transmitarray will be preferred for its analogy with reflectarray antennas.

In contrast, we exclude from this family frequency-selective surfaces which are designed primarily to achieve a selected frequency response for filtering applications without specific requirement on the transmission phase distribution across the aperture. We exclude also metasurfaces or metamaterial structures which are composed of periodic cells having a very small period as compared to the wavelength, making them equivalent to a homogeneous material with effective permittivity and permeability. Fresnel lenses are also well-known structures not considered in this chapter as they belong to a different class of fabrication technology although a lot of analogy can be found in their design, especially because transmitarrays implement phase-correcting Fresnel zones to focus a radiation beam [1, 2].

Transmitarray antennas currently developed in the literature actually inherit from a long history of large-size transmitting antenna structures such as constrained lenses and bootlace antennas developed in the 1940s–1980s for space and military applications (Fig. 6.1) [3–7]. Grid arrays belong also to this family of transmitting apertures and were investigated for beam-steering linearly polarized waves [8]. Transmitting structure gained a renewed interest in the 1990s for power combining applications with amplifiers integrated inside the transmitting panel in order to overcome the power limitations of integrated solid-state amplifiers [9].

Transmitarray antennas are able to achieve high directivity and gain, wideband performance and good power efficiency. These are mainly outcomes of the spatial feeding configuration which represents a major difference as compared to traditional phased arrays with corporate feed networks; in contrast, phased arrays usually enable a much higher flexibility in phase/magnitude control of each radiating element and therefore more accurate radiation control but the design of high-gain (e.g., above 30 dBi) phased arrays in the millimeter-wave domain remains highly challenging. In contrast to dielectric lenses, transmitarray antennas are spatially fed in free space instead of a dielectric material, and the beam collimating is achieved through sub-wavelength circuits instead of propagation; as a result, they generally offer a wider bandwidth, a higher efficiency, a lower weight, and more opportunities for electronic reconfiguration of the radiation properties (polarization, beam shape, and direction) at microwave and low millimeter-wave frequencies. Beyond 100–200 GHz, dielectric lenses certainly offer better efficiency and manufacturability while their weight and size become a minor issue. Unlike reflectarrays, transmitarray antennas do not suffer from masking effect from the focal source. This is an important point as there is a strong interest in integrating focal sources and active circuits (e.g., amplifiers, frequency conversion) at millimeter-wave frequencies, resulting in large systems of several wavelengths in size. Like other antenna array structures, transmitarrays offer also excellent polarization properties with a capability to generate linear or circular polarization with very low cross-polarization. They can be implemented, usually, on standard printed-circuit-board technologies which can be planar, conformed, or flexible, in order to guarantee low manufacturing costs in large volumes with good reproducibility. Some of the shortcomings of transmitarray antennas are their size (area and height), implementation in the upper millimeter-wave range or submillimeter wave range, and the implementation of large reconfigurable aperture with fine phase control due to the transmission topology which leaves little room for biasing circuits and biasing lines routing.

The electronic reconfiguration of transmitarray antennas to achieve beam-forming and beam-steering functions is certainly one of today's main challenges. It can be implemented at the focal source level using a reconfigurable focal array, which enables limited beam-steering capabilities but is still suitable for some applications, or it can be implemented at the transmitarray panel level, i.e., reconfiguration of each unit cell of the transmitarray, which enables a high radiation flexibility. In this case, the integration of active devices (switches, varactors, phase-shifters) significantly impacts the losses, fabrication complexity, and cost of these structures, especially at millimeter-wave frequencies.

Transmitarray antennas are attractive solutions for many current or emerging applications such as follows:

- Point-to-point links in telecommunication networks for long-haul and short-haul wireless links, where very high transmission capacity is needed. These markets cover a wide range of frequency bands from 20 GHz up to 90 GHz currently and additional bands beyond 90 GHz under investigation. Typical gain levels of 30–50 dBi are required and first industrial prototypes of transmitarray antennas developed at 60 GHz indicate a good level of maturity of this technology.

- Base stations in future mobile networks providing millimeter-wave mobile access; these systems will require compact and cost-effective base station antennas with beam-steering capabilities over a wide angular sector, multi-beam or beam-forming capability and gain levels in the range of 20–30 dBi.
- Satellite communications in Ku–Ka-bands, in particular at the level of mobile (vehicles) or portable terminals where lightweight compact and low-cost structures are needed.
- Automotive radars: today a variety of short-range or long-range radars equip modern vehicles for security (crash avoidance), comfort (auto-cruise), or autonomous driving. Gain ranges from 10 to 40 dBi and wide or narrow beam-scanning capabilities are needed depending on the application.
- Imaging and security applications where the use of millimeter-wave frequency bands becomes more and more popular for imaging (security screening in public transportations), surveillance (radars), and industrial control.

6.1.2 General Overview of the State of the Art

Many passive (fixed-beam) transmitarrays have been demonstrated to date in a wide range of frequency bands from C- to W-bands in linear, circular, or dual polarization [10–15]. Most of these designs are based on standard PCB fabrication which is well suited to low-cost, large-area, multilayer printed circuits in these frequency bands where standard line/slot resolution of about 100 μm is appropriate. First commercial products are expected soon for V- and E-band point-to-point communications.

Most devices in the past works were designed using geometric optics between the focal source and the transmitarray panel to calculate the appropriate phase correction of each unit cell to synthesize the linear phase distribution required for simple pencil-beam patterns. In fact, transmitarray antennas can be quite easily modeled analytically and simulated at low computation cost as shown in Sect. 6.2 [16–18]. Full-wave electromagnetic simulations are possible using current commercial softwares and standard computer stations. However, the simulation time can reach several hours, which is acceptable in a validation phase but hardly in an iterative design or optimization procedure. Analytical models can be easily coupled with optimization algorithms to customize radiation properties and soon, researchers demonstrated transmitarrays with synthesized beams, e.g., shaped beams [19], multiple beams [20], and near-field focusing.

Transmitarrays are based on elementary unit cells implementing a selected time delay or phase shift with low transmission/reflection losses. However, the implementation constraints usually limit the feasibility of any transmission phase value so that unit cell designs often exhibit a limited phase range (i.e., smaller than 360°) and/or a limited phase resolution (i.e., phase quantization). Several authors discussed these design issues and highlighted in particular the trade-offs between the phase variation range, the transmission loss, and the complexity of multilayer transmitarrays [21, 22]. Phase quantization, i.e., number of phase states available in a 360° range, is also an important parameter, especially in reconfigurable transmitarrays. Hence, the losses, as well as the complexity, cost, and power consumption, are increasing with the number of switching active devices and therefore the number of phase states.

Achieving a wide transmission bandwidth is often a major requirement to meet the needs of practical communication and radar applications. Most transmitarrays based on resonating radiators—e.g., half-wavelength patches, slots, dipoles, etc.—and phase-shifters, typically achieve a 3-dB bandwidth in the range of 10–20% suitable for many applications. However, some applications like imaging systems or multifunction radars require a much wider bandwidth, e.g., in order to transmit broadband pulsed signals. In this domain, multilayer sub-wavelength metallo-dielectric structures with true-time delay responses across a wideband have shown to operate over 30–50% bandwidth [23–25]. Dual-band transmitarrays have been demonstrated as well, a typical application being satellite communications (SATCOM) where transmit and receive frequency bands are separated by a large gap (e.g., 20/30 GHz for K/Ka-band systems) [26].

The focal distance between the focal source(s) and the transmitarray panel is often seen as a drawback in many applications where thin and low-profile requirements are important. Multiple options to mitigate this issue have been investigated by several groups, such as the use of multiple focal sources, guided feeds, and metasurfaces; several examples are discussed in Sect. 6.3.5.

The electronic reconfiguration of antennas radiation properties is clearly a major technical challenge and enabler for future millimeter-wave applications in order to cope with mobility and system reconfigurability needs with lower cost, mass, and size than mechanically steered systems. In the case of transmitarray antennas, reconfiguration can be implemented either at focal source or transmitarray panel levels with different capabilities in terms of scanning range or beam-forming flexibility. The reconfiguration of the focal source enables beam-scanning over narrow angular sectors around the nominal beam direction and can be implemented with simple switching circuits at low cost and with high efficiency; applications include automotive long-range radar or backhaul/fronthaul point-to-point systems where auto-alignment of the antenna beam is valuable to save time in installation phase and to mitigate small displacements due to wind or other causes. In contrast, the reconfiguration at the transmitarray panel enables much higher flexibility of the radiation properties including beam-scanning over a large angular sector, beam-forming adapted to arbitrary spatial constraints in terms of gain, sidelobes, null-steering, etc. Such flexibility comes at the cost of higher complexity, cost, and losses but is still valuable in many applications including SATCOMs, millimeter-wave access in future mobile networks, point-to-multi-point communications, radars, imaging, etc. Reconfigurable transmitarrays were considered and investigated since the early days of this antenna technology. Most of the investigations reported in the literature are performed in frequency bands from 4 to 15 GHz where few applications exist but technical implementation is easier. A few other examples are demonstrated around 30 GHz with the growing interest for SATCOM and 5G mobile communication networks. Several examples are discussed in Sect. 6.3.4. These works are based on a wide variety of reconfiguration technologies:

- PIN diodes [27–29] are currently the best switching technology available for millimeter-wave frequency bands in terms of transmission losses, power handling, linearity, and switching time. They lead to the best results demonstrated so far in terms of antenna efficiency despite the lower phase resolution (phase resolution is often limited to 90°–180°).
- MEMS switches have a potential for very low losses at high millimeter-wave frequencies; as compared to PIN diodes, higher power handling and linearity can be expected but switching times are orders of magnitude higher (in the range of microseconds). To date, this technology is still limited in terms of maturity and reliability [30] so that few demonstration works exist and practical applications seem far away.
- Varactor diodes are a widely available semiconductor technology enabling continuously tunable circuits. They enable a continuous phase control and, in that respect, can provide more accurate radiation capabilities. Their tuning range is often limited but they can be combined with a switching technology, such as PIN diodes, to extend the phase-shifting range. Several demonstrations were performed based on this technology at 2–15 GHz [31–37]. However, varactor diodes are characterized with a limited quality factor, i.e., significant losses at millimeter-wave frequencies. MEMS varactors are an attractive alternative in principle but have not been applied to transmitarray antennas to date, probably due to lack of a reliable and mature technology. Ferroelectric materials are promising for the implementation of tunable capacitors with better quality factor than semiconductor devices at millimeter-wave frequencies. First demonstrations were presented with attractive performance at 12 GHz [37].
- MMIC phase-shifters based on silicon or III-V semiconductor technologies are an obvious option to integrate phase-shifting capabilities inside transmitarray unit cells. The design of low-loss integrated phase-shifters is a very active field for many decades, driven mostly by phased-array developments in defense and spatial applications and more recently for civilian wireless applications. One of the challenges is the integration of these integrated circuits within the limited area of a transmitarray unit cell (typically $\lambda_0/2 \times \lambda_0/2$ in size). This approach is investigated in defense applications but few examples are available in the open literature [38]. In contrast to stand-alone devices like switches or varactors, integrated phase-shifters suffer from higher costs but enable more accurate phase control and can be combined with attenuation/amplification circuits.
- Microfluidics is a technology relying on the displacement of liquid metal in microchannels in order to reconfigure a radiating element or phase-shifting circuit. It has been investigated quite recently, e.g., at 8.8 GHz [39], and remains an exploratory area so far with little application potential. Hence, microfluidics is characterized by a certain fabrication and actuation complexity, as well as

very low switching times. On the upside, they can operate in static mode with zero power consumption and therefore may address applications with very low power consumption requirements and low reconfiguration rate.

- Liquid crystal materials exhibit a tunable dielectric permittivity which may be interesting for the design of reconfigurable transmitarrays. They are lossy at microwave frequencies and exhibit a small tuning range, but they offer interesting properties at millimeter-wave frequencies. To date, they were successfully used for reflectarrays but not for transmitarrays to our knowledge.

6.2 Modeling and Design

Transmitarray antennas are electrically large structures so their full-wave electromagnetic simulation is usually challenging in time and computer resources; most developments to date are based on simplified models at least in the first design steps and full-wave simulations are limited to validation and verification of main performances but with limited accuracy.

Transmitarray antennas can be modeled analytically to derive first-order performance metrics and guide the main design choices; such basic modeling is presented briefly in the following subsections. More accurate simulations can be performed using a mix of analytical modeling and full-wave electromagnetic simulations; such a methodology is presented in the following where focal source(s) and each individual unit cell are simulated with a 3D electromagnetic simulation software and these results are combined analytically to extract the global transmitarray antenna performances. Finally, some design guidelines are given on the two main families of transmitting structures, namely designs based on coupled layers or separate Tx–Rx antennas.

6.2.1 Theoretical Model

A generic configuration of a transmitarray antenna is shown in Fig. 6.2a. It is illuminated by a focal source with a complex radiation pattern $\vec{H}_{FS}(\theta,\phi)$. Each unit cell n is modeled by two antennas, of radiation patterns $\vec{H}_{1n}(\theta,\phi)$ and $\vec{H}_{2n}(\theta,\phi)$, connected through a phase-shifter or time delay device represented by a 2-port network $[S_n]$. In this work, the radiation patterns $\vec{H}$ are normalized to the square root of the isotropic gain so that the antenna efficiency η is given by

$$\eta = \frac{1}{4\pi}\iint \left|\vec{H}(\theta,\phi)\right|^2 \sin\theta d\theta d\phi. \tag{6.1}$$

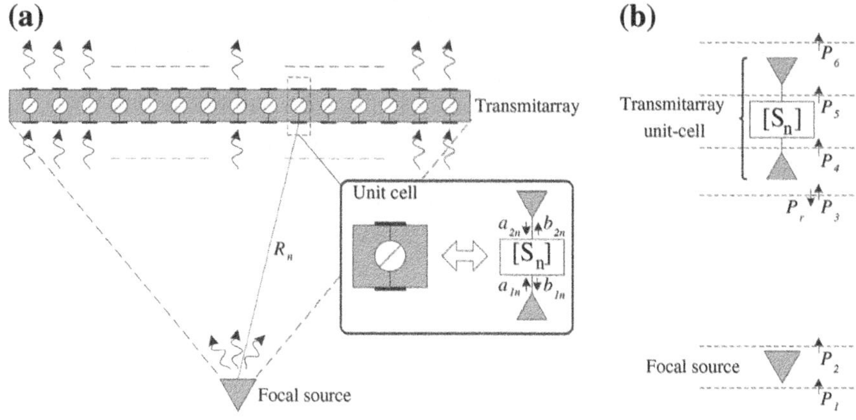

Fig. 6.2 Principle of a transmitarray antenna and generic model of a unit cell (**a**), power budget (**b**)

Assuming the focal source is fed with the input power P_1, the incident wave a_{1n} received by unit cell n is given by

$$a_{1n} = \sqrt{P_1}\frac{\lambda e^{-j2\pi R_n/\lambda}}{4\pi R_n}\vec{H}_{FS}(\theta_n, \phi_n) \cdot \vec{H}_{1n}(\theta_n, \phi_n), \tag{6.2}$$

where R_n is the distance between the focal source and (θ_n, ϕ_n) is the direction of the unit cell n with respect to the focal source. The reflected and transmitted waves b_{1n} and b_{2n} are computed as

$$b_{1n} = S_{11n}a_{1n} \text{ and } b_{2n} = S_{21n}a_{1n}. \tag{6.3}$$

Finally, the radiation pattern of the transmitarray $\vec{H}_{TA}(\theta, \phi)$ is computed by summing the contribution of each unit cell:

$$\vec{H}_{TA}(\theta, \phi) = \sum_n b_{2n}\vec{H}_{2n}(\theta, \phi). \tag{6.4}$$

The optimization and analysis of a transmitarray is best understood through a detailed power budget of the structure which is represented schematically in Fig. 6.2b. We can identify seven power figures:

- P_1: input power of the focal source.
- P_2: power radiated by the focal source.
- P_3: incident power from the focal source on the transmitarray.
- P_4: power collected by the transmitarray.

- P_5: power transmitted through the 2-port networks of the unit cells.
- P_6: power radiated by the transmitarray.
- P_r: power reflected by the transmitarray.

P_2 and P_6 can be computed by integration of the radiation patterns of the focal source and the transmitarray, respectively. P_3 can be computed by the integration of the focal source radiation pattern over the area of the transmitarray. P_4 and P_5 are the sum of the incident (a_{1n}) and transmitted (b_{2n}) waves, respectively. There is no rigorous technique to compute the reflected power P_r analytically since part of the incident radiation from the focal source is directly reflected by the transmitarray structure without going through the unit cell receiving antenna, i.e., the receiving antenna effective aperture is lower than the unit cell area ($P_4 < P_3$). However, if the losses in the receiving antennas are low, it may be approximated by summing the power not captured by the unit cells ($P_3 - P_4$) and the sum of the reflected waves b_{1n}:

$$P_r \approx P_3 - P_4 + \sum_n |b_{1n}|^2. \tag{6.5}$$

Based on these power figures, the total efficiency of the transmitarray can be defined as

$$\eta = \frac{P_6}{P_1} = \eta_{FS} \cdot \eta_{SO} \cdot \eta_{CA} \cdot \eta_{IL} \cdot \eta_{AL}, \tag{6.6}$$

where

- $\eta_{FS} = P_2/P_1$ is the efficiency of the focal source taking into account the resistive losses and mismatch losses.
- $\eta_{SO} = P_3/P_2$ is the illumination efficiency taking into account the spill-over losses.
- $\eta_{CA} = P_4/P_3$ is the efficiency of the receiving array. This term shall take into account resistive, impedance mismatch, and polarization mismatch losses. It can be considered as the effective aperture of the receiving array.
- $\eta_{IL} = P_5/P_4$ is the efficiency of the 2-port networks taking into account their insertion losses.
- $\eta_{AL} = P_6/P_5$ is the efficiency of the emitting array. This term shall take into account the resistive losses and impedance mismatch losses.

The total efficiency of the transmitarray can be computed classically by integration of the radiation patterns $\vec{H}_{TA}(\theta, \phi)$ given by Eq. 6.4.

The phase-shift values $ph(S_{21n})$ of each unit cell are selected to generate a phase distribution of the emerging waves $ph(b_{2n})$ corresponding to a desired radiation pattern:

$$ph(b_{2n}) = ph(a_{1n}) + ph(S_{21n}) = \varphi_{FS}(\theta_n, \phi_n) - k_0 R_n + \varphi_{1n}(\theta_{FS}, \phi_{FS}) + ph(S_{21n}), \tag{6.7}$$

where $\varphi_{FS}(\theta_n, \phi_n)$ is the phase of the focal source radiation pattern in the direction of unit cell n and $\varphi_{1n}(\theta_{FS}, \phi_{FS})$ is the phase of the unit cell n radiation pattern in the direction of the focal source. In many cases, a linear phase distribution is chosen to generate a pencil beam in a direction (θ_0, ϕ_0) and reach the maximum directivity:

$$ph(b_{2n}) = -k_0 \vec{r} \cdot \vec{r_n} = -k_0 \sin\theta_0 \cos\phi_0 x_n - k_0 \sin\theta_0 \sin\phi_0 y_n, \tag{6.8}$$

where $\vec{r}$ is the unit vector in the direction (θ_0, ϕ_0) of the beam and $\vec{r_n} = (x_n, y_n)$ is the coordinate vector of unit cell n.

The model described above is generic and can be used to investigate most, if not all, transmitarray configurations. A critical point is the implementation of the transmitting unit cells; as discussed in the following (Sect. 6.2.4), two main families can be identified. In the first one, the unit cells are composed of separate receive and transmit antennas with an intermediate interconnection network and correspond to the equivalent model presented in Fig. 6.2a. In these cases, each subpart of the unit cell can be designed and simulated independently in order to extract its characteristic parameters. Such structures usually include a ground plane isolating each side of the transmitarray panel. Some examples are the works presented in [28, 34, 36]. In the second family, unit cells are composed of coupled resonators or radiating elements (slots, patches, strips) and such unit cells cannot be divided into independent subblocks; typical examples are structures presented in [10, 11, 30]. The equivalent model of Fig. 6.2a can still be used by considering fictitious receive and transmit antennas having the same radiation patterns as the actual unit cell but ideal efficiency and impedance matching, while the intermediate 2-port network includes the S-parameters of the unit cell. These radiation patterns and S-parameters are computed from a Floquet analysis of the unit cell with periodic conditions and incident plane waves. We can note in this case that $P_5 = P_6$.

It is also worth to mention the case of multisource or focal-array-fed transmitarrays. As shown further (Sect. 6.3), the illumination of the transmitarray by multiple sources can be of interest to reduce the focal distance and make it more compact for instance, or to implement multi-beam functionalities. The same model can be applied readily in these cases using the superposition principle by summing the incident radiation from each focal source at the level of each unit cell.

6.2.2 Theoretical Results

A theoretical analysis of transmitarray antennas can be easily performed to derive generic performance results for a better understanding of the basic trade-offs and provide design guidelines to reach particular requirements.

Table 6.1 Maximum directivity and half-power beamwidth of an ideal focal source in $\cos^n\theta$

Index n	1	4	10	15	49
Directivity (dBi)	6	10	13.4	15	20
Beamwidth (3 dB)	120°	65.5°	42°	34.6°	19.2°

The far-field radiation of a focal source antenna, very often, can be approached by a $cos^n\theta$ function where the directivity $D(\theta, \phi)$ and half-power beamwidth $\Delta\theta_{3dB}$ are determined by the index n:

$$D(\theta, \phi) = 2(n+1)cos^n\theta \text{ and } \Delta\theta_{3dB} = 2\cos^{-1}\left(\sqrt[n]{\frac{1}{2}}\right). \tag{6.9}$$

Typical values of the maximum directivity and half-power beamwidth are given in Table 6.1. For such focal source, the spill-over losses can be easily calculated in the case of a circular transmitarray of diameter D_t and focal distance F by integration of the directivity over the array aperture:

$$\eta_{so} = 1 - cos^{n+1}\theta_0 \text{ with } \theta_0 = \tan^{-1}\left(\frac{D_t}{2F}\right). \tag{6.10}$$

This expression is important as one can note from Eq. 6.6 that the antenna total efficiency η is equal to the spill-over efficiency η_{so} if we assume a lossless focal source and transmitarray.

In [40], the theoretical directivity of an aperture illuminated by a $\cos^n\theta$ focal source is derived:

$$D = \frac{2F^2}{\lambda^2}(n+1)\left|\int_0^{\theta_0}\int_0^{2\pi}\sqrt{\frac{cos^{n+2}\theta}{1+tan^2\theta sin^2\phi}}\frac{sin\theta}{cos^3\theta}d\theta d\phi\right|^2. \tag{6.11}$$

This expression can be integrated numerically for a continuous aperture or transformed into a discrete sum in the (usual) case of a transmitarray composed of individual unit cells with a size in the order of 0.2–$1\lambda_0$. Very close numerical results are obtained using both approaches, showing that the discretization of the aperture into unit cells has no significant influence on the achievable directivity as long as the unit cell size remains below about λ_0.

From the directivity, one can easily compute the aperture efficiency of the transmitarray:

$$\eta_{ap} = \frac{D\lambda^2}{4\pi A} = D\left(\frac{\pi D_t}{\lambda}\right)^2, \tag{6.12}$$

where A is the aperture area of the transmitarray and D_t is its diameter. In the case of a lossless transmitarray, the aperture efficiency η_{ap} is also the product of the spill-over efficiency η_{so}, accounting for the power radiated by the focal source and not captured by the transmitarray, and the taper efficiency η_{taper}, accounting for the reduction of directivity due to a nonuniform amplitude distribution across the aperture; it follows that η_{taper} can be computed from Eqs. 6.10 and 6.12. In the case of a lossy transmitarray, the average loss of the unit cells can be taken into account to extract the taper efficiency.

Figure 6.3 reports theoretical plots of the spill-over efficiency, aperture efficiency, and taper efficiency for an ideal (lossless) circular planar transmitarray antenna illuminated by a $\cos^n\theta$ focal source. It clearly shows the trade-off between taper and spill-over losses as a function of the focal distance and the maximum aperture efficiency as a function of the focal source index.

The frequency bandwidth of an antenna is usually an important requirement in most applications. Figure 6.4 presents the frequency response of an ideal planar transmitarray illuminated by a $\cos^n\theta$ focal source with phase-shifting implemented as true-time delay (TTD) lines or constant phase-shifters. In the former case, TTD lines perfectly match the electrical delay between the focal source and each unit cell so that the transmitarray achieves its highest gain value at any frequency. In the latter case, the constant phase response of phase-shifters can be tuned to the optimal

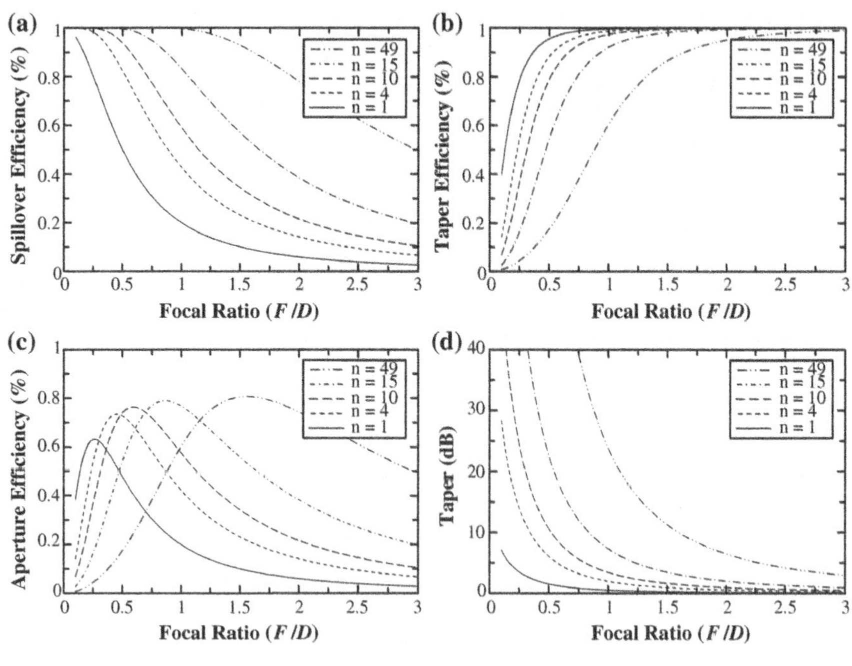

Fig. 6.3 Spill-over efficiency (**a**), taper efficiency (**b**), aperture efficiency (**c**) and edge amplitude taper (**d**) of an ideal transmitarray antenna illuminated by a $\cos^n\theta$ focal source

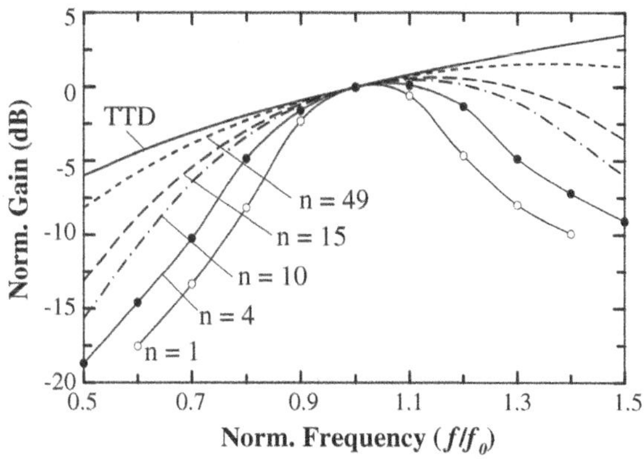

Fig. 6.4 Normalized frequency response of transmitarrays with true-time delays (TTD) or frequency-constant phase-shifters with $cos^n\theta$ focal sources with different indexes n (see Table 6.1 for gain versus index correspondence). Focal ratio *F/D* equals 0.25, 0.45, 0.7, 0.85, 1.55 for n = 1, 4, 10, 15, 49, respectively

electrical delays at the center frequency only and phase errors occurring at other frequencies limit the antenna bandwidth. This bandwidth reduction is more important for short focal distances (focal sources with a low index n in Fig. 6.4) since the path length difference between the center and edge cells is more important. True-time delay lines are a very attractive option theoretically in terms of bandwidth but have not been implemented in transmitarrays so far due to the very long delay values that are challenging to implement in practice.

6.2.3 Transmitarray Simulation

As with reflectarrays and other large-aperture antennas, full-wave electromagnetic simulation of transmitarray antennas is feasible but is usually not the method of choice in the design and optimization phase considering the considerable amount of computation resources necessary to reach a suitable accuracy. With computation times on the order of several hours at least using current mainstream computing stations and softwares, full-wave simulations can be very useful in the verification stage and for the investigation of the impact of the transmitarray environment (mechanical fixture, surrounding objects in the operational environment, etc.).

In contrast, the theoretical model presented in Sect. 6.2.1 enables a fast and accurate simulation of most transmitarray structures as it requires only a few electromagnetic full-wave simulations of small complexity (focal source(s) and unit cells) and derives accurate results on radiation patterns and main antenna characteristics through simple analytical calculations. A flow chart of the simulation procedure is presented in Fig. 6.5 and this procedure has been validated in numerous test cases presented in the articles of the author. A short simulation time is especially useful when optimization techniques with a large number of iterations are used, for instance, for radiation pattern synthesis. Several effects are not taken into account in this theoretical model such as the impact of spill-over radiation,

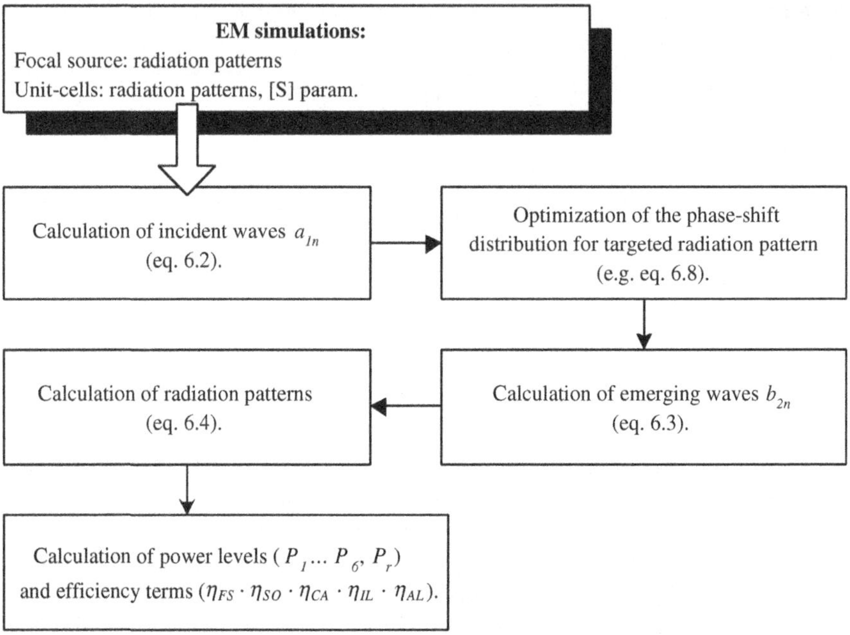

Fig. 6.5 Simulation procedure

diffraction, and scattering on the fixture or edges of the transmitarray; although these effects can be modeled, they would add a significant complexity to the simulation software and they usually can be mitigated through suitable design of the transmitarray structure (e.g., using shielding and absorbing materials).

Unit cells are simulated using periodic conditions and a Floquet analysis to derive the S-parameters and radiation patterns assuming an infinite array of identical cells. Such analysis is available in many electromagnetic simulation softwares and allows the investigation of large periodic structures through the simulation of a single unit cell. It is based on the theory established by G. Floquet to solve periodic linear differential equations. In the context of periodic electromagnetic structures excited by a plane wave, the fields' distribution is periodic as well and the reflected/transmitted waves are expressed as a sum of plane waves (Floquet modes) [41]. While the environment of each unit cell may generally be composed of different unit cells in the actual transmitarray, the assumption of infinite periodicity is generally good enough to allow for an accurate design. The size of unit cells is typically of $\lambda_0/2 \times \lambda_0/2$ so that only the first order of Floquet modes is relevant.

As mentioned in Sect. 6.2.1, in cases where unit cells are based on coupled layers and cannot be divided into independent subblocks (antennas and intermediate circuit), the Floquet simulation is used to derive the S-parameters of the unit cell which is then a 4×4 matrix taking into account the two orthogonal Floquet modes associated with the two polarization directions X and Y [42, 43]:

$$[S] = \begin{pmatrix} S_{11xx} & S_{11xy} & S_{12xx} & S_{12xy} \\ S_{11yx} & S_{11yy} & S_{12yx} & S_{12yy} \\ S_{21xx} & S_{21xy} & S_{22xx} & S_{22xy} \\ S_{21yx} & S_{21yy} & S_{22yx} & S_{22yy} \end{pmatrix}. \quad (6.13)$$

In some cases, unit cells are circularly polarized on either or both sides and circularly polarized S-parameters can be deduced from linearly polarized ones (Eq. 6.13) using classical expressions not reproduced here for brevity. Circular polarization can be also generated using sequential rotation schemes where the assumption of infinite periodicity of the individual unit cells is not met. In this case, it can be necessary to simulate sub-arrays to derive accurate radiation patterns of each unit cell and of the transmitarray [28].

Finally, the model, and in particular Eq. 6.2, is based on an important assumption that the transmitarray is located at far-field distance of the focal source since it assumes a 1/R field attenuation. The experience shows that a suitable accuracy was obtained with focal distances chosen for maximum aperture efficiency corresponding to typical applications. The replacement of Eq. 6.2 with a near-field model of the focal source radiation could be envisioned in specific cases with very short focal distances.

6.2.4 Transmitarray Unit Cells Design

A major challenge in the development of a transmitarray antenna is the design of unit cells with low transmission losses and a wide range of transmission phase, possibly up to 360° for short focal length arrays, beam-steering reconfigurable transmitarrays, or accurate beam synthesis. To reach this goal, one can identify two main strategies, which we will call separate Tx–Rx antennas and coupled-layers approaches, respectively. In the former, the unit cell can be clearly separated into two separate antennas on each side of the transmitarray panel and an intermediate circuit or interconnection as represented in Fig. 6.2a. In the latter, the multiple layers composing the unit cell operate collectively and the whole unit cell shall be simulated as a single structure.

Separate Tx–Rx antennas: the phase shift is generated in lumped or distributed circuits that have to be integrated within the transmitarray unit cell. As the area of the unit cell is usually limited to about $\lambda_0/2 \times \lambda_0/2$ and a small thickness is also desired, the main challenge lies in the miniaturization and minimization of losses of this circuit. This challenge is even more important if one would implement true-time delay unit cells which would be required for ultra-wideband operation. Hence, in large arrays, the time delay difference between unit cells in the center and at the edges amounts to several wavelengths (unless in very long focal transmitarrays); such an array has not been demonstrated so far to our knowledge.

Coupled-layers approach: this approach is inherited from frequency-selective surfaces (FSSs) which have been investigated for a long time for filtering purposes. It relies on the collective response of multiple layers of resonators or scatterers which exhibits a dispersive response much like a multistage filter. Small dimension changes or lumped capacitance/inductance changes shift the frequency response and result in a phase change. As it is often desired to minimize the number of layers for the sake of low manufacturing cost, it can be shown that a quarter-wavelength spacing between layers will maximize the phase-shift range and minimize reflection/insertion losses. The design principles and limitations of such structures have been investigated by several authors; in particular, the trade-off between the phase-shift range and transmission losses was investigated in [21] and it was shown that at least 3 and 4 layers were required for a 360° phase range within a 3- and 1-dB transmission loss limit, respectively. However, in [35], the authors showed that coupled-layers transmitarrays designed for beam-steering result in significant sidelobes in fixed directions dependent on the main beam direction but independent of the transmitarray size. This effect is attributed to higher order modes triggered in the structure at phase discontinuities and is emphasized by the large thickness (typically $\lambda_0/4$ between each layer) of the structure.

The approach described above based on resonant structures (half-wavelength scatterers, quarter-wavelength layer spacing) is usually narrowband due to a strongly dispersive frequency response; they may also exhibit a significant angular dispersion making the structure sensitive to incidence angles and prone to scanning losses. This lead several groups to investigate the use of non-resonant miniaturized (smaller than $\lambda_0/4 \times \lambda_0/4$) designs combining inductive and capacitive responses with reduced spacing between each layer to reach wideband responses. In this case, the phase shift per layer is reduced and a full 360° phase range would require a large number of layers resulting in a larger fabrication complexity. In [23], a 5-layer transmitarray with a total thickness of $0.08\lambda_0$ only was demonstrated with a 20% 3-dB gain bandwidth; in this example, the transmission phase range was about 180° only but this was sufficient for a transmitarray with a focal ratio of about 1. It can be shown that these structures usually result in a quasi-linear phase response or true-time delays which can be used to design ultra-wideband lenses. In [24], the authors demonstrated a 4-layer transmitarray with $0.24\lambda_0$ total thickness and a maximum true-time delay of about 40 ps at 8.5 GHz (equivalent to 123° phase shift). Although this delay is quite limited, this was suitable for a long focal lens (F/D = 1.5) with a very broad frequency response compatible with the transmission of pulsed signals with fractional bandwidth as large as 50%.

One can observe that such transmitarrays with a large number of layers come close to designs based on the effective-index approach, which indeed can provide very broadband performances but at a significant fabrication cost. The large number of layers makes very difficult the design of a reconfigurable transmitarray that would require either low-loss tunable materials or a very large number of tunable lumped devices distributed in several layers. So far, examples available in the literature have been limited to passive structures.

6.3 Practical Examples and State of the Art

A selection of transmitarrays taken from the literature is presented hereafter to illustrate the main achievements and most representative developments performed to date in the domains of passive transmitarrays, beam-steering or reconfiguration schemes, and feeding schemes.

6.3.1 Passive Transmitarrays

Numerous passive transmitarrays have been demonstrated to date in frequency bands ranging from 5 to 90 GHz. They are usually based on standard printed-circuit-board technologies and differ mainly by their phase-shifting schemes using either separate Tx–Rx antennas or coupled layers. Overall, most structures exhibit similar performances with 3-dB fractional bandwidths of 10–20%, aperture efficiencies of 25–45% and total efficiencies of 50–70%.

In [44], a 30-GHz circularly polarized transmitarray using the separate Tx–Rx antennas scheme was demonstrated on organic substrates with a thickness of 1.3 mm (0.13 λ_0) and three metal layers (Fig. 6.6). The transmitarray is a $10\lambda_0$ $10\lambda_0$ square array of unit cells having an insertion loss of 0.2 dB only. Each unit cell is linearly polarized on the focal source side and circularly polarized on the free space side. The total gain is 22.8 dBi which corresponds to an aperture efficiency of 15% only due to the 180° quantization used in the phase compensation scheme (about 3.1 dB of quantization loss) and to a significant taper loss due to the short

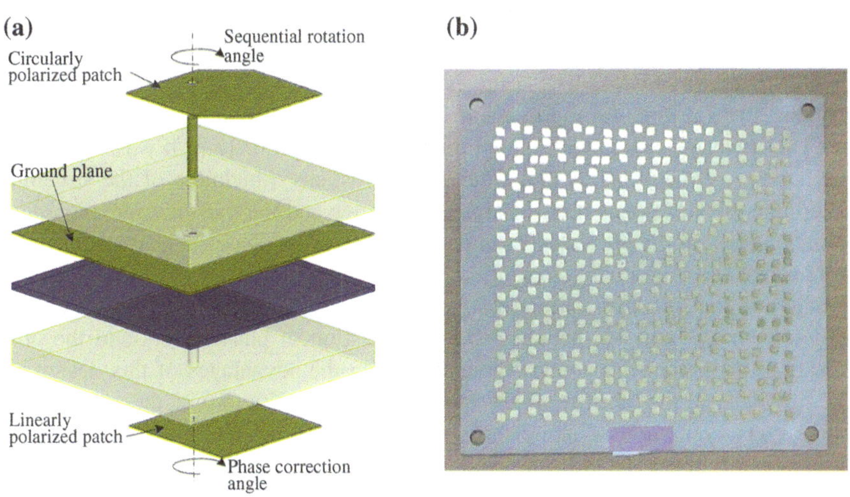

Fig. 6.6 30-GHz circularly polarized transmitarray unit cell (**a**) and prototype with sequential rotations (**b**) [44]

focal distance (*F/D* = 0.52). However, the antenna efficiency is very good (63.1%) for this frequency band and the bandwidth properties are excellent (3-dB bandwidth of 20%, 3-dB axial ratio bandwidth of 24%).

In [17], linearly and circularly polarized transmitarrays were demonstrated for the first time at 60 GHz. The advantages of the structure are its simplicity, made of only three metal layers, leading to very low loss (about 0.5 dB insertion loss) and wide bandwidth (7% 1-dB bandwidth). One of the critical fabrication difficulties in high millimeter-wave bands is the realization of vertical via interconnection between metal layers, which impacts the fabrication cost as well. This issue was overcome in [45] where via-less unit cells were demonstrated at 60 GHz with still only three metal layers and low thickness (735 μm). A transmitarray composed of seven different unit cells was demonstrated. These seven unit cells effectively realize 2.8 bits of quantization, i.e., seven phases state from 45° to 315° with 45° increments. Hence, it was found that it was not possible to implement a 0° phase state with the selected design options (only three metal layers, no via). A fixed-beam antenna was demonstrated with a broadside gain of 32 dBi at 60 GHz, a 3-dB bandwidth of 20%, and a very low cross-polarization level. Recently, the same design was demonstrated in E-band (71–86 GHz) with similar performance [46].

In [16], a 35-GHz linearly polarized filter-lens array is presented based on the coupled-layer approach. It is fabricated on two glass wafers with a total thickness of 1 mm (0.12 λ_0), and three metal layers with a diameter of 7.8 cm ($\sim 9\lambda_0$). It is composed of bandpass unit cells inspired from frequency-selective surfaces with a stack of patch-slot-patch resonators exhibiting insertion losses in the range of 2–3 dBi; their dimensions are tuned to provide the required phase shift at the central frequency to focus the beam in the broadside direction. The 3-dB bandwidth is 8.2%, the maximum gain is 25.6 dBi, the aperture efficiency is 45%, and the total efficiency is estimated to 52%.

In [47], another transmitarray based on the coupled-layer approach was demonstrated at 30-GHz The array is composed of stacked parallel continuous metallic strips diffracting the incident wave. The structure is made of three metal and two dielectric layers with a total thickness of 1 mm ($0.1\lambda_0$). It was shown that the number of metal layers and their relative distance (thickness of the structure) determine the achievable transmission phase range and the transmission magnitude variation [10]. In the presented example, a transmission phase range of $\sim$300° and a transmission magnitude variation of $\sim$2 dB were obtained. A difficulty in this structure is that the transmission magnitude and phase are not independent and a trade-off has to be made between the minimization of phase errors and the optimization of the transmission coefficient. Among the advantages of this design are the simplicity of the fabrication process and the possibility to tune the transmission phase quasi-continuously along the strip directions, thereby minimizing aperture discretization losses. In [48], the authors reduced the structure to two metal layers only with a transmission phase range of about 180° and used the amplitude correction scheme of Fresnel Zone Plate lenses to complete the 360° phase correction range. They demonstrated a gain of 28.6 dBi and an aperture efficiency of 28.9%

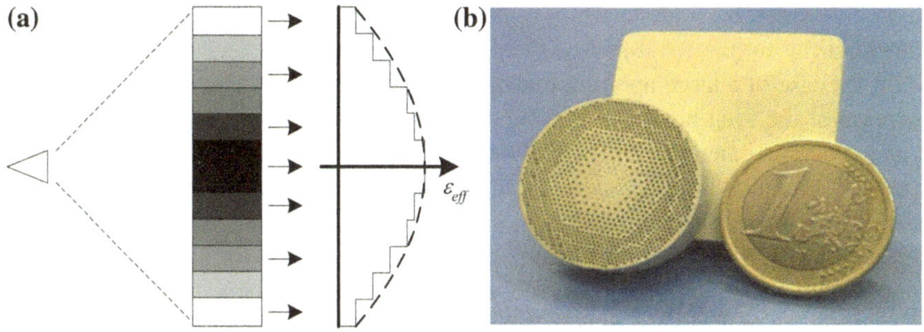

Fig. 6.7 Gradient-index dielectric flat lens: cross section and effective dielectric profile (**a**), prototype at 60 GHz (**b**) [11]

with a 3-dB bandwidth of 16.5%. In contrast, the same authors increased the number of metal layers to four in [19] to reach a more accurate phase and amplitude control and enable a beam-shaping capability with the demonstration of a flat-top beam.

In [11], transmitarrays implemented as gradient-index dielectric flat lenses were demonstrated at 60 and 77 GHz (Fig. 6.7). The lenses are made of an organic substrate with a permittivity $\varepsilon_r = 6.0$, a diameter of 25 mm ($5\lambda_0$ at 60 GHz), and a thickness of 7 mm ($1.4\lambda_0$). Holes are drilled in the dielectric substrate with a variable diameter and spacing in order to synthesize an equivalent permittivity tunable in the range of 2.25–6. The antenna gain is 18.3 dBi and 18.9 dBi at 60 GHz and 77 GHz, respectively, which corresponds to an aperture efficiency in the range of 19–27%. The antenna efficiency is estimated by the authors to 60–70%.

6.3.2 *Mechanical Beam-Steering*

Beam-steering is an important functionality in many applications and can be implemented mechanically. Instead of moving the whole antenna structure, transmitarray antennas offer the opportunity to keep the focal source(s) fixed and move only the transmitarray panel, which is interesting since the displacements of the RF and electronic interconnections of the focal source(s) can trigger significant complexity (rotary joint, flexible cables), cost, weight, and reliability issues, while the transmitarray panel does not have any electrical connection if it is passive.

As for dielectric lenses or reflectors, a common beam-steering technique is to rotate the transmitarray panel around its center so that the focal source remains on the focal arc (Fig. 6.8a) [49]. Two axes of rotations can be combined to steer in azimuth and elevation. This approach has been used in many works to demonstrate the capabilities of passive transmitarrays. It is usually limited to about ±30° because of increased spill-over losses and degraded radiation patterns (beam distortion, sidelobes) due to a non-optimal amplitude and phase illumination of the transmitarray by the focal source.

In the case of a large-aperture transmitarray, the rotation can induce a significant increase of the total height of the antenna system and a large volume requirement, which is inconvenient for many applications where low-profile antennas are desired such as vehicles or aircrafts. A similar beam-steering effect can be obtained by translation of the transmitarray panel above the focal source (Fig. 6.8b). In [50], a Ka-band circularly polarized transmitarray is demonstrated for satellite communications with in-plane translation of the transmitarray panel above the focal source to steer the beam in elevation over 0–50° with less than 3 dB scanning loss (Fig. 6.9). This translation is combined with a 360° in-plane rotation of the transmitarray panel to steer the beam in azimuth; hence, the focal source can be kept in fixed position because it is circularly polarized. The transmitarray is optimized for a beam tilted at 32.5° elevation angle when in center position and the 0–50° scanning range is covered for a translation of $[-0.3F, 0.6F]$, where F is the focal distance; this optimization takes into account the scan loss, the beam distortion, and the height requirements. The antenna aperture is 195×145 mm^2 and includes an extension to translate the panel without increasing spill-over losses, the focal ratio is $F/D \sim 0.55$, and the gain is 27.3 dBi.

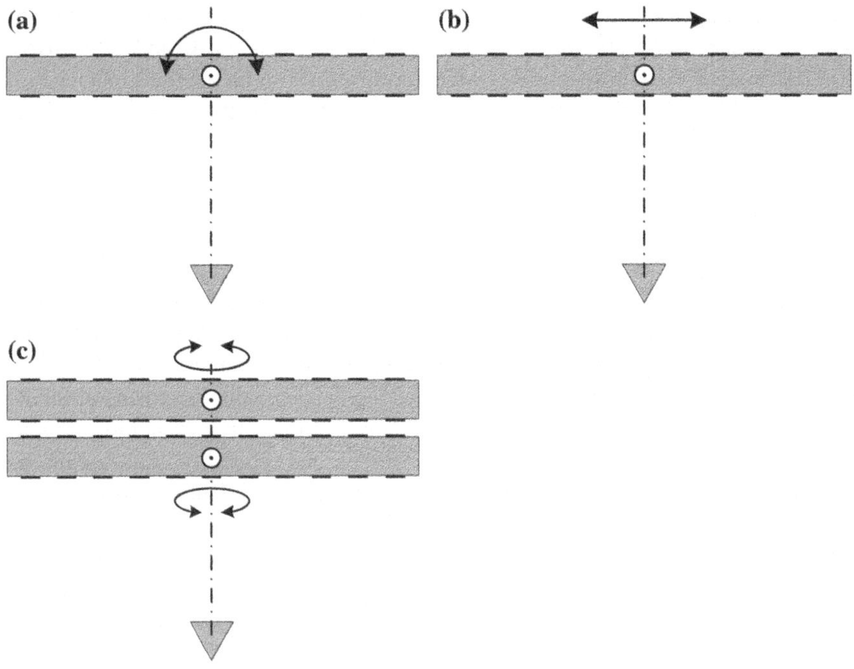

Fig. 6.8 Mechanical beam-steering schemes: rotation in the focal arc (**a**), translation (**b**), in-plane rotation (**c**)

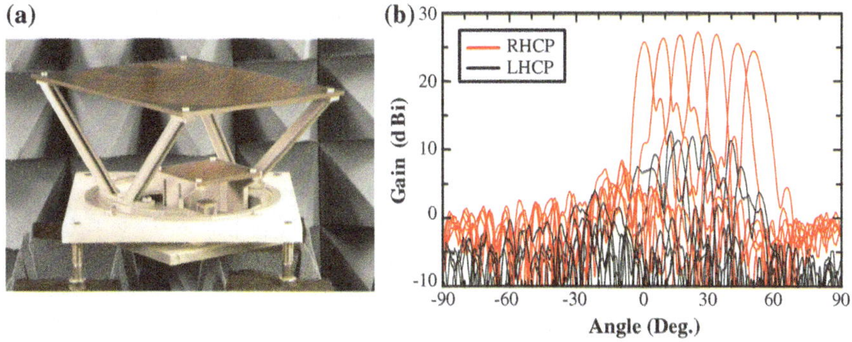

Fig. 6.9 Mechanical beam-steering through translation of the transmitarray; Ka-band prototype (**a**) and experimental radiation patterns with beam steered at 0–50° [50]

In-plane rotations of the transmitarray are easy to realize mechanically and do not degrade the overall size of the antenna. In [51], a 30-GHz linearly polarized transmitarray antenna composed of two rotating transmitting panels was presented (Fig. 6.8c). The panels are separated by about $\lambda_0/5$; each one exhibits a linear phase distribution; and the system is analogue to rotatable dielectric wedges, or Risley prisms, which were already known but seldom used in the microwave domain due to excessive thickness, weight, and reflection losses. This transmitarray mitigates these issues and beam-scanning capabilities up to 57° and 72° from the zenith were successfully demonstrated with scanning losses of 2.8 dB and 5.8 dB, respectively. As compared to a fixed-beam transmitarray of similar technology, the two-panel configuration results in about 1.8 dB additional losses at 30 GHz.

6.3.3 Switched-Beam Transmitarrays

In contrast to the mechanical beam-steering presented above, electronic switching or scanning offers much faster operation, which can be useful in many applications, especially radars and future mobile communication systems. In addition, it usually improves also the system in terms of cost, weight, and reliability. As previously known from dielectric and Fresnel lenses or reflectors, an array of sources in the focal plane can be used for a simple implementation of multi-beam and beam-switching capabilities (Fig. 6.10a). This approach is less complex and usually provides better efficiency than fully reconfigurable antennas (next section) at the cost of less radiation flexibility; these particularities make this configuration very attractive for millimeter-wave systems such as automotive radars or point-to-point communications.

This approach was first implemented at 60 GHz with circularly polarized small-aperture transmitarrays (11–14 dBi gain) targeting indoor short-range communications [52]. A focal array of five antennas integrated on high-resistivity

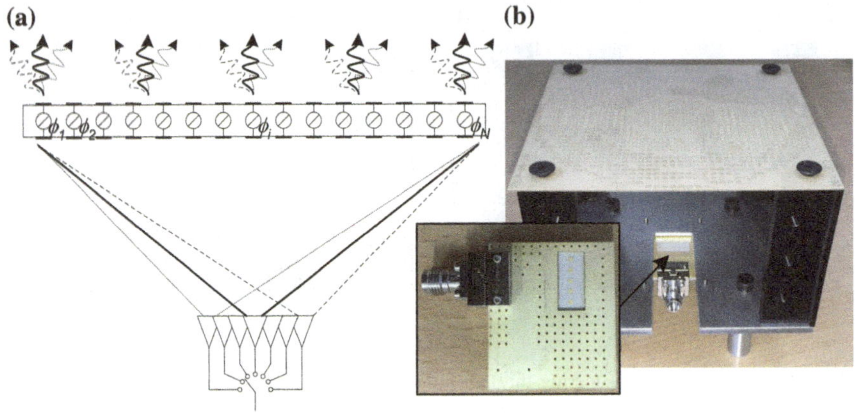

Fig. 6.10 Beam-switching transmitarray with switched focal sources (**a**), 60-GHz switched-beam transmitarray [45]

silicon was developed and a beam-switching capability of ±22° was demonstrated. In [45], a 60-GHz linearly polarized switched-beam transmitarray was developed approaching the specifications of point-to-point communications in wireless networks (Fig. 6.10b). A linear focal array of five patch antennas was designed associated to a SP5T switching network. The five sources are separated by 2.6 mm, the focal ratio *F/D* is 0.55, and this leads to five beams with a beamwidth of 3.2°, oriented at 0°, ±2.3°, ±4.6° and an inter-beam crossing level at about −1.5 dB; the overall −3 dB angular coverage is about ±6.1° in the steering plane. The maximum gain is 29.3 dBi and the 57–66 GHz band is covered with a gain above 25 dBi.

In the above examples, the transmitarrays were optimized for a source centered in the focal plane; off-center sources result in a steered beam but also in gain, beamwidth, and sidelobe degradation due to the oblique illumination. A specific optimization of the transmitarray can be considered to mitigate this issue and find a performance trade-off between broadside and steered beam [53].

6.3.4 Reconfigurable Transmitarrays

Reconfigurable transmitarray antennas correspond to structures where switching or tuning capabilities are integrated inside each unit cell in order to change the transmitting phase/amplitude/polarization and thereby reconfigure the antenna radiation patterns to steer the beam, change its spatial distribution (null-steering, sidelobe level control, beamwidth, multiple beams) or change its polarization. Most works to date have been focused on beam-steering performance which is the main applicative need and reconfigurable transmitarrays offer the perspective for more accurate beam-steering and larger scanning range than using switched focal sources (i.e., previous section), and faster steering than mechanical schemes. The design of reconfigurable unit cells with a high efficiency, small size ($0.5\lambda_0 \times 0.5\lambda_0$ in area, low thickness), low power consumption, and simple control has been the focus of many works for the last 15 years in frequency bands between 5 and 40 GHz. Targeting the highest frequency bands is also a challenge in terms of integration and efficiency. Few full-size reconfigurable transmitarrays have been demonstrated to date with attractive performances and two examples are described hereafter.

In [35], a 6 × 6-element transmitarray at 5 GHz using varactor diodes as tuning devices was demonstrated (Fig. 6.11). The array unit cell is composed of patch antennas on both sides of the panel with a phase-shifter circuit in an intermediate level [34]; this phase-shifter uses six varactor diodes but a single-biasing voltage. The unit cell exhibits a 400° phase range and a 3.6 dB average insertion loss (1.7–5.2 dB). One interest of this structure from the fabrication point of view is the lack of interlayer via interconnections; the signal transmission between layers is performed using slot coupling or proximity coupling. In contrast, coupling techniques can result in parasitic radiation degrading the radiation performance of the array in particular in terms of cross-polarization as observed by the authors. The authors mention that each unit cell was individually characterized in transmission phase to generate per-element lookup tables to control accurately the bias voltages of each varactor diode. The array generates a peak gain of 15 dBi, a scanning window of 100° × 100° and a fractional bandwidth of about 10% for 2 dB gain variation.

In [28], Di Palma et al. demonstrated a 20 × 20-element electronically reconfigurable transmitarray operating in Ka-band (27.4–31.7 GHz) with a beam-steering range of ±60°, full polarization reconfiguration (linear H/V, circular left/right), a broad 3-dB bandwidth (14.6%) and a good radiation efficiency of 58% (Fig. 6.12). The array is based on linearly polarized unit cells with 180° phase switching (1 bit phase quantization using PIN diodes) and the unit cells are arranged with a random 0°/90°/180°/270° sequential rotation. The random distribution mitigates the issue of

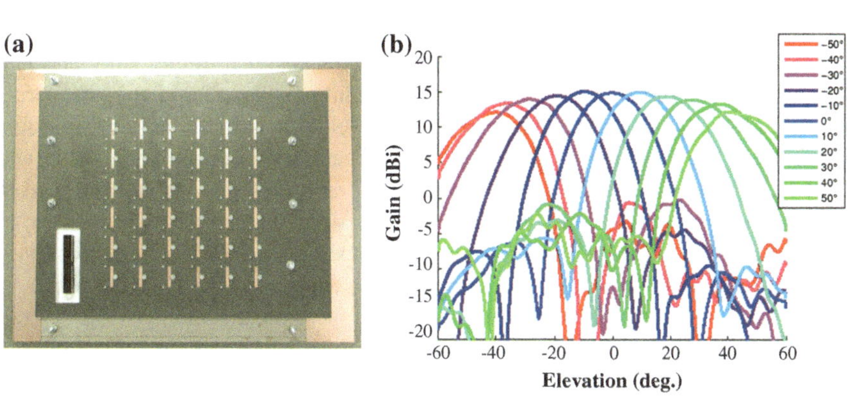

Fig. 6.11 6 × 6-element 5-GHz reconfigurable transmitarray (**a**) based on varactor diodes phase-shifters and radiation patterns in the E-plane with beam steered at ±50° (**b**) [35]

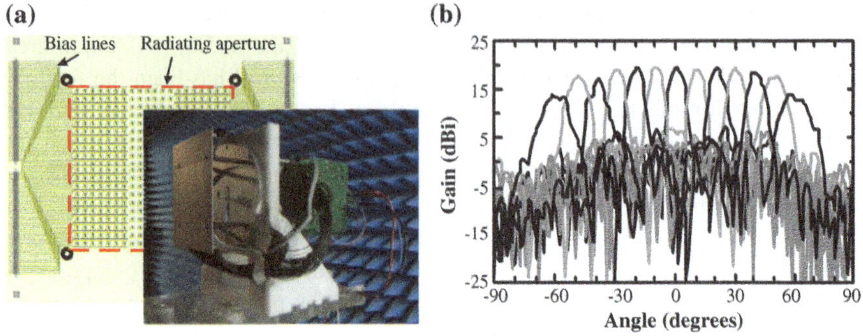

Fig. 6.12 Ka-band 20 × 20-element reconfigurable transmitarray (**a**), measured radiation patterns (realized gain) at 29 GHz for right-hand circularly-polarized beams in the horizontal plane (**b**) [28]

spurious grating lobes occurring in periodic arrays when steering the beam off the broadside direction. This sequential rotation scheme associated with the 180° phase switching enables a full polarization reconfigurability. In the case of circular polarization, the axial ratio remains lower than 2 dB over the full operating bandwidth of the antenna. The beam can be steered in all azimuth planes up to ±60° with 5 dB scanning loss and sidelobe levels below −12 dB. The array was demonstrated first with a 10 dBi focal horn source with a focal ratio *F/D* = 0.60 and next with a 2 × 2 planar focal array of SIW slot sub-arrays enabling a focal ratio reduction down to *F/D* = 0.36.

While the two selected examples above are based on varactor diodes or PIN diodes, it is important to note also the works performed using less mature technologies like RF-MEMS [30], ferroelectric [37], or microfluidics [39].

6.3.5 Focal Distance and Thickness Reduction

The space-feeding architecture of transmitarrays makes them quite bulky and is a drawback as compared to low-profile antenna solutions like phased arrays in terms of integration in applicative environments (e.g., vehicles, aircraft, building, portable systems, etc.). Depending on the focal source beamwidth, the maximum gain and aperture efficiency are reached for a typical focal ratio *F/D* of 0.3–1 as illustrated in Fig. 6.3 and therefore the focal distance is proportional to the transmitarray area. In this respect, the thickness of high-gain, large-aperture antennas can become significantly large and the idea emerged to divide the array into multiple sub-arrays of smaller area, so shorter focal distance, to make the structure thinner (Fig. 6.13a). The concept was introduced in the domain of Fresnel lenses, which suffer from the same issue [54] and was implemented in transmitarrays at X band with four focal sources and a focal distance reduction of about 1.8 [55]. As a rule of thumb, since the transmitarray aperture is divided into *N* sub-arrays (*N* focal sources), the focal

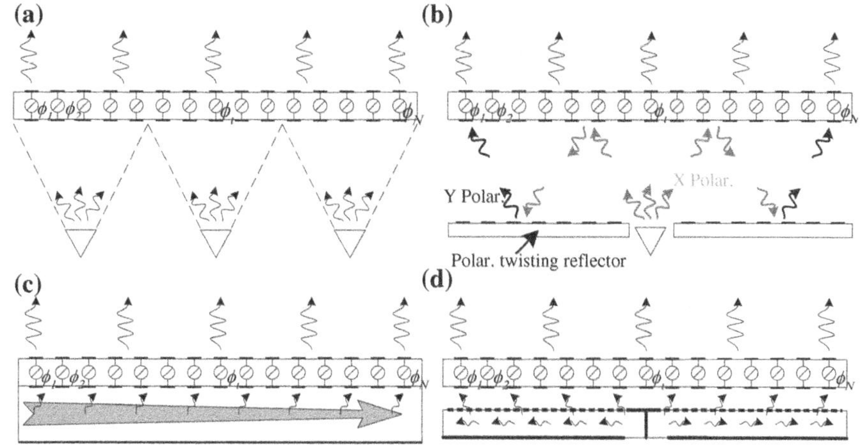

Fig. 6.13 Different schemes for reducing the focal distance and thickness of transmitarray antennas: multisource architecture (**a**), folded transmitarray (**b**), waveguide feed (**c**), metasurface feed (**d**)

distance can be reduced by about $\sqrt{N}$. In principle, the number N of focal sources can be increased as necessary to reduce dramatically the structure height but the benefit of a quasioptically fed structure will be reduced as well by the requirement of a feed network of the focal sources which need appropriate equal amplitude and phasing. It can be shown that the multiple focal source configuration can result in an increase of the transmitarray bandwidth since the distance between each unit cell and its closest focal source is reduced as compared to a single-source structure of equivalent area and therefore the differential time delays are reduced as well. It is worth to note that a multisource configuration can serve other purposes such as power combining (reducing the output power requirement of each focal source to reach a targeted radiated power level) or implementing multi-beam schemes for monopulse radars [56], imaging systems or spatial multiplexing schemes in communication systems.

Another approach is to illuminate the transmitarray indirectly by using a reflection in a so-called folded transmitarray architecture, which is similar to the folded reflectarray presented in Chap. 5. The focal source radiates in a polarization orthogonal to the transmitarray polarization so that the signal is reflected toward a polarization-twisting reflector. In contrast to the folded reflectarray, the phase correction to focus the beam is implemented in the transmitting surface and not in the reflector. To our knowledge, this approach has not been demonstrated so far with transmitarrays but only with a dielectric lens at 60 GHz in [57] where the focal distance has been reduced by a factor of about 3.

In [58], the transmitarray panel is no longer fed through a quasioptic radiation from the source but through guided waves (Fig. 6.13c). A feed network is composed of metallic hollow rectangular waveguides operating in the fundamental

TE10 mode. The unit cells are on the narrow side of the rectangular waveguides and each waveguide feeds in series a line of unit cells by coupling (leaking) a small amount of power to each of them. This approach is demonstrated with a reconfigurable transmitarray of 6 × 6 elements at 4.8 GHz and the performances are shown to be very similar in gain and efficiency to the same array previously demonstrated with a quasioptic feed [35]. This feeding scheme allows to reduce the thickness of the array to $0.75\lambda_0$ only. This comes at the cost of a significant increase in design complexity to ensure a proper phasing and power distribution across the aperture. The series feeding scheme leads also to a reduced bandwidth and a more significant beam-squint effect as in series-fed phased array. This architecture, composed of a guided feed network, phase-shifters, and radiating elements, comes close to classical phased arrays with the same difficulties in terms of scaling in size, weight, design complexity, and scaling to the millimeter-wave domain.

In [59, 60], the feed source is a metasurface antenna, of similar area as the transmitarray, generating a leaky-wave which illuminates the transmitarray panel located in its near-field (Fig. 6.13d). The metasurface source is fed in its center, the guided substrate mode propagates radially, and the surface is designed to generate a Bessel beam, i.e., with a nearly uniform amplitude, with a radial polarization. The transmitarray is also a metasurface and is designed to convert the radially polarized incident wave into the desired (linear or circular) polarization for the radiated beam. This scheme enables a very thin structure of $0.25\lambda_0$ only ($0.14\lambda_0$ separation between the transmitarray and the metasurface feed). The antenna efficiency and aperture efficiency for a passive (fixed-beam) transmitarray at 9.9 GHz were about 40% and 25%, respectively. As in the previous example (waveguide feed), the main challenge in this structure is the accurate design and control of the phase and amplitude distribution across the aperture.

6.4 Conclusion and Perspectives

Transmitarray antennas offer very promising opportunities for future systems, especially in the millimeter-wave range with a wealth of emerging applications in the telecommunication sector as well as radar and imaging systems. They have seen a lot of developments over the last 15 years related to the design of low-loss structures, especially in the millimeter-wave domain, the investigation of bandwidth limitations, and development of beam-steering schemes and fully reconfigurable antennas.

First commercial exploitation of passive transmitarray antennas is expected soon in the domain of point-to-point millimeter-wave communications but there are still many innovations and challenges expected to expand their capabilities, such as dual polarization, dual-band, and multiband operation, as well as filtering properties. The integration of transmitarrays with their environment is also a topic that has hardly been addressed so far: codesign with radomes, conformation on nonplanar surfaces, packaging and codesign of custom focal sources to reduce/cancel spill-over radiation and optimize the aperture efficiency. New and more complex transmitarray architectures are also a promising path for innovation: some of them have been described in Sect. 6.3.5 for size reduction like folded transmitarrays or distributed sources. The emergence of applications in the high millimeter-wave and submillimeter-wave ranges is also a good opportunity for transmitarrays to replace bulky and expensive dielectric lenses and mirrors; such development will challenge current manufacturing technologies to realize multilayer, thin, large-area, metal-dielectric structures with feature resolutions in the 10-μm range or below.

Most existing works are based on simple modeling and simulation principles which are good enough for proof of concept but lack the accuracy required for large-aperture antennas with synthesized beams in some challenging applications, like satellite communications. In this domain, new simulation and optimization tools based on multiscale electromagnetic modeling techniques will certainly be required. Such tools will certainly play a major role in the development of the complex architectures mentioned above.

Fully reconfigurable transmitarrays are certainly a major challenge for the future as current structures remain limited in performances and scalability in large apertures. This requires the design of transmitting unit cells with low loss, accurate phase control over a full 360° range, and simple and scalable control schemes. In the microwave and low millimeter-wave range, semiconductor switches and tunable devices will certainly remain the technology of choice to implement such systems. In the high millimeter-wave range, there are certainly important opportunities for innovating reconfiguration technologies likes MEMS, NEMS, ferroelectric devices, or tunable materials.

References

1. Y.J. Guo, S.K. Barton, *Fresnel Zone Antennas*, (Kluwer academic publishers, 2002)
2. I.V. Minin, O.V. Minin, *Basic Principles of Fresnel Antenna Arrays*, (Springer, 2010)
3. D.T. McGrath, Planar three-dimensional constrained lenses. IEEE Trans. Antennas Propag. **34**(1), 46–50 (1986)
4. W.E. Kock, Metal lens antennas. Proc. I.R.E. **34**(11), 826–836 (1946)
5. D.M. Pozar, Flat lens antenna concept using aperture-coupled microstrip patches. Electron. Lett. **32**(23), 2109–2111 (1996)
6. J. Ruze, Wide angle metal-plate optics. Proc. I.R.E. **38**(1), 53–59 (1950)
7. L. Schwartzman, L. Topper, Analysis of phased array lenses. IEEE Trans. Antennas Propag. **16**(6), 628–632 (1968)
8. C. Chekroun, D. Herrick, Y.M. Michel, R. Pauchard, P. Vidal, Radant—new method of electronic scanning. Microw. J. **24**, 45–53 (1981)
9. Z. Popovic, A. Mortazawi, Quasi-optic transmit/receive front ends. IEEE Trans. Microw. Theory Technol. **46**(11), 1964–1975 (1998)
10. N. Gagnon, A. Petosa, D.A. McNamara, Comparison of 2-, 3- and 4-Layer Phase Shifting Surface Lens Antennas, in *2013 IEEE AP-S International Symposium (APS URSI)*, (Orlando, FL, July 7–13, 2013)

11. M. Imbert, A. Papió, F. De Flaviis, L. Jofre, J. Romeu, Design and performance evaluation of a dielectric flat lens antenna for millimeter-wave applications. IEEE Antennas Wirel. Propag. Lett. **13**, 342–345 (2014)
12. P. Padilla, A. Muñoz-Acevedo, M. Sierra-Castañer, Passive microstrip transmitarray lens for Ku band, in *4th European Conference on Antennas and Propagation (EuCAP)*, (Barcelona, Spain, 12–16 April 2010)
13. R.H. Phillion, Flat lenses for circularly polarized electromagnetic waves, in *PhD Thesis*, (University of Calgary, AL, Canada, Dec 2010)
14. M.K. Taher Al-Nuaimi, W. Hong, Fabrication and experimental validation of high gain lens antenna for 71–86 GHz band, in *2015 IEEE Inernational Wireless Symposium (IWS)*, (Shenzhen, China, 30 March–1 April, 2015)
15. S.B. Yeap, X. Qing, Z.N. Chen, 77-GHz dual-layer transmit-array for automotive radar applications. IEEE Trans. Antennas Propag. **63**(6), 2833–2837 (2015)
16. A. Abbaspour-Tamijani, K. Saranbandi, G.M. Rebeiz, A millimeter-wave bandpass filter-lens array. IET Microw. Antennas Propag. **1**(2), 388–395 (2007)
17. H. Kaouach, L. Dussopt, J. Lantéri, T. Koleck, R. Sauleau, Wideband low-loss linear and circular polarization transmit-arrays in V-band. IEEE Trans. Antennas Propag. **59**(7), 2513–2523 (2011)
18. E.G. Plaza, G. Leon, S. Loredo, F. Las-Heras, A simple model for analyzing transmitarray lenses. IEEE Antennas Propag. Mag. **57**(2), 131–144 (2015)
19. N. Gagnon, A. Petosa, and D.A. Mc Namara, Electrically thin free-standing phase and amplitude shifting surface for beam shaping applications. Microw. Opt. Technol. Lett. **54**(7), 1566–1571 (2012)
20. A.H. Abdelrahman, P. Nayeri, A.Z. Elsherbeni, F. Yang, Design of single-feed multi-beam transmitarray antennas, in *IEEE AP-S International Symposium (APS URSI)* (Memphis, TN, 7–11 July 2014)
21. A.H. Abdelrahman, A.Z. Elsherbeni, F. Yang, Transmission phase limit of multilayer frequency-selective surfaces for transmitarray designs. IEEE Trans. Antennas Propag. **62**(2), 690–697 (2014)
22. H. Nematollahi, J.-J. Laurin, J.E. Page, J.A. Encinar, Design of broadband transmitarray unit cells with comparative study of different numbers of layers. IEEE Trans. Antennas Propag. **63**(4), 1473–1481 (2015)
23. M.A. Al-Joumayly, N. Behdad, Wideband planar microwave lenses using sub-wavelength spatial phase shifters. IEEE Trans. Antennas Propag. **59**(12), 4542–4552 (2011)
24. M. Li, N. Behdad, Wideband true-time-delay microwave lenses based on metallo-dielectric and all-dielectric lowpass frequency selective surfaces. IEEE Trans. Antennas Propag. **61**(8), 4109–4119 (2013)
25. M. Li, N. Behdad, Broadband true-time-delay microwave lenses based on miniaturized element frequency selective surfaces. IEEE Trans. Antennas Propag. **61**(3), 1166–1179 (2013)
26. S.H. Zainud-Deen, S.M Gaber, H. Malhat, K.H Awadalla, Single feed dual-polarization dual-band transmitarray for satellite applications, in *30th National Radio Science Conference*, (Cairo, Egypt, 16–18 April 2013)
27. C. Huang, W. Pan, X. Ma, X. Luo, 1 bit reconfigurable circularly polarized transmit-array in X band. IEEE Antennas Wireless Propag. Lett. **15**, 448–451 (2016)
28. L. Di Palma, A. Clemente, L. Dussopt, R. Sauleau, P. Potier, P. Pouliguen, Circularly-polarized reconfigurable transmitarray in Ka-band with beam scanning and polarization switching capabilities. IEEE Trans. Antennas Propag. **65**(2), 529–540 (2017)
29. W. Pan, C. Huang, X. Ma, X. Luo, A dual linearly-polarized transmitarray element with 1-bit phase resolution in X-band. IEEE Antennas Wireless Propag. Lett. **14**, 167–170 (2015)
30. C. Chih-Chieh, B. Lakshminarayanan, A. Abbaspour-Tamijani, A programmable lens-array antenna with monolithically integrated MEMS switches. IEEE Trans. Microw. Theory Technol. **7**(8), 1874–1884 (2009)

31. L. Boccia, I. Russo, G. Amendola, G. Di Massa, Multilayer antenna-filter antenna for beam-steering transmit-array applications. IEEE Trans. Microw. Theory Tech. **60**(7), 2287–2300 (2012)
32. C. Huang, W. Pan, X. Ma, B. Zhao, J. Cui, X. Luo, Using reconfigurable transmitarray to achieve beam-steering and polarization manipulation applications. IEEE Trans. Antennas Propag. **63**(11), 4801–4810 (2015)
33. A. Iturri-Hinojosa, J.I. Martinez-Lopez, A.E. Martynyuk, Analysis and design of E-plane scanning grid arrays. IEEE Trans. Antennas Propag. **58**(7), 2266–2274 (2010)
34. J.Y. Lau, S.V. Hum, A wideband reconfigurable transmitarray element. IEEE Trans. Antennas Propag. **60**(3), 1303–1311 (2012)
35. J.Y. Lau, S.V. Hum, Reconfigurable transmitarray design approaches for beamforming applications. IEEE Trans. Antennas Propag. **60**(12), 5679–5689 (2012)
36. P. Padilla, A. Muñoz-Acevedo, M. Sierra-Castaner, M. Sierra-Pérez, Electronically reconfigurable transmitarray at Ku band for microwave applications. IEEE Trans. Antennas Propag. **58**(8), 2571–2579 (2010)
37. M. Sazegar, Y. Zheng, C. Kohler, H. Maune, M. Nikfalazar, J.R. Binder, R. Jakoby, Beam steering transmitarray using tunable frequency selective surface with integrated ferroelectric varactors. IEEE Trans. Antennas Propag. **60**(12), 5690–5699 (2012)
38. T. Chaloun, C. Hillebrand, C. Waldschmidt, W. Menzel, Active transmitarray submodule for K/Ka band satcom applications, in *2015 German Microwave Conference (GeMIC)*, (Nuremberg, Germany, 16–18 March 2015)
39. E. Erdil, K. Topalli, N.S. Esmaeilzad, O. Zorlu, H. Kulah, O.A. Civi, Reconfigurable nested ring-split ring transmitarray unit cell employing the element rotation method by microfluidics. IEEE Trans. Antennas Propag. **63**(3), 1163–1167 (2015)
40. C.A. Balanis, Antenna theory, analysis and design, 2nd edn. (Wiley, 1997)
41. A.K. Bhattacharyya, *Phased Array Antennas: Floquet Analysis, Synthesis, BFNS and Active Array Systems*, (Wiley-Interscience, 2006)
42. R.C. Hall, R. Mittra, K.M. Mitzner, Analysis of multilayered periodic structures using generalized scattering matrix theory. IEEE Trans. Antennas Propag. **36**(4), 511–517 (1988)
43. M. Steer, *Microwave and RF Design: A Systems Approach*, (SciTech Publishing, 2010)
44. L. Di Palma, A. Clemente, L. Dussopt, R. Sauleau, P. Potier, P. Pouliguen, Circularly-polarized transmitarray with sequential rotation in Ka-band. IEEE Trans. Antennas Propag. **63**(11), 5118–5124 (2015)
45. A. Moknache, L. Dussopt, J. Säily, A. Lamminen, M. Kaunisto, J. Aurinsalo, T. Bateman, J. Francey, A switched-beam linearly-polarized transmitarray antenna for V-band backhaul applications, in *10th European Conference on Antennas and Propagation (EuCAP)*, (Davos, Switzerland, 11–15 April 2016)
46. L. Dussopt, A. Moknache, T. Potelon, R. Sauleau, Switched-beam E-band transmitarray antenna for point-to-point communications, in *11th European Conference on Antennas and Propagation (EuCAP)*, (Paris, France, 19–24 March 2017)
47. N. Gagnon, A. Petosa, D.A. McNamara, Thin microwave quasi-transparent phase-shifting surface (PSS). IEEE Trans. Antennas Propag. **58**(4), 1193–1201 (2010)
48. N. Gagnon, A. Petosa, D.A. Mc Namara, Printed hybrid lens antenna. IEEE Trans. Antennas Propag. **60**(5), 2514–2518 (2012)
49. M.K. Taher Al-Nuaimi, W. Hong, On the beam scanning capabilities of discrete dielectric flat lens at 73.5 GHz and 83.5 GHz, in *2015 IEEE International Wireless Symposium (IWS)*, (Shenzhen, China, 30 March–1 April, 2015)
50. E.B. Lima, S.A. Matos, J.R. Costa, C.A. Fernandes, N.J.G. Fonseca, Circular polarization wide-angle beam steering at Ka-band by in-plane translation of a plate lens antenna. IEEE Trans. Antennas Propag. **63**(12), 5443–5455 (2015)
51. N. Gagnon, A. Petosa, Using rotatable planar phase shifting surfaces to steer a high-gain beam. IEEE Trans. Antennas Propag. **61**(6), 3086–3092 (2013)
52. J. Zevallos Luna, L. Dussopt, A V-band switched-beam transmit-array antenna. Int. J. Microw. Wireless Technol. **6**(1), 51–56 (2014)

53. P. Nayeri, F. Yang, A.Z. Elsherbeni, Design of multifocal transmitarray antennas for beamforming applications, in *2013 IEEE AP-S International Symposium (APS URSI)*, (Orlando, FL, 7–13 July 2013)
54. A. Petosa, S. Thirakoune, I.V. Minin, O.V. Minin, Array of hexagonal Fresnel zone plate lens antennas. IEE Electron. Lett. **42**(15), 834–836 (2006)
55. L. Di Palma, A. Clemente, L. Dussopt, R. Sauleau, P. Potier, P. Pouliguen, Design and experimental characterization of a reconfigurable transmitarray with reduced focal distance. Int. J. Microw. Wireless Technol. **8**(03), 447–454 (2016)
56. L. Di Palma, A. Clemente, L. Dussopt, R. Sauleau, P. Potier, P. Pouliguen, Radiation pattern synthesis for monopulse radar applications with a reconfigurable transmitarray antenna. IEEE Trans. Antennas Propag. **64**(9), 4148–4154 (2016)
57. O. Soykin, A. Artemenko, A. Mozharovskiy, A. Myskov, R. Maslennikov, Millimeter-wave high gain lens antenna with reduced focus distance, in *European Conference on Antennas and Propagation (EuCAP)*, (Davos, Switzerland, 10–15 April 2016)
58. J.G. Nicholls, S.V. Hum, Full-space electronic beam-steering transmitarray with integrated leaky-wave feed. IEEE Trans. Antennas Propag. **64**(8), 3410–3422 (2016)
59. C. Pfeiffer, A. Grbic, Millimeter-wave transmitarrays for wavefront and polarization control. IEEE Trans. Microw. Theory Technol. **61**(12), 4407–4417 (2013)
60. C. Pfeiffer, A. Grbic, Planar lens antennas of subwavelength thickness: collimating leaky-waves with metasurfaces. IEEE Trans. Antennas Propag. **63**(7), 3248–3253 (2015)

Chapter 7
Fabry-Perot Cavity Antennas

Alexandros Feresidis, Konstantinos Konstantinidis and Peter Gardner

Abstract Fabry-Perot Cavity Antennas (FPAs) are a type of highly directive planar antennas that offer a promising alternative to standard planar microstrip patch arrays or waveguide slot array antennas. They offer significant advantages in terms of low fabrication complexity, particularly at mm wave frequencies, high radiation efficiency, and good radiation pattern performance. These advantages, in conjunction with a renewed interest in periodic surfaces and meta-surfaces, led to a reinvigoration of international research on this antenna type. This chapter reports recent advances on the design and implementation of FPAs at mm-wave bands. The main concept of FPAs, their operating principles and analysis approaches are briefly introduced. The basic types of FPAs are summarized following a historical account of various implementations until recent years. The main body of this chapter provides an overview of recent designs with a main focus on mm-wave bands and the advantages of the reported antennas for high-frequency implementations.

This chapter reports recent advances on the design and implementation of Fabry-Perot Cavity-type antennas (FPAs) at mm-wave bands. The main concept of FPAs, their operating principles, and analysis approaches are briefly introduced. The basic types of Fabry-Perot cavity antennas are summarized following a historical account of various implementations until recent years. The main body of this chapter provides an overview of recent designs with a main focus on mm-wave bands and the advantages of the reported antennas for high-frequency implementations.

A. Feresidis (✉) · K. Konstantinidis · P. Gardner
Department of Electronic Electrical and Systems Engineering,
University of Birmingham, Birmingham B15 2TT, UK
e-mail: a.feresidis@bham.ac.uk

P. Gardner
e-mail: p.gardner@bham.ac.uk

A. Boriskin and R. Sauleau (eds.), *Aperture Antennas for Millimeter and Sub-Millimeter Wave Applications*, Signals and Communication Technology,
DOI 10.1007/978-3-319-62773-1_7

7.1 Introduction

7.1.1 Fabry-Perot Cavity Antenna Concept

The term Fabry-Perot Cavity Antenna (FPA) was coined in recent years for a relatively simple planar antenna structure that yields high directivity and gain, offering a promising alternative to standard planar microstrip patch array or waveguide slot array antennas. FPAs offer significant advantages in terms of low fabrication complexity, particularly at mm-wave frequencies, high radiation efficiency, and good radiation pattern performance [1]. These advantages, in conjunction with a renewed interest in periodic surfaces and meta-surfaces, led to a reinvigoration of international research on this antenna type that led to a surge of journal and conference publications since the early 2000s.

The basic concept of FPAs, although the term was not used at the time, was first introduced by Trentini [2] in 1956. This work demonstrated how several types of periodic surfaces, acting as partially reflective surfaces (PRSs), can increase the directivity of a single waveguide aperture surrounded by a ground plane. In mid-1980s, the directivity and gain enhancement produced by dielectric superstrates over printed antennas was reported independently [3, 4]. Around the same time, Trentini's work was revisited by [5], producing highly directive antennas with multiple periodic superstrates. However, it was not until 2001, when the similarities of the operating principle of these antennas with the Fabry-Perot interferometer widely used in optics were reported [1] and a study on the effect of the periodic superstrate (or PRS) characteristics on the antenna bandwidth and directivity was performed. A large number of papers have been published since, with the term Fabry-Perot Cavity antenna used in many of those. However, other terms have also been used for this type of antennas, such as Electromagnetic Band-Gap (EBG) resonator antennas, high-gain leaky-wave antennas, or PRS antennas.

The main FPA structure is formed by a partially reflective surface (PRS) placed at a distance of about half-wavelength and in parallel to a ground plane thus forming an open Fabry-Perot type resonant cavity. An excitation source is typically included within the cavity, e.g., a dipole, a microstrip patch or a slot in the ground plane. The gain and bandwidth of the FPA depend on the reflection (amplitude and phase) from the PRS as well as the distance from the ground plane. The PRS can be either a passive periodic array, such as Frequency-Selective Surfaces (FSSs) which are customarily used for filtering electromagnetic waves, or a metasurface with sub-wavelength and potentially nonuniform unit cells. Various implementations have been proposed in recent years, most of them employing 2D metallo-dielectric arrays of conducting elements [1, 2, 5–12] or metallic periodic arrays of apertures in a conducting sheet [13–15]. However, different configurations have also been proposed employing one or more dielectric layers [3, 4, 16–19] with different dielectric constants and thicknesses functioning as the PRS layer. The use of dielectric superstrates as the PRS may offer some advantages in terms of reduced fabrication complexity, however it reduces the degrees of freedom in the design.

7.1.2 General Overview of FPA Analysis Techniques

The work of Trentini [2] first employed a geometric optics approximation to describe mathematically the antenna operation by means of the multiple reflections of plane waves between the ground plane and the PRS. The analysis was adopted to describe the paths of transmitting and reflecting rays from the PRS. This approximate method assumes an infinite extent of the structure and thus is not taking into account diffraction effects while it also ignores higher order mode coupling. The approach gives an estimation of the directivity performance after calculating the reflection characteristics of the PRS structures under plane-wave incidence assuming infinite size structures. The operating frequency is defined by the cavity distance, which is typically close to half-wavelength in order to achieve constructive interference of the waves bounced between the PRS and the ground.

Another analysis technique for FP antennas is based on the principle of reciprocity. This technique is based on the principle that a source placed at a point A in the far-field will provoke an electric field strength at a point B inside the structure equal to the one that would be provoked at point A, from a source at point B (Fig. 7.1). Thus, the technique reduces the estimation of the far-field radiation characteristics to the calculation of the fields at an observation point inside the antenna when illuminated by a plane wave of fixed magnitude and a varying angle of incidence. By sampling the relative field strength, at the observation point B, for various angles of incidence the (receiving) radiation pattern of the antenna can be obtained. This approach is described in [20, 21] for antennas consisting of a single periodic layer and also employed in [22, 23] for FP antennas with multiple periodic layers. The technique can be combined with any other method that allows the calculation of the near field at point B. For more accurate results, the calculation of the near field at point B can be done using a rigorous full-wave modeling technique, for example as in [22].

A more rigorous analysis and design of FP antennas can be performed using leaky-wave theory. This is due to the partial electromagnetic transparency of the PRS at frequencies near the resonance and the associated leaky-modes that are excited. The main information about the operation of leaky-wave antennas can be provided by investigating the complex wavenumber, $k = \beta - j\alpha$, of the leaky-modes. Obtaining the values of the real part of k gives an estimation of the

Fig. 7.1 Schematic diagram of reciprocity technique

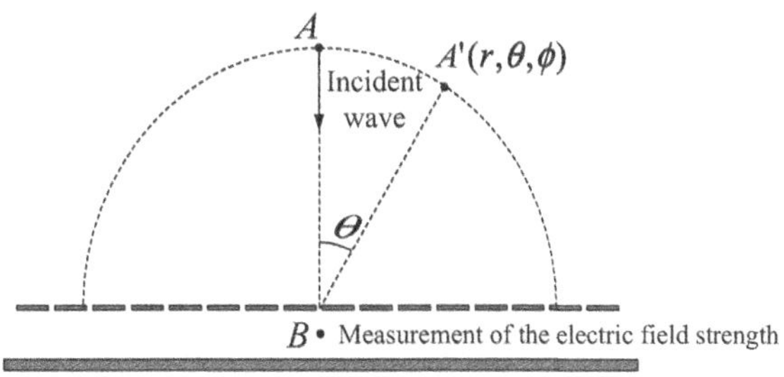

angle of the main radiated beam, while the imaginary part of k gives information about the beamwidth of the radiated beam for the corresponding leaky mode. In [24], a technique based on the transverse equivalent network (TEN) method has been applied to study periodic patches printed on a grounded slab via a dispersion analysis, which can give great insight into the operation of FP leaky-wave antenna. The procedure is employed to extract the dispersion equation of the surface and leaky-wave modes from the plane-wave reflection data. The homogenization process leads to an equivalent admittance for the periodic surface, whose poles and zeros are derived from the reflection coefficients in the fast-wave region.

More recently, the finite-difference time-domain (FDTD) method has been used to model the unit cell and rigorously derive the dispersion properties for 2-D FP antennas [25]. Moreover, LW analysis of FP antennas based on a PRS and a HIS has been carried out using FDTD [26]. The LWAs under investigation are studied using a periodic FDTD methodology, initially proposed in [27, 28]. According to this methodology, any periodic leaky-wave structure can be rigorously modeled by simulating its unit cell within a FDTD computational domain, terminated with periodic boundary conditions (PBCs). This type of modeling allows for the computationally efficient calculation of the electromagnetically converged complex propagation constants of the leaky-modes supported by the infinitely long structures under consideration.

7.2 Basic Types of FPA Antennas

7.2.1 Metallic and Metallo-Dielectric FP Antennas

As mentioned above, the first and most common implementation of an FP antenna is the one introduced in [1] which has been based on earlier work [2]. It consists of a doubly periodic array of metallic elements, printed on a dielectric substrate and placed at approximately half-wavelength distance over a ground plane (Fig. 7.2). The antenna excitation is a waveguide-fed slot in the ground plane.

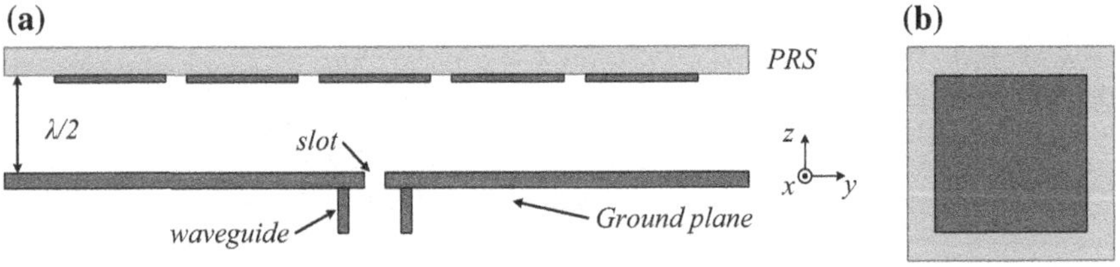

Fig. 7.2 Typical implementation of a Fabry-Perot antenna with metallo-dielectric PRS over a ground plane and a waveguide-fed slot, **b** front view of a square element unit cell

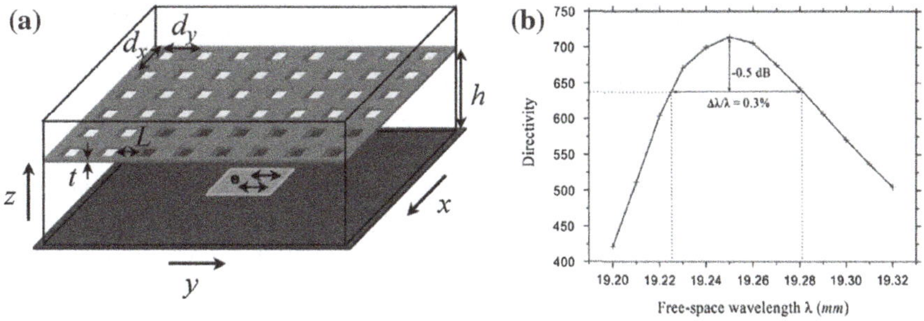

Fig. 7.3 a Schematic representation of the antenna based on a Fabry-Perot cavity, L = 3.06 mm, t = 0.14 mm, $d_x = d_y$ = 5.8 mm, and h = 9 mm, b directivity versus wavelength (taken from [13])

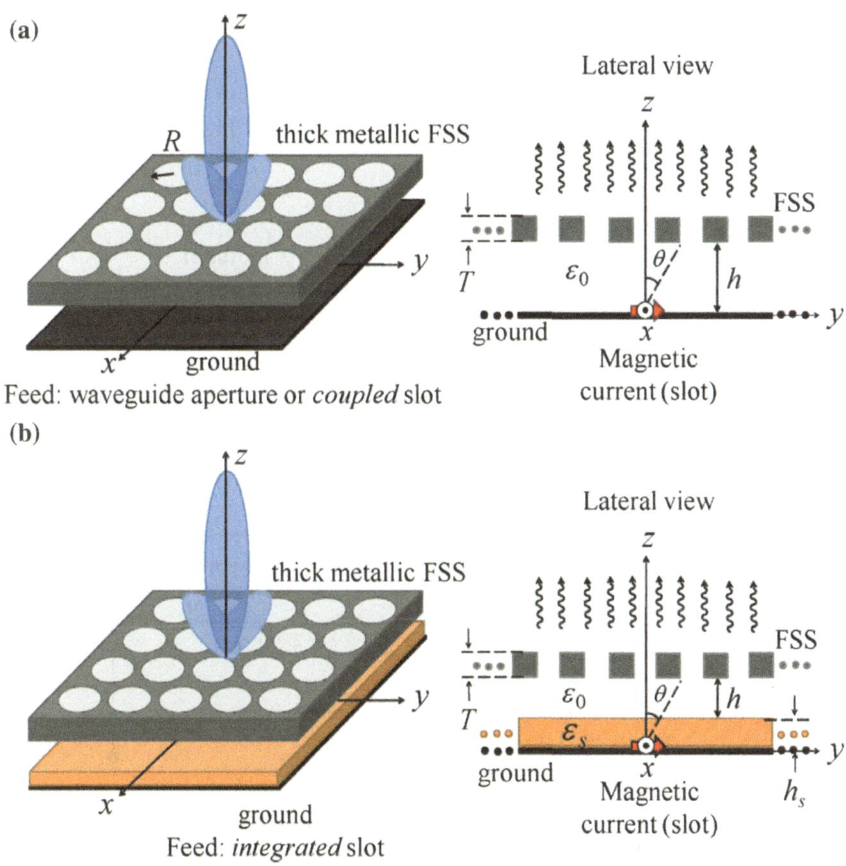

Fig. 7.4 a 3D (*left*) and lateral (*right*) view of a single-layer FPC antenna formed by a thick metallic FSS with periodic circular holes fed by a magnetic current on its ground. b Similar to (a) but for the dual-layer cavity antenna (taken from [14])

This idea has been extensively investigated, employing different types of primary feeds and geometries for the periodic array. Moreover, a lot of work has been carried out at mm-wave frequencies, in particular at 44 GHz [8], 60 GHz [9, 10] and 94 GHz [11].

At the same time, all-metallic PRSs have been introduced [13, 14] as the one shown in Fig. 7.3. This particular design achieves high directivity performance but with narrow bandwidth operating at *Ku*-band (Fig. 7.3b). The PRS is formed by an array of square apertures on a 1 cm-thick metallic sheet and the antenna is fed by a square patch (Fig. 7.3a). All the dimensions of the structure are shown in the figure. Another implementation of FP antennas employing all-metallic PRS is presented in Fig. 7.4 [14]. In this case, the antenna is designed to operate around 60 GHz and it consists of a PRS with circular apertures.

7.2.2 Dielectric FP Antennas

The use of single [16] or multiple dielectric layers [3, 4, 17–19] as superstrates for FP antennas has been investigated since the mid-1980s, Fig. 7.5 [3]. Since the

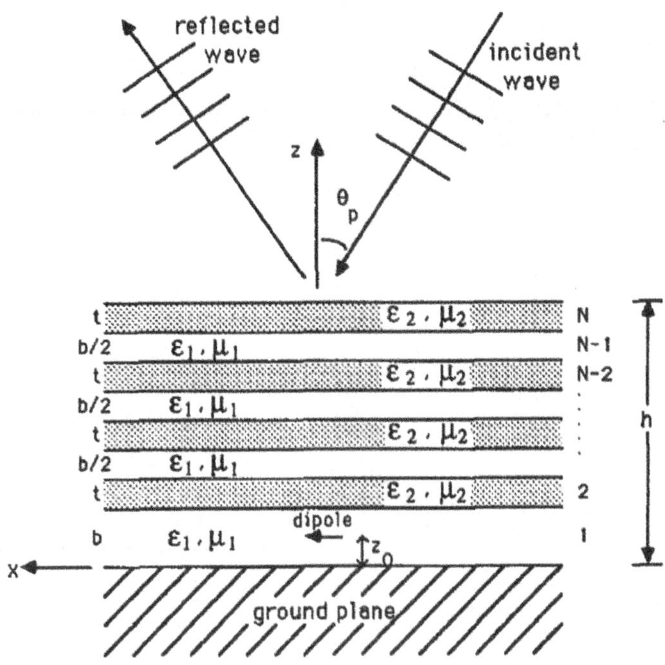

Fig. 7.5 FP antenna structure, consisting of dielectric layers of alternating thickness and material constants stacked above a ground plane. The structure is fed by a simple dipole source (taken from [3])

operation of FP antennas is based on the reflection characteristics of the PRS, high dielectric constant values were chosen in order to achieve high reflectivity values which in turn leads to high antenna directivity and gain. Moreover, a large number of layers are necessary in order to obtain comparable directivity values with the conventional metallo-dielectric PRS antenna. In the design shown in the figure, dielectric layers of alternating thickness and permittivity values are stacked on top of each other. The larger the number of layers used, the greater the achieved directivity.

7.3 Mm-Wave FPA Implementations

In this section, a review of recent Fabry-Perot antenna implementations at mm-wave frequencies is presented. Mm-wave antenna designs and implementation are typically limited by the fabrication process employed and the associated tolerances. Some of the highest mm-wave frequency implementations of FPAs are reported in this section. Furthermore, the reduction of the profile of FPAs has been studied extensively, by replacing the ground plane with an Artificial Magnetic Conductor (AMC) surface [29, 30], which leads to quarter-wavelength or even sub-wavelength profile highly directive antennas [22, 31–38]. Moreover, numerous implementations with multiple PRS layers have been presented for bandwidth enhancement of FP antennas [23, 37–43]. Finally, different feeding techniques [8, 10, 44–46] have been studied resulting in alternative configurations providing enhanced radiation performance.

7.3.1 Higher mm-Wave and Submm-Wave FPAs

Recently, a first report of a submm-wave Fabry-Perot cavity antenna has been presented employing a fully metallized structure [15]. In this case, the choice of material was made to avoid dielectric losses which may be detrimental at these frequencies. The complete antenna structure is shown in Fig. 7.6a and b and consists of six layers. The first three layers form a waveguide H-bend used in order to facilitate the interconnection of the antenna with the waveguide flange, while the remaining three layers form the FPA. The antenna has been fabricated using a micromachining process based on SU-8 photoresist that has been developed at the University of Birmingham. The process offers the advantages of high-dimensional accuracy, high achievable structure aspect ratio, and a capability of large-scale inexpensive production. Figure 7.6c shows the directivity/gain performance along with the S_{11} response. It can be observed that high directivity of 23 dBi has been achieved at 284 GHz with the proposed design.

Another implementation of FP antenna employing dielectric superstrate is obtained by using woodpile Electromagnetic Band-Gap (EBG) structures.

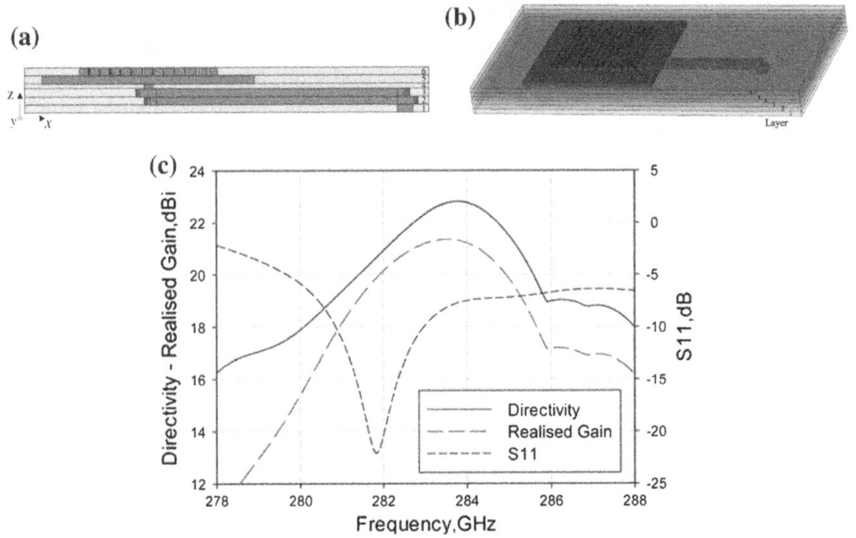

Fig. 7.6 **a** Cross section and **b** perspective of the structure (the hollow space is illustrated in *blue color* while the surrounding conductors were made transparent to offer a clearer view). **c** Simulated directivity, realized gain, and return loss of the proposed antenna (taken from [15])

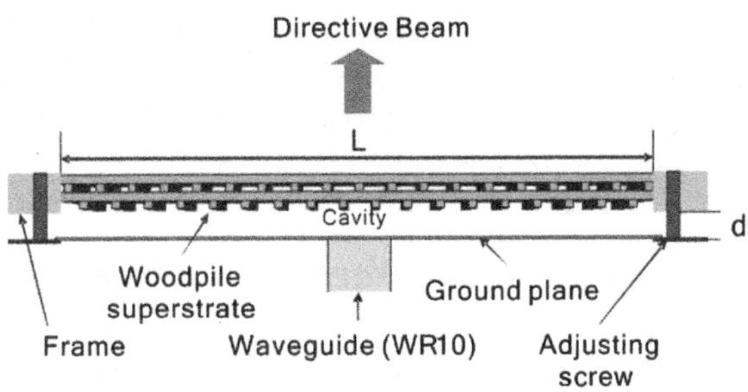

Fig. 7.7 Geometry of woodpile superstrate antenna operating around 90 GHz (taken from [18])

A representative example of this category is the one presented in [18] (Fig. 7.7). In the particular design, 3D EBG structures are formed by using ceramic material of high dielectric constant. The band-gap of the EBG is designed to be around the desired operating frequency, which in this case is ~90 GHz. The woodpile structure is placed at a distance d = λ/2 over a ground plane as illustrated in Fig. 7.6. The response of the antenna depends on the geometrical parameters of the woodpile structure and can be modified by varying the filament width (w) and spacing (a). The directivity performance of the antenna for different w and fixed a is shown in Fig. 7.8a while for fixed w and different a is shown in Fig. 7.8b. From the figure, it can be observed that the directivity-bandwidth performance is highly dependent on these two geometrical parameters.

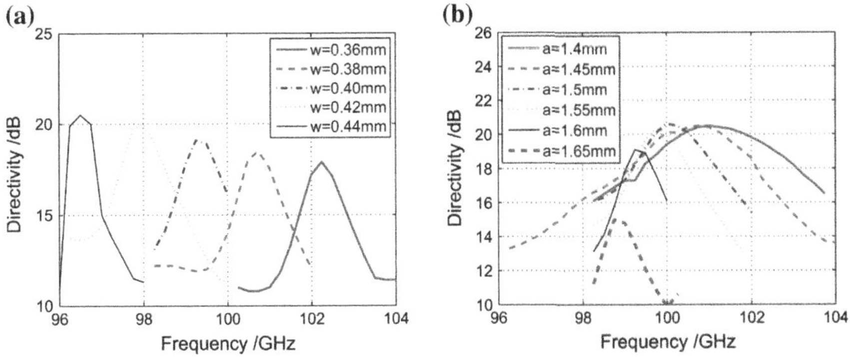

Fig. 7.8 Directivity versus frequency for: **a** different filament width *w* with a = 1.6 mm and **b** different filament spacing *a* with w = 0.4 mm (taken from [18])

7.3.2 Reduced Profile FP Antennas Using HIS

High Impedance Surfaces (HIS) are artificial metamaterial structures introduced by Sievenpiper [29, 30] that exhibit a reflection phase of zero at a specific frequency acting as Artificial Magnetic Conductors (AMC). They typically employ 2D periodic arrays of elements on a grounded dielectric substrate. This property has been exploited to design FP antennas with reduced profile.

More specifically, various implementations have been reported, where the ground plane of the FP antenna has been replaced by an HIS structure. The concept is based on altering the resonant condition of the cavity due to the zero reflection phase of the HIS ground. This has been initially proposed in [31, 32], introducing a high-gain FP antenna with quarter-wavelength profile. The idea has then been extensively investigated [22, 33–38] to achieve even further profile reduction. For example, in [36] a $\lambda/60$ profile FP antenna is proposed based on an HIS ground and a composite surface as a superstrate, acting as both a PRS and an AMC (Fig. 7.9). The composite PRS-AMC surface consists from a square patch periodic array and an inductive grid printed on either sides of a dielectric substrate. The choice of the dimensions of the HIS ground and the PRS-AMC surface provides the suitable reflection phases that result in the desired ultrathin cavity. The antenna is designed to operate at around 10 GHz achieving a directivity of 18.9 dB. However, this design has the disadvantage of a narrowband performance due to the rapid reflection phase variation of both surfaces forming the structure.

Moreover, HIS structures have also been employed for producing circularly polarized (CP) Fabry-Perot cavity antennas as the one presented in [34]. The antenna incorporates a double-sided PRS placed above an anisotropic HIS ground plane (Fig. 7.10) and the cavity is excited by a linearly polarized source rotated by 45°. The CP Fabry-Perot antenna operates at 15 GHz with a 21.1 dB gain. The circular polarization is obtained by controlling the reflection characteristics of the HIS and the PRS enforcing the resonance condition for both polarizations along

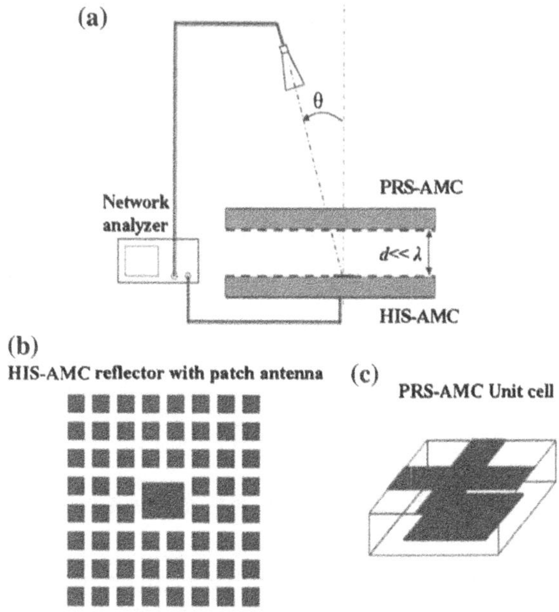

Fig. 7.9 **a** Schematic diagram of the antenna and measurement setup, **b** *top view* of the HIS ground plane with the patch antenna employed to feed the structure and **c** perspective view of the PRS-AMC unit cell (taken from [36])

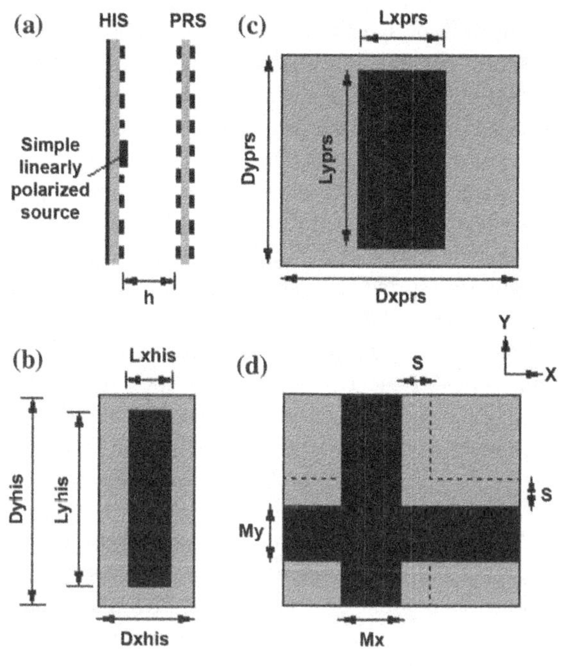

Fig. 7.10 **a** Schematic diagram of the antenna, **b** *top view* of the HIS array unit cell, **c** *top* and **d** *bottom view* of the PRS unit cell (taken from [34])

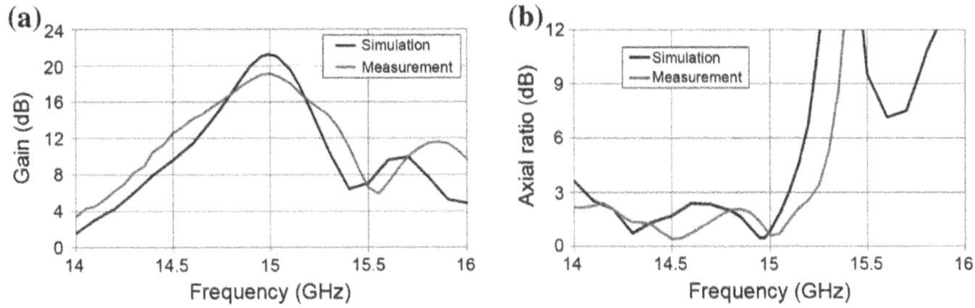

Fig. 7.11 **a** Gain (simulated and measured) of finite antenna with a patch antenna source, **b** Axial ratio (simulated and measured) of finite antenna with a patch antenna source (taken from [34])

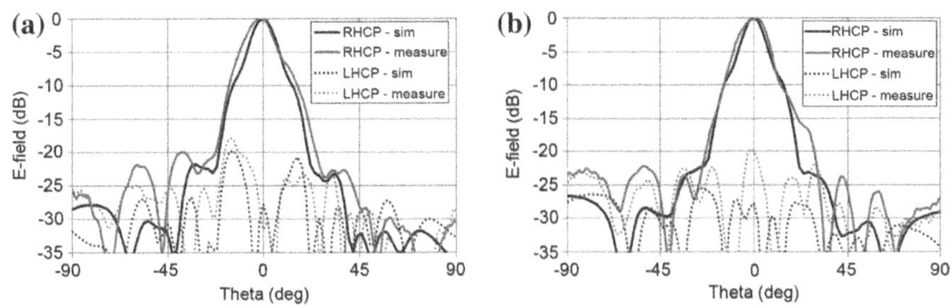

Fig. 7.12 Simulated and measured radiation patterns at 15 GHz for **a** $\varphi = 0°$, **b** $\varphi = 90°$ (taken from [34])

with a 90° phase difference between them. The simulated and measured gain and axial ration of the antenna is shown in Fig. 7.11a and b respectively. The axial ratio at 15 GHz is 0.6 dB with the measured 3-dB axial ratio bandwidth being 9%. In Fig. 7.12, the radiation patterns for the two principal planes are presented. It can be observed that highly directive patterns are obtained for both the planes, while the cross polarization level is less than −20 dB in both planes.

7.3.3 Broadband Multi-layer FP Antennas

The implementations presented thus far, demonstrate some of the advantages of FP antennas, such as high-gain performance, simple design and feeding techniques. Moreover, low-profile FP antennas have been obtained with the incorporation and proper design of HIS structures. However, all these configurations suffer from an inherit narrowband performance. To address this issue, a technique for bandwidth improvement was introduced [39] and further investigated, based on the coupling between two [23] or more [40] PRS layers of dissimilar array dimensions. The concept exploits the different reflectivity values of the PRS layers to obtain a

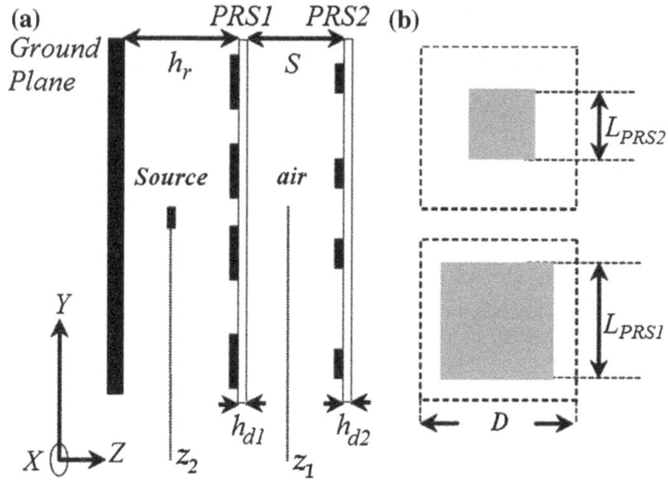

Fig. 7.13 Cross-section of **a** double-layer free-standing dipole PRS leaky-wave antenna and **b** unit cell dimensions of the double-layer dipole PRS (taken from [23])

positive reflection phase gradient with frequency that satisfies the resonance condition of the antenna cavity for a wide frequency range [1].

In [23], the proposed technique has been applied, leading to a double-layer PRS design with half-wavelength separation between the PRSs and the ground plane (Fig. 7.13), achieving a significant bandwidth enhancement compared to a single-layer FP PRS antenna [1]. The PRS layers are formed by 2D periodic arrays of square patches with the same periodicity but different size, printed on dielectric substrates (Fig. 7.13b). In Fig. 7.14a, the simulated and measured gain and S11 of the proposed antenna is depicted. It can be observed that high gain is obtained between 13.4 and 14.1 GHz, with a maximum value of 17.4 dBi and a 3 dB bandwidth of 6%. Furthermore, the simulated and measured radiation patterns at the central frequency for H- and E-plane are presented in Fig. 7.14b, showing good agreement and validating the concept.

Another approach of the same technique has been adopted in [42], where a positive PRS reflection phase gradient is achieved employing dipoles printed on both sides of a dielectric substrate. The unit cell and the complete antenna structure are illustrated in Fig. 7.15a and b respectively. Three designs were presented with different dipole dimensions, leading to different performances. Depending on the obtained reflection phase and magnitude of each double-sided structure, a dual-band antenna (Fig. 7.16a), and two broadband antennas with different bandwidths and maximum gain (Fig. 7.16b and c) were obtained.

More recently, a new concept for designing broadband, yet sub-wavelength profile Fabry-Perot-type antennas has been introduced [37]. The idea is based on combining the bandwidth enhancement technique (using multiple layers) and the profile reduction technique (using HIS structures as described in the previous section).

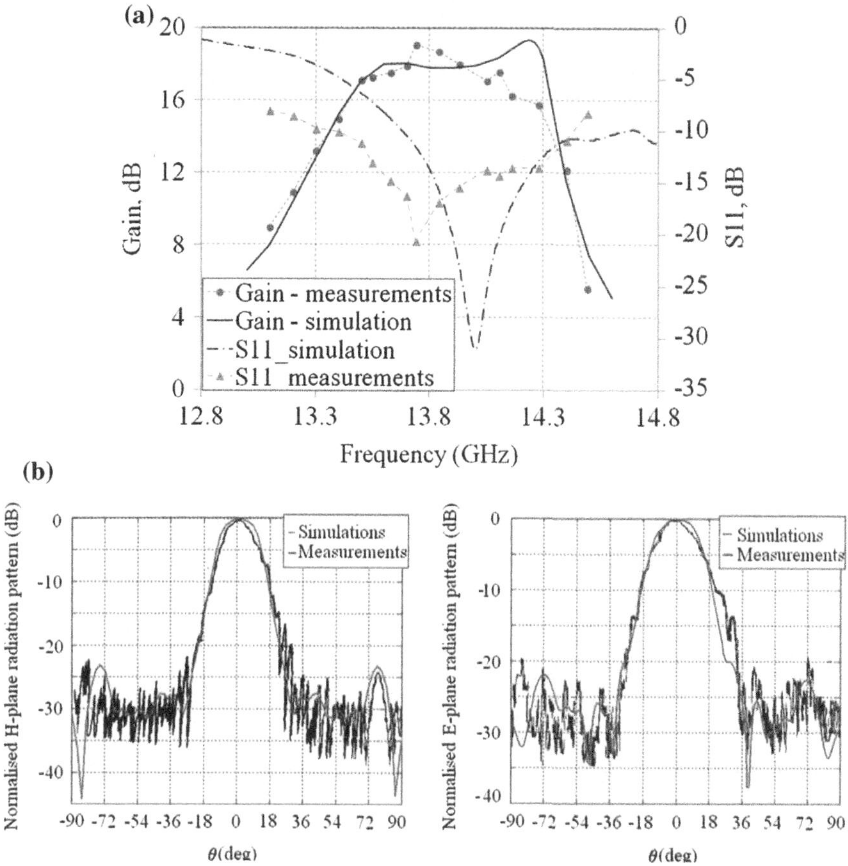

Fig. 7.14 Measured and simulated **a** gain and S11 response and **b** H-plane (*left*) and E-plane (*right*) radiation patterns at 13.9 GHz (taken from [23])

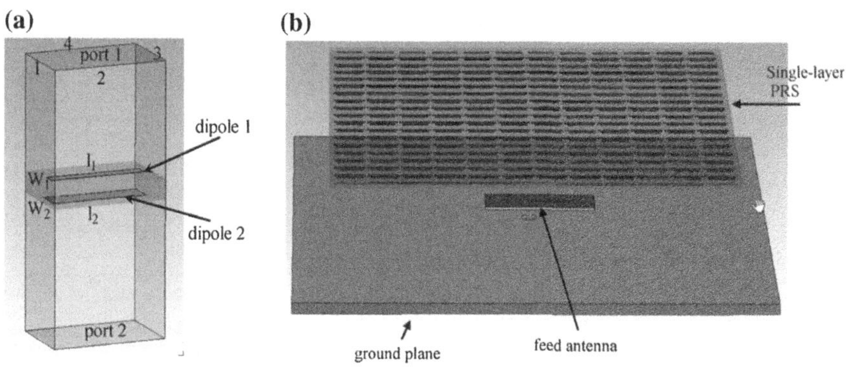

Fig. 7.15 Perspective view of **a** the unit cell of the proposed structure and **b** the complete antenna with the PRS, the ground plane and the primary source taken from [42]

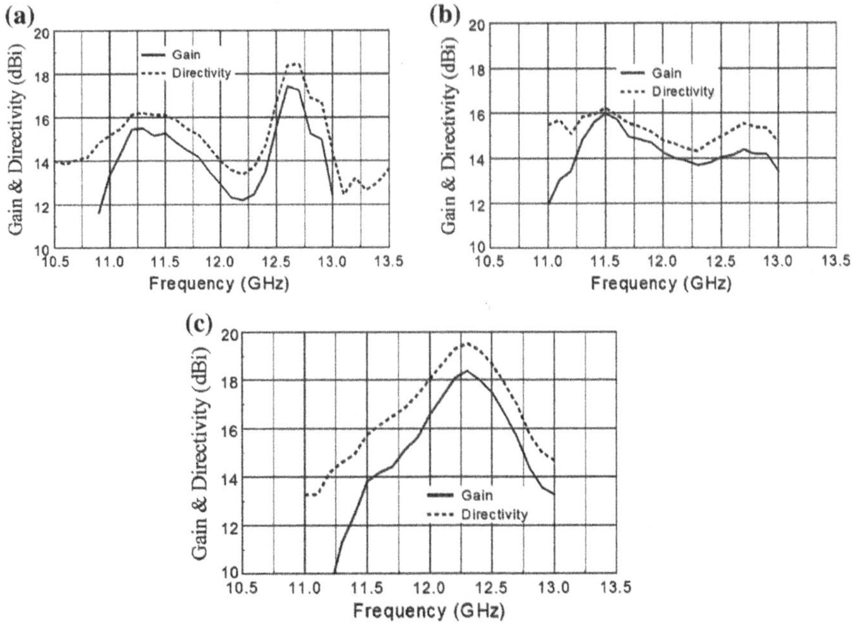

Fig. 7.16 Measured gain and directivity for the three antennas (taken from [42])

More particularly, two composite PRS-AMC layers are employed creating two sub-wavelength profile Fabry-Perot cavities that significantly enhance the bandwidth performance of the resulting low-profile antenna (Fig. 7.17). The composite layers are designed to achieve reflection phases that satisfy the resonance condition for

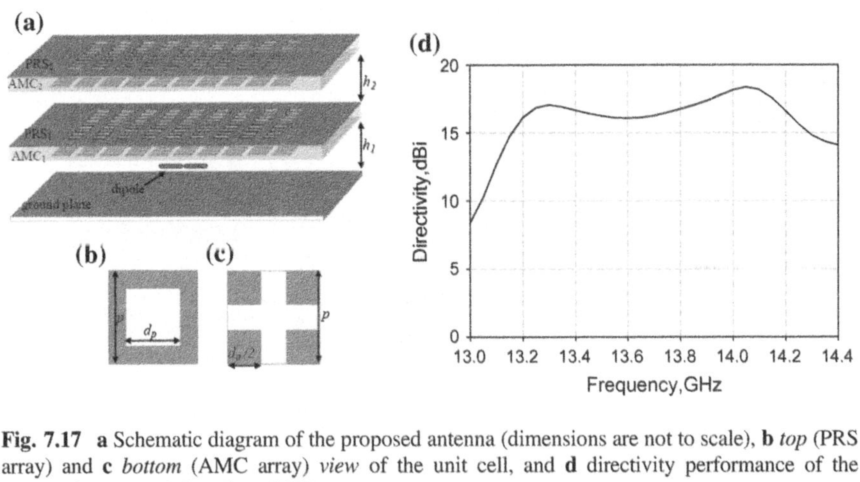

Fig. 7.17 **a** Schematic diagram of the proposed antenna (dimensions are not to scale), **b** *top* (PRS array) and **c** *bottom* (AMC array) *view* of the unit cell, and **d** directivity performance of the proposed antenna (taken from [37])

quarter-wavelength cavities. Thus, they are placed at about quarter-wavelength from a ground plane and from each other creating two quarter-wave air cavities with a total profile of about $\lambda/2$ (Fig. 7.17). The simulated directivity of the proposed antenna is illustrated in Fig. 7.17d showing a broadband performance. This concept can be extended to achieve further profile reduction keeping the increased bandwidth performance. This is obtained using multiple composite layers designed to create $\lambda/6$ air cavities [38].

7.3.4 Feeding Techniques for FP Antennas

The design of broadband high-gain FP antennas remains an important challenge. Over the past few years, several feeding techniques have been proposed to improve the radiation performance of these antennas. For example, in [45] an array of multiple sources was employed replacing the single feed in the ground plane, in order to increase the radiating aperture of the antenna. This led to an increase of the bandwidth for a fixed gain value and therefore an increase of the gain-bandwidth product.

In [46] a dual-slot feeding technique is proposed for broadband multi-layer periodic Fabry-Perot cavity antennas achieving enhanced directivity as well as broadband radiation and broadband input matching. A microstrip line is printed on a dielectric substrate with the ground plane printed on the other side of the substrate and the feeding slots etched off it Fig. 7.18a. The S_{11} response for a single slot feeding of the antenna is compared with the proposed dual-slot feeding (Fig. 7.18b). It can be seen that the introduction of the second slot yields a more broadband input matching that covers all the operational frequency range. Moreover, the electric near-field distribution for the three-layer antenna with one and two feeding slots shown in Fig. 7.18c and d demonstrates that the field distribution is significantly improved.

In [8] a metallo-dielectric FPC antenna fabricated on a quartz disk is presented operating at 44 GHz. The choice of the substrate has been made to guarantee mechanical stability which is particularly important to obtain the expected response. A planar feed is introduced which is depicted in Fig. 7.19a. The feeding structure is made of three parts; the Back-Gap (BG) structure for ease of probing, the 50 O CPW line and a magnetic dipole (slot). The simulated and measured reflection coefficient of a prototype is shown in Fig. 7.19b. A similar feeding technique has been used in [10] for operation at 60 GHz.

In [6], a plane-parallel Fabry-Perot resonator comprising two uniform and nonuniform inductive metal meshes excited by a horn antenna is presented (Fig. 7.20a). The structure performs beam focusing employing the nonuniform mirror, whose geometry is chosen so that it behaves as a spherical equiphase surface with a desired radius of curvature. The directivity of the antenna is controlled by the curvature of the synthesized wavefront. This concept is successfully validated in the

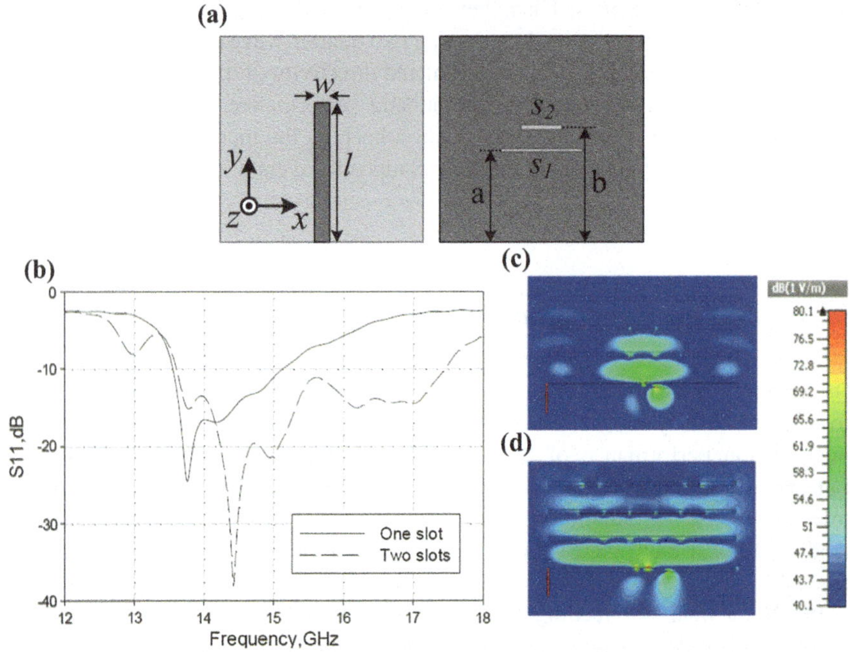

Fig. 7.18 **a** *Back* and *front view* of ground plane with the *microstrip line* and *slots*. Cross section of the electric field distribution (YZ plane) for the three layer Fabry-Perot antenna at 14 GHz **b** one slot **c** two slots **d** S_{11} of one and two microstrip fed slots of the proposed three-layer Fabry-Perot antenna taken from [46]

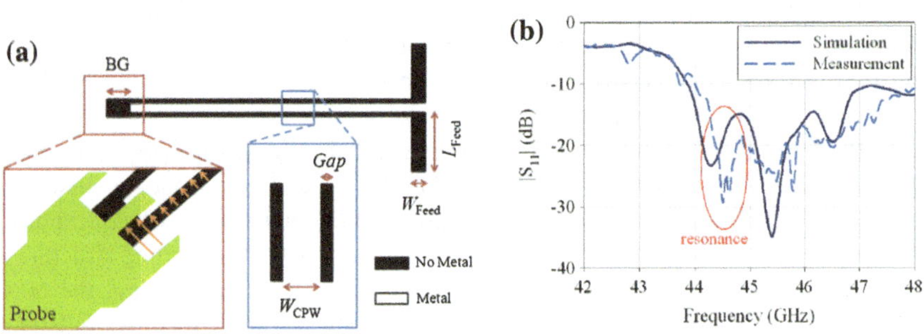

Fig. 7.19 **a** Feeding structure of the antenna, **b** the reflection coefficient of the antenna, a comparison between the simulation and measurement results (taken from [8])

60 GHz band. The radiation patterns of the proposed antenna are Gaussian with very low sidelobes (Fig. 7.20b).

Recently, a grid-based multiple-layer FPA has been presented where the layers are separated by air-filled cavities (Fig. 7.21a) [12]. The FP antenna is fed by a microstrip patch antenna and operates at 60 GHz. The simulations of these

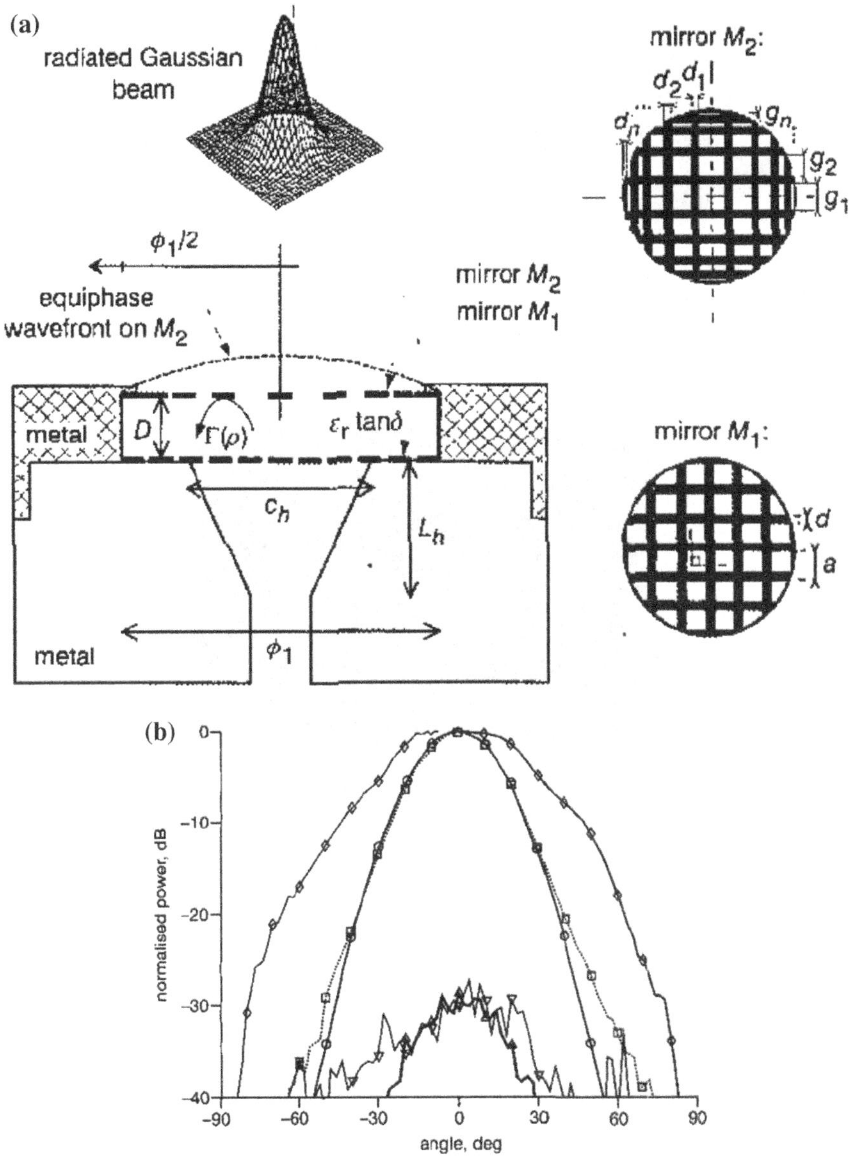

Fig. 7.20 **a** Geometry of dielectric-loaded plane-parallel Fabry-Perot cavity excited by horn antenna, **b** radiation patterns in H-plane at 57.38 GHz (taken from [7])

superstrate designs showed antenna gains as high as 19 dBi (Fig. 7.21b). These structures were fabricated on sheets of a flexible, low-loss foam, copper laminate material, and their performances were measured. Due to the nonuniformity in the thickness of the foam layers, as well as other material and fabrication issues, the

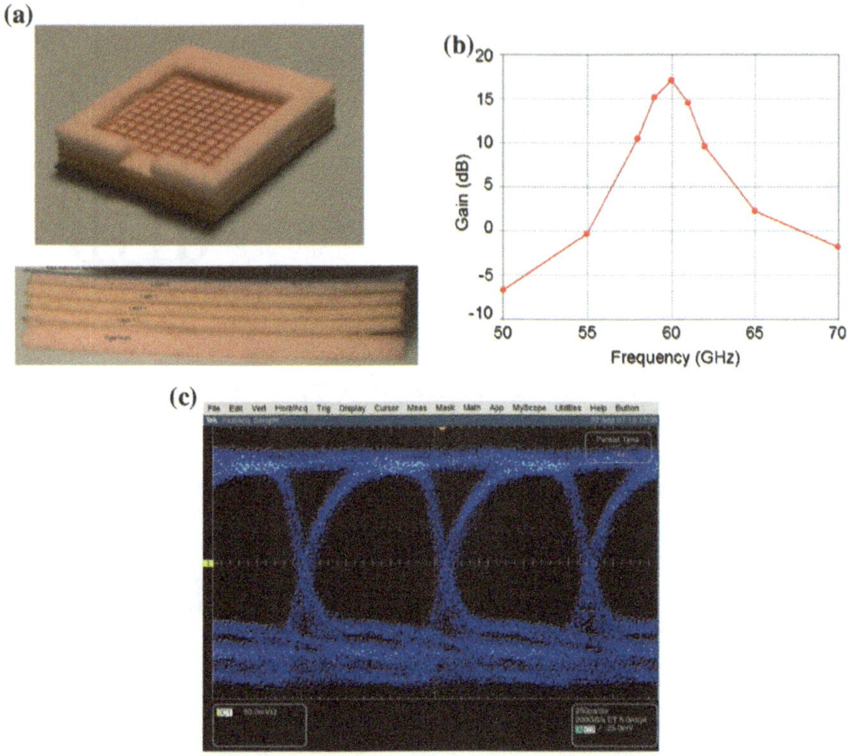

Fig. 7.21 **a** Four-layer grid-based superstrate structure with microstrip patch antenna feeding, **b** simulation results for an 11 × 11 grid structure on two foam layers, optimized in CST with lossless materials, **c** measured eye pattern of the proposed antenna excited by a 1.25-Gb/s signal. (taken from [12])

highest gain achieved, was just over 11 dBi which is significantly lower than the predicted one. This demonstrates the importance of high accuracy fabrication processes for the implementation of FP cavity-type antennas at mm-waves.

Importantly, the fabricated antenna prototypes were connected to a practical 60 GHz wireless system in order to measure the bit error rate (BER) of the system in the presence of the proposed antenna [12]. The maximum data rate of the system was measured and demonstrated by producing the "eye" diagrams (time-domain response of the system) (Fig. 7.21c). The results showed that the proposed antennas produce a significant increase in the system gain for gigabit per second signals in the 60-GHz band, supporting a maximum error free data rate of 3 Gb/s in this specific example.

7.4 Discussion and Future Trends

FPAs have attracted increasing research interest over the past decade and this trend seems to continue. Although the research on FPAs started at lower microwave frequencies, where there are now already applications on real practical systems [47], there is now significant effort toward developing FPA-type antennas at higher mm-wave and THz frequencies. These efforts are taking advantage of emerging microfabrication technologies as well as new materials that are becoming available. Some of these efforts have been reported in this chapter, ranging from frequencies of 10 GHz up to approximately 300 GHz.

References

1. A.P. Feresidis, J.C. Vardaxoglou, High-gain planar antenna using optimized partially reflective surfaces. IEE Proc. Microw. Antennas Propag. **148**(6) (2001)
2. G.V. Trentini, Partially reflecting sheet array. IRE Trans. Antennas Propag. **AP-4**, 666–671 (1956)
3. D.R. Jackson, A.A. Oliner, A. Ip, Leaky-wave propagation and radiation for a narrow-beam multiple-layer dielectric structure. IEEE Trans. Antennas Propag. **41**, 344–348 (1993)
4. D.R. Jackson, N.G. Alexopoulos, Gain enhancement methods for printed circuit antennas. IEEE Trans. Antennas Propag. **33**, 976–987 (1985)
5. J.R James, S.J.A Kinany, P.D. Peel, G. Andrasic, Leaky-wave multiple dichroic beamformers. Electron. Lett. **25**, 1209–1211 (1989)
6. R. Sauleau, P. Coquet, T. Matsui, J.-P. Daniel, A new concept of focusing antennas using plane-parallel Fabry-Perot cavities with nonuniform mirrors. IEEE Trans. Antennas Propag. **51**(11), 3171–3175 (2003)
7. R. Sauleau, Ph Coquet, D. Thouroude, J.-P. Daniel, Beam focusing using 60-GHz Fabry-Perot resonators with uniform and non-uniform metal grids. Electron. Letters **39**(4), 341–342 (2003)
8. S.A. Hosseini, F. Capolino, F.D. Flaviis, A 44 GHz single-feed Fabry-Perot cavity antenna designed and fabricated on quartz, in *IEEE Antennas and Propagation Society (AP-S) International Symposium*, pp. 1285–1288, Spokane, Washington, USA, July 3–8, 2011
9. S.A. Hosseini, F. Capolino, F. De Flaviis, *Design of a single feed all-metal 63 GHz Fabry-Perot cavity antenna using a TL and a wideband circuit model* (IEEE Int. Symp. Antennas Propag., Charleston, SC, 2009)
10. S.A. Hosseini, F. De Flaviis, F. Capolino, A highly-efficient single-feed planar Fabry-Pérot cavity antenna for 60 GHz technology, in *IEEE International Symposium on Antennas Propagation*, Chicago, IL, USA, July 2012
11. K. Konstantinidis, A.P. Feresidis, P.S. Hall, M.J. Lancaster, Design of Fabry-Perot cavity antenna at 94 GHz, in *Antennas and Propagation Conference (LAPC)*, 2012 Loughborough, pp. 1–4, 12–13 November 2012
12. S.J. Franson, R.W. Ziolkowski, Gigabit per second data transfer in high-gain metamaterial structures at 60 GHz. IEEE Trans. Antennas Propag. **57**, 2913–2925 (2009)
13. N. Guerin, S. Enoch, G. Tayeb, P. Sabouroux, P. Vincent, H. Legay, A metallic Fabry-Perot directive antenna. IEEE Trans. Antennas Propag. **54**(1), 220–224 (2006)
14. R. Gardelli, M. Albani, F. Capolino, Gain enhancement of a V-band antenna using a Fabry-Pérot Cavity with a self-sustained all-metal cap with FSS. IEEE Trans. Antennas Propag. **63**(3) (2015)

15. K. Konstantinidis, A.P. Feresidis, M.J. Lancaster, P.S. Hall, Micromachined terahertz Fabry-Perot cavity highly directive antennas. IET Microw. Antennas Propag. **9**(13), 1436–1443 (2015)
16. H. Ostner, E. Schmidhammer, J. Detlefsen, D.R. Jackson, Radiation from dielectric leaky-wave antennas with circular and rectangular apertures. Electromagnetics **17**, 505–535 (1997)
17. A.R. Weily, L. Horvath, K.P. Esselle, B.C. Sanders, T.S. Bird, A planar resonator antenna based on a woodpile EBG material. IEEE Trans. Antennas Propag. **53**(1), 216–223 (2005)
18. Y. Lee, X. Lu, Y. Hao, S. Yang, J.R.G. Evans, C.G. Parini, Low-profile directive millimetre wave antennas using free-formed three-dimensional (3-D) electromagnetic bandgap structures. IEEE Trans. Antennas Propag. **57**(10), 2893–2903 (2009)
19. H.Y. Yang, N.G. Alexopoulos, Gain enhancement methods for printed circuit antennas through multiple superstrates. IEEE Trans. Antennas Propag. **35**, 860–863 (1987)
20. T. Zhao, D.R. Jackson, J.T. Williams, H.-Y.D. Yang, A.A. Oliner, 2-D periodic leaky-wave antennas-part I: metal patch design. IEEE Trans. Antennas Propag. **53**(11), 3505–3514 (2005)
21. T. Zhao, D.R. Jackson, J.T. Williams, H.-Y.D. Yang, A.A. Oliner, 2-D periodic leaky-wave antennas-part II: slot design. IEEE Trans. Antennas Propag. **53**(11), 3515–3524 (2005)
22. C. Mateo-Segura, G. Goussetis, A.P. Feresidis, Sub-wavelength profile 2-D leaky-wave antennas with two periodic layers. IEEE Trans. Antennas Propag. **59**(2), 416–424 (2011)
23. C. Mateo-Segura, A.P. Feresidis, G. Goussetis, Bandwidth enhancement of 2-D leaky-wave antennas with double-layer periodic surfaces. IEEE Trans. Antennas Propag. **62**(2), 586–593 (2014)
24. S. Maci, M. Caiazzo, A. Cucini, M. Casaletti, A pole-zero matching method for EBG surfaces composed of a dipole FSS printed on a grounded dielectric slab. IEEE Trans. Antennas Propag. **53**(1), 70–81 (2005)
25. P. Kosmas, A.P. Feresidis, G. Goussetis, Periodic FDTD analysis of a 2-D leaky-wave planar antenna based on dipole frequency selective surfaces. IEEE Trans. Antennas Propag. **55**(7), 2006–2012 (2007)
26. J.R. Kelly, T. Kokkinos, A.P. Feresidis, Analysis and design of sub-wavelength resonant cavity type 2-D leaky-wave antennas. IEEE Trans. Antennas Propag. **56**(9), 2817–2825 (2008)
27. T. Kokkinos, C.D. Sarris, G.V. Eleftheriades, Periodic FDTD analysis of leaky-wave structures and applications to the analysis of negative-refractive-index leaky-wave antennas. IEEE Trans. Microw. Theory Techn. **54**(4), 1619–1630 (2006)
28. T. Kokkinos, C.D. Sarris, G.V. Eleftheriades, Periodic finite-difference time-domain analysis of loaded transmission-line negative-refractive-index metamaterials. IEEE Trans. Microwave Theory Tech. **53**(4), 1488–1495 (2005)
29. D. Sievenpiper, High-impedance electromagnetic surfaces. Ph.D. dissertation, Dept. Elect. Eng., Univ. California at Los Angeles, Los Angeles, CA, 1999
30. D. Sievenpiper, L. Zhang, R.F.J. Broas, N.G. Alexopoulos, E. Yablonovitch, High impedance electromagnetic surfaces with a forbidden frequency band. IEEE Trans. Microw. Theory Tech. **47**(11), 2059–2074 (1999)
31. S. Wang, A.P. Feresidis, G. Goussetis, J.C. Vardaxoglou, Low-profile resonant cavity antenna with artificial magnetic conductor ground plane. Electron. Lett. **40**(7), 405–406 (2004)
32. A.P. Feresidis, G. Goussetis, S. Wang, J.C. Vardaxoglou, Artificial magnetic conductor surfaces and their application to low-profile high-gain planar antennas. IEEE Trans. Antennas Propag. **53**(1), 209–215 (2005)
33. J.R. Kelly, T. Kokkinos, A.P. Feresidis, Analysis and design of sub-wavelength resonant cavity type 2-d leaky-wave antennas. IEEE Trans. Antennas Propag. **56**(9), 2817–2825 (2008)
34. R. Orr, G. Goussetis, V. Fusco, Design method for circularly polarized Fabry-Perot cavity antennas. IEEE Trans. Antennas Propag. **62**(1), 19–26 (2014)
35. L. Zhou, H. Li, Y. Qin, Z. Wei, C.T. Chan, Directive emissions from subwavelength metamaterial-based cavities. Appl. Phys. Lett. **86**, 1011011 (2005)

36. A. Ourir, A. de Lustrac, and J.-M. Lourtioz, All-metamaterial-based subwavelength cavities for ultrathin directive antennas. Appl. Phys. Lett. **88**(8), 84103-1–84103-3 (2006)
37. K. Konstantinidis, A. Feresidis, P. Hall, Dual subwavelength Fabry-Perot cavities for broadband highly directive antennas. IEEE Antennas Wirel. Propag. Lett. **13**, 1184–1186 (2014)
38. K. Konstantinidis, A.P. Feresidis, P.S. Hall, Broadband sub-wavelength profile high-gain antennas based on multi-layer metasurfaces. IEEE Trans. Antennas Propag. **63**(1), 423–427 (2015)
39. A.P. Feresidis, J.C. Vardaxoglou, A broadband high-gain resonant cavity antenna with single feed, in *Proceedings of EuCAP*, Nice, France, 2006
40. K. Konstantinidis, A.P. Feresidis, P.S. Hall, multilayer partially reflective surfaces for broadband Fabry-Perot cavity antennas. IEEE Trans. Antennas Propag. **62**(7), 3474–3481 (2014)
41. A. Zeb, Y. Ge, K.P. Esselle, Z. Sun, M.E. Tobar, A simple dual-band electromagnetic band gap resonator antenna based on inverted reflection phase gradient. IEEE Trans. Antennas Propag. **60**(10), 4522–4529 (2012)
42. Y. Ge, K.P. Esselle, T.S. Bird, The use of simple thin partially reflective surfaces with positive reflection phase gradients to design wideband, low-profile EBG resonator antennas. IEEE Trans. Antennas Propag. **60**(2), 743–750 (2012)
43. M. Al-Tarifi, D. Anagnostou, A. Amert, K. Whites, Bandwidth enhancement of the resonant cavity antenna by using two dielectric superstrates. IEEE Trans. Antennas Propag. **61**, 1898–1908 (2013)
44. R. Sauleau, Ph Coquet, T. Matsui, Low-profile directive quasi-planar antennas based on millimeter wave Fabry-Perot cavities. IEE Proc. Microw. Antennas Propag. **150**(4), 274–278 (2003)
45. R. Gardelli, M. Albani, F. Capolino, Array thinning by using antennas in a Fabry–Perot cavity for gain enhancement. IEEE Trans. Antennas Propag. **54**(7) (2006)
46. K. Konstantinidis, A.P. Feresidis, P.S. Hall, Dual-slot feeding technique for broadband Fabry–Perot cavity antennas. IET Microw. Antennas Propag. **9**(1) (2015)
47. J.C. Iriarte, I. Ederra, R. Gonzalo, Y. Brand, A. Fourmault, Y. Demers, L. Salgetti-Drioli, P. de Maagt, EBG superstrate array configuration for the WAAS space segment. IEEE Trans. Antennas Propag. **57**(1), 81–93 (2009)

Chapter 8
Near-Field Focusing by Non-diffracting Bessel Beams

Mauro Ettorre, Santi Concetto Pavone, Massimiliano Casaletti, Matteo Albani, Agnese Mazzinghi and Angelo Freni

Abstract This chapter illustrates the capabilities of non-diffractive Bessel beams for near-field focusing. After a brief introduction of the non-diffractive phenomenon and its origin, the generation of non-diffractive Bessel beams by inward cylindrical traveling waves is analyzed in detail. A ray interpretation is proposed for such beams for infinite and finite radiating apertures. Their main radiation capabilities and limitations in terms of focusing, bandwidth, and operating range are discussed. In particular, it is shown that inward cylindrical traveling wave aperture field distributions can generate non-diffractive Bessel beams over a large bandwidth. As a practical implementation of the discussed theory, a class of launchers based on radial waveguides loaded by metallic gratings or slots is proposed, for which an efficient design and optimization technique is described. In addition, a different and unconventional leaky-wave approach is also adopted for the design of radial waveguides loaded by metasurfaces. The chapter ends by outlining the future research activities and possible applications of non-diffractive beams.

8.1 Introduction

Diffraction is an always-present phenomenon limiting the focusing capabilities of any radiating wave. The quest for possible solutions overcoming diffraction spreading has been a topic of interest since a long time, with preliminary works

---

M. Ettorre (✉)
Institut d'Electronique et de Télécommunications de Rennes (IETR),
UMR CNRS 6164, Université de Rennes 1, Rennes, France
e-mail: Mauro.Ettorre@univ-rennes1.fr

S.C. Pavone · M. Albani
University of Siena, Siena, Italy

M. Casaletti
Université Pierre and Marie Curie (UPMC), Paris, France

A. Mazzinghi · A. Freni
University of Florence, Florence, Italy

A. Boriskin and R. Sauleau (eds.), *Aperture Antennas for Millimeter and Sub-Millimeter Wave Applications*, Signals and Communication Technology,
DOI 10.1007/978-3-319-62773-1_8

dating back to the beginning of the last century [1, 2]. However, the seminal works by Durnin in 1987 gave a completely new light and interest to non-diffractive waves, especially within the optics and physics communities. Durnin demonstrated theoretically [3] and verified experimentally [4], the existence of a particular class of solutions to the scalar wave equation. These solutions, called Bessel beams, remain confined and do not undergo diffractive spreading.

They are of the form $E(\rho, z) = J_0(k_\rho \rho) e^{-jk_z z}$, where $k_\rho^2 + k_z^2 = k_0^2$. The free-space propagation constant is denoted by k_0, (ρ, z) are the observation point coordinates in the cylindrical coordinate system, and $J_0(.)$ is the zeroth-order Bessel function of the first kind. The name Bessel beam stems from the transverse distribution of the electric field, as clear from the previous expression. This expression also shows that the field profile does not change with z, and thus does not exhibit diffractive spreading. In the ideal case, such beams have a finite energy density, but are not square integrable. Another peculiar feature of such beams is that their spatial spectrum consists of a single ring (annulus). In other words, an ideal Bessel beam can be thought as a superposition of plane waves with propagation directions lying on a cone. The transverse propagation constant of the beam k_ρ controls the null-to-null beamwidth (NNBW) of the beam $(NNBW \sim 4.81/k_\rho)$. In other words, the beamwidth can be arbitrarily reduced by increasing the transverse propagation constant. For $0 \leq k_\rho \leq k_0$, the beam is propagating (along z) whereas for $k_\rho > k_0$ is evanescent [5]. It is worth mentioning that Bessel beams are not the only non-diffractive solution of the wave equation. During the years, other invariant solutions have been proposed and generalized to the polychromatic case [1, 2, 6–8]. However, this chapter will mainly consider the generation of zeroth-order Bessel beams at micro- and millimeter waves.

The generation of an ideal Bessel beam requires an infinite amount of energy and unlimited radiating apertures. In practice, Bessel beams have been generated over a finite area (aperture size) and did not undergo diffraction over a limited range within the Fresnel zone of the structure [4, 9]. This operating range is called the *non-diffractive range* of the Bessel beam launcher.

At optical frequencies, Bessel beams are typically generated by using axicons [4, 9–11] which are conical lenses able to convert an impinging Gaussian beam into a Bessel one in a bi-conical region close to the axicon. Other methods considered holograms [12], localized modes [13], Fabry–Perot cavities [14, 15], just to mention a few. In optics, Bessel beams have been used in a wide range of applications including laser machining, biosensing, optical trapping of nanoparticles, etc. [16, 17].

At radio frequencies, Bessel beams have recently received an increased attention and several experimental implementations of Bessel beam launchers have been reported from the micro to the Terahertz frequency range. These implementations include axicons [18], holograms [19], metallic circular gratings [20, 21], parabolic reflectors [22], planar arrays [23], open circular waveguides [24], metamaterials and metasurfaces [25–32], near-field plates [33], and radial line slot arrays (RLSA) [34, 35]. Among the cited works, this chapter will focus the attention on the Bessel launchers using radial waveguides loaded with slots (RLSA) or metallic gratings and by metasurfaces. This chapter is organized as follows.

Section 8.2 provides an accurate analysis of the non-diffraction radiation generated by inward cylindrical traveling waves. This analysis completes the general approach followed in literature where standing cylindrical waves are considered for the generation of the non-diffractive radiation [1, 2]. The Geometrical Optics (GO) and Space Wave (SW) contributions to the field radiated by infinite radiating apertures are given in closed-form expressions (Sect. 8.2.1). The SW contribution is expressed by using incomplete Hankel functions. It is clearly shown that the GO contribution creates a non-diffractive Bessel beam close to the axis of symmetry of the generating aperture, where the SW is negligible. Therefore, non-diffractive radiation is not only generated by standing cylindrical waves (Bessel-like distributions) thus simplifying the beam generator synthesis. In contrast, outward cylindrical waves over an infinite radiating aperture cannot provide a non-diffractive beam. The finite aperture case is then considered in Sect. 8.2.2. The radiated field is decomposed in its GO and diffractive (D) contributions. The GO term defines the region in which non-diffractive Bessel beams can be generated, whereas the nonuniform description of the edge-diffracted field gives a correction to the GO field and rigorously shows why the non-diffractive radiation is limited by the non-diffractive range. In addition, it is found that the interface between the GO and the D contributions produces field amplitude ripples over the longitudinal axis as described in Sect. 8.2.3.

Section 8.3 considers the practical generation of non-diffractive radiation. An equivalent current approach is proposed in Sect. 8.3.1 to implement the cylindrical aperture field distributions discussed in Sect. 8.2 with RLSAs and radial waveguides loaded with metallic gratings. Section 8.3.2 presents the slot layout for RLSA Bessel beam launchers and its main design parameters. An innovative and fast optimization tool is described in Sect. 8.3.3 for the synthesis of the required aperture current distribution with RLSA launchers and metallic gratings. The design procedure is used in Sect. 8.3.4 to design a circular-polarized RLSA Bessel launcher operating at 15 GHz. The same procedure is adopted for the Bessel launcher presented in Sect. 8.3.5 operating at 30 GHz and made by a radial waveguide loaded with metallic gratings.

Section 8.4 presents the generation of Bessel beams by cylindrical leaky-wave modes. The motivations behind the proposed design are introduced in Sect. 8.4.1. It is shown that a radial waveguide loaded with a homogenous impedance sheet can radiate non-diffractive Bessel beams as long as the structure support leaky-wave modes. A detailed field analysis is thus presented in Sect. 8.4.2, where the main design criteria and limitations are provided. In particular, it is shown that the non-diffractive radiation is achieved as a combination of inward and outward traveling leaky-wave modes within the radial waveguide. Section 8.4.3 introduces the dispersion analysis of the proposed launcher and give the main design equations. Section 8.4.4 presents two prototypes of the proposed launcher operating in X- and Ka-band. The prototypes are based on a radial waveguide loaded by a uniform impedance sheet. The impedance sheets are synthesized by uniform metasurfaces (please refer to Chap. 11 for more details on metasurfaces). An innovative leaky-wave launcher based on a modulated metasurface is also presented in Sect. 8.4.5.

Finally, Sect. 8.5 concludes the chapter by outlining the current applications and novel research venues of non-diffractive Bessel beams. Particular attention is devoted to Bessel beam launchers operating over a large bandwidth for the generation of non-diffractive pulses at millimeter waves.

8.2 Non-diffractive Radiation

8.2.1 Analytical Formulation for Infinite Apertures

The geometry of the problem is shown in Fig. 8.1. The origin of the reference system is at the aperture center with the z-axis normal to the aperture. For the sake of convenience, the observation point is expressed either in cylindrical (ρ, ϕ, z) or in spherical (ρ, θ, ϕ) coordinates; vectors are bold and a hat denotes a unit vector. In the following discussion, Transverse Magnetic (TM) modes with respect to the z-direction will only be considered. However, a similar procedure can be applied to Transverse Electric (TE) modes. An inward cylindrical wave is assumed for the magnetic current distribution on the radiating aperture, namely

$$\mathbf{M}(\rho, \phi, z = 0) = M(\rho, z = 0)\hat{\boldsymbol{\phi}} = H_1^{(1)}(k_{\rho a}\rho)\hat{\boldsymbol{\phi}}, \tag{8.1}$$

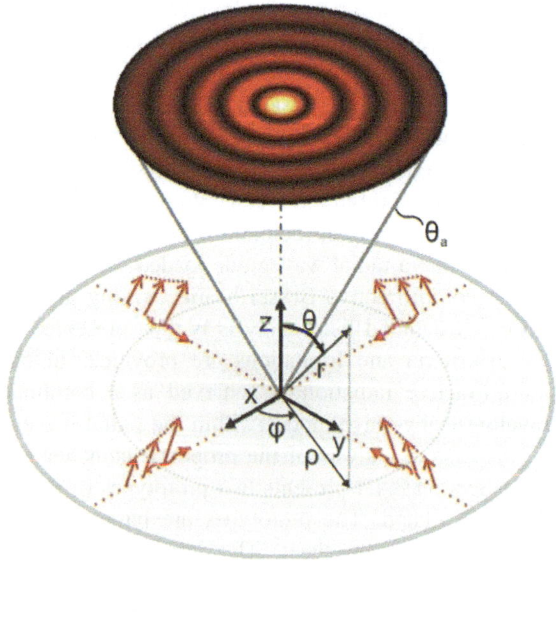

Fig. 8.1 Schematic view of the considered configuration for the generation of a Bessel beam. The aperture field distribution at z = 0 is a cylindrical inward wave with radial propagation constant $k_{\rho a}$. The non-diffractive zone is limited by the cone with angle $\theta_a = \arcsin(k_{\rho a}/k)$, where k is the free-space wavenumber

where $k_{\rho a}$ is the radial propagation constant of the cylindrical inward wave and $H_n^{(i)}(.)$ is the n-th order Hankel function of the i-th kind.

A time dependence $e^{j\omega t}, \omega = 2\pi f$ being the angular frequency, is assumed and suppressed. The electric field radiated by the aperture is given by [36]

$$\mathbf{E}(\rho, z) = \frac{1}{4\pi} \int_{\infty e^{-j\pi}}^{+\infty} \left[\frac{k_\rho}{k_z} H_0^{(2)}(k_\rho \rho)\hat{\mathbf{z}} + jH_1^{(2)}(k_\rho \rho)\hat{\boldsymbol{\rho}} \right] \widetilde{M}(k_\rho) e^{-jk_z z} k_\rho dk_\rho, \tag{8.2}$$

$$\widetilde{M}(k_\rho) = -2\pi j \int_0^{+\infty} E_t(\rho, z=0) J_1(k_\rho \rho) \rho d\rho, \tag{8.3}$$

where k_ρ, and $k_z = \sqrt{k^2 - k_\rho^2}$ are the transverse and longitudinal spectral variables, k is the wavenumber in free space, whereas $J_n(.)$ and $H_n^{(i)}(.)$ are the n-th order Bessel and Hankel functions of the i-th kind, respectively. Equation (8.3) is the Hankel transform of the magnetic current distribution over the aperture and, for the assumed inward cylindrical wave distribution, it is given by

$$\widetilde{M}(k_\rho) = \frac{-4k_\rho}{k_{\rho a}(k_\rho^2 - k_{\rho a}^2)}. \tag{8.4}$$

The radiated electric field can be exactly expressed as the superposition of the GO and SW contributions [35], namely

$$\mathbf{E}(\mathbf{r}) = \mathbf{E}^{GO}(\mathbf{r}) + \mathbf{E}^{SW}(\mathbf{r}), \tag{8.5}$$

with

$$\begin{aligned} \mathbf{E}^{GO}(\mathbf{r}) = {} & 2\left[\frac{k_{\rho a}}{jk_{za}} J_0(k_{\rho a}\rho)\hat{\mathbf{z}} + J_1(k_{\rho a}\rho)\hat{\boldsymbol{\rho}}\right] e^{-jk_{za}z} \mathrm{U}(\theta_a - \theta) \\ & + \left[\frac{k_{\rho a}}{jk_{za}} H_0^{(1)}(k_{\rho a}\rho)\hat{\mathbf{z}} + H_1^{(1)}(k_{\rho a}\rho)\hat{\boldsymbol{\rho}}\right] e^{-jk_{za}z} \mathrm{U}(\theta - \theta_a), \end{aligned} \tag{8.6}$$

$$\begin{aligned} \mathbf{E}^{SW}(\mathbf{r}) = {} & -\frac{k_{\rho a}}{jk_{za}} \left[sgn(w_0^-) H_0^{(2)}(k_{\rho a}\rho, |w_0^-|) e^{-jk_{za}z} - H_0^{(2)}(k_{\rho a}\rho, w_0^+) e^{jk_{za}z} \right] \hat{\mathbf{z}} \\ & - \left[sgn(w_0^-) H_1^{(2)}(k_{\rho a}\rho, |w_0^-|) e^{-jk_{za}z} + H_1^{(2)}(k_{\rho a}\rho, w_0^+) e^{jk_{za}z} \right] \hat{\boldsymbol{\rho}} - \frac{2}{j\pi k_{\rho a}} \left(\hat{\mathbf{z}} + cot\,\theta \hat{\boldsymbol{\rho}} \right) \frac{e^{-jkr}}{r} \\ & \sim \frac{2 \sin\theta}{j\pi k_{\rho a}(cos^2\theta - \cos^2\theta_a)} \frac{e^{-jkr}}{r} \hat{\boldsymbol{\theta}}, \end{aligned} \tag{8.7}$$

with $k_{za} = \sqrt{k^2 - k_{\rho a}^2}$ denoting the Bessel beam normal propagation constant. In (8.7), $H_n^{(i)}(\Omega, w_0)$ are the n-th order incomplete Hankel functions of i-th kind with the second argument $w_0^{\pm} = tanh^{-1}cos\,\theta \pm tanh^{-1}cos\theta_a$ [37], whereas $sgn(.)$ and $U(.)$ are the sign and Heaviside step function, respectively. The GO contribution exhibits the well-known Bessel beam shape (first term in (8.6)) within the cone $\theta < \theta_a$ where $U(\theta - \theta_a) = 1$, whereas becomes an inward Hankel beam outside such a cone where $U(\theta - \theta_a) = 0$.

By considering the Bessel function as the superposition of two Hankel functions $2J_n = H_n^{(1)} + H_n^{(2)}$, GO is interpreted as the superposition of two ray contributions, namely, an inward Hankel beam conical wave is associated with a ray arising from a point on the aperture, and an outward Hankel beam conical wave is associated with a ray arising from a point on the aperture which is diametrically opposite with respect to the observation point. Indeed, the launched inward conical wave becomes an outward conical wave beyond the caustic at the z-axis. While the former ray reaches the observation point regardless its location, the latter exists only when observing inside the cone $\theta < \theta_a$ (Fig. 8.2a). At the GO discontinuity cone $\theta = \theta_a$, the SW contribution (8.7) exhibits an opposite abrupt discontinuity which renders the total field smooth and continuous. Such a transitional behavior is described by the sign and incomplete Hankel functions when $w_0^- \to 0$. Outside the transition region near the discontinuity cone, the SW contribution exhibits a ray-optical behavior which is derived from the asymptotic expression of incomplete Hankel functions for large arguments [37] and reported in the last line of (8.7). Indeed, the SW is a transverse (i.e.,$\hat{\theta}$ polarized) spherical wave associated with a ray launched at the aperture center (origin of the reference system in Fig. 8.1), with a radiation null on the aperture symmetry axis $\theta = 0$.

As an example, Fig. 8.3 shows the amplitude of the z (Fig. 8.3a, d, and g) and ρ (Fig. 8.3b, e and h) components and the total electric field (Fig. 8.3c, f and i) radiated by a standard Bessel distribution (Fig. 8.3a, b and c), by an inward cylindrical wave (Fig. 8.3d, e and f), and by an outward cylindrical wave

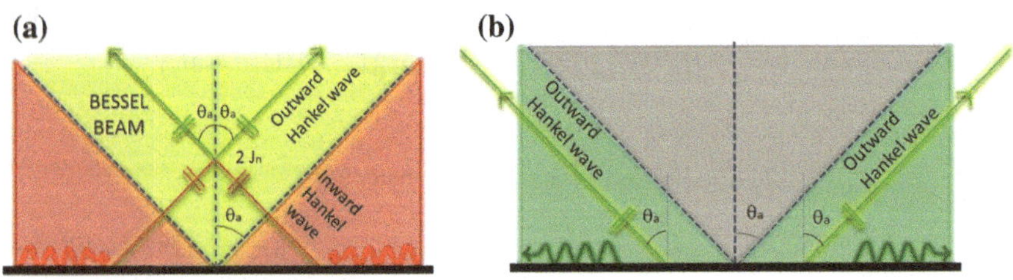

Fig. 8.2 Ray interpretation of the GO field for an infinite aperture. **a** Inward Hankel aperture distribution: the GO field comprises an inward Hankel beam ray (in *red*), present throughout the space, and an outward Hankel beam ray (in *green*) bounded inside the cone $\theta < \theta_a$; the superposition of the two rays inside the cone (*yellow area*) creates a Bessel beam. **b** Outward Hankel aperture distribution: the GO field comprises only an outward Hankel beam ray (in *green*) bounded outside the cone $\theta < \theta_a$ [38]

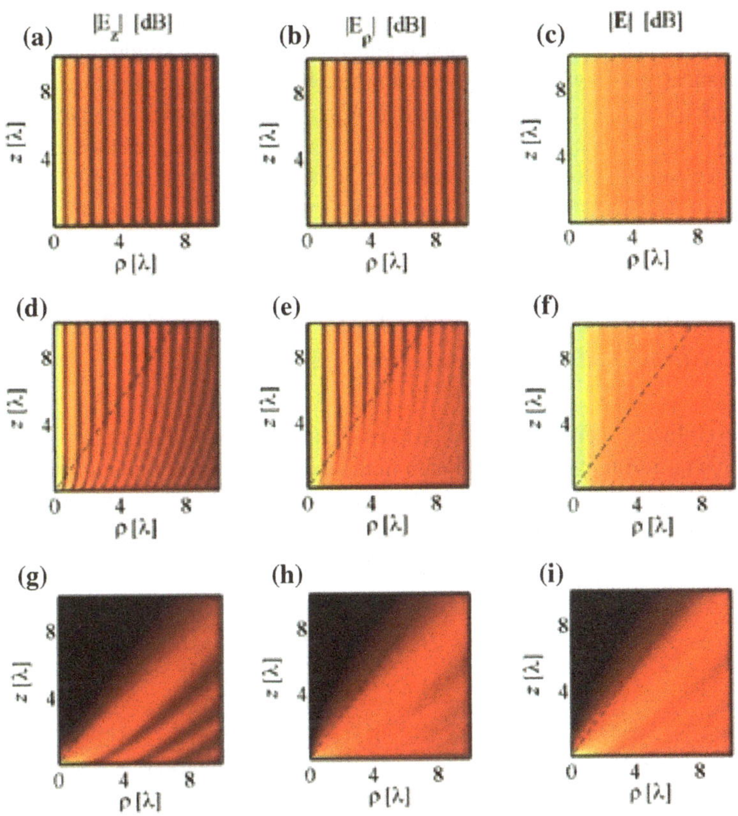

Fig. 8.3 Electric field radiated by an infinite aperture. $|E_z|$ (**a**, **d**, **g**), $|E_\rho|$ (**b**, **e**, **h**) and total electric field amplitude $|E|$ (**c**, **f**, **i**). Standard Bessel beam reference field (**a**, **b**, **c**), field radiated by an inward Hankel distribution (**d**, **e**, **f**), and field radiated by an outward Hankel distribution (**g**, **h**, **i**). The axes are normalized with respect to the wavelength (λ) at the operating frequency. The *dashed line* marks the GO boundary $\theta = \theta_a$

(Fig. 8.3g, h and i), with $k_{\rho a} = 0.6k$ in the vertical ρz plane. As clear from the previous results, the non-diffractive behavior of the field radiated by an inward cylindrical wave can be appreciated within a cone with angle $\theta_a \approx 37°$ (dotted line) where the various components of the electric field (Fig. 8.3d, e, and f) recover the respective components of a standard Bessel beam (Fig. 8.3a, b, and c). By repeating the same formulation for the case of an outward traveling wave over the radiating aperture $\mathbf{M}(\rho, \phi, z = 0) = H_1^{(2)}(k_{\rho a}\rho)\hat{\boldsymbol{\varphi}}$, the radiated field is found again as in Eq. (8.5) with the following expression for the GO contribution:

$$\mathbf{E}^{GO}(\mathbf{r}) = \left[\frac{k_{\rho a}}{jk_{za}} H_0^{(2)}(k_{\rho a}\rho)\hat{\mathbf{z}} + H_1^{(2)}(k_{\rho a}\rho)\hat{\boldsymbol{\rho}}\right] e^{-jk_{za}z} \mathrm{U}(\theta - \theta_a), \tag{8.8}$$

and a SW contribution which is the negative of that in (8.7). In (8.8), differently from (8.6), the GO field contribution is constituted by only a single outward Hankel beam conical wave, which however vanishes within the cone delimited by $\theta < \theta_a$. Its ray interpretation is shown in Fig. 8.2. Again, the SW contribution discontinuity perfectly matches the GO jump at the shadow boundary cone, thus providing a smooth continuous total field. Therefore, it is apparent that an outward Hankel distribution cannot produce a Bessel beam, as also clear from Fig. 8.3g, h, and i.

8.2.2 *Nonuniform Asymptotic Field Evaluation for Finite First Kind Hankel Aperture Distributions*

Once established that an infinite inward Hankel current distribution is able to generate a non-diffractive Bessel beam in the longitudinal component of the electric field, let us assume a finite equivalent magnetic current distribution of radius a and directed along the ϕ-axis over a planar aperture. The Hankel transform of such equivalent magnetic current distribution has the form

$$\tilde{M}(k_\rho) = \frac{-4}{k_\rho^2 - k_{\rho a}^2}\left[\frac{k_\rho}{k_{\rho a}} + \frac{\pi a}{2j}\left(k_\rho H_1^{(1)}(k_{\rho a}a)J_0(k_\rho a) - k_{\rho a}H_0^{(1)}(k_{\rho a}a)J_1(k_\rho a)\right)\right]. \tag{8.9}$$

To render $\tilde{M}(k_\rho)$ more suitable for an asymptotic evaluation, it is convenient to recast its expression as the sum of three terms

$$\tilde{M}(k_\rho) = \tilde{M}_\infty(k_\rho) + \tilde{M}_+(k_\rho) + \tilde{M}_-(k_\rho), \tag{8.10}$$

where

$$\tilde{M}_\infty(k_\rho) = -\frac{k_\rho}{k_{\rho a}}\frac{4}{k_\rho^2 - k_{\rho a}^2}, \tag{8.11}$$

is the Hankel transform of the magnetic current distribution over an infinite aperture (refer to (8.4)), whereas

$$\tilde{M}_\pm(k_\rho) = \frac{j\pi a}{k_\rho^2 - k_{\rho a}^2}[k_\rho H_1^{(1)}(k_{\rho a}a)H_0^{(2,1)}(k_\rho a) - k_{\rho a}H_0^{(1)}(k_{\rho a}a)H_1^{(2,1)}(k_\rho a)], \tag{8.12}$$

in which the Hankel functions of second or first kind apply to the upper or lower sign, respectively, and arise because of the aperture truncation whose radius is denoted by a.

By using (8.12), the S scalar potential introduced in the appendix of [38] is split into three terms, namely $S_{\infty,+,-}$, in which the corresponding $\tilde{M}_{\infty,\pm}$ are used. Such

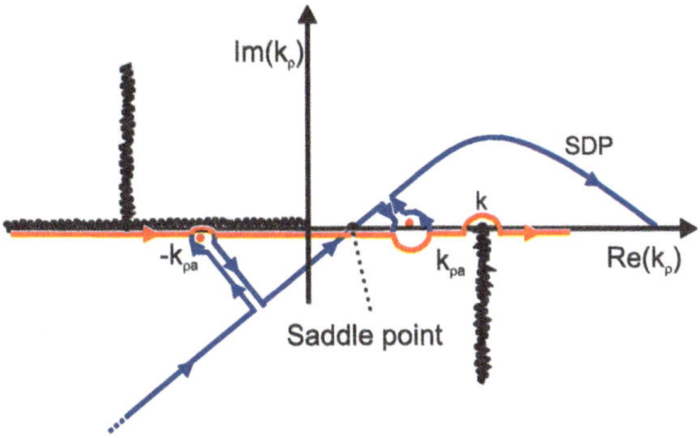

Fig. 8.4 Topology of the integrand function in (8.4). The deformation of the initial integration path (*red line*) into the SDP (*blue line*) is shown for completeness. The *black lines* correspond to the brunch cuts associated with the Hankel functions and the normal propagation constant k_z

integrals can be asymptotically evaluated by using the Steepest Descent Path (SDP) method. As far as S_∞ is considered, its spectral integral representation exhibits a saddle point located at $k_\rho = k\sin\theta$ and pole singularities at $k_\rho = \pm k_{\rho a}$, which are detoured by the integration contour as in Fig. 8.4. The integration path is therefore deformed onto the SDP through the saddle point, and the pole residue contribution has to be taken into account when poles are captured during the path deformation. Hence, the infinite aperture TM scalar potential asymptotically reduces to the sum of a Geometrical Optics (GO) and a diffracted (D) rays,

$$S_\infty(\mathbf{r}) = S_\infty^{GO}(\mathbf{r}) + S_\infty^{D}(\mathbf{r}) \tag{8.13}$$

corresponding to the residue

$$S_\infty^{GO}(\mathbf{r}) = \frac{e^{-jk_{za}z}}{jk_{\rho a}k_{za}}\left[H_0^{(2)}(k_{\rho a}\rho)U(\theta_a - \theta) + H_0^{(1)}(k_{\rho a}\rho)\right], \tag{8.14}$$

with $k_{za} = \sqrt{k^2 - k_{\rho a}^2}$, and the SDP

$$S_\infty^{D}(\mathbf{r}) = \frac{1}{4\pi}\int_{SDP} \tilde{M}_\infty(k_\rho)H_0^{(2)}(k_\rho\rho)\frac{e^{-jk_z z}}{k_z}dk_\rho, \tag{8.15}$$

contributions, respectively. In (8.14), $U(.)$ is the unit step function, which vanishes for negative argument, i.e., in the GO shadow region $\theta > \theta_a = \arcsin(k_{\rho a}/k)$ where the pole at $k_\rho = k_{\rho a}$ in Fig. 8.4 is not captured.

The diffractive contribution given in (8.15) can be asymptotically evaluated as the saddle point contribution, leading to a closed-form Geometrical Theory of Diffraction (GTD) ray description

$$S_\infty^{D}(\mathbf{r}) \sim \frac{2}{j\pi k^3\sin\theta_a\left(\sin^2\theta - \sin^2\theta_a\right)}\frac{e^{-jkr}}{r}. \tag{8.16}$$

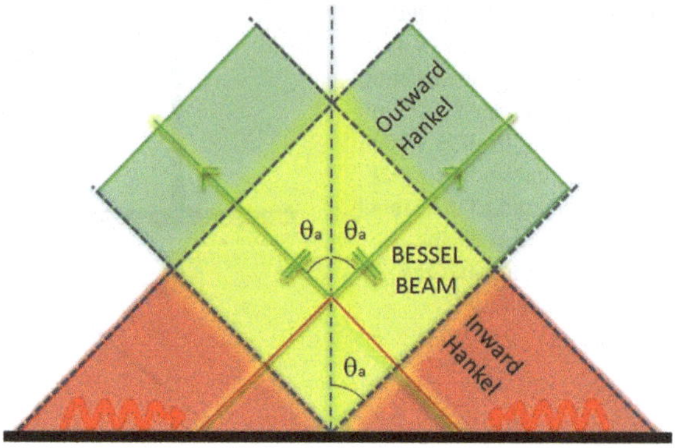

Fig. 8.5 Aperture geometry and map of the GO contribution to the z-component of the electric field radiated by a finite inward cylindrical traveling wave aperture distribution. The Bessel beam is generated in the *yellow region* around the longitudinal z-axis

The above expression reveals that the diffractive contribution of the infinite aperture S_∞ asymptotically behaves like a spherical wave (space wave) arising from the aperture center. According to the spherical wave spreading, S^D decays as $1/r$, unlike S^{GO}, whose amplitude does not decay, thus the diffractive contribution becomes more and more negligible with respect to the GO contribution moving away from the aperture. In addition, S^D_∞ exhibits an angular pattern with a zero on the longitudinal axis ($\theta = 0$). Therefore, its contribution close to the axis is particularly weak. The nonuniform GTD asymptotic expression (8.16) fails at the GO shadow boundary $\theta = \theta_a$. A more sophisticated uniform asymptotics might cure such a singularity, but it is beyond the purpose of the present chapter.

An analogous procedure can be followed for the calculation of the contributions to the potentials $S_\pm$ corresponding to $\tilde{M}_\pm$. Their spectral integral representation exhibits a saddle point at $k_\rho = k \sin\theta_\pm$, with $\theta_\pm = \arctan[(\rho \pm a)/z]$ (refer to Fig. 8.5), and one pole singularity at $k_\rho = \pm k_{\rho a}$. By resorting to the path deformation onto the SDP, one obtains

$$S_\pm(\mathbf{r}) = S^{GO}_\pm(\mathbf{r}) + S^D_\pm(\mathbf{r}), \tag{8.17}$$

in which

$$S^{GO}_\pm(\mathbf{r}) = \frac{1}{jk_{\rho a}k_{za}} H_0^{(2,1)}(k_{\rho a}\rho)e^{-jk_{za}z}U(\theta_a \mp \theta_\pm), \tag{8.18}$$

where the Hankel functions of second or first kind, respectively, apply to the upper or lower sign, and

$$S^D_\pm(\mathbf{r}) = \frac{1}{4\pi}\int\limits_{SDP} \tilde{M}_\pm(k_\rho)H_0^{(2)}(k_\rho\rho)\frac{e^{-jk_z z}}{k_z}dk_\rho. \tag{8.19}$$

Also in this case, the GTD asymptotic evaluation of (8.19) provides insight about the ray nature of these wave constituents

$$S_{\pm}^{D}(\mathbf{r}) \sim \frac{je^{\pm j\frac{3\pi}{4}}\sqrt{a}H_0^{(1)}(k_{\rho a}a)}{\sqrt{2\pi k}k^2\sin\theta_{\pm}(\sin\theta_{\pm} \mp \sin\theta_a)}\frac{e^{-jkr_{\pm}}}{\sqrt{\rho r_{\pm}}}, \tag{8.20}$$

with $r_{\pm} = \sqrt{(\rho \pm a)^2 + z^2}$. Indeed $S_{\pm}^{D}$ appears in the form of a cylindrical edge-diffracted wave, arising from the aperture rim, as can be inferred from the $e^{-jkr_{\pm}}/\sqrt{r_{\pm}}$ factor. Such a GTD expression fails at the corresponding GO shadow boundary ($\theta_{\pm} = \theta_a$) and at the edge-diffracted wave caustic on the longitudinal z-axis ($\rho = 0$), as expected. After gathering all the GO and D terms, the scalar potential can be finally arranged in the following form:

$$S(\mathbf{r}) = S^{GO}(\mathbf{r}) + S_{\infty}^{D}(\mathbf{r}) + S_{+}^{D}(\mathbf{r}) + S_{-}^{D}(\mathbf{r}) \tag{8.21}$$

in which the overall GO term is

$$S^{GO}(\mathbf{r}) = \frac{e^{-jk_{za}z}}{jk_{\rho a}k_{za}}\{H_0^{(2)}(k_{\rho a}\rho)[U(\theta_a - \theta) - U(\theta_a - \theta_+)] + H_0^{(1)}(k_{\rho a}\rho)$$
$$U(-\theta_- - \theta_a)\}. \tag{8.22}$$

It is worth noting that in the case of an infinitely large aperture $a \to +\infty$, $\theta_{\pm} \to \pm\pi/2$, and in turn $U(\theta_a \mp \theta_{\pm}) \to 0$, whence, in (8.18), $S_{\pm}^{GO} \to 0$. Also, $r_{\pm} \to +\infty$ and since in (8.20) $\sqrt{a}H_0^{(1)}(k_{\rho a}a)$ remains limited, it follows that $S_{\pm}^{D} \to 0$. Therefore, it is easily proved that as $a \to +\infty$, then the scalar potential reduces to that of the infinite aperture case reported in [38], namely $S \to S_{\infty}^{GO} + S_{\infty}^{D}$, as expected. Note that such a beam is not a pure Bessel beam throughout the space, but it comprises a Bessel beam, which is confined within the conical region $\theta < \theta_a$, and a diffractive space wave component, though it is negligible close to, and at the z-axis.

From the scalar potential, the GO contribution to the electric field can be exactly calculated by differentiating the GO contribution of the scalar potential (refer to the appendix of [38]) as follows:

$$\mathbf{E}^{GO}(\mathbf{r}) - \left[\frac{k_{\rho a}}{jk_{za}}H_0^{(2)}(k_{\rho a}\rho)\hat{\mathbf{z}} + H_1^{(2)}(k_{\rho a}\rho)\hat{\boldsymbol{\rho}}\right][U(\theta_a - \theta) - U(\theta_a - \theta_+)]e^{-jk_{za}z}$$
$$+ \left[\frac{k_{\rho a}}{jk_{za}}H_0^{(1)}(k_{\rho a}\rho)\hat{\mathbf{z}} + H_1^{(1)}(k_{\rho a}\rho)\hat{\boldsymbol{\rho}}\right]U(-\theta_- - \theta_a)e^{-jk_{za}z}. \tag{8.23}$$

Focusing the attention on the z-component of the previous expression, it can be noticed that

- In the region defined by $0 < |\theta| < \theta_a, \theta_a < \theta_+ < \pi/2$, and $\theta_a < -\theta_- < \pi/2$, a Bessel beam can be established, since the GO contribution to the electric field is proportional to $J_0(k_{\rho a}\rho)e^{-jk_{za}z}$, as highlighted in Fig. 8.5 (yellow region).
- In the regions $0 < |\theta| < \theta_a, \theta_a < \theta_+ < \pi/2, -\theta_- < \theta_a$ or $0 < \theta < \theta_a, \theta_+ < \theta_a, \theta_a < -\theta_- < \pi/2$, the field is described by an outward cylindrical traveling wave $H_0^{(2)}(k_{\rho a}\rho)e^{-jk_{za}z}$, as highlighted in Fig. 8.5 (green regions).
- In the regions delimited by $\theta_a < |\theta| < \pi/2, \theta_a < \theta_+ < \pi/2$, and $\theta_a < -\theta_- < \pi/2$, the field is described by an inward cylindrical traveling wave $H_0^{(1)}(k_{\rho a}\rho)e^{-jk_{za}z}$, as highlighted in Fig. 8.5 (red regions).

The diffractive contributions to the radiated field can be evaluated by differentiating the respective potential terms $S^D_{\infty,+,-}$ [20] which, in the asymptotic limit, lead to the space wave ray fields

$$\mathbf{E}^D_\infty(\mathbf{r}) \sim \frac{2j \sin\theta}{\pi k \sin\theta_a\left(\sin^2\theta - \sin^2\theta_a\right)} \frac{e^{-jkr}}{r} \hat{\boldsymbol{\theta}} \tag{8.24}$$

$$\mathbf{H}^D_\infty(\mathbf{r}) \sim \frac{2j \sin\theta}{\pi k \eta \sin\theta_a\left(\sin^2\theta - \sin^2\theta_a\right)} \frac{e^{-jkr}}{r} \hat{\boldsymbol{\phi}}, \tag{8.25}$$

and the edge-diffracted ray fields

$$\mathbf{E}^D_\pm(\mathbf{r}) = \frac{e^{\pm j\frac{\pi}{4}}\sqrt{a}H_0^{(1)}(k_{\rho a}a)}{j\sqrt{2\pi k}(\sin\theta_\pm \mp \sin\theta_a)} \frac{e^{-jkr_\pm}}{\sqrt{\rho r_\pm}} \hat{\boldsymbol{\theta}}_\pm \tag{8.26}$$

$$\mathbf{H}^D_\pm(\mathbf{r}) = \frac{e^{\pm j\frac{\pi}{4}}\sqrt{a}H_0^{(1)}(k_{\rho a}a)}{j\sqrt{2\pi k}\eta(\sin\theta_\pm \mp \sin\theta_a)} \frac{e^{-jr_\pm}}{\sqrt{\rho r_\pm}} \hat{\boldsymbol{\phi}}. \tag{8.27}$$

The various ray constituents (Fig. 8.5) of the diffracted fields inherit the same shape and behavior of the respective contributions of the potential. According to ray-optical approximation, field polarization is transverse to the ray, i.e., tangent to the wavefront. The diffracted magnetic field ray contributions $\mathbf{H}^D_{\infty,\pm}$ are all polarized along $\hat{\boldsymbol{\phi}}$ orthogonally to the picture plane in Fig. 8.5, whereas the diffracted electric field ray contributions $\mathbf{E}^D_{\infty,\pm}$ are polarized along the local unit vectors $\hat{\boldsymbol{\theta}}, \hat{\boldsymbol{\theta}}_\pm$ transverse to the ray and parallel to the picture plane in Fig. 8.5.

The above ray format expressions for the various wave constituents provide a neat physical insight to the Bessel beam generation by an inward cylindrical traveling wave aperture distribution and clearly show that its capabilities are similar to those of commonly used Bessel aperture distributions. The above expressions fail close to the z-axis, which is of particular interest for Bessel beams. In the next section we provide alternative expressions valid in such region.

8.2.3 Evaluation of the Electromagnetic Field Along the Longitudinal Z-Axis

Let us focus our attention on the z-axis, corresponding to $\rho = 0$. Here it is convenient to rewrite the TM scalar potential introduced in [38], to calculate the electromagnetic field radiated by Hankel aperture distributions in terms of Bessel functions, namely

$$S(\mathbf{r}) = \frac{1}{2\pi}\int_0^{+\infty} \tilde{M}(k_\rho) J_0(k_\rho \rho) \frac{e^{-jk_z z}}{k_z} dk_\rho. \tag{8.28}$$

By differentiating the previous expression [20], the electric field can be evaluated on the longitudinal z-axis and it is clear that E_ρ and H_ϕ vanish, as expected due to the cylindrical symmetry of the radiating aperture, whereas the longitudinal component of the electric field can be evaluated as

$$\mathbf{E}(z\hat{\mathbf{z}}) = \frac{1}{2\pi}\int_0^{+\infty} k_\rho^2 \tilde{M}(k_\rho) \frac{e^{-jk_z z}}{k_z} dk_\rho \hat{\mathbf{z}}. \tag{8.29}$$

As in Sect. 8.2.2, the Hankel transform of the equivalent magnetic current is split into the three terms and the resulting three integrals are evaluated by deforming the integration path onto the SDP. The residue contributions corresponding to the GO terms are the same as in (8.23) once evaluated at $\rho = 0$, thus reducing to

$$\mathbf{E}^{GO}(z\hat{\mathbf{z}}) = \frac{2k_{\rho a}}{jk_{za}} e^{-jk_{za}z} U(\theta_a - \theta_0)\hat{\mathbf{z}}, \tag{8.30}$$

in which $\theta_0 = \arctan(a/z)$ is the limiting value of θ_+ and $-\theta_-$, when $\rho \to 0$ (refer to Fig. 8.5). The saddle point contribution relevant to the infinite aperture exhibits now a quadratic zero of the integrand function at the saddle point itself. Therefore, its asymptotic value is very weak and corresponds to a longitudinally polarized ray contribution

$$\mathbf{E}_\infty^D(z\hat{\mathbf{z}}) \sim -\frac{4k}{\pi k_{\rho a}^3} \frac{e^{-jkz}}{z^2} \hat{\mathbf{z}}, \tag{8.31}$$

indeed the leading asymptotic term of the space wave (8.31) vanishes on the axis, where $\theta = 0$ and $r = z$. The other two saddle point contributions provide the expression for the edge-diffracted rays at their caustic on the z-axis

$$\mathbf{E}_+^D(z\hat{\mathbf{z}}) + \mathbf{E}_-^D(z\hat{\mathbf{z}}) \sim \frac{H_0^{(1)}(k_{\rho a}a)\sin^2\theta_0}{j(\sin\theta_0 - \sin\theta_a)} e^{-jkr_0}\hat{\mathbf{z}}, \tag{8.32}$$

where (8.26) fails. In (8.32), $r_0 = \sqrt{a^2+z^2}$ is the limiting value of r_+ and r_-, when $\rho \to 0$ (refer to Fig. 8.5).

The above GTD asymptotic expression is nonuniform with respect to the pole singularity which might lie in the neighborhood or at the saddle point and fails at the GO shadow boundary $\theta_0 = \theta_a$, or equivalently $z = z_{NDR} = a \cot \theta_a$, where the GO contribution (8.30) abruptly disappears. z_{NDR} is commonly referred as non-diffractive range.

In conclusion, on the z-axis the total electric field results from the summation of the three wave constituents, namely GO (8.30), space wave (8.31), and edge-diffracted wave (8.32). As the observation point moves away from the aperture, the space wave contribution rapidly decays and becomes negligible. Therefore, in addition to the ideal Bessel beam GO contribution, a spurious diffractive contribution arises from the aperture truncation. This diffractive contribution interferes with the dominant term and creates an amplitude ripple in the total field. Such an effect is present also in the field radiated by a standard Bessel distribution and it is unavoidably associated to the aperture finiteness. It can be mitigated by increasing the aperture size, as it is apparent from the dependence on a in (8.32).

8.3 Generation of Non-diffractive Radiation

8.3.1 Equivalent Aperture Field Distribution

According to the Schelkunoff theorem, the aperture field distributions discussed in Sect. 8.2 are equivalently generated in the half-space $z > 0$ by a distribution of magnetic surface currents, radiating on a perfect electric conductor, of the form

$$\mathbf{M} = \mathbf{E} \times \hat{z}\big|_{z=0}. \tag{8.33}$$

In particular for the generation of a circularly polarized Bessel beam such current is given by [34, 39]:

$$\mathbf{M} = \mathbf{E} \times \hat{z}\big|_{z=0} = A_0 J_0(k_\rho \rho)\hat{\mathbf{p}}_{RH,LH}, \tag{8.34}$$

where $\hat{\mathbf{p}}_{RH,LH} = (\hat{x} \mp j\hat{y})/\sqrt{2}$ for right-hand and left-hand circularly polarized beams.

In the following sections, it will be discussed the possibility to synthetize such current distributions or equivalently the associate aperture field by using a radial waveguide loaded with concentric circular slits or sub-resonant slots. A holographic approach will be adopted. Holography predicts that a certain antenna aperture equivalent current distribution is produced by the interference between the field map generated by the input feeding wave alone and the hologram. In our cases the latter is the field generated by the array radiating elements [40].

The case of sub-resonant slots distribution will be treated in detail in the next section for the design and optimization of a circularly polarized radial line slot array (CP-RLSA) antenna [41–43].

8.3.2 *Slot Layout for Radial Line Slot Array Bessel Launchers*

A CP-RLSA is usually composed of several hundreds of circular-polarized radiating elements, each one comprising two slots $\lambda_d/4$ apart, making an angle of $\pm 45°$ with respect to the radial direction (Fig. 8.6). Since the two slots in the pair are orthogonal to each other and fed in phase quadrature by the feeding wave in the PPW, each element radiates a CP wave in the broadside direction. The arrangement of these CP elements on the radiating aperture determines the antenna radiation pattern.

Since the elements of the array are closely spaced, their radiation pattern is strongly related to that of the associated continuous aperture current distribution, of which the array constitutes a spatial sampling. Such an aperture distribution is chosen specifically to achieve the desired radiation pattern by following one of the techniques present in the literature. Among them, here the alternate projection method [44, 45] has been adopted. As it is well known, the radiation pattern of the slot array will sufficiently approximate the continuous aperture reference pattern if the following two conditions are verified: the array element excitation corresponds to the continuous aperture sampling, and, according to the Schelkunoff sampling criterion, the inter-element distance is sufficiently tightly linked. It is worth pointing out that an array pattern synthesis technique assuming given element positions cannot be adopted here. Indeed, it is necessary to resort to a continuous aperture distribution because the specific slot positions are used to calibrate the radiating element phase and cannot be given a priori.

According to Bethe theory [46], the radiation from a slot cut into a metallic plane can be modeled by an equivalent magnetic dipole sitting on a ground plane, with dipole moment proportional to the magnetic field of the feeding wave in the PPW

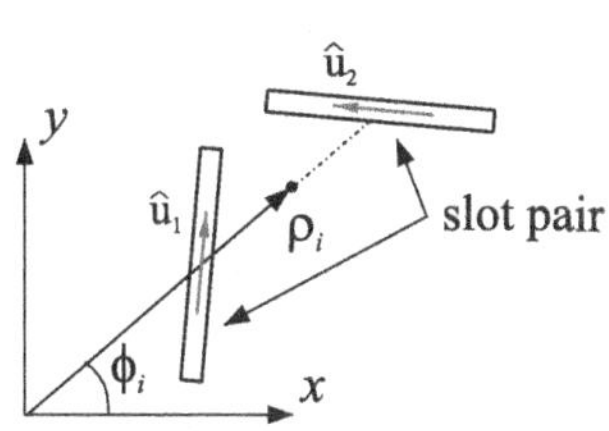

Fig. 8.6 Geometry of a slot pair element radiating a RHCP when illuminated by an outward traveling wave. The distance between the slot centers is set to $\lambda_d/4$, where λ_d is the wavelength in the PPW. The LHCP is achieved by mirroring the slots with respect to $\hat{\rho}$

$$M = \underline{\alpha}_m \cdot H^{\text{inc}}, \tag{8.35}$$

through a dyadic polarizability $\underline{\alpha}_m$. Although Bethe theory also introduces an equivalent electric dipole, its radiation effect is negligible with respect to that of the magnetic dipole for slots with a length much greater than their width. Since here we are dealing with such kind of elongated slots the effect of the equivalent electric dipoles is not considered.

For a single slot oriented along $\hat{\mathbf{u}}$, the dyadic polarizability is given by $\underline{\alpha}_m = \alpha_m \hat{\mathbf{u}}\hat{\mathbf{u}}$, where the amplitude of α_m grows by increasing the slot length up to the resonance length. Therefore, the slot length can be adjusted in order to match the amplitude of the target aperture distribution, while the phase of the equivalent magnetic dipole of each slot can be controlled by adjusting its position along the feeding wave propagation direction, thus varying the phase of the incident wave excitation $\mathbf{H}^{\text{inc}}$.

Since the basic element of the CP-RLSA is a slot pair, it is worthwhile introducing an analogous description for the whole slot pair. Thus, by summing the magnetic dipoles relevant to the two slots in the pair, and normalizing them to the incident magnetic field at the slot pair center, a polarizability $\underline{\alpha}_m^p$ for the whole slot pair is obtained. Note that $\underline{\alpha}_m^p$ depends on the phase progression of the incident feeding wave in the PPW, which establishes the phase shift between each slot and the pair center reference point.

In the following, since the Bessel aperture distribution $A(\rho)$ is rotationally symmetric, we consider only the class of CP-RLSAs with a rotationally symmetric radiation pattern. Also, without loss of generality, we focus on a right-hand CP (RHCP) antenna. The left-hand CP (LHCP) can be achieved similarly. Under the above hypothesis, the ideal target aperture magnetic current will be in the form $\mathbf{M}_0(\boldsymbol{\rho}_i) = A(\rho)\hat{\mathbf{p}}_{RH,LH}$, where $\hat{\mathbf{p}}_{RH,LH} = (\hat{\mathbf{x}} \mp j\hat{\mathbf{y}})/\sqrt{2}$ denotes the polarization unit vector and the upper/lower sign applies to the RHCP/LHCP.

To match the target aperture distribution, one has to assure that, for each slot pair element, the equivalent magnetic dipole $\mathbf{M}(\boldsymbol{\rho}_i) = \underline{\alpha}_m^p(\boldsymbol{\rho}_i) \cdot \mathbf{H}^{\text{inc}}(\boldsymbol{\rho}_i)$ is parallel to $\mathbf{M}_0(\boldsymbol{\rho}_i)$, i.e., $\mathbf{M}(\boldsymbol{\rho}_i) = Ce^{j\gamma}\mathbf{M}_0(\boldsymbol{\rho}_i)$, where $\boldsymbol{\rho}_i$ is the position of the i-th slot pair element center, and γ, C are arbitrary real constants. Such condition is satisfied only if, for each slot pair, it holds

$$\left|\underline{\alpha}_m^p(\boldsymbol{\rho}_i) \cdot \mathbf{H}^{\text{inc}}(\boldsymbol{\rho}_i)\right| = C|A(\rho_i)|, \tag{8.36}$$

and

$$\angle\left\{\mathbf{M}_0^*(\boldsymbol{\rho}_i) \cdot \underline{\alpha}_m^p(\boldsymbol{\rho}_i) \cdot \mathbf{H}^{\text{inc}}(\boldsymbol{\rho}_i)\right\} = \gamma. \tag{8.37}$$

For a given γ only specific positions of the slot pair elements verify (8.37). Hence, (8.37) is the key equation controlling the correct slot pair positioning which ensures the target aperture phase distribution. On the other hand, (8.36) imposes the correct amplitude distribution on the radiating aperture which can be achieved by properly choosing the slot length. In fact, a variation in slot length mainly affects the associated magnetic moment magnitude, while the phase change is a higher order effect; similarly, a variation in slot position mainly affects the associated magnetic moment phase while the magnitude remains almost the same. Such a quasi-orthogonality property permits independent actions on the two sets of variables (i.e., the slot pair lengths and positions) to determine the holographic slot layout.

For example, for the slot pair arrangement in Fig. 8.6, excited by a radial outward feeding wave, by adopting the simple Bethe model and neglecting the slot coupling, the slot pair polarizability can be estimated as follows:

$$\underline{\alpha}_m^p = \alpha_m \left[e^{j\pi/4} \hat{\mathbf{u}}_1 \hat{\mathbf{u}}_1 + e^{-j\pi/4} \hat{\mathbf{u}}_2 \hat{\mathbf{u}}_2 \right] = \alpha_m \left[e^{-j\phi} \hat{\mathbf{p}}_{LH} \hat{\boldsymbol{\rho}} + j e^{j\phi} \hat{\mathbf{p}}_{RH} \hat{\boldsymbol{\phi}} \right]. \tag{8.38}$$

By considering a PPW excited by a $TM^z_{0,0}$ cylindrical mode, (i.e., $\mathbf{H}^{\rm inc}(\boldsymbol{\rho}) = H^{inc} H_1^{(2)}(k_d \rho) \hat{\boldsymbol{\phi}}$, where $H_1^{(2)}$ denotes the first-order Hankel function of second kind, and $k_d = 2\pi/\lambda_d$ the wavenumber inside the PPW, (refer to Sect. 9.4.1 in [47] for modes definition), Eq. (8.36) reduces to

$$|\alpha_m(\rho)| = \frac{C|A(\rho)|}{\left| H_1^{(2)}(k_d \rho) \right|}, \tag{8.39}$$

which is exploited so as to calculate the slot length profile. Indeed, a parametric analysis of an isolated slot pair provides α_m as function of the slot length ℓ [42], from which a slot length profile $\ell = \ell(\rho)$ satisfying (8.39) can be obtained. Note that the constant C does not affect the aperture amplitude shape, but has to be chosen to keep the slot length within a proper range, and to calibrate the total aperture radiating power. Namely, if the value of the C constant is too large, the power of the feeding wave is entirely radiated by the slots close to the antenna center, and the slots at the antenna rim are fed too weakly to provide the desired amplitude, even assuming the maximum resonant length. On the other hand, when C is too small, the desired aperture shape is easily achieved but a considerable amount of power is not radiated and remains trapped in the PPW beyond the last peripheral slots. Although the residual power can be absorbed with an absorbing rim [48–50], this results in a poor antenna radiation efficiency. Summarizing, C must be optimized as a tradeoff between fine aperture amplitude shape reconstruction and high radiation efficiency.

Next, from (8.37), the centers of the slot pairs have to be located on the curve

$$\angle\left\{ A^*(\rho)\alpha_m(\rho) e^{j\phi} H_1^{(2)}(k_d \rho) \right\} = \gamma. \tag{8.40}$$

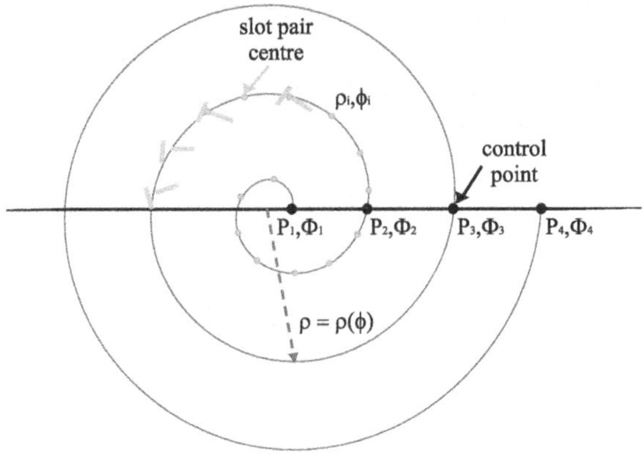

Fig. 8.7 Slot pair positioning. Slot pair centers $\boldsymbol{\rho}_i = (\rho_i, \phi_i)$ (*turquoise dots*) are initially chosen by sampling the curve $\rho = \rho(\phi)$ defined by (8.40), which ensures that the slot pair elements match the target aperture phase distribution, according to the simplified model based on Bethe theory. Next, in the optimization process the slot positioning curve $\rho = \rho(\phi)$ is represented by a thinned set of control points $\mathbf{P}_i = (\mathrm{P}_i, \Phi_i)$ (*black dots*) [40]

It is worth noting that for a uniform distribution (i.e., $A(\rho) = \text{const.}$), by using the asymptotic large argument expansion $\angle H_1^{(2)}(k_d\rho) \sim -k_d\rho + 3\pi/4$ and neglecting the phase variation for the polarizability of slots having different lengths, Eq. (8.40) results in the classic CP-RLSA spiral slot arrangement $\phi - k_d\rho = \text{const}$[42].

Likewise, in any general CP-RLSA design, Eq. (8.36) allows the calculation of the slot length profile $\ell(\rho)$, and then Eq. (8.37) provides the definition of a curve where the slot pair centers are to be distributed. Such a curve will be in general of the form $\rho = \rho(\phi)$ (Fig. 8.7). The slot pair centers are then distributed uniformly along such a curve by choosing the sampling points ϕ_i via $\phi_i = \phi_{i-1} + \Delta s/\rho(\phi_{i-1})$. The sampling step Δs along the curve is arbitrary, and its value is a design parameter which is constrained between a minimum value, to avoid slot overlapping, and a maximum one, according to the Schelkunoff sampling criterion. Given the sampling points ϕ_i, slot pair centers will be located at $\boldsymbol{\rho}_i = (\rho_i, \phi_i)$, with $\rho_i = \rho(\phi_i)$, and the length of the slots in the pair will be $\ell_i = \ell(\rho_i)$. Note that the total number of slot pairs is, in our approach, not a priori determined. In fact, we keep adding slots until they remain within the assigned antenna size.

The above-mentioned algorithm completely neglects mutual coupling among slot pairs, which can be accounted for only with a full-wave analysis. However, it is sufficient for initializing the optimization process described in the next section.

8.3.3 Optimization Procedure

When a not uniform field aperture distribution, such as a Bessel one, is employed or the geometrical requirements on the antenna radius determine a number of slot pair elements in the order of several thousands, the use of an optimization algorithm to synthesize the required distribution is necessary. A fine optimization cannot be performed by using simplified models as in [51, 52], since they do not describe accurately the global slot coupling. An accurate but efficient full-wave numerical method is then mandatory for the optimization of the antenna. A successful example is the technique described in [53] which reduces the return loss in a linear polarized RLSA configuration, even if the antenna is recursively optimized ring by ring and not globally. Moreover, several papers point out that, when a careful antenna design is applied and the efficiency of the antenna is maximized, the amplitude of higher modes excited in the PPW by the slots distribution itself is no longer negligible. Indeed, it results in an appreciable degradation of the radiation pattern and/or of the reflection coefficient of the antenna. For example, in [54] the authors have recently proved that the distortion of the rotational symmetry of the radiation pattern and the unbalanced sidelobe levels typical of a spiral RLSA are related to these higher order modes, and in particular to the spurious $TM^{z}_{0,-2}$ mode of the PPW.

The optimization procedure here used is tailored on the particular physical mechanism that governs the RLSA's operation, as explained in [40]. Specifically, it acts on the position and the length of each slot so as to obtain a matching between the actual antenna aperture distribution, calculated by the full-wave analysis, and its target profile. However, the optimization cannot act individually on each slot; otherwise the enormity of the problem would result in an instable procedure that does not converge at all. Instead, the antenna parameterization, here proposed, involves some global parameters the tuning of which is performed resorting to specific fitness functions. Such fitness functions are introduced on the basis of the physical picture behind the working mechanism of the antenna, and are chosen in such a way that each fitness function provides the optimization feedback for only one geometrical parameter. Therefore, the optimization scheme of the large multivariable problem is reduced to the parallel arrangement of some single-parameter problems. Such size reduction and quasi-orthogonality property ensure a very fast convergence.

For a given antenna geometry, by using a full-wave method, the dipole moment of each slot is easily evaluated. Specifically, $\mathbf{M}_1(\boldsymbol{\rho}_i)$ and $\mathbf{M}_2(\boldsymbol{\rho}_i)$ denote the magnetic dipole moments of the inner and outer slots in the pair, respectively, while $\mathbf{M}(\boldsymbol{\rho}_i) = \mathbf{M}_1(\boldsymbol{\rho}_i) + \mathbf{M}_2(\boldsymbol{\rho}_i)$ is the dipole moment of the i-th slots pair of center $\boldsymbol{\rho}_i$.

For each slot pair a complex fitness functions F_i is then introduced as follows:

$$F_i = \frac{\hat{\mathbf{p}}_{RH} \cdot \mathbf{M}(\boldsymbol{\rho}_i)}{A(\rho_i)} \frac{\bar{A}}{\bar{M}}, \tag{8.41}$$

where $\bar{A}=\sum_{i=1}^{N_p}|A(\rho_i)|/N_p$ and $\bar{M}=\sum_{i=1}^{N_p}|\hat{\mathbf{p}}_{RH}\cdot\mathbf{M}(\boldsymbol{\rho}_i)|/N_p$ are the target and the realized average co-polar dipole moment amplitude, respectively, with N_p denoting the total number of slot pairs.

Note that $\angle F_i$ and $|F_i|$ indicate the phase and amplitude errors between the target and the actual dipole moment distribution, which can be compensated by adjusting the slot pair radial position ρ_i and the slot length ℓ_i, respectively.

In principle, to minimize the overall error we could act on each slot pair independently. However, the strong mutual coupling between the slots renders this way of operating instable, and after a few steps the slots start breaking the order of the initial layout and overlapping.

To overcome this drawback a set of N control points $\mathbf{P}_n=(\mathrm{P}_n,\Phi_n)$ is introduced. Since N is much lower than the total slot pair number $(N\ll N_p)$, the control point set represents a coarse sampling of the slot pair distribution curve $\rho=\rho(\phi)$ (Fig. 8.7). Analogously, the slot length is described by the thinned set of N length samples L_n. For a given set of angular control point locations Φ_n, the antenna geometry is generated from the reduced sets P_n and L_n by calculating slot pair positions and lengths through a linear interpolation as

$$\rho(\phi)=\sum_{n=1}^{N}P_n b_n(\phi) \tag{8.42}$$

$$l(\phi)=l(\rho(\phi))=\sum_{n=1}^{N}L_n b_n(\phi), \tag{8.43}$$

where

$$b_n(\phi)=\begin{cases}\frac{\phi-\Phi_{n-1}}{\Phi_n-\Phi_{n-1}} & \Phi_{n-1}<\phi\leq\Phi_n,\\ \frac{\Phi_{n+1}-\phi}{\Phi_{n+1}-\Phi_n} & \Phi_n\leq\phi<\Phi_{n+1},\\ 0 & \text{elsewhere.}\end{cases} \tag{8.44}$$

Equations (8.42) and (8.43) represent a parameterized version of the position and length curves and allow the automatic generation of the antenna layout. In Fig. 8.7 and in the examples provided in Sect. 8.3.4, we have chosen $\Phi_n=2n\pi$ (i.e., a control point each spiral turn), though different choices are possible.

The initial set of parameters is chosen by sampling the initial design curves obtained with the simplified approach illustrated in Sect. 8.3.2 as $\mathrm{P}_n=\rho(\Phi_n), L_n=\ell(\Phi_n)$. The antenna layout can next be adjusted by varying both the values of the control points P_n (thus smoothly moving the slot pairs), and the values of the samples L_n (thus smoothly controlling the slots length). This smooth deformation of the layout geometry avoids a significant variation of the distance between adjacent slots, thus minimizing the risk of overlapping, unlike the scheme where each slot moves independently of its neighbor. The updating law is described in the following.

Optimization Loop

At each step s of the optimization process, when the antenna layout is parameterized by the values of the parameters $\mathrm{P}_n^{(s)}, L_n^{(s)}$, a full-wave analysis of the whole antenna is performed and the equivalent magnetic dipole moments $\mathbf{M}(\boldsymbol{\rho}_i)$ are calculated. In addition, the antenna spillover efficiency

$$\eta_{\mathrm{so}}^{(s)} = \frac{P_{\mathrm{acc}} - P_{\mathrm{so}}}{P_{\mathrm{acc}}} \tag{8.45}$$

is also calculated, where P_{acc} is the power accepted by the antenna at the feeding port, whereas P_{so} is the total power trapped in the PPW beyond the slots distribution. It is worth noting that when losses in the PPW are negligible η_{so} equals the radiation efficiency. Then, a linearly weighted version of the complex fitness function (8.41) is defined at each step s as

$$\bar{F}_n^{(s)} = \sum_{i=1}^{N_p} F_i^{(s)} b_n(\phi_i) \tag{8.46}$$

which expresses the local average error on the slot pairs in the neighborhood of the n-th control point. In fact, each n-th value of the fitness function is associated with the relevant n-th control point, but it averages the local error at each i-th slot pairs in the sub-spiral around the control point. The minimization at each control point of the magnitude and the phase of the fitness function leads to the optimum slot pair length and position, respectively.

Then, the control parameters $\mathrm{P}_n^{(s)}, L_n^{(s)}$ are updated according to

$$\mathrm{P}_n^{(s+1)} = \mathrm{P}_n^{(s)} - \chi_\rho \angle \bar{F}_n^{(s)} / k_d \tag{8.47}$$

and

$$L_n^{(s+1)} = \left[1 + \chi_\ell \left(1 - |\bar{F}_n^{(s)}| \sqrt{\eta_{\mathrm{so}}^{(s)}}\right)\right] L_n^{(s)} \tag{8.48}$$

with $0 < \chi_\rho, \chi_\ell \leq 1$ suitable damping factors (typically $\chi_\rho \simeq 0.5$ and $\chi_\ell \simeq 0.5$) which prevent the optimization loop from oscillating. The updating rule (8.47) corrects the slot position minimizing the local average phase error; indeed slots are moved inward/outward to compensate for a local aperture illumination phase delay/advance. The updating rule (8.48) corrects the slot length minimizing the local amplitude error; indeed, slots are shortened/lengthened to compensate for excessive/scarce local aperture illumination amplitude. Also, the global factor

$\sqrt{\eta_{\mathrm{so}}^{(s)}}$ has been introduced in (8.48) so that the slot length profile of the entire antenna is adjusted to maximize the radiation efficiency. Indeed, all the slots are globally shortened/lengthened, without changing the aperture distribution shape, to guarantee that most of the power injected in the PPW at the feeding point is radiated by the slots before reaching the antenna rim.

The loop is repeated until the spillover efficiency η_{so} is over a fixed threshold (usually $\eta_{\mathrm{so}} > 95\%$), and the average aperture distribution error

$$\varepsilon_{\Delta} = \frac{1}{N_p}\sum_{i=1}^{N_p}\left|\frac{\hat{\boldsymbol{\rho}}_{RH}\cdot\mathbf{M}(\boldsymbol{\rho}_i)}{\bar{M}} - \frac{A(\rho_i)}{\bar{A}}\right| = \frac{1}{N_p}\sum_{i=1}^{N_p}\left|\frac{A(\rho_i)}{\bar{A}}\right||F_i - 1| \tag{8.49}$$

is sufficiently small (e.g., $\varepsilon_{\Delta} < 0.01$).

Minimization of Spurious Modes Excited by the Slots Arrangement

Another effect, which is neglected in the simplified model used for the first design of the CP-RLSA, is the perturbation of the incident feeding wave due to the field scattered in the PPW by the slots etched in the upper plate. Despite the field scattered inside the PPW by a single-slot pair is relatively weak, the collective radiation of all the slot pairs generates a higher order of PPW modes, some of which can be of appreciable amplitude with respect to the feeding. Since the PPW height h is chosen less than half wavelength $(h < \lambda_d/2)$, only $TM_{n,m}^z$ modes with $n = 0$ can propagate (where n and m tag the radially propagating mode eigenvalues along z and ϕ, respectively). When the antenna is fed by an outgoing $TM_{0,0}^z$ mode, excited higher order $(m \neq 0)$ modes vanish at the feeding point $\rho = 0$, and therefore their occurrence does not result in an increased antenna reflection loss. Nevertheless, since they exhibit a phase variation $\exp(jm\phi)$, their presence in the PPW creates an interference with the $TM_{0,0}^z$ feeding mode resulting in a not uniform total field impinging on the slots. As a matter of fact, the feeding wave loses the rotational symmetry and exhibits an amplitude ripple and a phase oscillation along ϕ. In turn, the realized antenna aperture distribution loses the rotational symmetry with a degradation of the radiation pattern. Therefore, in the case of Bessel aperture distribution, this effect results in a circular asymmetry of the pseudo-Bessel beam around the propagation axis which needs to be kept under control during the design.

To minimize a given undesired $TM_{0,m}^z$ mode one can adjust the distance $\Delta\rho_i$ between the two slots in a pair, and their length difference $\Delta\ell_i$. In particular, the inner (tagged by $q = 1$) and the outer (tagged by $q = 2$) slots in the i-th pair will be placed at $\boldsymbol{\rho}_{iq} = (\rho_{iq}, \phi_{iq})$, with $\rho_{iq} = \rho_i + (-1)^q \Delta\rho_i/2$, and their respective length will be $\ell_{iq} = \ell_i + (-1)^q \Delta\ell_i/2$. Note that the azimuthal coordinate of the inner/outer slot ϕ_{iq} might be displaced with respect to the slot pair center ϕ_i to avoid slot overlapping [41]. To optimize the design parameters $\Delta\rho_i$ and $\Delta\ell_i$, for each slot pair a complex fitness function

$$
{}^{m}F_{i} = -\frac{H_{m}^{(2)\prime}(k_{d}\rho_{i1})e^{jm\phi_{i1}}\hat{\boldsymbol{\phi}}_{i1}\cdot\mathbf{M}_{1}(\boldsymbol{\rho}_{i})}{H_{m}^{(2)\prime}(k_{d}\rho_{i2})e^{jm\phi_{i2}}\hat{\boldsymbol{\phi}}_{i2}\cdot\mathbf{M}_{2}(\boldsymbol{\rho}_{i})} \tag{8.50}
$$

is introduced, where m tags the undesired mode $TM^{z}_{0,m}$ to which the fitness function is related, and $H_{m}^{(2)\prime}$ denotes the derivative of the m-th order Hankel function of second kind. In (8.50) $\hat{\boldsymbol{\phi}}_{iq}$ is the azimuthal unit vector at inner/outer slot center $\boldsymbol{\rho}_{iq}$. The fitness function in (8.50) is the negative of the ratio between the amount of $TM^{z}_{0,m}$ individually excited by each slot in the pair. When ${}^{m}F_{i} = 1$ the contributions of the two slots to the $TM^{z}_{0,m}$ mode perfectly cancel out each other. Hence, the amplitude $|{}^{m}F_{i}|\gtrless 1$ and the phase $\angle {}^{m}F_{i}\gtrless 0$ can be used to feedback $\Delta\ell_{i}$ and $\Delta\rho_{i}$, respectively. Namely, $|{}^{m}F_{i}|$ can be reduced/increased by increasing/reducing $\Delta\ell_{i}$, so that the scattering contribution of the inner/outer slot is weakened while that of the outer/inner slot is reinforced. Also, $\angle {}^{m}F_{i}$ can be increased/reduced by reducing/increasing $\Delta\rho_{i}$, so that the phase delay between the scattering contribution of the two slots is reduced/increased.

Again, to introduce smooth variation and avoid optimization instability, we resort to a linear variation of the design parameters

$$
\Delta\rho(\phi) = \sum_{n=1}^{N}\Delta \mathrm{P}_{n} b_{n}(\phi) \tag{8.51}
$$

$$
\Delta\ell(\phi) = \sum_{n=1}^{N}\Delta L_{n} b_{n}(\phi) \tag{8.52}
$$

which are calculated in terms of a thinned set of samples $\Delta \mathrm{P}_{n}$ and ΔL_{n} defined at the control points Φ_{n}. Such parameters are initialized with their nominal values $\Delta \mathrm{P}_{n} = \lambda/4, \Delta L_{n} = 0$ according to the ideal simplified model introduced in Sect. 8.3.2. Next, they are updated in the optimization loop by resorting to a linearly weighted version of the complex fitness function (8.50) associated to each control point n

$$
{}^{m}\bar{F}_{n}^{(s)} = \sum_{i=1}^{N_{p}} {}^{m}F_{i}^{(s)}\, b_{n}(\phi_{i}). \tag{8.53}
$$

In particular, the updating rules are defined as follows:

$$
\Delta \mathrm{P}_{n}^{(s+1)} = \Delta \mathrm{P}_{n}^{(s)} - \chi_{\Delta \mathrm{P}}\angle {}^{m}\bar{F}_{n}^{(s)}/k_{d} \tag{8.54}
$$

and

$$\Delta L_n^{(s+1)} = \Delta L_n^{(s)} + \chi_{\Delta L}\left[1 - \left|{}^m\bar{F}_n^{(s)}\right|\right]\Delta \tag{8.55}$$

with $0 < \chi_{\Delta P}, \chi_{\Delta L} \leq 1$ suitable damping factors (usually $\chi_{\Delta P} \simeq 0.1$ and $\chi_{\Delta L} \simeq 0.1$) and $\Delta = \max\left(L_n^{(s)}/20, \Delta L_n^{(s)}\right)$. The ratio between the amplitude of the undesired $TM_{0,m}^z$ mode in the PPW and the amplitude of the feeding mode can be monitored at each step s and introduced into the stop criterion of the optimization procedure. Therefore, also this second fitness function is calculated at each control point and the minimization of its magnitude and phase, respectively, leads to the optimum slot length difference inside the pair, and the optimum displacement $\Delta\rho$ of each slot in the pair with respect to the pertinent pair center $\boldsymbol{\rho}_i$.

Note that the parameters pair $\Delta\rho_i$ and $\Delta\ell_i$ might alternatively be adjusted in order to minimize the cross-polarized radiation of the slot pair. However, the CP-RLSA spiral arrangement presents a broadside null of the cross-polarized radiation pattern in the array factor. Therefore, the polarization level of the antenna combining a CP element and a CP array factor is usually excellent and does not require a cross-polarization level improvement.

8.3.4 *Circular-Polarized Bessel Beam Launcher Made by a RLSA Structure*

The optimization procedure presented in Sect. 8.3.3 converges to the optimum solution in a very limited number of steps, allowing the use of a full-wave analysis of the entire antenna [40] that, in turn, guarantees an accurate prediction of the antenna behavior. In the full-wave analysis, at least five rectangular waveguide type modes are required to accurately model the shape of the magnetic current along each slot, because of the strong coupling between the slots [53]. In the present design, unlike [53], the slots couple significantly both in the internal (PPW) and in the external (grounded half-space) regions, because the dielectric constant is almost the same in the two regions.

Consider now the design of a CP-RLSA antenna working at the nominal frequency of 15 GHz for the generation of a Pseudo-Bessel beam characterized by a half-power beamwidth of less than 12 cm and to need a focalization distance in between 4 and 8 m.

Hence, the antenna has to generate, on the plane $z = 0$, an aperture field distribution $A(\rho) = A_0 J_0(k_B \rho 0)$, as specified in Eq. (8.34), where the constant k_B is related to the half-power beamwidth (HPBW) through $k_B = 2.235/HPBW$ [3]. Thus, the HPBW requirement is satisfied by choosing $k_B = 20\,\mathrm{m}^{-1}$. Moreover, assuming reasonable a depth of field $z_{\max} = 9\,\mathrm{m}$, through the relation $z_{\max} = R/\sqrt{(k_0/k_B)^2 - 1}$, it is obtained an antenna radius of 57 cm. In the following, by presuming a slight degradation of the actual antenna performances,

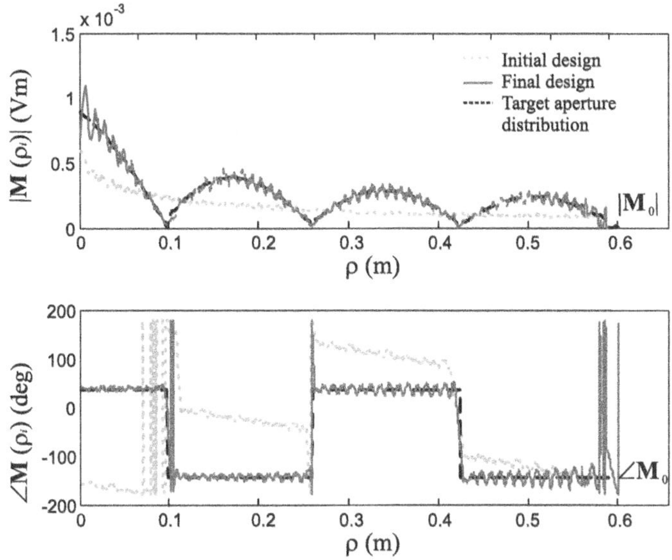

Fig. 8.8 Evolution of the equivalent dipole moment of each slot pair along the optimization steps: magnitude (*above*), and phase (*below*). Initial design (*dotted green line*), final design (*red line*), and target aperture distribution $\mathbf{M}_0(\boldsymbol{\rho}_i)$ (*dashed-black line*)

an antenna radius of 60 cm will be considered. Due to the large dimension of the antenna, a hollow PPW cannot be employed. Thus, to guarantee mechanical robustness and to maintain its thickness constant, the PPW is supposed filled by a foam substrate (ROHACELL WF51, $\varepsilon_r = 1.07$ and $\tan\delta = 0.0035$ @10 GHz).

The optimization process converges in 16 steps. The comparison between the target aperture distribution and the synthesized one is shown in Fig. 8.8, where one can trace how the equivalent dipole moment of each slot pair $\mathbf{M}(\boldsymbol{\rho}_i) = A(\rho_i)\widehat{\mathbf{p}}$ evolves during the optimization process from the first draft design (green dotted line) and it converges to the target (black dashed line) $\mathbf{M}_0(\boldsymbol{\rho}_i)$, at the last step (red solid line). It is also noticeable a rippled behavior of the synthetized dipole moment due to the excitation of spurious higher circumferential modes in the PPW, and especially modes $TM^z_{0,-2}$. If no action had been taken at each optimization step when increasing the efficiency of the antenna (i.e., the length of the slots and consequently its interaction), the ripple would have become unacceptable. However, this phenomenon was contrasted by adjusting the relative position $\Delta\rho$ of each slot pair that constitutes the elementary circular-polarized element of the array through Eq. (8.53). Introducing this degree of freedom in the optimization process, we limit the amplitude of the undesired $TM^z_{0,\pm 2}$ modes to less than 15% with respect to the amplitude of the feeding $TM^z_{0,0}$ mode. Thus, the effect of the ripples is mitigated.

The discretization of the aperture distribution by slots necessarily introduces a degradation of the Bessel beam performances as the reduction of the maximum

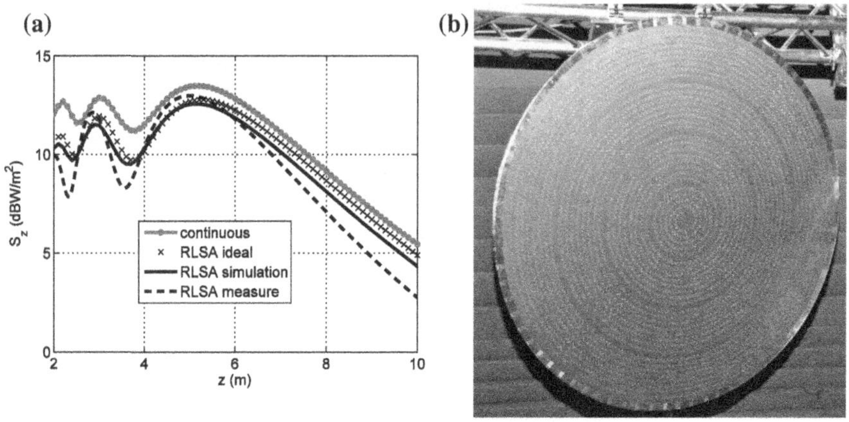

Fig. 8.9 **a** the axial z-component of the radiated power density s_z along the z-axis for the target continuous distribution (*black red line*), for the target distribution sampled on the slot pair positions (*gray line*), for the synthesized RLSA distribution (*black solid line*), and for the measured antenna prototype (*black dashed line*). **b** A photograph of the RLSA prototype

radiated power density, but not of the beam shape. In Fig. 8.9 the axial z-component of the radiated power density associated to the ideal continuous distribution (red solid line) is compared to that associated to the ideal distribution (gray line) sampled on the slot pair positions of the synthesized RLSA (the power accepted by the antenna is set equal to 1 W). The small reduction of the maximum radiated power density associated to the synthesized antenna (solid black line), with respect to the one associated to the sampled ideal distribution, demonstrates the effectiveness of the optimization process. The RLSA is expected to have the maximum power density at a distance from the radiating surface of 5.1 m instead of 5.2 m of the continuous aperture, with a −6 dB depth of field of $430\lambda_0$ instead of $455\lambda_0$.

By defining the focusing efficiency as the ratio between the power focalized in the target spatial xy-spot of 12 cm diameter at the maximum radiated power distance for the actual antenna and the ideal (continuous) pseudo-Bessel beam distribution, the expected efficiency for the synthesized RLSA is about 84%.

The point of maximum power density was expected at z = 5.1 m (continuous red line in Fig. 8.9a). However, it has been measured at z = 5.0 m. The shift of the maximum power density distance toward the antenna is more evident in Fig. 8.9 where the dashed-black curve represents the measured power density along z. This deviation is compatible with the manufacturing error associated with the laser cutting of the slots in the metallic plate of the antenna. However, it can be considered a very good result for that electrically large antenna (60 λ_0 at 15 GHz) with almost ten thousand slots (i.e., 9154), as shown in Fig. 8.9b.

The measured radiated power density (normalized to its relative maximum value) in the xz-plane is shown in Fig. 8.10a. The −3 dB beam remains confined in the specified spatial region of diameter 12 cm for more than 2 m in the entire band of frequency. In Fig. 8.10b, the simulated and measured normalized power densities

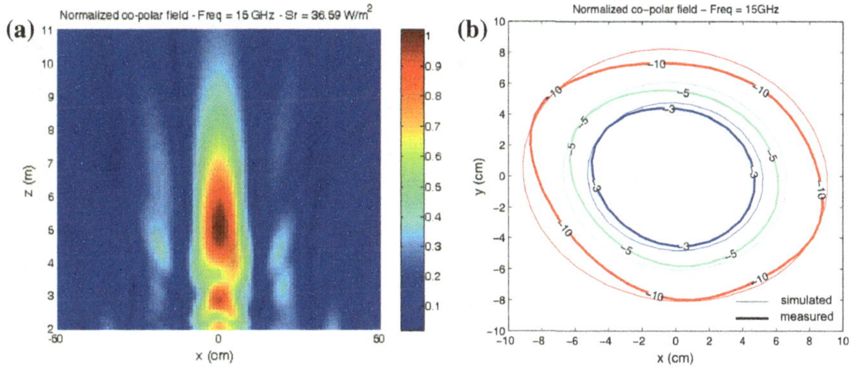

Fig. 8.10 Power density at 15 GHz, normalized to its maximum value, in the xz-plane (**a**) and in the xy-plane for a distance from the radiating surface of 5.0 m (**b**)

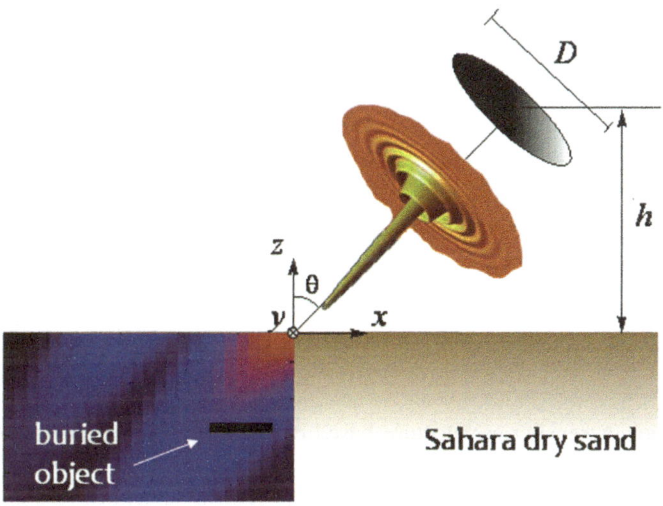

Fig. 8.11 Geometry of the problem: an antenna that produces a Bessel beam is used to detect a buried metallic plate. Specifically, we are considering a noncontact detection of buried objects at 15 GHz when a circularly polarized antenna is on-board of a vehicle. Practical observation suggests focalizing with a half-power beamwidth (HPBW) of a few centimeters (<12 cm) and with a not well-defined focalization distance in between 4 and 8 m (i.e., $200 \div 400\lambda_0$)

are compared at 5 m distance from the radiating surface. The agreement between the simulated and measured beam shapes is very good.

The antenna here described has been used for the specific application for non-contact microwave detection of buried mines, as sketched in Fig. 8.11.

In conventional ground penetrating radar, the detection cannot be performed in remote conditions, because of the very large clutter contribution that degrades radar

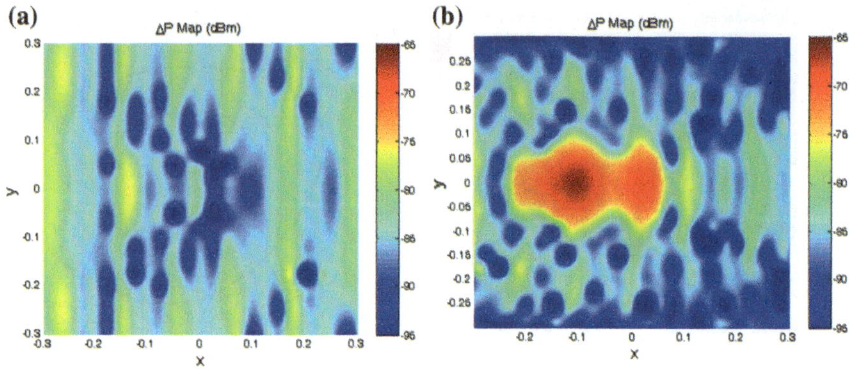

Fig. 8.12 Echo map relevant to a 10 cm × 10 cm metal plate buried 15 cm deep in Sahara dry sand when a classical pencil beam (**a**) or a Bessel beam (**b**) are used

Fig. 8.13 Radial waveguide loaded with metallic gratings and fed by a coaxial fed

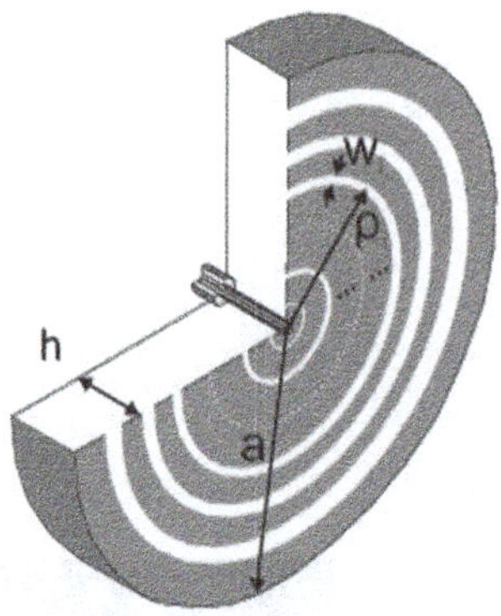

performance. A sensing architecture based on the concept of localized waves has been proposed in [34] to dramatically increase the signal-to-noise ratio. The advantage of using a Bessel beam instead of a classical pencil beam in such kind of ground penetrating radar is evident in the comparison of the relevant echo maps shown in Fig. 8.12.

8.3.5 Bessel Launcher Made by a Radial Waveguide Loaded with Metallic Gratings

The practical implementation of a Bessel beam launcher exploiting an inward-traveling wave aperture distribution is shown in Fig. 8.13. The structure is made by a parallel-plate waveguide (PPW) with annular slots etched on the top plate. A coaxial feed at the center excites an outward cylindrical radial wave inside the PPW. An in-house Method of Moments (MoM) [20] is used to take into account

Table 8.1 Geometrical sizes of the Bessel beam launcher: radial positions (ρ_i) and widths (w_i) of the annular slots

Slot	ρ_i (mm)	w_i (mm)	Slot	ρ_i (mm)	w_i (mm)
1	1.5	0.55	7	110	2.3
2	29.8	0.94	8	126.5	2.6
3	45.7	1.1	9	143	3
4	61.5	1.5	10	159.5	3.5
5	76.4	1.2	11	175.4	3.8
6	94.1	3	12	192.4	4

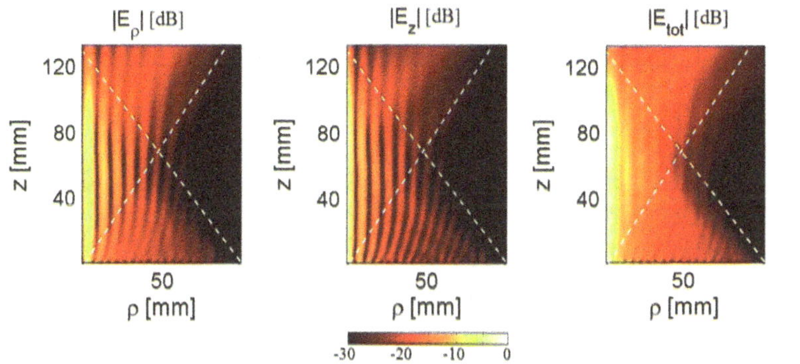

Fig. 8.14 Electric field radiated by the launcher in Fig. 8.13, calculated using COMSOL Multiphysics. From left to right, $|E_z|$, $|E_\rho|$ and $|\mathbf{E}|$. The axes are normalized with respect to the wavelength (λ) at the operating frequency. The dashed-white lines mark the GO boundaries

the mutual coupling amongst the various slots. The design follows the same steps outlined in Sect. 8.3.3 to tailor the magnetic current distribution over the top plate in the shape of an inward cylindrical wave. This is accomplished by controlling the positions along the radial direction ρ_i and sizes w_i of the circular slots (refer to Fig. 8.13).

Indeed, slots are positioned at those points where the phase of the outward cylindrical feeding wave inside the PPW matches that of the target inward cylindrical aperture distribution, according to a holographic criterion.

The slot width is used to modulate the amplitude of the aperture distribution. In addition the design procedure assures that the total energy launched by the coaxial feed in the PPW is radiated, thus avoiding any spurious radiation by the edges of the structure. The geometrical details of the structure are provided in Table 8.1. COMSOL multiphysics has been used to analyze the structure. Field maps of the electric field above the launcher are shown in Fig. 8.14, which are found very similar to their theoretical counterparts in Sect. 8.2.1. To allow a clearer quantitative comparison, we also show in Fig. 8.15 the E_z component of the electric field at the distance $z = 4.667\lambda$ from the aperture, corresponding to the maximum transverse extension of the rhomboidal region where the Bessel beam is created (refer to Fig. 8.14),

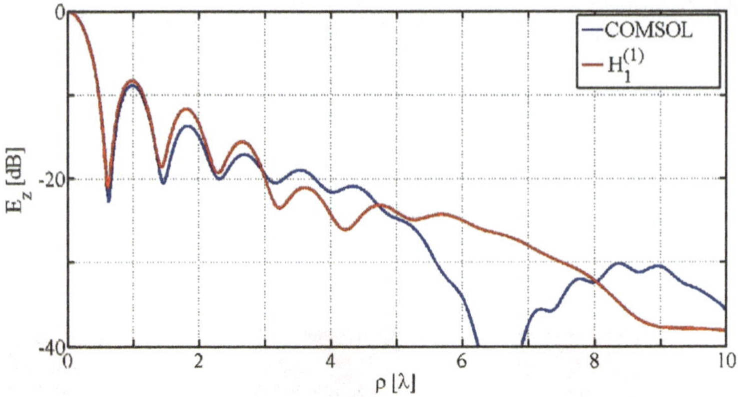

Fig. 8.15 Normalized $|E_z|$ component of the electric field at $z = 4.667\lambda$ for the structure in Fig. 8.13 simulated with COMSOL Multiphysics. The field radiated by a truncated inward Hankel distribution is shown for comparison

compared to the ideal truncated ($a = 7\lambda$) inward Hankel aperture distribution. The good agreement reveals the accuracy of the launcher design. It is worth noting that the traveling nature of the synthesized aperture distribution guarantees a wideband operation as a difference with resonant Bessel designs [9, 40, 55].

8.4 Leaky-Wave Bessel Beam Launchers

8.4.1 The Leaky Radial Waveguide: Motivations

In this section, we consider a propagating transverse magnetic (TM) Bessel beam [56]. The z-directed (in the direction of the structure's surface normal) electric field component takes on the following form: $E_z(\rho, z) = J_0(k_\rho \rho)e^{-jk_z z}$. Since the wave is TM-polarized, the other components of the electromagnetic field can be expressed as [57]

$$E_\phi = H_z = H_\rho = 0, \tag{8.56}$$

$$E_\rho = j\frac{k_z}{k_\rho}J_1(k_\rho \rho)e^{-jk_z z}, \tag{8.57}$$

$$H_\rho = j\frac{k_0}{\eta_0 k_\rho}J_1(k_\rho \rho)e^{-jk_z z}, \tag{8.58}$$

where η_0 is the characteristic impedance of free space and the wavenumbers obey the separation relation: $k_0^2 = k_\rho^2 + k_z^2$.

Fig. 8.16 A leaky radial waveguide loaded with a capacitive sheet impedance

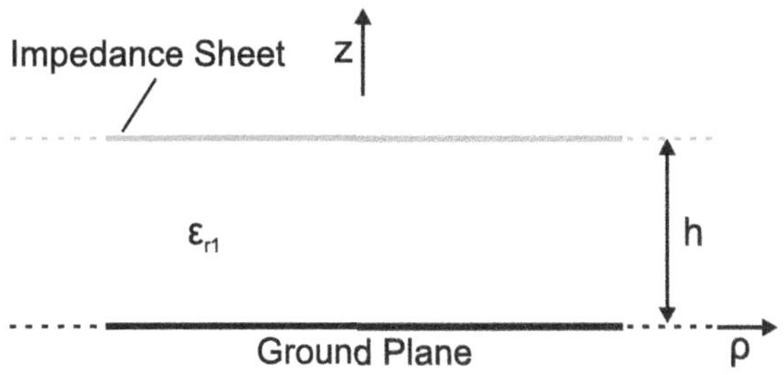

Equations (8.56)–(8.58) show that the components of the electric and magnetic field are Bessel functions, and do not change their profile with respect to z. Since the Bessel beam is propagating, the field components possess a transverse propagation constant k_ρ smaller than k_0, the free-space wavenumber. As a result, an open structure exciting a propagating Bessel beam supports an azimuthally invariant fast/leaky mode in the transverse direction. Since the structure supports a leaky mode, the normal wavenumber is complex: $k_z = \beta_z + j\alpha_z$ and the structure's surface impedance becomes

$$\eta_{surf} = -\frac{E_\rho}{H_\phi} = -\frac{\eta_0}{k_0}\left(\beta_z + j\alpha_z\right). \tag{8.59}$$

From (8.59), it is clear that a structure with a capacitive surface impedance is required to support a TM-polarized, cylindrical leaky mode [58]. A radiating system that supports such a leaky mode is the radial waveguide loaded with a capacitive impedance sheet, depicted in Fig. 8.16. This capacitively loaded radial waveguide will be studied in more detail in the next sections.

It is worth mentioning that the argument above for TM modes can be also extended to transverse electric (TE) modes. In the TE case, the z-component of the magnetic field presents a Bessel profile, and an inductive impedance sheet is required to generate an azimuthally invariant leaky mode inside the radial waveguide.

8.4.2 *Field Derivation for the Leaky Radial Waveguide*

The considered radial waveguide is shown in Fig. 8.16. It consists of a capacitive impedance sheet at a height h above a metallic ground plane. A ϕ-invariant TM mode with respect to the z-direction is considered. A magnetic vector potential with only a z-component (**A**) is used to derive the field radiated and within the radial waveguide. In particular, for a radial waveguide filled with a homogenous media with permittivity ε_{r1}, the vector potential can be expressed in cylindrical coordinates as [57]

$$A = \cos(k_{z1}z)\left[C_1 H_0^{(1)}(k_\rho \rho) + D_1 H_0^{(2)}(k_\rho \rho)\right]\hat{z}, \tag{8.60}$$

where k_{z1} and k_ρ are the normal and transverse propagation constants, and $H_0^{(1)}(.)$ and $H_0^{(2)}(.)$ are the zero-order Hankel functions of the first and second kind, respectively. The variables C_1 and D_1 denote amplitude coefficients stipulated by the boundary conditions and excitation.

The components of the electric and magnetic field can now be derived as

$$E_\phi = H_z = H_\rho = 0, \tag{8.61}$$

$$E_\rho = \frac{-j\eta_1}{k_1}\frac{\partial A}{\partial \rho \partial z}, \tag{8.62}$$

$$E_z = \frac{-j\eta_1}{k_1}\left(\frac{\partial^2}{\partial_z^2} + k_1^2\right)A, \tag{8.63}$$

$$H_\phi = -\frac{\partial A}{\partial \rho}, \tag{8.64}$$

where A, η_1, and k_1 are the z-components of the magnetic potential, characteristic impedance, and propagation constant within the radial waveguide, respectively. The fields within the radial waveguide can therefore be expressed as

$$E_\rho = \frac{-j\eta_1 k_{z1} k_\rho}{k_1}\sin(k_{z1}z)\left[C_1 H_1^{(1)}(k_\rho \rho) + D_1 H_1^{(2)}(k_\rho \rho)\right], \tag{8.65}$$

$$E_z = \frac{-j\eta_1 k_\rho^2}{k_1}\cos(k_{z1}z)\left[C_1 H_0^{(1)}(k_\rho \rho) + D_1 H_0^{(2)}(k_\rho \rho)\right], \tag{8.66}$$

$$H_\phi = k_\rho \cos(k_{z1}z)\left[C_1 H_1^{(1)}(k_\rho \rho) + D_1 H_1^{(2)}(k_\rho \rho)\right]. \tag{8.67}$$

Equations (8.65)–(8.67) show that the field components consist of two cylindrical waves: one traveling outward $\left(H_0^{(2)}(.) \text{ or } H_1^{(2)}(.)\right)$, and the other inward $\left(H_0^{(1)}(.) \text{ or } H_1^{(1)}(.)\right)$. The field component E_z assumes a J_0 profile when $C_1 = D_1$. Such field profiles can be achieved by placing a circular metallic rim at a distance $\rho = \rho_1$, which corresponds to a null of the J_0 function. Therefore, the cancelation of the z-directed electric field within the radial waveguide leads to

$$C_1 = -D_1 \frac{H_0^{(2)}(k_\rho \rho_1)}{H_0^{(1)}(k_\rho \rho_1)}. \tag{8.68}$$

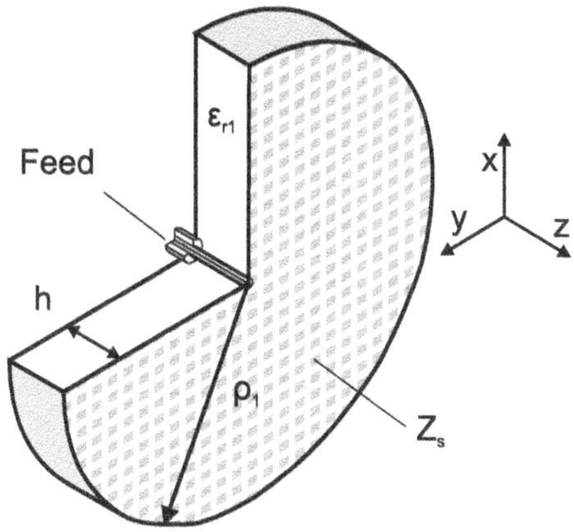

Fig. 8.17 The leaky radial waveguide with a metallic rim at $\rho = \rho_1$. An array of patches represents the impedance surface Z_s. The launcher is fed by a coaxial probe at the center

The corresponding structure is shown in Fig. 8.17.

Assuming a complex wavenumber along $\rho(k_\rho = \beta_\rho - j\alpha_\rho)$ and large rim radius ρ_1, the Hankel functions can be approximated as $H_0^{(1),(2)}(k_\rho \rho) \sim \sqrt{2/\pi k_\rho \rho} e^{\pm j(k_\rho \rho - \pi/4)}$. As a result, two conditions can be derived from (8.68) to ensure $C_1 = D_1$:

$$\beta_\rho \rho_1 = -\frac{\pi}{4} + q\pi, \tag{8.69}$$

$$\alpha_\rho \rho_1 \ll 1, \tag{8.70}$$

where q is an integer. Equation (8.69) requires that the inward and outward modes are out of phase at $\rho = \rho_1$, one of the zeros of the zeroth-order Bessel function. Equation (8.70) states that the amplitudes of the backward and forward leaky-wave modes are nearly identical. Note that in practical leaky-wave antennas, the leaky-wave mode is almost zero at the end of the structure, in order to avoid an additional sidelobe due to backward traveling leaky-wave modes. Here, the structure is resonant in the ρ−direction and the backward traveling mode is critical to establishing the Bessel beam.

Once Eqs. (8.69) and (8.70) are satisfied, the field components (8.65)–(8.67) can be rewritten as

$$E_\rho = C_1 \frac{-j\eta_1 k_{z1} k_\rho}{k_1} \sin(k_{z1} z) J_1(k_\rho \rho), \tag{8.71}$$

$$E_z = C_1 \frac{-j\eta_1 k_\rho^2}{k_1} \cos(k_{z1} z) J_0(k_\rho \rho), \tag{8.72}$$

$$H_\phi = -C_1 k_\rho \cos(k_{z1} z) J_1(k_\rho \rho). \tag{8.73}$$

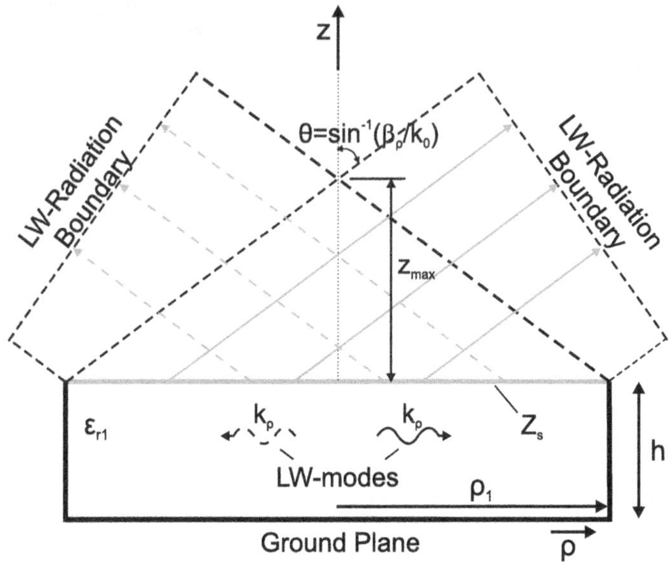

Fig. 8.18 Non-diffractive range of the leaky radial waveguide

As expected, the z-component of the electric field assumes a J_0 profile.

The radius ρ_1 of the Bessel beam launcher (leaky radial waveguide) is also directly related to the non-diffractive range of the truncated Bessel beam [3] (see Fig. 8.18):

$$z_{\max} = \rho_1 \sqrt{\left(\frac{k_0}{\beta_\rho}\right)^2 - 1}. \tag{8.74}$$

As shown in Fig. 8.18, the non-diffractive range is defined by the intersection of the leaky-wave radiation boundaries.

Two characteristics of the Bessel beam radiation become evident from Fig. 8.18 and (8.74): (1) larger non-diffractive ranges can be achieved with leaky-wave modes pointing closer to broadside; (2) the far-field radiation pattern is conical and points at $\theta = \sin^{-1}(\beta_\rho/k_0)$. As a result, the non-diffractive range can be located in the Fresnel zone or near-field of the structure by changing the transverse propagation constant from a value close to zero to k_0. In addition, larger non-diffractive ranges correspond to Bessel beams with wider null-to-null beamwidths, where the null-to-null beamwidth is given by $NNBW \sim 4.81/\beta_\rho$. Therefore, a tradeoff exists between beam size and non-diffractive range, where the leaky complex wavenumber is determined by (8.69), (8.70), and (8.74).

Finally, to completely define the fields within the leaky radial waveguide, the coefficient D_1 in (8.68) needs to be found. This can be done by choosing a particular excitation for the structure. Here, the coaxial excitation shown in Fig. 8.17 is selected. It is assumed that the total current flowing on the inner conductor of the coaxial probe's surface is uniform and equal to $I = 2\pi a J_s$, where a is the radius of the inner conductor and J_s is the surface current density. The assumption is valid for very thin radial waveguides [57], $h \ll \lambda_0(\lambda_0$, free-space wavelength at the operating frequency $f_0)$, and small values of the impedance sheet Z_s. Imposing the continuity of the ϕ-component of the magnetic field along the coaxial probe's surface and for small probes' radii, the coefficient D_1 can be found:

$$D_1 = \frac{I}{4j\left(1 + \frac{H_0^{(2)}(k_\rho \rho_1)}{H_0^{(1)}(k_\rho \rho_1)}\right)}. \tag{8.75}$$

8.4.3 Dispersion Analysis of the Leaky Radial Waveguide

In order to design the leaky radial waveguide, closed-form expressions for Z_s and h as a function of the transverse propagation constant are needed. These expressions can be found using the equivalent transmission line model of the radial waveguide shown in Fig. 8.19. The dispersion equation of the radial waveguide will be derived first. The parameters Z_s and h will then be expressed in terms of the normal propagation constant of the leaky-wave modes within the radial waveguide. Using the transverse resonance technique, the dispersion equation can be expressed as

$$Y_{\text{air}} + Y_s - jY_1^{TM} \cot(k_{z1} h) = 0, \tag{8.76}$$

with $Y_{\text{air}} = 1/Z_{\text{air}}, Y_s = 1/Z_s$ and $Y_1^{TM} = \varepsilon_{r1} k_0/\eta_0 k_{z1}$ denoting the characteristic admittance for a TM mode in free space. It is worth noting that the expression provided in [59] can be used for Z_{air}. However, in order to derive closed-form

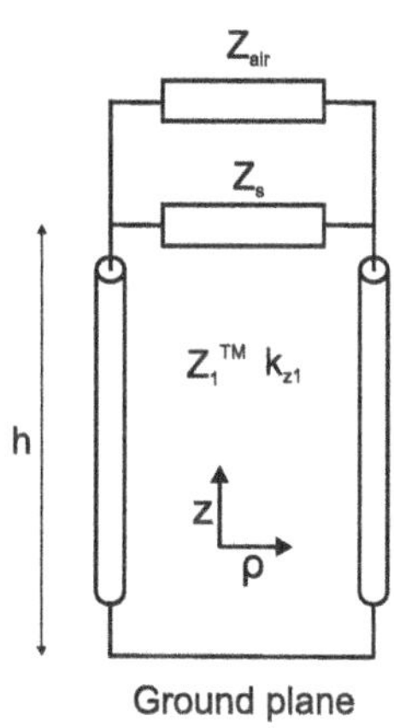

Fig. 8.19 Equivalent transmission line model for the leaky radial waveguide in Fig. 8.17

expressions for Z_s and h, in the following it is assumed Z_s equal to the free-space characteristic impedance of a TM mode [56].

Considering $Z_s = jX_s$, and using the small argument expression for the cotangent function, (8.76) becomes

$$\frac{k_0}{\eta_0 k_z} - j\frac{1}{X_s} - \frac{\varepsilon_{r1} k_0}{\eta_0 k_{z1}} \frac{1}{k_{z1} h} = 0. \quad (8.77)$$

Equation (8.77) can be further simplified for a hollow radial waveguide $(\varepsilon_{r1} = 1)$. Under this assumption, (8.77) simplifies to

$$k_z h - j\left(\frac{1}{X_s} \eta_0 h \frac{k_z^2}{k_0} + 1\right) = 0. \quad (8.78)$$

Given that the normal wavenumber is complex $(k_z = \beta_z + j\alpha_z)$ for leaky-wave modes, (8.78) can be separated into real and imaginary parts. After some algebrical simplifications, Z_s and h can be written as a functions of k_z:

$$X_s = \frac{-2\alpha_z}{k_0} \eta_0, \quad (8.79)$$

$$h = \frac{2\alpha_z}{\alpha_z^2 + \beta_z^2}. \quad (8.80)$$

Therefore, a unique solution can be found for Z_s and h once the transverse or normal propagation constant is fixed, assuming the structure is operating in its fundamental mode. The sheet impedance Z_s is found to be capacitive, as needed for exciting TM-leaky-wave modes [58]. Equations (8.79) and (8.80) show that the impedance sheet values are determined by the attenuation constant of the leaky-wave mode, whereas the height h is dependent on both the propagation and attenuation constants. In particular, lower values of h correspond to faster leaky-wave modes, at a given frequency of operation. Finally, the expressions for the general case of dielectric filled higher order (not operating on the principal mode) leaky-wave radial waveguide can be found in [60]. Reference [60] also defines the operating bandwidth of Bessel launchers based on radial leaky waveguides.

8.4.4 *Leaky Radial Waveguide: Experimental Results*

In this section two prototypes are presented for the leaky-wave Bessel beam launchers. The final prototypes have been designed using the procedure outlined in the previous sections. More details about the structures and designs can be found in [25, 60].

Fig. 8.20 Leaky-wave Bessel beam launcher operating in X-band

The first prototype is shown in Fig. 8.20. It operates at $f_0 = 10\,\text{GHz}$ using the principal leaky-wave mode within the radial waveguide. The launcher is fed by a coaxial probe at the center of the structure.

The normalized transverse complex wavenumber of the excited leaky-wave mode is $k_\rho/k_0 = 0.8 - j0.0167$. The aperture radius is fixed equal to 88.89 mm (2.96 λ_0, λ_0 is the wavelength at f_0). The sheet impedance and radial waveguide height are given by (8.79) and (8.80): $Z_s = -16.77j\,\text{Ohm}, h = 0.588\,\text{mm}$. The required impedance sheets for the two designs shown in this section are realized by printing periodic square patch elements on both sides of a thin dielectric substrate, as shown in Fig. 8.20, [25].

The E_z component of the electric field above the prototype was measured using a short electric probe attached to an automatically controlled 3D translation stage. The electric probe was approximately $\lambda_0/10$ (λ_0 is the free-space wavelength at f_0 = 10 GHz). Figure 8.21a and b reports the normalized E_z component of the electric field of the Bessel beam launcher at z = 0.75 λ_0. The measured fields are at f = 9.54 GHz [25]. This frequency shift with respect to the initial operating frequency is attributed to tolerances inherent to the milling and etching fabrication processes used for the prototype. Close agreement is observed between the simulated and measured results. The cylindrical symmetry of the measured Bessel beam is evident in the 2D plot of Fig. 8.21b. The dark and light rings correspond to nulls and maxima of the Bessel beam. The 2D Fourier transform is shown in Fig. 8.21c. The spectral ring of the Bessel beam is prominent confirming the generation of the Bessel beam in free space. Figure 8.21d provides the measured E_z component of the electric field along the *xz*-plane within the non-diffractive range equal to z_{max} 2.1 λ_0. The main beam, zeros, and shape of the field do not change for distances up to z = 2 λ_0, further verifying the Bessel beam operation. However, at z = 2 λ_0 higher sidelobes appear at the edges of the measurements.

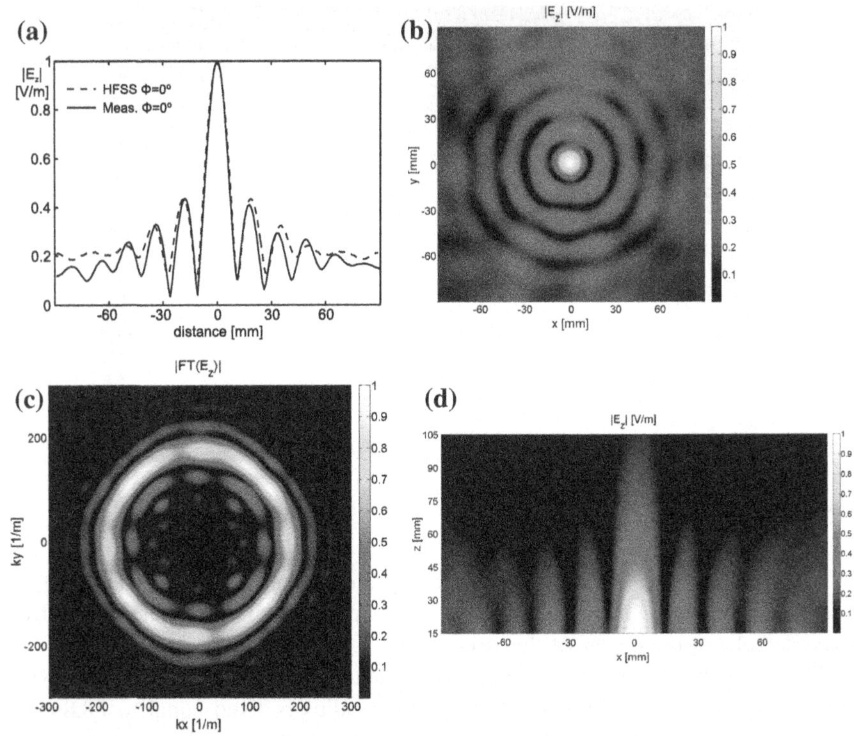

Fig. 8.21 Measured and normalized E_z component of the electric field. **a** Comparison between measured and simulated results at $z = 0.75\ \lambda_0$ from the impedance surface. **b** 2D plot at $z = 0.75\ \lambda_0$ from the impedance surface. **c** Normalized amplitude of the Fourier transform of the measured E_z component of the electric field at the operating frequency at $z = 0.75\ \lambda_0$ from the impedance surface. **d** 2D plot along the xz-plane

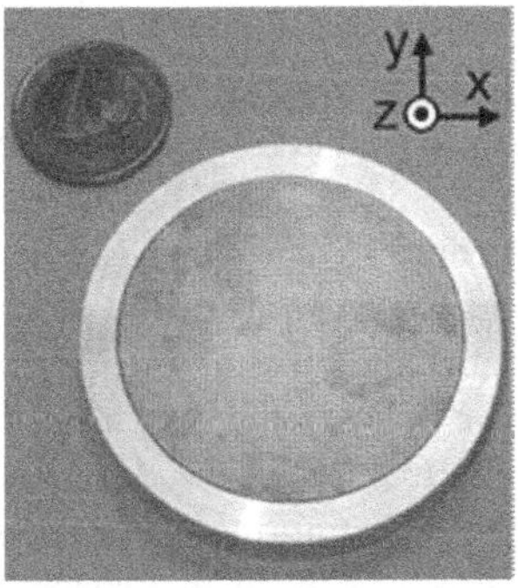

Fig. 8.22 Leaky-wave Bessel beam launcher operating in Ka-band

The second prototype is shown in Fig. 8.22. The aim of this prototype is to use higher order leaky-wave modes for the generation of Bessel beams [60]. As a difference with the structure in Fig. 8.20, the present launcher is suited to Printed

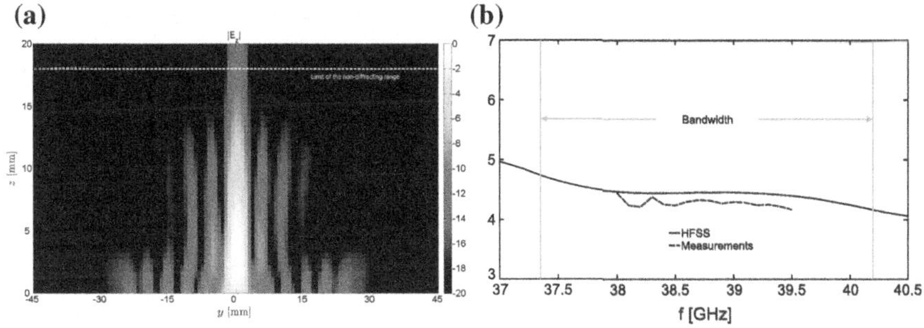

Fig. 8.23 **a** Measured E_z component of the normalized electric field at f = 38.3 GHz along the xz-plane. The *white-dashed line* corresponds to the limit of the non-diffractive range. **b** Variation of the HPBW versus frequency for the Bessel beam launcher at $z = \lambda_0$ from the radiating aperture. The *vertical lines* mark the extrema of the maximum available bandwidth as defined in [60]

Circuit Board (PCB) technology. The operating frequency is $f = 39.3\,\text{GHz}$ with a final radiating aperture of about 44.6 mm and a thickness smaller than 3.5 mm. The launcher is still fed by a simple coaxial probe at the center of the structure.

Also in this case, the E_z component of the electric field above the prototype was measured using a short electric probe attached to an automatically controlled 3D translation stage. Figure 8.23 provides the measured E_z field component along the xz-plane. As expected, the half-power beamwidth (HPBW) of the launched Bessel beam is preserved within the non-diffractive range with an upper limit at z = 16.4 mm (marked with a dashed-white line). Above this distance, diffraction prevails, and the beam starts to widen.

The variation of the measured HPBW with the frequency at a fixed distance ($z = \lambda_0$, with λ_0 is the free-space wavelength at f = 40 GHz) from the radiating aperture is shown in Fig. 8.23b. Simulation results obtained with Ansys HFSS are also proved for comparison. Due to the dispersive behavior of leaky-wave mode generating the beam, both the simulated and the measured HPBWs slowly decrease within the band 37.3–40.3 GHz. However, the measured HPBW remains almost constant within the frequency range 38–39.5 GHz. This result proves that the proposed launcher can generate a Bessel launcher over a relative bandwidth of about 4%.

8.4.5 Bessel Launcher Using a Modulated Metasurface

This section presents an implementation of Bessel beam launchers using modulated metasurfaces [32]. The modulated metasurface (surface impedance) is used to transform a surface wave (SW), launched by a feeder, into a LW mode by using a locally periodic modulation. In the following, we suppose that the feeder launches a cylindrical incident SW propagating along a metasurface placed in the xy-plane. For a lossless non-modulated inductive scalar surface impedance $Z_s = j\bar{X}_s(\bar{X}_s > 0)$, the dominant SW mode has a TM_z polarization (where z is the normal to the surface, and TM stands for transverse magnetic). The magnetic field on the surface is assumed to be of the form

$$\mathbf{H}^{inc}_{aperture}(\boldsymbol{\rho}') = A^{TM}(\boldsymbol{\rho}')H_0^{(2)}(k_t\rho)\hat{\boldsymbol{\rho}}, \tag{8.81}$$

where $\boldsymbol{\rho}' = x'\hat{\mathbf{x}} + y'\hat{\mathbf{y}}$ is a general point on the antenna surface, A^{TM} the amplitude, and k_t is obtained by solving the transverse dispersion problem [61].

The field radiated by the locally modulated metasurface can be expressed as the field radiated by the equivalent surface magnetic currents, defined as

$$\begin{aligned}\mathbf{M}(\mathbf{r}') &= 2\mathbf{E}(\mathbf{r}') \times \hat{\mathbf{z}} = 2\left[Z_s(\boldsymbol{\rho}')\hat{\mathbf{z}} \times \mathbf{H}^{inc}_{aperture}(\boldsymbol{\rho}')\right] \times \hat{\mathbf{z}} \\ &= -2A^{TM}(\boldsymbol{\rho}')Z_s(\boldsymbol{\rho}')H_0^{(2)}(k_t\rho)\hat{\boldsymbol{\rho}}. \end{aligned} \tag{8.82}$$

A Bessel beam polarized along z-direction can be obtained by a magnetic current aperture distribution placed in $z = 0$ of the form (8.34) polarized along $\hat{\rho}$. The radial wavenumber k_ρ and the radius of the metasurface $\rho_{\max}$ are directly related to the non-diffractive range of the Bessel beam. The needed phase behavior can be obtained from (8.82) using the following surface impedance [32]:

$$Z_s(\boldsymbol{\rho}') = j\bar{X}_s\left(1 + M(\boldsymbol{\rho}')\cos\left[-k_t\rho' + \frac{3}{4}\pi - \pi\chi\left(-\cos\left(\beta_\rho^B\rho' - \frac{3}{4}\pi\right)\right)\right]\right), \tag{8.83}$$

where $\chi(x) = \begin{cases} 1\, x \geq 0 \\ 0\, x < 0 \end{cases}$ is the step function.

It can be noted that (8.83) can also assure a similar amplitude law with respect to (8.34) since the Bessel and the Hankel functions have the same amplitude factor in their asymptotic expansion for large arguments. The two amplitude laws match exactly only in the case where $k_\rho = k_t$. This latter condition could not be satisfied since $k_\rho < k_0$ while $k_t > k_0$. The effect of this small amplitude error (tapering) is shown in Fig. 8.24a where a circular aperture of radius 3λ illuminated with a Bessel function with $k_\rho = 0.6k_0$ has been considered. As can be seen, the above amplitude difference gives rise to small differences on the secondary lobes while the focusing effect is still present. However, it can be noticed that a perfect amplitude matching could be obtained by locally changing radially the average impedance $\bar{X}_s(\rho)$.

An alternative formulation can be obtained by using an inward cylindrical wave distribution as in (8.1). Also for this case, a tapered distribution law can be obtained (amplitude and phase) using the following expression [32]:

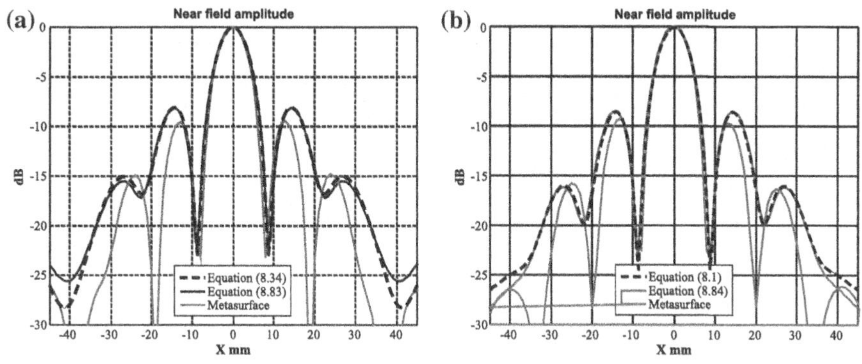

Fig. 8.24 **a** Near-field comparison at a distance of 2λ from the antenna between an ideal Bessel beam, the field generated by an aperture described by Eq. (8.34) and its metasurface antenna implementation (8.83). **b** Same comparison using the aperture distribution (8.1) and its metasurface implementation (8.84)

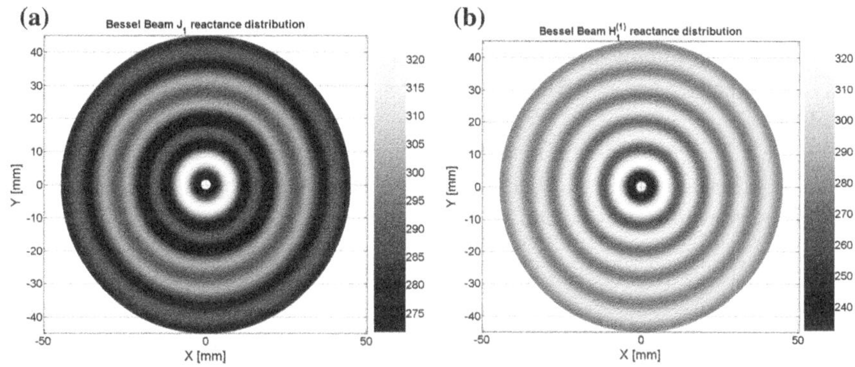

Fig. 8.25 Reactance chart for a Bessel beam polarized along z, generated with a J_1 distribution or an $H_1^{(1)}$ distribution. The average impedance is Z_s = j 290 Ω and modulation index M = 0.17

$$Z_s(\boldsymbol{\rho}') = j\bar{X}_s\left(1 + M(\boldsymbol{\rho}')\mathrm{Re}\left\{e^{jArg\left(\mathrm{H}_1^{(2)}(k_t\rho')\right)}e^{-jArg\left(\mathrm{H}_1^{(1)}(\beta_\rho^B\rho')\right)}\right\}\right). \tag{8.84}$$

Similar consideration on the aperture field amplitudes holds also for this configuration as shown in Fig. 8.24b.

Figure 8.24 shows a comparison between the ideal Bessel beam, the tapered ideal beam, and the field radiated by the metasurface antenna at 2λ from the antenna obtained with HFSS for both configurations. A good agreement is observed between theory and full-wave results. Reactance chart examples for both formulations are shown in Fig. 8.25. An operating frequency of 20 GHz is considered in all cases.

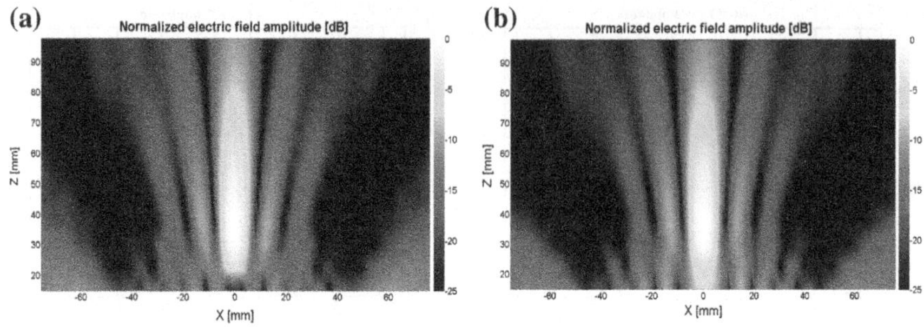

Fig. 8.26 Normalized near-field scanning in dB for a Bessel beam antenna **a** using the aperture distribution (8.34). **b** Alternative aperture distribution (8.1)

The two presented configurations have been implemented using a circular panel of radius 3λ. The metasurface impedance is realized by printing squared patches with different dimensions on a squared lattice of side 1.25 mm on a grounded dielectric slab (Rogers 4350b with a relative permittivity of 3.66 and a thickness of 1.52 mm). A higher modulation index has been used close to the periphery of the antenna in order to prevent reflections [32]. The parameter for the Bessel functions has been selected as $k_\rho = 0.6k_0$; the corresponding non-diffracting zone is given by $z \leq 4\lambda$.

Figure 8.26 represents the map of the near-field generated by the metasurface launcher. The non-diffractive behavior of the generated beam can be recognized within the non-diffracting region of the launcher.

8.5 Conclusions and Future Directions

This chapter has briefly introduced the concept of non-diffractive radiation with a special focus on Bessel beams. The possibility to generate Bessel beams with inward-traveling waves has been presented. It has been proved that inward-traveling waves can generate non-diffractive Bessel beams over a rhomboidal region in front of the radiating aperture. In addition, the non-diffractive radiation is achieved over a large bandwidth with respect to classical methods based on resonant radiating aperture. The wideband feature of such launchers is extremely appealing for the generation of non-diffractive pulses and in particular of X-waves [1, 2, 62]. Besides, planar radiating structures are preferred at millimeter waves to cumbersome axicon lenses adopted at optical frequencies [9, 10].

An efficient design procedure has then been presented for planar Bessel launchers based on radial waveguides loaded with radiating slots (RLSA antennas) or metallic gratings. The procedure allows synthetizing a defined aperture field distribution by controlling the location and size of the slots or metallic gratings of the radial waveguide. The procedure has been adopted for designing a RLSA Bessel launcher operating in Ku-band with circular polarization and a deep of focus of several meters. Such launcher can find application in remote sensing for the detection of buried objects [34] or as a field concentrator for drilling operations in planetary explorations [63].

The same design procedure has been used for a Bessel launcher using inward-traveling waves. The launcher is made by a radial waveguide loaded with metallic gratings. The structure is suited to low-cost printed circuit board fabrication processes. The wideband behavior, low loss, and low cost of such launcher are attractive features for imaging applications at millimeter waves [64]. Besides, the proposed structures may find applications for future short-range wireless links for next-generation communication networks in D-band (120 GHz) and even beyond 300 GHz [65]. On the other hand, the design procedure proposed here for non-diffractive launcher can be adopted for the design of antenna systems with an arbitrary near-field profile [66].

The chapter has also outlined the possibility to generate Bessel beams with leaky-wave modes. The non-diffractive beam is obtained by the radiation of an inward and outward leaky-wave mode traveling within a radial waveguide. These modes are excited by a coaxial probe at the center of the radial waveguide loaded with a uniform metasurface. The propagation constant of the leaky-wave modes controls the non-diffractive range and beamwidth of the generated beam. In particular, it has been shown that resonant apertures are needed for the generation of the non-diffractive radiation. Despite the resonant nature of the radiating aperture, Bessel launchers based on leaky-wave modes are a novel paradigm for the generation of non-diffractive radiation.

The combination of this concept with the recent achievements on inward-traveling waves presented in this chapter may pave the way to extremely efficient launcher at millimeter waves. Recent works are already pursuing such possibility by using metasurfaces [27, 30, 32]. The possible applications of such launchers include low-profile antennas [67], wireless power transfer systems [68], microparticle manipulations with tractor beams [69], and the generation of surface plasmon polaritons [70].

References

1. H.E. Hernandez-Figueroa, M. Zamboni-Rached, E. Recami, *Localized Waves* (J. Wiley, NY, 2008)
2. H.E. Hernandez-Figueroa, E. Recami, M. Zamboni-Rached, *Non-diffractive Waves* (J. Wiley-VCH, Berlin, 2013)
3. J. Durnin, Exact solutions for nondiffracting beams. I. The scalar theory. J. Opt. Soc. Am. A **4**, 651–654 (1987)
4. J. Durnin, J.J. Miceli Jr., J.H. Eberly, Diffraction-free beams. Phys. Rev. Lett. **58**, 1499 (1987)
5. S. Ruschin, A. Leizer, Evanescent Bessel beams. J. Opt. Soc. Amer. A **15**(5), 1139–1143 (1998)

6. M.A. Bandres, J.C. Gutierrez-Vega, S. Chàvez-Cerda, Parabolic nondiffracting optical wave fields. Opt. Lett. **29**, 44–46 (2004)
7. J.C. Gutierrez-Vega, M.D. Iturbe-Castillo, S. Chàvez-Cerda, Alternative formulation for invariant optical fields: mathieu beams. Opt. Lett. **25**, 1493–1495 (2000)
8. J. Salo, J. Fagerholm, A.T. Friberg, M.M. Salomaa, Unified description of nondiffracting X and Y waves. Phys. Rev. A **62**, 4261 (2000)
9. M.R. Lapointe, Review of non-diffracting Bessel beam experiments. Opt. Laser Tech. **24**(6), 315 (1992)
10. J. Arlt, K. Dholakia, Generation of high-order Bessel-beams by use of an axicon. Opt. Commun. **177**(4), 297–301 (2001)
11. O. Brzobohaty, T. Cizmàr, P. Zemànek, High quality quasi-Bessel beam generated by round-tip axicon. Opt. Express **16**, 12688–12700 (2008)
12. J. Turunen, A. Vasara, A.T. Friberg, Holographic generation of diffraction-free beams. Appl. Opt. **27**(19), 3959–3962 (1988)
13. J. Williams, J.B. Pendry, Generating Bessel beams by use of localized modes. J. Opt. Soc. Am. A **22**(5), 992–997 (2005)
14. G. Indebetouw, Nondiffracting optical fields: some remarks on their analysis and synthesis. J. Opt. Soc. Am. A. **6**(1), 150–152 (1989)
15. A.J. Cox, D.C. Dibble, Nondiffracting beam from a spatially filtered Fabry-Perot resonator. J. Opt. Soc. Am. A. **9**(2), 282–286 (1992)
16. Q. Zhan, Cylindrical vector beams: from mathematical concepts to applications. Adv. Opt. Photon. **1**, 1–57 (2009)
17. D. McGloin, K. Dholakia, Bessel beams: diffraction in a new light. Contemp. Phys. **46**, 15–28 (2005)
18. S. Monk, J. Arlt, D.A. Robertson, J. Courtial, M.J. Padgett, The generation of Bessel beams at millimetre-wave frequencies by use of an axicon. Opt. Commun. **170**, 213–215 (2000)
19. J. Salo, J. Meltaus, E. Noponen, J. Westerholm, M.M. Salomaa, A. Lönnqvist, J. Säily, J. Häkli, J. Ala-Laurinaho, A.V. Räisänen, Millimetre-wave Bessel beams using computer holograms. Electron. Lett. **37**(12), 834–835 (2001)
20. S.C. Pavone, M. Ettorre, M. Albani, Analysis and design of Bessel beam launchers: longitudinal polarization. IEEE Trans. Antennas Propag. **64**, 2311–2318 (2016)
21. Z. Li, K.B. Alici, H. Caglayan, E. Ozbay, Generation of an axially asymmetric Bessel-like beam from a metallic subwavelength aperture. Phys. Rev. Lett. **102**, 143901 (2009)
22. D. Mugnai, A. Ranfagni, R. Ruggeri, Observation of superluminal behaviors in wave propagation. Phys. Rev. Lett. **84**, 4830–4833 (2000)
23. P. Lemaitre-Auger, S. Abielmona, C. Caloz, Generation of Bessel beams by two-dimensional antenna arrays using sub-sampled distributions. IEEE Trans. Antennas Propag. **61**, 2645–2653 (2012)
24. M.A. Salem, A.H. Kamel, E. Niver, Microwave Bessel beams generation using guided modes. IEEE Trans. Antennas Propag. **59**(6) part 2, 2241–2247 (2011)
25. M. Ettorre, S.M. Rudolph, A. Grbic, Generation of propagating Bessel beams using leaky-wave modes: experimental validation. IEEE Trans. Antennas Propag. **60**, 2645–2653 (2012)
26. C. Pfeiffer, A. Grbic, Metamaterial Huygens' surfaces: tailoring wave fronts with reflectionless sheets. Phys. Rev. Lett. **110**, 197401 (2013)
27. C. Pfeiffer, A. Grbic, Controlling vector Bessel beams with metasurfaces. Phys. Rev. Appl. **2**, 044012 (2014)
28. Y.B. Li, B.G. Cai, X. Wan, T.J. Cui, Diffraction-free surface waves by metasurfaces. Opt. Lett. **39**, 5888–5891 (2014)
29. F. Aieta, P. Genevet, M.A. Kats, N. Yu, R. Blanchard, Z. Gaburro, F. Capasso, Aberration-free ultrathin flat lenses and axicons at telecom wavelengths based on plasmonic metasurfaces. Nano Lett. **12**, 4932–4936 (2012)
30. B.G. Cai, Y.B. Li, W.X. Jiang, Q. Cheng, T.J. Cui, Generation of spatial Bessel beams using holographic metasurface. Opt. Express **23**, 7953—7601 (2015)

31. M.Q. Qi, W.X. Tang, T.J. Cui, A broadband Bessel beam launcher using metamaterial lens. Sci. Rep. **5**, 11732 (2015)
32. M. Casaletti, M. Śmierzchalski, M. Ettorre, R. Sauleau, N. Capet, Polarized beams using scalar metasurfaces. IEEE Trans. Antennas Propag. **AP-64**(8), 3391–3400 (2016)
33. M.F. Imani, A. Grbic, Generating evanescent Bessel beams using near-field plates. IEEE Trans. Antennas Propag. **60**, 3155–3164 (2012)
34. A. Mazzinghi, M. Balma, D. Devona, G. Guarnieri, G. Mauriello, M. Albani, A. Freni, Large depth of field pseudo-Bessel beam generation with a RLSA Antenna. IEEE Trans. Antennas Propag. **62**(15), 3911–3919 (2014)
35. M. Ettorre, S.C. Pavone, M. Casaletti, M. Albani, Experimental validation of Bessel beam generation using an inward Hankel aperture distribution. IEEE Trans. Antennas Propag. **63** (6), 2539–2544 (2015)
36. L.B. Felsen, N. Marcuvitz, *Radiation and Scattering of Waves. Electromagnetic Wave Theory* (IEEE Press, Piscataway, NJ, 1994)
37. R. Cicchetti, A. Faraone, Incomplete Hankel and modified Bessel functions: a class of special functions for electromagnetics. IEEE Trans. Antennas Propag. **52**, 3373–3389 (2004)
38. M. Albani, S.C. Pavone, M. Casaletti, M. Ettorre, Generation of non-diffractive Bessel beams by inward cylindrical traveling wave aperture distributions. Opt. Express **22**, 18354–18364 (2014)
39. S.C. Pavone, M. Ettorre, M. Casaletti, M. Albani, Transverse circular-polarized Bessel beam generation by inward cylindrical aperture distribution. Opt. Express **24**, 11103–11111 (2016)
40. M. Albani, A. Mazzinghi, A. Freni, Automatic design of CP-RLSA antennas. IEEE Trans. Antennas Propag. **60**, 5538–5547 (2012)
41. M. Ando, K. Sakurai, N. Goto, K. Arimura, Y. Ito, A radial line slot antenna for 12 GHz satellite TV reception. IEEE Trans. Antennas Propag. **AP-33**(12), 1347–1353 (1985)
42. M. Takahashi, J. Takada, M. Ando, N. Goto, A slot design for uniform aperture field distribution in single-layered radial line slot antennas. IEEE Trans. Antennas Propag. **39**(7), 954–959 (1991)
43. T. Yamamoto, M. Ando, N. Goto, and M. Takahashi, Application of radial line slot antenna to mil-limiter-wave systems at 60 GHz band, in *Proceedings of the 1996 IEEE AP-S International Symposium*, vol. 1, pp. 62–65, 21–26 July 1996
44. H. Elmikati, A. Elsohly, Extension of projection method to nonuniformly linear antenna arrays. IEEE Trans. Antennas Propagat. **32**(5), 507–512 (1984)
45. O.M. Bucci, G. D'Elia, G. Mazzarella, G. Panariello, Antenna pattern synthesis: a new general approach. Proc. IEEE **82**(3), 358–371 (1994)
46. H.A. Bethe, Theory of diffraction by small holes. Phys. Rev. **66**(7–8), 163–182 (1944)
47. C.A. Balanis, *Advanced Engineering Electromagnetics*, Chapter 9 (Wiley, NY, 1989)
48. J. Takada, M. Ando, N. Goto, A reflection canceling slot set in a linearly polarized radial line slot antenna. IEEE Trans. Antennas Propag. **AP-40**(4), 433–438 (1992)
49. P.W. Davis, M.E. Bialkowski, Experimental investigation into a linearly polarized radial slot antenna for DBS TV in Australia. IEEE Trans. Antennas Propag. **AP-45**(7), 1123–1129 (2007)
50. M.E. Bialkowski, P.W. Davis, Analysis of a circular patch antenna radiating in a parallel-plate radial guide. IEEE Trans. Antennas Propag. **AP-50**(2), 180–187 (2002)
51. A. Akiyama, T. Yamamoto, M. Ando, N. Goto, Numerical optimisation of slot parameters for a concentric array radial line slot antenna. IEE Proc. Microw. Antennas Propag. **145**(2), 141–146 (1998)
52. J.I. Herranz, A. Valero-Nogueira, F. Vico, V.M. Rodrigo, Optimization of beam-tilted linearly polarized radial-line slot-array antennas. IEEE Antennas Wirel. Propag. Lett. **9**, 1165–1168 (2010)
53. J.I. Herranz-Herruzo, A. Valero-Nogueira, M. Ferrando-Bataller, Optimization technique for linearly polarized radial-line slot-array antennas using the multiple sweep method of moments. IEEE Trans. Antennas Propag. **AP-52**(4), 1015–1023 (2004)

54. H. Ueda, J. Hirokawa, M. Ando, M. Albani, A coaxial feeder with two pairs of parasitic pins for realizing rotationally symmetric aperture illumination in spiral array radial line slot antennas. IECIE Trans. Commun. **E93-B**, 2554–2561 (2010)
55. S. Chàvez-Cerda, A new approach to Bessel beams. J. Mod. Opt. **46**, 923–930 (1999)
56. M. Ettorre, A. Grbic, Generation of propagating Bessel beams using leaky-wave modes. IEEE Trans. Antennas Propag. **60**, 3605–3613 (2012)
57. R.F. Harington, *Time-Harmonic Electromagnetic Fields* (Wiley-IEEE Press, New York, 2001)
58. R.E. Collin, F.J. Zucker, *Antenna Theory* (McGraw-Hill, New York, 1969)
59. N. Marcuvitz, *Waveguide Handbook* (McGraw-Hill, New York, 1951)
60. W. Fuscaldo, G. Valerio, A. Galli, R. Sauleau, A. Grbic, M. Ettorre, Higher-order leaky-mode Bessel-Beam launcher. IEEE Trans. Antennas Propag. **60**(3), 904–913 (2016)
61. A. Oliner, A. Hessel, Guided waves on sinusoidally-modulated reactance surfaces. IRE Trans. Antennas Propag. **7**(5), 201–208 (1959)
62. W. Fuscaldo, S.C. Pavone, G. Valerio, A. Galli, M. Albani, M. Ettorre, Analysis of limited-diffractive and limited-dispersive X-waves generated by finite radial waveguides. J. Appl. Phys. **119**, 194903 (2016)
63. S. Costanzo, G. Di Massa, A. Borgia, A. Raffo, T.W. Versloot, L. Summerer, Microwave Bessel beam launcher for high penetration planetary drilling operations, in *10th European Conference on Antennas and Propagation*, Switzerland (2016)
64. H. Meng, J. Zhang, B. Xiang, W. Dou, Y. Yu, *Millimeter Wave Imaging with Bessel Beams* (IEEE MTT-S International Microwave Symposium Digest, USA, 2013)
65. P. Smulders, The road to 100 Gb/s wireless and beyond: basic issues and key directions. IEEE Commun. Mag. **51**(12), 86–91 (2013)
66. M. Ettorre, M. Casaletti, G. Valerio, R. Sauleau, L. Le Coq, S.C. Pavone, M. Albani, On the near-field shaping and focusing capability of a radial line slot array. IEEE Trans. Antennas Propag **62**, 1991–1999 (2012)
67. C. Pfeiffer, A. Grbic, Planar lens antennas of subwavelength thickness: collimating leaky-waves with metasuraces. IEEE Trans. Antennas Propag. **AP-63**(7), 3248–3253 (2015)
68. M. Ettorre, M. Casaletti, A. Grbic, Power link budget for propagating Bessel beams, in *IEEE Antennas and Propagation Society International Symposium (APSURSI)* (2013), pp. 960–961
69. C. Pfeiffer, A. Grbic, Generating stable tractor beams with dielectric metasurfaces. Phys. Rev. **B 91**, 115408 (2015)
70. J. Lin, J. Dellinger, P. Genevet, B. Cluzel, F. de Fornel, F. Capasso, Cosine-Gauss plasmon beam: a localized long-range nondiffracting surface wave. Phys. Rev. Lett. **109**, 093904 (2012)

Chapter 9
Metasurface Antennas

Gabriele Minatti, Marco Faenzi, Mario Mencagli, Francesco Caminita, David González Ovejero, Cristian Della Giovampaola, Alice Benini, Enrica Martini, Marco Sabbadini and Stefano Maci

Abstract This chapter reports design and analysis methods for planar antennas based on modulated metasurfaces (MTSs). These antennas transform a surface wave (SW) into a leaky wave by means of the interaction with a MTS having a spatially modulated equivalent impedance. The basic concept is that the MTS imposes the impedance boundary conditions (BCs) seen by the SW, and therefore the MTS controls amplitude, phase, and polarization of the aperture field. Thus, MTS antennas are highly customizable in terms of their performances, by simply changing the MTS and without affecting the overall structure. Several technological solutions can be adopted to implement the MTS, from sub-wavelength patches printed on a grounded slab at microwave frequencies, to a bed of nails structure in the millimetre and sub-millimetre wave range: in any case, the resulting device has light weight and a low profile. The design of the MTS is based on a generalized form of the Floquet wave theorem adiabatically applied to curvilinear locally periodic BCs. The design defines the continuous BCs required for reproducing a

---

G. Minatti · M. Faenzi · M. Mencagli · F. Caminita · C.D. Giovampaola · A. Benini · E. Martini · S. Maci (✉)
Department of Information Engineering and Mathematics, University of Siena, Via Roma 56, 53100 Siena, Italy
e-mail: macis@diism.unisi.it

G. Minatti
e-mail: gabriele.minatti@unisi.it

M. Faenzi
e-mail: faenzi@diism.unisi.it

M. Mencagli
e-mail: mencagli@diism.unisi.it

C.D. Giovampaola
e-mail: dellagiovampaola@diism.unisi.it

A. Benini
e-mail: benini@diism.unisi.it

E. Martini
e-mail: enrica.martini@wave-up.it

A. Boriskin and R. Sauleau (eds.), *Aperture Antennas for Millimeter and Sub-Millimeter Wave Applications*, Signals and Communication Technology, DOI 10.1007/978-3-319-62773-1_9 desired aperture field, and it is verified by a fast full-wave solver for impedance BCs. Next, the continuous BCs are discretized and implemented by a distribution of electrically small printed metallic elements in a regular lattice, like pixels in an image. The final layout is composed of tens of thousands of pixels and it is analyzed by a full-wave solver which makes use of entire domain basis functions combined with a fast-multipole algorithm. Examples of design and realizations of MTS antennas are shown, proving the effectiveness of the concept.

9.1 Introduction

Metasurfaces (MTSs) are thin metamaterial layers characterized by unusual reflection properties of plane waves and/or dispersion properties of surface/guided waves. Metamaterials are artificial materials that can be formed by periodic arrangements of many small inclusions in a dielectric host environment, for achieving macroscopic electromagnetic or optical properties that cannot be found in nature.

MTSs have found several applications in designing antennas and microwave components and they may be distinguished in *penetrable* and *impenetrable*. A penetrable MTS (sometimes called *metafilm*) is constituted by a planar distribution of small periodic elements in a very thin host medium [1]. A ground plane is instead a constitutive element for impenetrable MTSs, which are the ones treated in this chapter. In the microwave regime, impenetrable MTSs are usually formed by a dense texture of small elements in a regular lattice, printed on a grounded slab with or without shorting vias. Impenetrable MTSs can be also simply constituted by a dense distribution of pins [2] on a ground plane. This solution is usually adopted at the optical frequencies to overcome the issue of dielectric losses.

The MTS is referred to as *uniform* when the elements in the periodic lattice are identical all over the MTS plane [3]. Here, however, we deal with *aperiodic* elements having geometrical parameters gradually changing from cell-to-cell: in this case the MTS is referred to as *nonuniform* or *modulated*. Nonuniform MTSs allow one to change the phase velocity and/or propagation path of the guided wave sustained by the MTS.

F. Caminita · E. Martini
Wave Up s.r.l, Via Edimburgo 6, 50126 Florence, Italy
e-mail: francesco.caminita@wave-up.it

D. González Ovejero
Institut d'Électronique et de Télécommunications de Rennes - UMR CNRS 6164, 263 Avenue du Général Leclerc, Rennes 35042, France
e-mail: david.gonzalez-ovejero@univ-rennes1.fr

M. Sabbadini
European Space Agency, Keplerlaan 1, 2201 AZ Noordwijk, The Netherlands
e-mail: marco.sabbadini@esa.int

The applications of modulated MTSs can be divided in four classes: (a) MTSs for aperture antennas [4–12], (b) MTSs for the control of surface-waves (SWs) wavefront [13–16], (c) MTS screens for the control of field transmission [17–19], (d) MTSs for EBG [20]. Here, we focus the attention on the first class of applications.

The subject of surface wave control through MTSs has been treated in several recent papers ([4–33]), some of them addressing the relevant phenomena in the framework of Transformation Optics (see [14, 23–29] and Chap. 11 of this book). On the other hand, the control of leaky-wave (LW) radiation through modulation of boundary conditions (BCs) has led to a new class of modulated MTS antennas [4–9] (referred to as MoMetAs in [6], which stands for Modulated Metasurface Antennas). This kind of antennas transforms a bounded SW into a radiative LW through the SW interaction with a modulated MTS. In the microwave range, the MTS antennas we are dealing with are typically constituted by sub-wavelength patches of various shapes, printed on a grounded dielectric substrate [4–10]. They exhibit low profile, low weight, reduced complexity, and low cost. Neither external protruding or backing feed arrangements nor (sub-) reflectors are required, since the feeding element is embedded in the MTS; this constitutes an advantage with respect to other types of printed antennas, like reflectarrays, especially for space applications [6]. Although modulated MTS antennas are based on a LW mechanism, their behavior offers an enormous flexibility with respect to conventional LW antennas. The key features of modulated MTS antennas, making them very appealing mostly for space applications but also for a wide range of ground applications, can be summarized as follows:

- Unique decoupling of electrical and thermo-mechanical design.
- Unique capability of low-complexity (dynamic) beam control (shaping, pointing, and scanning).
- Unique support for late electrical performance refinement and design reusability.
- Complete and simple on-surface control of medium-to-large size apertures.
- Low power losses.
- Low mass and low envelope.
- Implementation amenable to very different technologies.
- Simple low-cost manufacturing.

Some prototypes of MTS antennas are shown in Fig. 9.1. At the frequency of operation, the elements that implement the BCs are small in terms of the wavelength (between $\lambda/5$ and $\lambda/10$, with λ being the free-space wavelength), so that the interacting SW sees the interface BC as a continuum. These elements behave like pixels in a black and white printed image, whose gray scale is realized by changing the dimension of the pixel elements inside a regular lattice. One can also control the polarization by using asymmetric pixel elements (this can be done, e.g., by adding a feature like a cut or a slot to a circular patch, or using an elliptical patch), and rotating them inside the lattice.

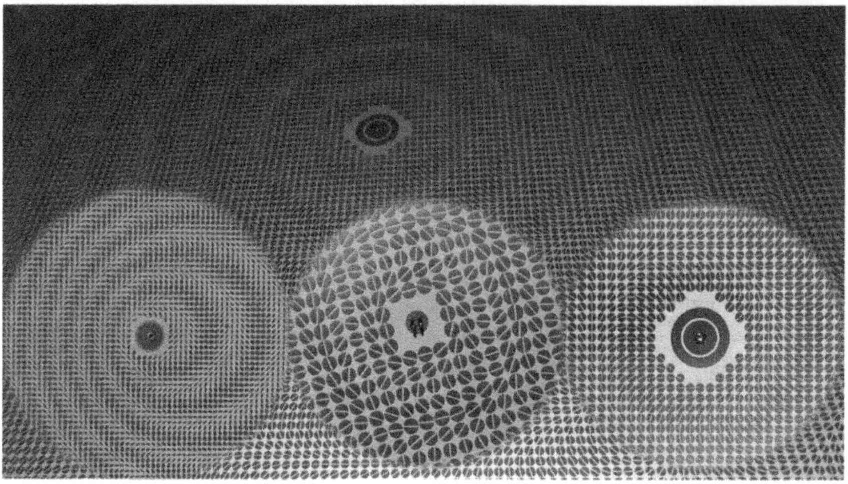

Fig. 9.1 Examples of realized prototypes of MTS antennas. The SW is excited by a single probe (*left* and *right insets*) or by 4 probes (*central inset*). In the three examples, the feeding region is located at the center of the MTS

Table 9.1 summarizes a number of modulated MTS designs, several of which have been built and measured, while the most recent ones are yet to be manufactured. Nevertheless, the extremely good accuracy of the design tools in use makes it possible to rely on the calculated performance figures.

The model of these antennas is inherently multi-scale (see Fig. 9.2). To convey the idea, we adopt a terminology used in physics (Fig. 9.2), to denote the scales in terms of (a) "macroscopic scale", (b) "mesoscopic scale", (c) "molecular scale", and (d) "atomic scale". The largest dimension of the antenna can be several wavelengths (e.g., 15λ, Fig. 9.2a). The local period of the global modulation is normally of the order of one wavelength (Fig. 9.2b). Zooming in the scale, one sees the lattice of the texture, which consists of pixel elements with a size of approximately $\lambda/10$ (Fig. 9.2c). The specific features added to the patches to control polarization (e.g., the slots) can be of the order of $\lambda/50$, which is approximately the size of the discretization triangles required in a Method of Moment mesh (Fig. 9.2d). It is, hence, clear that a conventional global full-wave analysis of the overall structure would require the solution of a million-unknowns problem. For this reason, a single model cannot cover all the scales, and the synthesis procedure cannot be solely based on a direct full-wave solver. Note that the number of unknowns is even larger if the computational approach is based on other methods such as FEM or FDTD. Here, we only deal with MoM methods for solving these structures, as the use of other methods is out of the scope of this chapter. In the following, we introduce a design process for modulated MTS antennas that makes use of different models for the different scales, corresponding to different levels of accuracy.

Table 9.1 Spot-beam modulated metasurface antennas—design and performances summary

	Configurations												
Date	2010	2014		2015	2015	2015	2015	2015	2015	2015		2016	2016
Application	–	Space exploration		–	–	–	Telecom	EO	–	Space exploration		Space exp	Telecom
Type	Disk	Disk		Disk	Disk	Disk	Disk	Disk	Disk	Disk		Disk	Disk
Verification	Test	Test		Test	Test	Analysis	Analysis	Analysis	Analysis	Analysis		Model	Model
Frequency (GHz)	17	8.3	7.1	28	28	13.25	29.5	26.3	25.1	7.1	8.4	7–8	29.5 GHz
Diameter	200 mm	540 mm	300 mm	86 mm	86 mm	120 mm	560 mm	270 mm	270 mm	890 mm		1 m	1 m
Directivity (dBi)	25	28.5	21.4	23.8	23.3	26	43.0	36.2	24	27.5	32.0	36	48.4
Gain (dBi)	24.5	28.0	21.1	23.5	23.0	–	–	–	–	–	–	35.5	47.4
Pointing angle (°)	0	0	0	30	0	0	0	0	0	0		0	0
Polarization	CP	CP	CP	CP	LP	Dual-LP	CP	CP	CP	CP		CP	CP
Cross-polar level	−13 dB	−25 dB	−20 dB	−25 dB	−25 dBi	−25 dB	−30 dB	−30 dB	–	−25 dB	−30 dB	−35 dB	−35 dB
−3 dB beamwidth (°)	5.5	4.6	13.3	10.3	8.1	6.0	1	2.7	–	3.5	3.0	2.5	0.7
Bandwidth @ −1 dB (%)	3	4	6	–	–	–	–	–	12.5	3	3	17	5
Bandwidth @ −3 dB (%)	6	8	15	–	–	–	–	–	25	6	6	34	10
Efficiency[a] (%)	25	32	27	43	34	40	65	70	–	–	–	75	73
Return loss (dB)	−20	−23	−20	−20	−22	–	–	–	–	–	–	−25	−25
Isolation	–	>60 dB		–	–	–	–	–	–	–	–	–	–
Remarks	First design attempt	Split-aperture solution		Low accuracy prototype	Low accuracy prototype			Efficiency optimisat.	Bandwidth enlargem.	Dual-band solution		Low accuracy modeling	Low accuracy modeling

[a]Efficiency = Aperture efficiency × Spillover efficiency (reduction of aperture illumination w.r.t uniform law × amount of surface wave power converted into radiation)

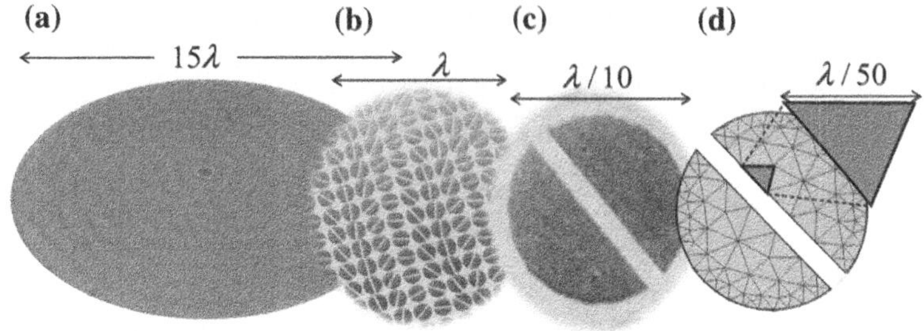

Fig. 9.2 Multiscale features of MTS antennas: "macroscopic" scale (**a**), "mesoscopic" scale (**b**), "molecular" scale (**c**), "atomic" scale (**d**)

The chapter is organized as follows. Section 9.2 provides a global picture of the design process, whose blocks are described more in detail in the following sections. In particular, Sect. 9.3 introduces the Flat Optics model for the description of fields and currents on the modulated MTS. Section 9.4 describes the synthesis scheme for the design of an impedance BC providing a desired radiating aperture field. Section 9.5 discusses the full-wave analysis method based on a continuous impedance BC model. Section 9.6 presents the technique used for the synthesis of the BC, based on the extraction of the equivalent impedance of the single pixels. Section 9.7 illustrates an efficient MoM approach for the full-wave analysis of textured MTS antennas consisting of elliptical patches. An example of design process, including all the previously mentioned steps, is presented in Sect. 9.8. Also, other designs for highly efficient antennas or wide band capability are discussed in Sect. 9.8. The issue of the efficiency for these antennas is discussed in Sect. 9.9. Finally, practical examples of MTS antennas are reported in Sect. 9.10, and conclusions are given in Sect. 9.11.

9.2 Overview of the Design Process

The global synthesis scheme is shown in Fig. 9.3 [12]. It is composed of three main blocks, each of them working on a different model scale. The first block synthesizes the continuous impedance surface, which, when excited by a reference SW produces the objective aperture field. Then, within the second block, the continuous impedance surface is discretized and implemented through a dense texture of small patches. We will refer to them as "pixels" since they resemble pixels in a black and white printed image. Finally, the third block analyzes the textured antenna layout through a global full-wave solver. Since the number of pixels is quite large (typically several thousands), the global solver must be engineered with smart numerical strategies to overcome the computational complexity arising from the huge number of unknowns of the linear problem.

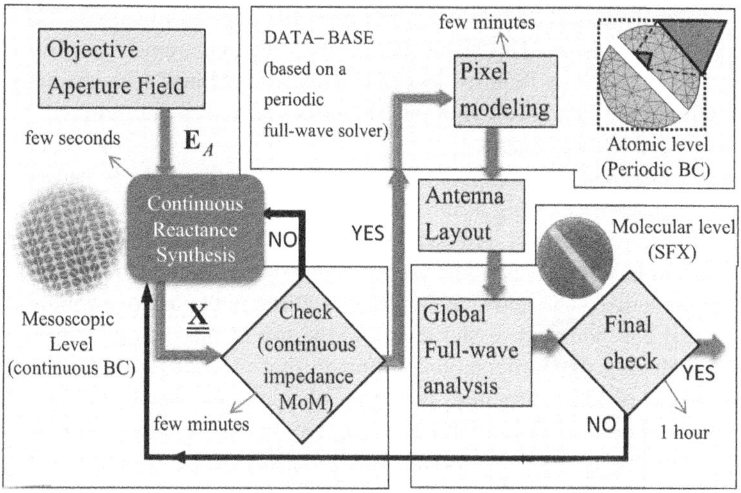

Fig. 9.3 Block diagram of the synthesis process

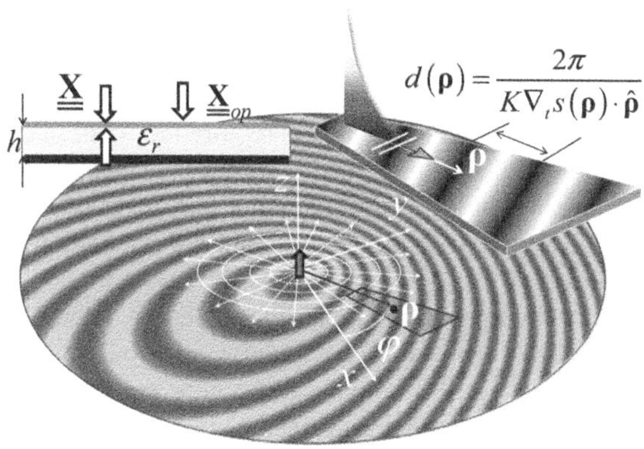

Fig. 9.4 Geometry of the problem. The sub-wavelength printed patches are modeled by an anisotropic tensor reactance, which defines through (9.1) discontinuous BCs over a grounded dielectric slab (inset on the *left corner*). The inset on the *right top corner* sketches the local periodic problem for the definition of the adiabatic Floquet modes

The design process starts assuming that the radiative components of the objective aperture field are known. We do not treat here the process to synthesize a continuous aperture field from the pattern requirements, since this is the subject of a rich literature, see for instance [34] and the references therein. Within the first block, the synthesis of the BCs is performed at "mesoscopic" level (area on the left in Fig. 9.3), namely, assuming that the patterned metallic cladding can be treated as a continuous, space-dependent reactance tensor $\underline{\underline{\mathbf{X}}}$, which models continuous anisotropic BCs related to the discontinuity of the magnetic field (Fig. 9.4). The block in red is the core of this task, and it is based on the analytical method introduced in [35, 36] that we denote succinctly as "Flat Optics" (FO) and synthesized in the next paragraph. The time required to produce $\underline{\underline{\mathbf{X}}}$ is a few seconds using a good notebook *independently* of the antenna size. After the synthesis has been performed, a full-wave analysis still based on a continuous BC approach, is used to check the accuracy of the synthesis. This step is performed with an extremely fast formulation described in [37], which requires less than 10 s for antennas with diameter of 20 wavelengths. If the first check with this solver is successful, one proceeds with the synthesis of the MTS elements.

The element synthesis (called pixel modeling in Fig. 9.3) is carried out at "atomic" level, namely, using a full-wave code with RWG basis functions. A local "micro-periodicity" concept is used in this phase: it is assumed that the local element is immersed in a periodic environment of equal elements with the periodicity of the lattice. The above assumption legitimizes the use of periodic BCs on the elemental cell boundaries, and therefore the use of the periodic Green's function, thus reducing the computational effort to that of a single periodic cell. The analysis of the periodic structure, which is inherently extremely fast, is repeated several times in order to construct a database that relates the different geometries of the unit cell element with the corresponding entries of the anisotropic reactance tensor. The database is built before the entire synthesis process, and possibly reused for other syntheses. Actually, only a few samples of the parameter-space are directly calculated and interpolation is next used to create a denser database.

The database construction requires few tens of minutes for a 2D database (i.e., obtained by varying 2 geometrical parameters) and few hours for a 3D database (3 geometrical parameters). The database is normally constructed at the antenna operating frequency; its extension around the frequency of operation can be important when the antenna is required to work over a large band. To this purpose, the pole-zero matching method [38] or the analytical model presented in [32] can be used. The output of the "atomic" part of the design is a complete layout of the antenna, where the dimensions and the orientation of any single pixel-element are defined.

The final phase of the design process is a detailed analysis of the layout, which requires a global full-wave integral equation model. This model is implemented at "molecular level", namely using few entire domain basis functions per pixel-element and incorporating them into a fast-multipole algorithm to accelerate the solution. In case the final check is not successful, a repetition of the continuous model synthesis may be required. Actually, as the analytic core of the synthesis, i.e., the "continuous reactance synthesis" block in Fig. 9.3, is powerful and accurate, the overall process rarely requires a feedback, which is typically needed only for very demanding requirements, like large bandwidth or very high efficiency.

The synthesis step that we are going to introduce in the next paragraph is based on a description of the MTS in terms of "transparent" BC in place of "opaque" BC (see left-side inset of Fig. 9.4). With the first terminology, introduced in [39], we denote the relationship between the tangential electric field and the discontinuity of the magnetic field at the metallic cladding level. "Opaque" BCs relate instead the tangential electric field to the limit of the magnetic field at the interface between the MTS and free space. The motivation of this choice is that the transparent BC is less spatially dispersive than the opaque BC [32,33].

In the next section, we present the main issue concerning each of the blocks of the design scheme in Fig. 9.3. In the last part of the chapter, we summarize the presented design workflow through an applicative example.

9.3 Adiabatic Floquet Wave Expansion

The synthesis of the continuous reactance (red block in Fig. 9.3) is based on a theoretical model we have developed to describe currents and fields on a class of synthesizable reactances that we will refer to as "constant-average non uniform reactances". Such reactances describe the transparent locally periodic modulated BCs formed by the patch cladding on a grounded dielectric slab. We have found that currents and fields on such a locally periodic environment are well described by an adiabatic Floquet Wave (FW) expansion matched to the local periodicity.

Let us refer to the configuration shown in Fig. 9.4. The MTS is defined on a circular surface of radius a, centered at the origin of a reference system described in cylindrical coordinates (ρ, φ), with unit vectors $(\hat{\boldsymbol{\rho}}, \hat{\varphi})$. Accordingly, the observation point on the surface at the interface between the metal cladding and free space will be indicated by $\boldsymbol{\rho} = \rho \cos \varphi \hat{\mathbf{x}} + \rho \sin \varphi \hat{\mathbf{y}}$. Vectors will be denoted by bold characters and tensors by bold characters underlined by a double bar; k and ζ identify free-space wavenumber and impedance, respectively. Transverse magnetic (TM) and transverse electric (TE) modes will be referred to the normal $\hat{\mathbf{z}}$ to the surface. We will restrict our analysis to MTSs composed by a lossless grounded dielectric slab of relative permittivity ε_r and thickness h, on which perfectly electric conducting subwavelength patches are printed. We will assume that the SW is excited by an electric vertical dipole located at the center of the reference system. In practice, the feeder shall be designed to maximize the amount of SW power and minimize the amount of power directly radiated in the free-space. Therefore, to maximize the SW efficiency, some strategies are adopted that can make the feeding system more complex than a simple small vertical dipole. Despite this, the excited SW can still be considered azimuthally symmetric, as the one of the vertical dipole, and this is enough for giving validity to our simplified dipole assumption.

9.3.1 *A Class of Synthesizable Continuous Reactances*

We model the printed MTS by anisotropic "transparent" BCs [35, 39] identified by

$$\mathbf{E}_t = j\underline{\underline{\mathbf{X}}} \cdot \hat{\mathbf{z}} \times \left(\mathbf{H}_t|_{0^+} - \mathbf{H}_t|_{0^-} \right) = j\underline{\underline{\mathbf{X}}} \cdot \mathbf{J} \tag{9.1}$$

where $\underline{\underline{\mathbf{X}}}$ is the "transparent" reactance and $\mathbf{J}$ is the electric current flowing in the metallic homogenized cladding. In practical realizations, the transparent reactance mainly depends on the patch loading. In absence of losses, $\underline{\underline{\mathbf{X}}}$ is a Hermitian tensor [29] (or the *impedance* $j\underline{\underline{\mathbf{X}}}$ is anti-Hermitian); the eigenvectors $\hat{\mathbf{e}}_1, \hat{\mathbf{e}}_2$ of $\underline{\underline{\mathbf{X}}}$ are therefore orthogonal, and the eigenvalues are real.

The class of transparent *lossless* reactance tensors we would like to synthesize is given by

$$\underline{\underline{\mathbf{X}}} = \underline{\underline{\mathbf{X}}}^{(0)} + \underline{\underline{\mathbf{X}}}^{(-1)} + \underline{\underline{\mathbf{X}}}^{(+1)} \tag{9.2}$$

$$\underline{\underline{\mathbf{X}}}^{(0)} = \bar{X}_\rho \hat{\boldsymbol{\rho}}\hat{\boldsymbol{\rho}} + \bar{X}_\varphi \hat{\boldsymbol{\varphi}}\hat{\boldsymbol{\varphi}} \tag{9.3}$$

$$\underline{\underline{\mathbf{X}}}^{(\mp 1)} = \frac{1}{2} e^{\pm jKs} \left[m_\rho \left(\bar{X}_\rho \hat{\boldsymbol{\rho}}\hat{\boldsymbol{\rho}} - \bar{X}_\varphi \hat{\boldsymbol{\varphi}}\hat{\boldsymbol{\varphi}} \right) e^{\pm j\Phi_\rho} + m_\varphi \bar{X}_\rho \left(\hat{\boldsymbol{\rho}}\hat{\boldsymbol{\varphi}} + \hat{\boldsymbol{\varphi}}\hat{\boldsymbol{\rho}} \right) e^{\pm j\Phi_\varphi} \right] \tag{9.4}$$

where $\bar{X}_\rho$ and $\bar{X}_\varphi$ are negative (capacitive) quantities and K is a large $\boldsymbol{\rho}$-independent constant such that $K|\nabla_t s(\boldsymbol{\rho})| \gg |\nabla_t \Phi_{\rho,\varphi}(\boldsymbol{\rho})|$. The above reactance is written in terms of exponentials, which combined together lead to cosine functions. Therefore, the above class is constituted by symmetric tensors, namely, a special case of Hermitian tensors which are representative of MTSs constituted by elements with two orthogonal symmetry axes. The above conditions identify reactances with rapidly oscillating entries, whose average tensor possesses principal axes aligned with $(\hat{\boldsymbol{\rho}}, \hat{\boldsymbol{\varphi}})$. We denote this class as "constant-average MTSs" and $\underline{\underline{\mathbf{X}}}$ in (9.2) as "constant-average reactance". In (9.3), $\underline{\underline{\mathbf{X}}}^{(0)}$ denotes the $\boldsymbol{\rho}$-independent average reactance, whereas $\underline{\underline{\mathbf{X}}}^{(\mp 1)}$ in (9.4) are the rapidly oscillating contributions which contain the modulation functions $m_{\varphi,\rho}(\boldsymbol{\rho})$. Within the entries of $\underline{\underline{\mathbf{X}}}^{(\mp 1)}$ one can identify three factors: (i) a rapidly varying phase factor $\exp(\pm jKs(\boldsymbol{\rho}))$, which provides the main interaction with the exciting SW; (ii) the slowly varying phase factors $\exp(\pm j\Phi_{\rho,\varphi}(\boldsymbol{\rho}))$, which mainly control the polarization; (iii) the modulation indexes $m_{\varphi,\rho}(\boldsymbol{\rho})$, which mainly control the amplitude of the field. The latter are considered small enough to avoid local changes of the nature of the transparent reactance from capacitive to inductive, that could lead to the excitation of a higher order (TE-dominant) SW mode. This is ensured by the condition $m_{\rho,\varphi}(\boldsymbol{\rho}) < 1$.

9.3.2 Analysis by Adiabatic Floquet Waves

One of the main properties of the *constant-average* reactance $\underline{\underline{\mathbf{X}}}$ in (9.2) is concerned with the fact that the *average* current flowing in it is quite similar to the SW-currents that the dipole source would excite in $\underline{\underline{\mathbf{X}}}^{(0)}$. An appropriate modification of these currents will constitute therefore the 0-index mode of an adiabatic FW expansion. Since $\underline{\underline{\mathbf{X}}}^{(0)}$ is $\boldsymbol{\rho}$-independent, the TM-dominant SW excited on it by a monopole is a purely TM cylindrical wave asymptotically given by

$$\mathbf{J}_0 = J_0 H_1^{(2)}(\beta_{sw}\rho)\hat{\boldsymbol{\rho}} \tag{9.5}$$

where $H_1^{(2)}$ is the Hankel function of second kind and first order. The propagation constant β_{sw} is solution of a dispersion equation which is affected by the $\bar{X}_\rho$ component only. An excellent closed-form approximation of β_{sw} is given in [32].

In presence of the modulation, the 0-indexed FW mode (simply denoted as 0-mode) is obtained from (9.5) by *locally* transforming the real unperturbed wavenumber β_{sw} in the complex radial wavenumber $k^{(0)}(\boldsymbol{\rho}) = \beta_{sw} + \beta_\Delta(\boldsymbol{\rho}) - j\alpha(\boldsymbol{\rho})$. It is, therefore, implicitly assumed that the 0-mode has a cylindrical, attenuated wavefront, where the *local* attenuation parameter $\alpha(\boldsymbol{\rho})$ accounts for the transfer of energy from the 0-mode to the −1 (leaky) mode during the propagation path. The parameters $\beta_\Delta(\boldsymbol{\rho})$ and $\alpha(\boldsymbol{\rho})$ are functions of the $\underline{\underline{\mathbf{X}}}$ entries. Since $k^{(0)}(\boldsymbol{\rho})$ is space dependent, its associated *global* phase $\tilde{k}^{(0)}(\boldsymbol{\rho})\rho$ from the reference point at the origin is obtained by integrating the local relation $\partial(\tilde{k}^{(0)}(\boldsymbol{\rho})\rho)/\partial\rho = k_0(\boldsymbol{\rho})$; i.e.,

$$\tilde{k}^{(0)}(\boldsymbol{\rho}) = \frac{1}{\rho}\int_0^\rho k^{(0)} d\rho' = \beta_{sw} + \frac{1}{\rho}\int_0^\rho [\beta_\Delta(\boldsymbol{\rho}') - j\alpha(\boldsymbol{\rho}')] d\rho' \tag{9.6}$$

This leads to the following global adiabatic FW expansion for the current

$$\mathbf{J} \approx \sum_n \mathbf{J}^{(n)} \tag{9.7}$$

$$\mathbf{J}^{(n)} = \mathbf{j}^{(n)} e^{-jnKs(\boldsymbol{\rho})} H_1^{(2)}\left(\tilde{k}^{(0)}\rho\right) \tag{9.8}$$

where $Ks(\boldsymbol{\rho})$ is the same as in the definition of the reactances (9.4). Although the adiabatic expansion contains an infinite number of terms, only three terms of the expansion are sufficient to describe the total current, whenever only the −1 mode belongs to the visible range. Therefore, we will retain only these three terms for a synthesis process. The complex terms $\mathbf{j}^{(n)} = J_\varphi^{(n)}\hat{\boldsymbol{\varphi}} + J_\rho^{(n)}\hat{\boldsymbol{\rho}}$ in (9.8) denote the slowly varying part of the n-th current mode and they are the unknowns of the problem. The 0-indexed mode has a dominant TM component $J_\rho^{(0)}$, whose amplitude is much larger than that of the TE component $J_\varphi^{(0)}$, and also much larger than both the components of the (±1)-indexed modes. This dominant behavior is more evident for small modulation indexes. Indeed, in the limit for vanishing modulation, $J_\rho^{(0)}$ should be the only component and should coincide with J_0 in (9.5).

Taking the asymptotic form of the Hankel function (valid for distances larger than about one wavelength from the source), it is apparent that each mode in (9.8) has a curvilinear-wavefront, which propagates with *local* n-indexed FW wavevector

$$\boldsymbol{\beta}^{(n)} = \mathrm{Re}\nabla_t\left[\tilde{k}^{(0)}\rho + nKs(\boldsymbol{\rho})\right] = (\beta_{sw} + \beta_{\Delta})\hat{\boldsymbol{\rho}} + nK\nabla_t s(\boldsymbol{\rho}) \tag{9.9}$$

and with n-independent local attenuation parameter $\alpha(\boldsymbol{\rho})$. The curvilinear-wavefront associated with each FW current mode is given by $\beta_{sw}\rho + nKs(\boldsymbol{\rho}) = const.$ The adiabatic FW electric fields can be obtained through the spectral Green's function (GF) of the grounded slab evaluated at the local wavevector $\boldsymbol{\beta}^{(n)}$ in (9.9), that is

$$\mathbf{E}_t(\boldsymbol{\rho}) = \sum_n \mathbf{E}^{(n)} \approx \sum_n \underline{\underline{\mathbf{Z}}}_{GF}^{(n)} \cdot \mathbf{J}^{(n)} \tag{9.10}$$

where $\underline{\underline{\mathbf{Z}}}_{GF}^{(n)}$ depends on the grounded slab spectral GF (see appendix in [36]). It can be seen that $\underline{\underline{\mathbf{Z}}}_{GF}^{(n)}$ is purely imaginary for any $n \neq -1$ while it is complex for n = −1. The −1 mode is actually the only mode of interest in order to design the modulated impedance that provides a given radiation pattern.

The basis on which we have expanded currents and fields does not satisfy Maxwell's equations, since it is based on an asymptotic, adiabatic approximation valid far from the point source. However, this basis locally recovers a Floquet wave expansion and may be used for obtaining an adiabatic solution. This solution is found by using (9.10) in the transparent boundary conditions (9.1).

$$\sum_{n=-2}^{2} \underline{\underline{\mathbf{Z}}}_{GF}^{(n)} \cdot \mathbf{J}^{(n)} = j\left(\underline{\underline{\mathbf{X}}}^{(0)} + \underline{\underline{\mathbf{X}}}^{(-1)} + \underline{\underline{\mathbf{X}}}^{(+1)}\right) \cdot \left(\mathbf{J}^{(1)} + \mathbf{J}^{(0)} + \mathbf{J}^{(-1)}\right) \tag{9.11}$$

This allows for finding an analytical solution and a local $\boldsymbol{\rho}$–dependent adiabatic dispersion equation. To this end, the terms in (9.11) with the same rapid phase variation are equalized and a set of equations is obtained. The latter is solved by substitutions, thus leading to a homogeneous system of type $j\underline{\underline{\chi}}(\boldsymbol{\rho}) \cdot \mathbf{J}^{(0)} = 0$. The latter admits a non trivial solution only for $\det\left[\underline{\underline{\chi}}(\boldsymbol{\rho})\right] = 0$ which is the *local dispersion equation*; this equation allows for determining the complex value of the 0-mode *local* wavenumber $k^{(0)}$. The final solution is therefore individuated in terms of the two TE/TM components of the 0-mode $\mathbf{J}^{(0)}$ (namely $J_{\varphi}^{(0)}$ and $J_{\rho}^{(0)}$) still to be determined. While their relative amplitude (i.e., their ratio) is found by using $k^{(0)}$ in the linear system $j\underline{\underline{\chi}}(\boldsymbol{\rho}) \cdot \mathbf{J}^{(0)} = 0$, the amplitude $J_{\rho}^{(0)}$ can be found by some considerations on power conservation [35, 36]. Indeed, within a zero-order approximation, the *average* current flowing in the *constant-average* reactance $\underline{\underline{\mathbf{X}}}$ can be identified with the SW-currents flowing in $\underline{\underline{\mathbf{X}}}^{(0)}$ in absence of modulation. Thus, $J_{\rho}^{(0)}$

is estimated by means of the radial Poynting vector associated with the surface wave excited on the uniform MTS [35].

9.4 Continuous Reactance Synthesis

Using the theoretical framework introduced in the previous paragraph, it is possible to set up the synthesis process of the BCs, namely the red block in Fig. 9.3. The aim of the design is to determine $\underline{\underline{\mathbf{X}}}$ in (9.2) that creates a far field as the one radiated by the objective circular aperture field. We describe the latter aperture field by the rather general form

$$\mathbf{E}_A = E_0 e^{-jk\ell(\boldsymbol{\rho})}\left[e_\rho(\boldsymbol{\rho})e^{j\gamma_\rho(\boldsymbol{\rho})}\hat{\boldsymbol{\rho}} + e_\varphi(\boldsymbol{\rho})e^{j\gamma_\varphi(\boldsymbol{\rho})}\hat{\boldsymbol{\varphi}}\right]\mathrm{U_A} \tag{9.12}$$

where U_A is a unit step function which is 1 inside a circle of radius a and zero elsewhere. In (9.12), $\gamma_\rho(\boldsymbol{\rho})$, $\gamma_\varphi(\boldsymbol{\rho})$, $e_\rho(\boldsymbol{\rho})$, $e_\varphi(\boldsymbol{\rho})$, $\ell(\boldsymbol{\rho})$ are real, weakly variable functions of space.

In order to determine $\underline{\underline{\mathbf{X}}}$, we identify $\mathbf{E}_A$ with $\mathbf{E}^{(-1)}$, namely, with the field associated with the adiabatic leaky-wave field. In doing this, we implicitly assume that the difference between the objective aperture field $\mathbf{E}_A$ and the $\mathbf{E}_t$ field in (9.10) is a dominantly reactive field contribution, with spectrum mainly concentrated outside the visible region. This assumption can be verified once the design procedure is completed and the aperture fields are assessed.

The determination of $\underline{\underline{\mathbf{X}}}$ involves the iterative process shown in Fig. 9.5 and thoroughly described in [36]. The process starts by setting the aperture field, the aperture size, substrate parameters, and operational frequency as input parameters. An initial guess of the modulation parameters (namely $Ks(\boldsymbol{\rho})$, $m_{\rho,\varphi}(\boldsymbol{\rho})$, $\Phi_{\rho,\varphi}(\boldsymbol{\rho})$) is found taking the 0-mode current and wavenumber as the ones in the uniform MTS with average reactance $\underline{\underline{\mathbf{X}}}^{(0)}$.

Next, this first initial guess of the MTS is analyzed to get the effective values of the complex displacement $\beta_\Delta(\boldsymbol{\rho}) - j\alpha(\boldsymbol{\rho})$ on the modulated MTS. Such displacement is then used to refine the estimation of the modulation parameters in the next iterative step. When the modulation parameters vary less than a chosen threshold, then the process is stopped and $\underline{\underline{\mathbf{X}}}$ is determined.

9.5 Full-Wave Homogenized Impedance Analysis

The continuous reactance obtained in Sect. 9.4 is verified at this step (see the block diagram in Fig. 9.3) using a full-wave solver for planar apertures that supports a space-dependent impedance BC. The synthesized apertures are typically large,

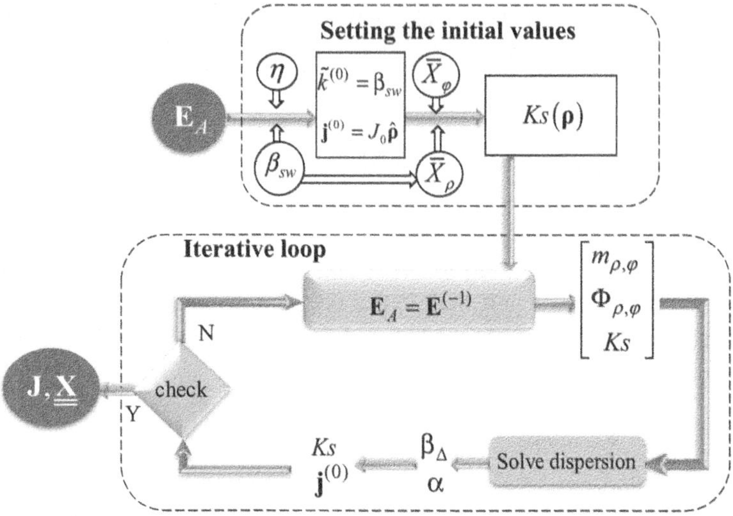

Fig. 9.5 Block diagram for synthesizing the continuous reactance $\underline{\underline{\mathbf{X}}}$ from the aperture field $\mathbf{E}_A$. The input parameters for setting the initial value are: aperture field, aperture size, substrate parameter, and frequency

and an efficient analysis is thus necessary to complete the "mesoscopic" (homogenized impedance) block of the scheme in Fig. 9.3. We solve the problem of a transparent impedance BC (9.2) imposed on top of a grounded dielectric slab of infinite extent in the x-y plane. An equivalent problem can be described only by horizontal electric currents at the air–dielectric interface. The imposed IBC leads to the following integral equation

$$\hat{\mathbf{z}} \times \left[\iint_{S'} \underline{\underline{\mathbf{G}}}^{\mathrm{EJ}}(\boldsymbol{\rho}, \boldsymbol{\rho}') \cdot \mathbf{J}(\boldsymbol{\rho}')\, \mathrm{d}S' - j\underline{\underline{\mathbf{X}}}(\boldsymbol{\rho}') \cdot \mathbf{J}(\boldsymbol{\rho}') \right] = -\hat{\mathbf{z}} \times \mathbf{E}^{\mathrm{inc}} \tag{9.13}$$

where $\hat{\boldsymbol{\rho}}$ and $\hat{\boldsymbol{\rho}}'$ are the source and observation points, respectively. $\underline{\underline{\mathbf{G}}}^{\mathrm{EJ}}$ represents the electric dyadic Green's function, $\mathbf{E}^{\mathrm{inc}}$ is the incident field due to the excitation and $\mathbf{J}$ is the unknown surface current density. The motivation for carrying out the MoM analysis with $\underline{\underline{\mathbf{X}}}$ instead of an opaque reactance is, on the one hand, to account for the spectral dispersion due to the thickness of the grounded dielectric slab [31]. Resorting to layered medium Green's functions, one can introduce the effect of the grounded slab in an efficient manner, since the unknowns in (9.13) will be just the electric currents $\mathbf{J}$. On the other hand, by using $\underline{\underline{\mathbf{X}}}$ along with the relevant Green's functions it is possible to avoid ill-conditioning issues that arise in the formulations based on opaque reactances for typical surface reactance values in MTS structures [39].

It has been shown in [37] that an appropriate choice of entire domain basis functions leads to an extremely fast solution of (9.13). Such basis functions, hereinafter referred to as Gaussian ring basis functions (GRBFs) [37], constitute a particularly convenient choice for representing $\mathbf{J}$ on circular apertures, and they can be written in the space-domain as

$$\mathrm{f}_{m,n}(\rho,\varphi) = \mathrm{e}^{-\mathrm{j}n\varphi}\,\Psi_{m,n}(\rho);\quad \Psi_{m,n}(\rho) = \frac{1}{2\sigma^2}\,\mathrm{e}^{-\frac{(\rho-\rho_m)^2}{4\sigma^2}}\,\Upsilon_n\left(\frac{\rho_m\rho}{2\sigma^2}\right) \tag{9.14}$$

where $\Upsilon_n(z) = \mathrm{I}_n(z)\ \mathrm{e}^{-|Re\{z\}|}$, with $\mathrm{I}_n(z)$ being the modified Bessel function of the first kind and order n. Equation (9.14) corresponds to a Gaussian-type ring with linear azimuthal phase. GRBFs peak at ρ_m and their half-maximum beamwidth (HMBW) is $4\sqrt{\ln 2}\sigma$. Figure 9.6a shows the shape of $\mathrm{f}_{m,n}(\rho, 0)$ for different values of σ. One can also fix σ to a constant value and choose $\rho_m = \rho_0 + m4\sqrt{\ln 2}\sigma$ ($m = 0, 1, 2, 3\ldots$) for numbering the basis functions in a progressive sequence of GRBFs that cross each other at half of their maximum (see Fig. 9.7a). Since the dependence of $\Psi_{m,n}(\rho)$ on n (Bessel function's order) is extremely weak for this particular choice of ρ_m, changing n maintains the sequence of functions unchanged.

On the other hand, the spectral-domain counterpart of (9.14) presents an analytical form:

$$\mathrm{F}_{m,n}(\mathbf{k}_\rho) = \mathrm{e}^{-\mathrm{j}n\alpha}2\pi\mathrm{j}^n\,\Phi_{m,n}(k_\rho);\quad \Phi_{m,n}(k_\rho) = \mathrm{e}^{-\sigma^2k_\rho^2}\,\mathrm{J}_n(\rho_m k_\rho) \tag{9.15}$$

where $\mathbf{k}_\rho = k_x\hat{\mathbf{x}} + k_y\hat{\mathbf{y}} = k_\rho\cos\,\alpha\hat{\mathbf{x}} + k_\rho\sin\,\alpha\hat{\mathbf{y}}$, with $k_\rho = \sqrt{k_x^2+k_y^2}$ and k_x, k_y being the spectral variables of the classical Fourier domain. Equation (9.15) can be used for obtaining closed-form expressions of the MoM impedance matrix elements. Examples of GRBFs spectra are shown in Figs. 9.6b and 9.7b.

The surface current density on the MTS plane can be expanded using the proposed basis functions as

$$\mathbf{J}(\boldsymbol{\rho}) = \sum_{m=0}^{M-1}\sum_{n=-N/2}^{N/2} i^x_{m,n}\mathrm{f}^x_{m,n}(\boldsymbol{\rho})\,\hat{x} + i^y_{m,n}\mathrm{f}^y_{m,n}(\boldsymbol{\rho})\,\hat{y} \tag{9.16}$$

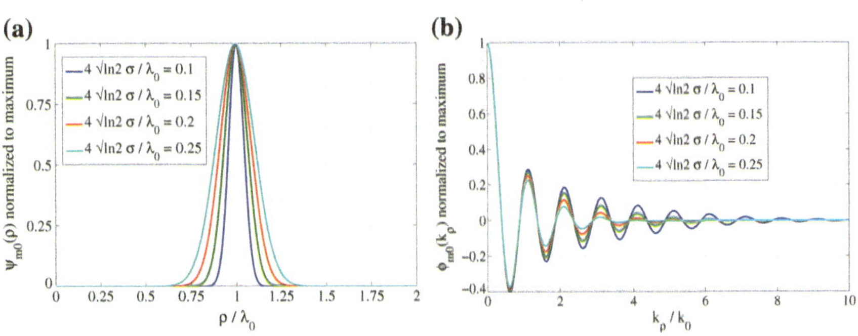

Fig. 9.6 Representation of the GRBF in the space-domain as a function of ρ/λ_0 **(a)** and in the spectral-domain as a function of k_ρ/k_0**(b)**, for different values of $4\sqrt{\ln 2}\sigma$ ($\rho_m/\lambda_0 = 1$, $n = 0$)

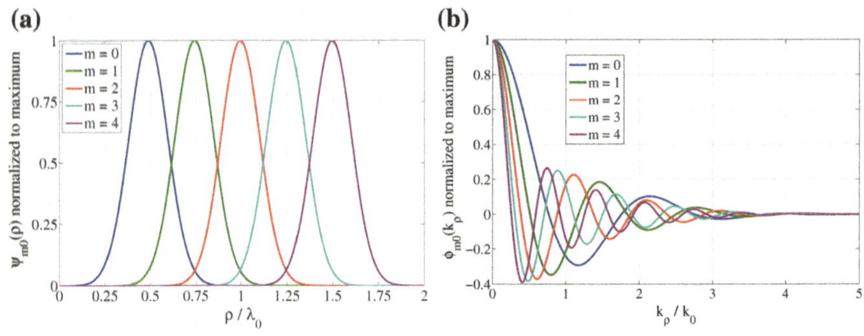

Fig. 9.7 Representation of the GRBF in the space-domain as a function of ρ/λ_0 **(a)** and in the spectral-domain as a function of k_ρ/k_0**(b)**, for $\rho_m = \rho_0 + m4\sqrt{\ln 2}\sigma$ $(m = 0, 1, 2, 3\ldots)$ with $\rho_0 = 0.5\lambda_0$ and $m = 0, 1, \ldots, 4$ $(\sigma = 0.25\lambda_0/(4\sqrt{\ln 2}),\ n = 0)$

where M and $N/2$ are the number of basis functions along ρ and the higher order azimuthal harmonic respectively. In (9.16), x and y-directed components are required to account for the vector nature of the current.

After testing with the complex conjugate of the basis functions and arranging the unknowns in the system of equations according to their direction, one arrives to an algebraic system $\mathbf{Z}\,\mathbf{i} = \mathbf{v}$, where $\mathbf{Z} = [\mathbf{Z}^{xx}, \mathbf{Z}^{xy}; \mathbf{Z}^{yx}, \mathbf{Z}^{yy}]$, with $\mathbf{Z}^{xx}$ being an $M(N+1) \times M(N+1)$ matrix. Similarly, $\mathbf{i} = [\mathbf{i}^x; \mathbf{i}^y]$ $\mathbf{v} = [\mathbf{v}^x; \mathbf{v}^y]$ with $\mathbf{i}^x$ and $\mathbf{v}^x$ being $M(N+1) \times 1$ vectors. A generic entry of the $\mathbf{Z}^{xx}$ sub-matrix can be written as the sum of

$$\mathrm{G}^{xx} = \int\limits_S \mathrm{f}_{m,n}^{x,\star}(\boldsymbol{\rho}) \int\limits_{S'} \mathrm{G}_{xx}^{\mathrm{EJ}}(|\boldsymbol{\rho} - \boldsymbol{\rho}'|)\, \mathrm{f}_{m',n'}^{x}(\boldsymbol{\rho}')\, \mathrm{d}S'\, \mathrm{d}S \tag{9.17}$$

and,

$$\chi^{xx} = -\int\limits_S \mathrm{f}_{m,n}^{x,\star}(\boldsymbol{\rho}) \mathrm{j}\mathrm{X}^{xx}(\boldsymbol{\rho}) \mathrm{f}_{m',n'}^{x}(\boldsymbol{\rho}) \mathrm{d}S \tag{9.18}$$

Both quantities depend implicitly on (m, n, m', n'). In (9.17), (9.18), $\star$ denotes the complex conjugate and $\mathrm{G}_{xx}^{\mathrm{EJ}}$ is the xx component of $\overline{\overline{\mathrm{G}}}^{\mathrm{EJ}}$. The term G^{xx} involves the relevant grounded slab Green's function, whereas χ^{xx} is related to the impedance boundary condition. It can be shown [37] that the integral in (9.17) can be written in closed-form after combining the spectral-domain approach and rational function fitting, whereas the use of an asymptotic expansion allows one to solve (9.18) without resorting to any numerical integration. The other sub-matrices can be found by using formal substitution of superscripts. Thus, the proposed basis functions allow one to write the MoM impedance matrix entries in a closed-form.

The typical structure of the $2M(N+1) \times 2M(N+1)$ MoM impedance matrix is symbolically shown in Fig. 9.8. The dashed lines in Fig. 9.8 limit the $(N+1) \times (N+1)$ block for $m = m' = 0$ and $n, n' \in [-N/2, N/2]$. The matrix with the Green's function contribution, shown in Fig. 9.8a, is composed by square tri-diagonal blocks of dimension $(N+1) \times (N+1)$, where all the elements with $|n - n'| > 2$ are equal to 0. The matrix containing the IBC contribution is shown in Fig. 9.8b. It is important to notice that here the color-bar is amplified by a factor of 10 with respect to Fig. 9.8a. Therefore, it is obvious that the IBC contribution (χ^{xx}) dominates that of the Green's function (G^{xx}). This is a general feature and suggests the use of IBC^{xx} as preconditioner for the iterative solution of very large problems (larger than 30 λ_0). The matrix IBC is dominantly block-diagonal, i.e., only the few off-diagonal blocks with $|m - m'| \leq 2$ are significant. This property, that derives from the fact that χ^{xx} is nonvanishing only for overlapping GRBFs, greatly helps the design phase when the profile of $\underline{\underline{\mathbf{X}}}$ should be optimized by successive iterations. It is also important to notice that the contribution G^{xx} is independent of the IBC and should be calculated only once in an optimization procedure. We have shown how the introduction of the MTS as a transparent impedance BC in the integral equation, and the closed-form spectra of the GRBFs enables an efficient computation of the MoM impedance matrix using the spectral-domain approach and an asymptotic expansion. More importantly, these basis functions represent the global evolution of the surface current density in an effective manner, which results in a significant reduction of the number of unknowns, if compared with sub-entire domain basis functions, like Rao–Wilton–Glisson functions defined on triangular domains.

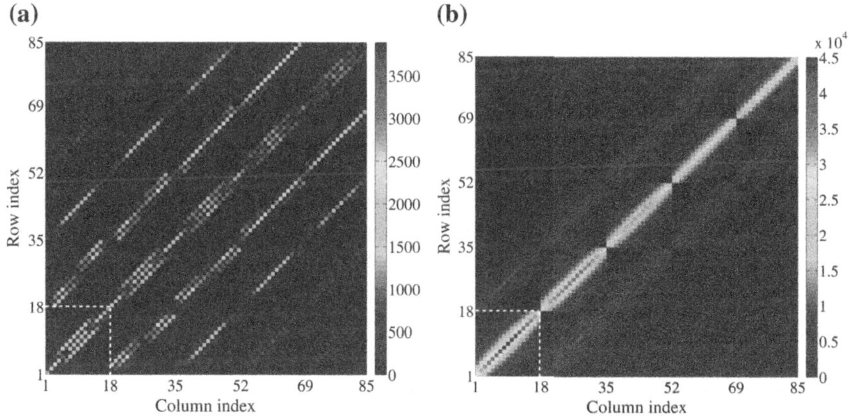

Fig. 9.8 Magnitude of the MoM impedance matrix elements for a typical structure with $M = 48$ and $N = 16$. For simplicity, the first 51 rows and columns are shown for: the matrix with the Green's function contribution **(a)** and the matrix with IBC contribution **(b)**

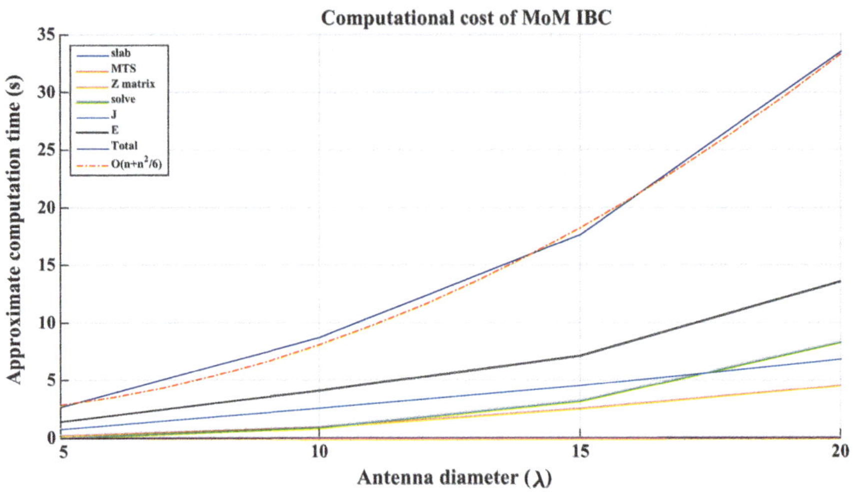

Fig. 9.9 Computational time as a function of the antenna diameter evaluated on a Quad-core laptop with 8 GB RAM. The overall computational time has a rather mild quadratic dependence with the size (*red dot-dashed line*)

The fast analysis framework obtained with this method is particularly important when used in an optimization procedure of the impedance profile. Figure 9.9 shows the timing as function of the antenna diameter evaluated on a Quad-core i7 with 8 GB RAM together with a fitting curve that shows a rather mild quadratic dependence with size. Antennas with diameter of 20 λ can be solved in about 35 s.

9.6 Pixel Modeling

At this point of the design process, the continuous reactance $\underline{\underline{\mathbf{X}}}$ that produces the objective aperture field is completely known all over the aperture domain. The next design step consists in sampling it and implementing it by a dense texture of electrically small patches. Typically, the reactance is distributed on a Cartesian lattice with square unit cells, whose dimensions range, as a rule of thumb, from λ/10 to λ/6. However, in principle, other lattices with non-square cells can be conveniently exploited. To emphasize that patches are small in terms of wavelength, sometimes we denote them and the cell that they occupy as "pixels" in the global aperture picture.

The periodicity of the lattice is always taken as a constant on the aperture and the variation of impedance is achieved by changing the geometrical parameters of the elements. To produce anisotropic impedances, the geometry of the element should have additional features in order to change the electromagnetic properties for propagation along different axes. Different geometrical shapes can be used.

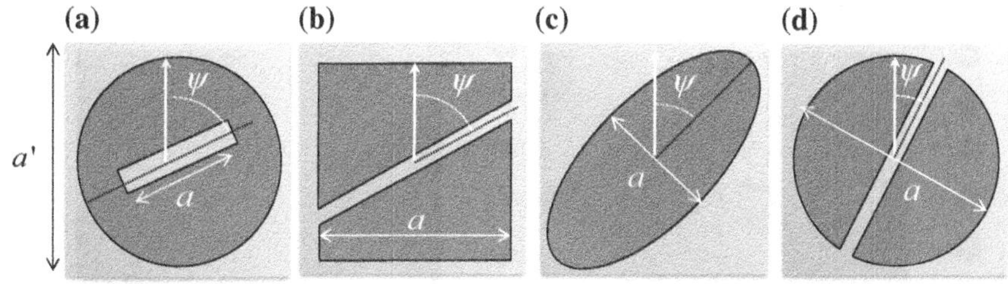

Fig. 9.10 Patch geometries for anisotropic impedance surface. Each geometry possesses two specific nondimensional parameters *a/a'* and ψ that are considered for constructing the reactance database

The solutions of Fig. 9.10b and Fig. 9.10d are the ones suggested in [7, 8], respectively. The elliptical shape in Fig. 9.10c can be analyzed with the quasi-analytical method proposed in [30]. All the investigated solutions exhibit two nondimensional parameters *a/a'* and ψ. Retrieval of the reactance values from the geometrical parameters is performed by resorting to a local periodicity concept; i.e., by periodic boundary conditions applied to a unit cell with parameters a/a' and ψ. This allows one to construct a database, which is next interpolated defining two continuous functions of the parameters a/a' and ψ. The values required in (9.2) are obtained pixel by pixel by a best fit process. This process is typically very fast after the database construction.

Figures 9.11 and 9.12 show two examples of databases, namely the impedance maps, relevant to the patch geometry of Fig. 9.10d and Fig. 9.10c, respectively, obtained through a periodic full-wave MoM solver. All of the examples are evaluated for a periodic square cell with side $a \approx \lambda/13$ on a dielectric slab with $\varepsilon_r = 13$ and thickness $\lambda/23$. The maps show the components of the opaque reactance tensor $X_{\rho\rho}, X_{\rho\varphi}, X_{\varphi\varphi}$ normalized to an impedance $\bar{X}_s = 300\,\Omega$ (for the Hermitian property of the reactance tensor it results $X_{\rho\varphi} = X_{\varphi\rho}$).

The database can be visualized in two different ways. The first one is presented in Fig. 9.11a, c, and e for slotted circular patches and in Fig. 9.11b, d and f for elliptical patches. The second one is presented in Fig. 9.12 and consists on isofrequency dispersion curves (IDCs) which represent the locus of the trajectory of the wavevectors end-point when changing the direction of propagation.

Although a systematic way to build the database is given by full-wave analysis [30, 38], the alternative approach presented in [33] can be used. It allows for constructing extensive databases for various shapes, minimizing the number of parameters to be stored. The formulation is restricted to elements with at least two symmetry axes and is valid for the dominant TM SW until the limit of the Floquet-Bloch region [32], which is the region where the higher order Floquet modes become significant. This generalization permits a closed-form representation of the isofrequency dispersion curves as a function of two parameters only; these are the equivalent quasi-static capacitances along the symmetry directions of the geometry.

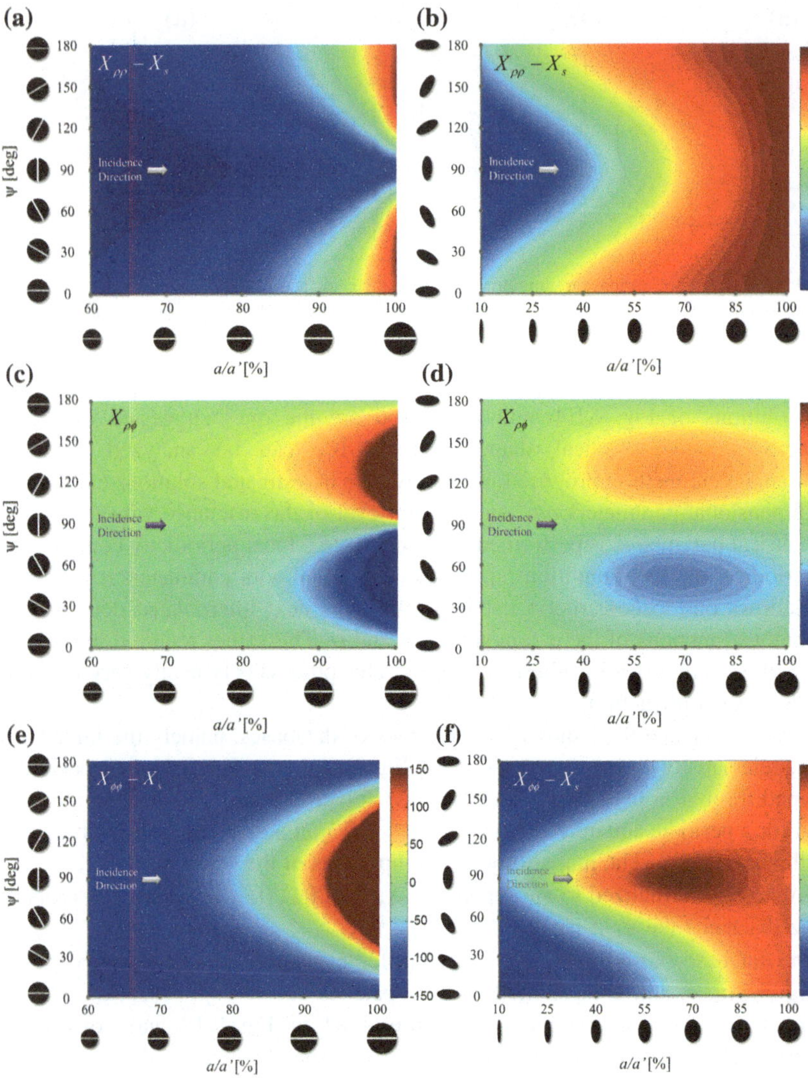

Fig. 9.11 Example of impedance maps for the patch geometry of Fig. 9.10d ((**a**) $X_{\rho\rho}$, (**c**) $X_{\rho\varphi}$, (**e**) $X_{\varphi\varphi}$) and for the patch geometry of Fig. 9.10c. ((**b**) $X_{\rho\rho}$, (**d**) $X_{\rho\varphi}$, (**f**) $X_{\varphi\varphi}$). All the maps (except for (**c**) and (**d**)) are normalized to $\bar{X}_s = 300\,\Omega$ and relevant to a cell size of λ/13 on a dielectric with thickness λ/23 and $\varepsilon_r = 13$

Once the latter are determined [32], the IDCs can be approximated by using the "two-circle approximation" [33].

Here, we provide some examples to show the accuracy of the analytical formulation in [33], considering various shapes of the printed elements: elliptical patch, rectangular patch, elliptical ring, and circular patch with rectangular slot.

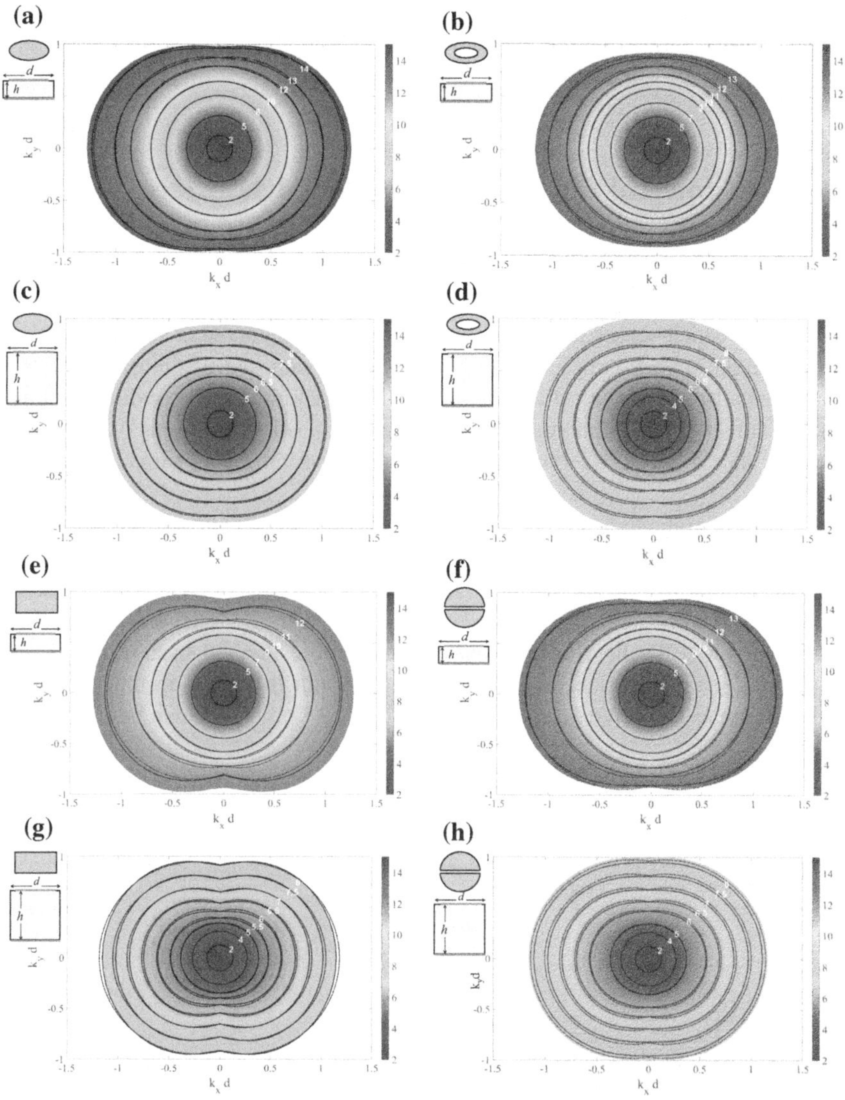

Fig. 9.12 IDCs for several patch geometries, solid lines are relevant to MoM results, dashed lines are relevant to the analytical formulation, curve labels are in GHz. Elliptical patch with $h = d/3$ (**a**) and with $h = d$ (**c**). Elliptical ring patch with $h = d/3$ (**b**) and with $h = d$ (**d**). Rectangular patch with $h = d/3$ (**e**) and with $h = d$ (**g**). Slotted circular patch with $h = d/3$ (**f**) and with $h = d$ (**h**)

The basic periodic cell is always square and with period $d = 3\,\text{mm}$. The substrate has relative permittivity $\varepsilon_r = 9.8$ and two different thicknesses have been considered ($h = 1\,\text{mm}$ and $h = 3\,\text{mm}$) for each configuration. Figure 9.12 shows the isofrequency dispersion diagrams for the above-mentioned patch geometries.

The two-parameters-dependent analytical formulation (dashed lines) is compared with the one obtained from a full-wave analysis (solid lines). The agreement is good up to the Floquet-Bloch frequency, as expected.

9.7 Fast Integral Equation Method for Metasurface Antennas

The global full-wave analysis of the textured layout of the MTS is the last step of the design, which serves as "virtual prototyping" of the antenna structure. This is not a straightforward issue, as the final layout is composed of thousands of electrically small patches with different shapes, thus yielding an ill-conditioned linear system, consisting of hundreds of thousands or even several millions of unknowns. Clearly, the solution of this problem cannot be afforded with conventional techniques.

The use of suitable patch geometries to texture the MTS allows for representing currents on them with entire domain basis functions, thus greatly reducing the number of unknowns of the problem. For instance, currents on the elliptical patches in Fig. 9.13 can be effectively represented using the entire domain basis functions defined in [30]. We have also found that two entire domain basis functions per patch are usually sufficient to describe currents with the same accuracy obtained with more than 100 RWG basis functions. The MoM procedure can be furthermore

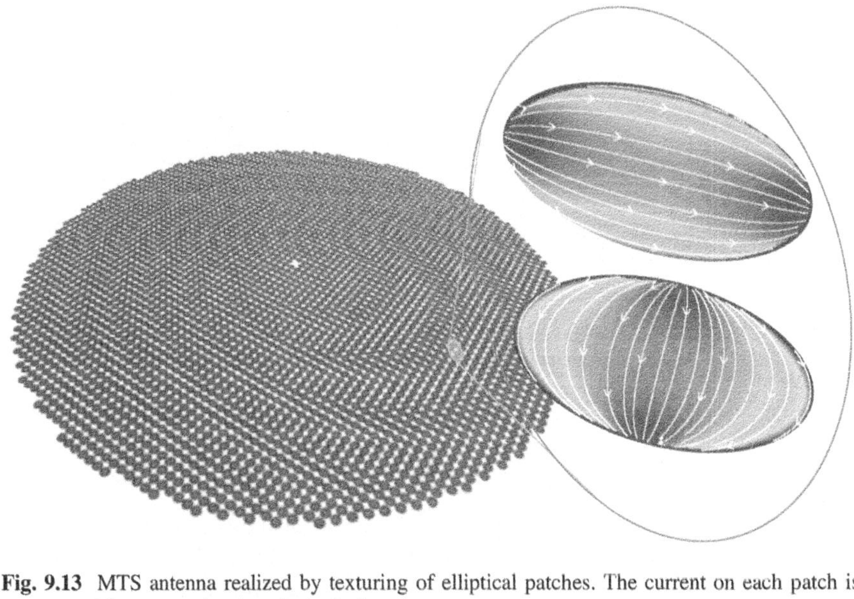

Fig. 9.13 MTS antenna realized by texturing of elliptical patches. The current on each patch is represented by two entire domain basis functions

engineered using an algorithm based on the fast multiple method (FMM) to accelerate the solution.

To set up the simulation, the metallic patches are replaced by unknown currents. The MoM is then applied to convert the electric field integral equation (EFIE) into a linear algebraic system of equations, expanding the unknown current distribution in a set of basis functions. A suitable couple of entire domain basis functions, able to represent the currents on elliptical patches, have been proposed in [30] where both spatial and spectral expressions are reported.

The spatial behavior of the two functions is illustrated in the inset of Fig. 9.13. Applying the Galerkin procedure, the standard EFIE-MoM linear system assumes the matrix form $\underline{\underline{\mathbf{Z}}} \cdot \mathbf{I} = \mathbf{V}$ where the computation of the elements of $\mathbf{V}$ depends on the nature of the source. As in the classical FMM [40, 41], the computational efficiency is increased by representing the MoM interaction matrix $\underline{\underline{\mathbf{Z}}}$ as superposition of a near interaction matrix $\underline{\underline{\mathbf{Z}}}_{\text{near}}$ and a far interaction matrix $\underline{\underline{\mathbf{Z}}}_{\text{far}}$. The 2D-FMM version reported in [40, 41] is used to accelerate the matrix-vector product relevant to $\underline{\underline{\mathbf{Z}}}_{\text{far}}$, while the near interaction matrix is processed using the standard MoM. In particular, the $\underline{\underline{\mathbf{Z}}}_{\text{near}}$ entries are calculated by using the spectral method of moment, implementing the efficient procedure in [42].

9.8 Examples of Antenna Design

In this paragraph, we recap the design process while showing some examples of modulated MTS antenna design. A wide range of aperture field distributions can be realized by properly shaping the MTS, spanning from the ones radiating a highly directive beam to the ones producing a shaped beam. In [43] a modulated MTS antenna has been designed and analyzed, which is capable to handle a broadside dual polarization. Also, here we will show that it is possible to design a MTS antenna with an operational bandwidth greatly wider than other periodic leaky-wave antennas.

9.8.1 Shaped Beam Antenna

As first example, we discuss the steps relevant to the design of a circularly polarized isoflux pattern antenna. Such shaped beam is of interest for data transmission antennas on satellite platforms for Earth observation missions. The shaped beam is conceived to provide a uniform power-flux density (isoflux) over a well-defined portion of the visible Earth surface, compensating for the rather large differential path loss between Nadir and grazing incidence on the Earth. The Earth coverage is normally obtained with conical isoflux beam [8]. To achieve higher gains as required to transmit higher data rates, the same overall coverage of the axially symmetric isoflux antenna can be obtained by a sector isoflux beam. An azimuthal rotation is needed to keep the beam pointing toward the ground station while the satellite moves along the orbit. For this example, the operational frequency is chosen at 26.25 GHz inside the Ka-band and the substrate thickness is 0.5 mm, with relative permittivity of 9.8. The size of the aperture is chosen arbitrarily as 9 λ and it allows to convert 90% of the input SW power into LW power. The target aperture distribution can be found through standard techniques, as stated in Sect. 9.2. For this example, we have extended to polarized aperture fields the method presented in [44]. Figure 9.14 shows phases and amplitudes of the objective aperture field distribution for both the radial component, ((a) and (b) respectively) and the azimuthal component ((c) and (d)). Once the input parameters for the continuous reactance synthesis block in Fig. 9.3 have been set, the impedance surface is found through the iterative process described in Sect. 9.3, assuming the impedance surface fed by a small vertical electric dipole located at the center of the

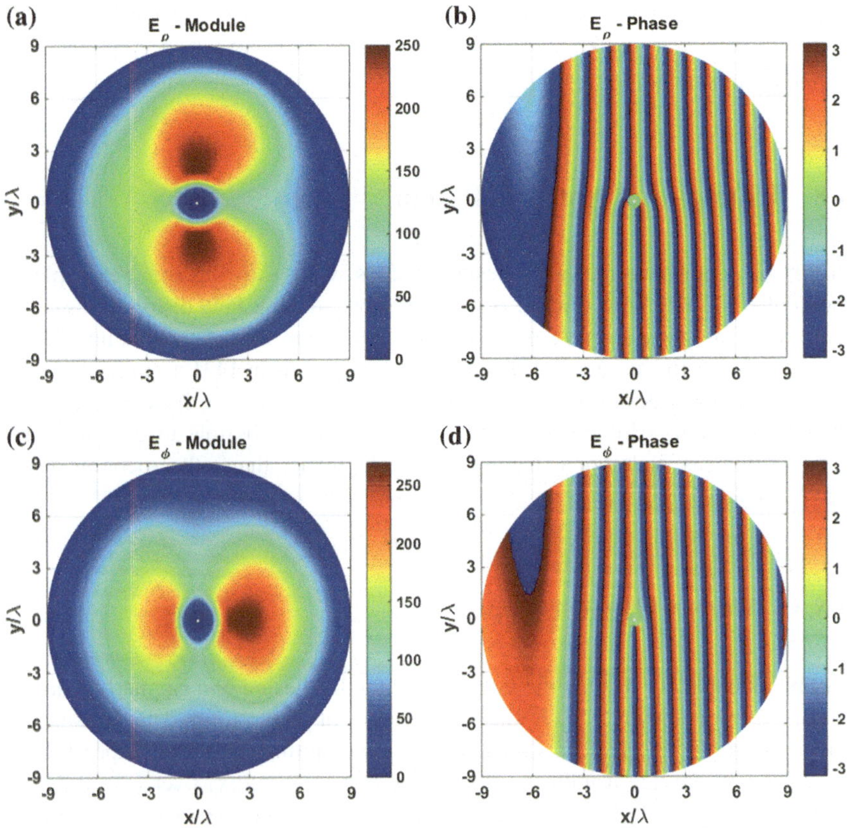

Fig. 9.14 Objective aperture field distributions for the isoflux-shaped beam antenna. Amplitude and phase of the radial component (**a, b**), and amplitude and phase of the azimuthal component (**c, d**)

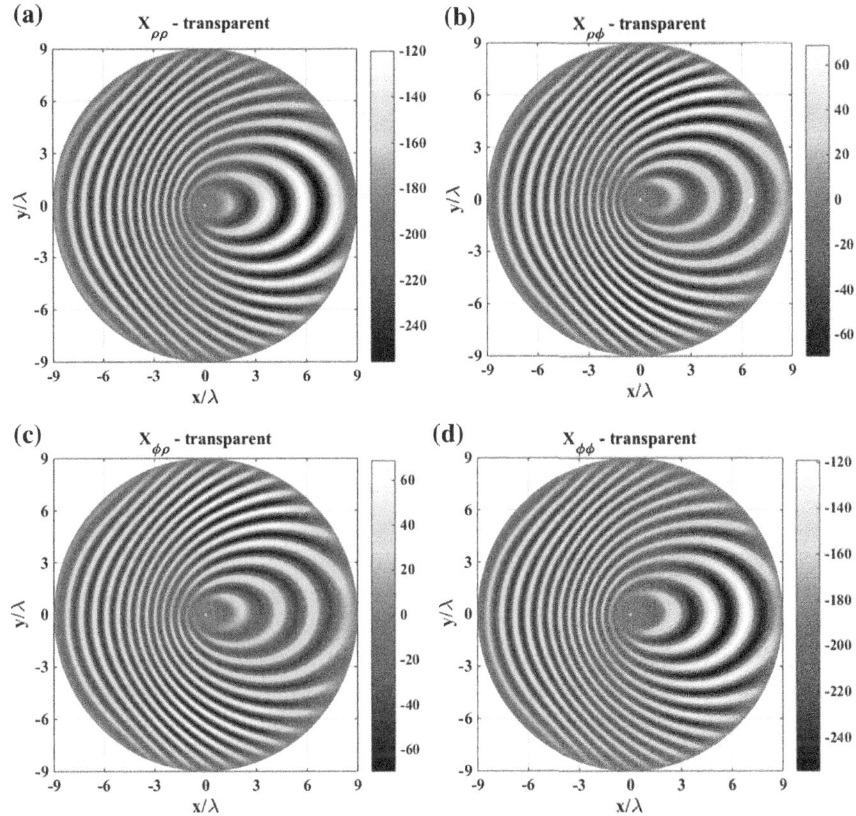

Fig. 9.15 Impedance maps of $\underline{\underline{\mathbf{X}}} = \hat{\rho}\hat{\rho}X_{\rho\rho} + (\hat{\rho}\hat{\varphi} + \hat{\varphi}\hat{\rho})X_{\rho\varphi} + \hat{\varphi}\hat{\varphi}X_{\varphi\varphi}$ resulting from the synthesis process. **a** $X_{\rho\rho}$, **b** $X_{\rho\varphi}$, **c** $X_{\varphi\rho}$, and **d** $X_{\varphi\varphi}$

surface. The entries of $\underline{\underline{\mathbf{X}}}$ for the concerned example, as a function of the position, are shown in Fig. 9.15. The impedance surface is analyzed through the continuous impedance boundary condition full-wave solver (IBC-MoM) [35].

If the design is unsatisfactory, which is something that rarely happens and only for very demanding requirements (for instance a very high efficiency), then the synthesis of $\underline{\underline{\mathbf{X}}}$ is repeated changing some of the design parameters (for instance, for a highly efficient beam one could act on the average impedance or on the aperture size).

When the agreement between the target far-field and the one radiated by the continuous impedance is judged satisfactory, the metasurface is textured with electrically small patches. In this example, we have chosen the elliptical patch geometry as the one depicted in Fig. 9.10c, so that we can use the entire domain basis functions described in [30] to create the database maps and to analyze the final layout with the global full-wave solver. To create the database, the parameters of

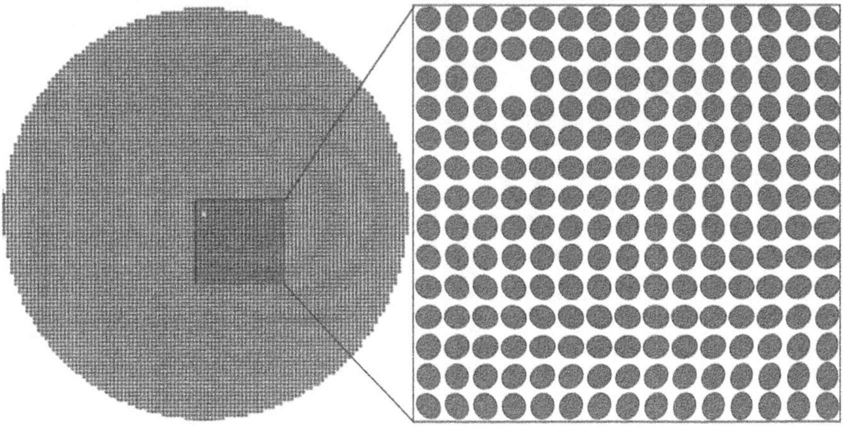

Fig. 9.16 Final layout of the antenna analyzed by the FMM-MoM: global picture and a detail of the patch texture close to the center of the antenna

the elliptic geometry are changed and the impedance associated with the geometry is retrieved by using a full-wave solver in which the concerned patch geometry is embedded into a periodic environment. The database allows for defining a function that maps a given patch geometry into the relevant tensorial impedance.

The final layout is shown in Fig. 9.16 and it is composed of elliptical patches characterized by geometrical parameters gradually variable with the position on the surface. The final structure is composed of 8936 patches and it is next analyzed using a global full-wave solver. The current on each patch is described by 2 entire domain basis functions and the MoM procedure is accelerated using a fast multipole algorithm. Figure 9.17 shows a comparison among the three methods described here for the normalized radiation diagrams: the Flat Optics solution (Sect. 9.3) the IBC-MoM solution (Sect. 9.5), which analyzes the continuous impedance BCs, the FMM-MoM solution (Sect. 9.7) which analyzes the textured layout of the antenna. Figure 9.18 shows the directive pattern and the spectral u-v plane coming from the IBC-MoM analysis.

9.8.2 High-Efficiency Antenna

Figure 9.19 shows the directive pattern of a broadside MTS antenna having a radius of 8 wavelengths whose impedance surface has been designed to produce an aperture field with very high tapering efficiency. The directivity predicted by MoM code solving the continuous impedance surface is around 33.3 dBi, corresponding to a tapering efficiency of about 85%. The tapering efficiency of the aperture distribution on the MTS, broadly speaking, is mainly controlled by tapering the modulation indexes which in turn control the leakage rate.

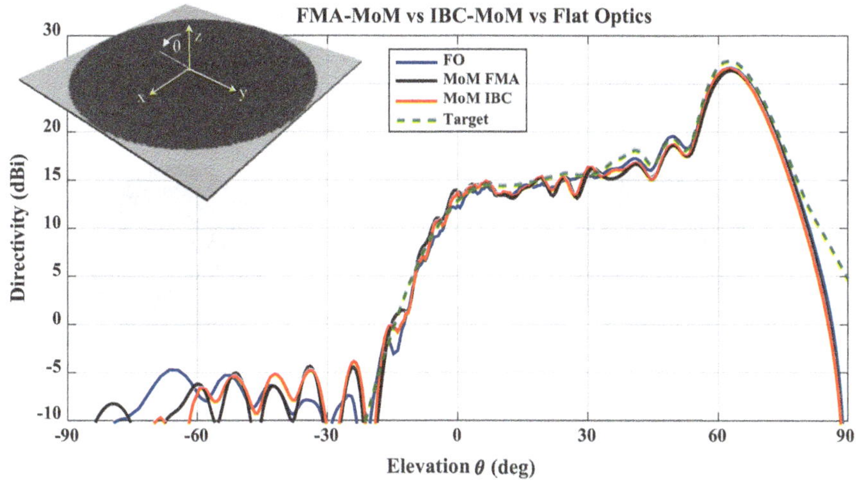

Fig. 9.17 Comparison between the copolar (RHCP) directivity diagrams. *Green-dashed line* is produced by the objective aperture distribution in Fig. 9.14. *Blue solid line* is the directivity diagram produced by the Flat Optics (FO) [35, 36] currents introduced in Sect. 9.3. *Red solid line* is resulting from the IBC-MoM analysis of the synthesized continuous impedance surface. *Black solid line* is the directive pattern resulting from the FMM-MoM analysis of the textured layout of the antenna

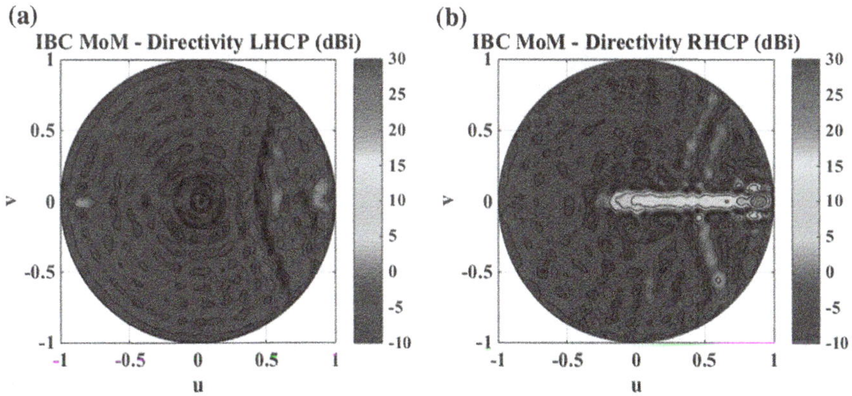

Fig. 9.18 Far field diagrams in the spectral u-v plane coming from the IBC-MoM analysis of the synthesized impedance surface, LHCP **(a)** and RHCP components **(b)**

The inset of Fig. 9.20 shows the $X_{\rho\rho}$ component of the modulated impedance surface, whilst the main frame shows the amplitude along the x-axis of the z-directed component of the Poynting vector, the modulation index m_ρ and the attenuation parameter α. It is seen that m_ρ (which is the same as m_φ for this example) increases with x, thus increasing the leakage rate α and therefore keeping

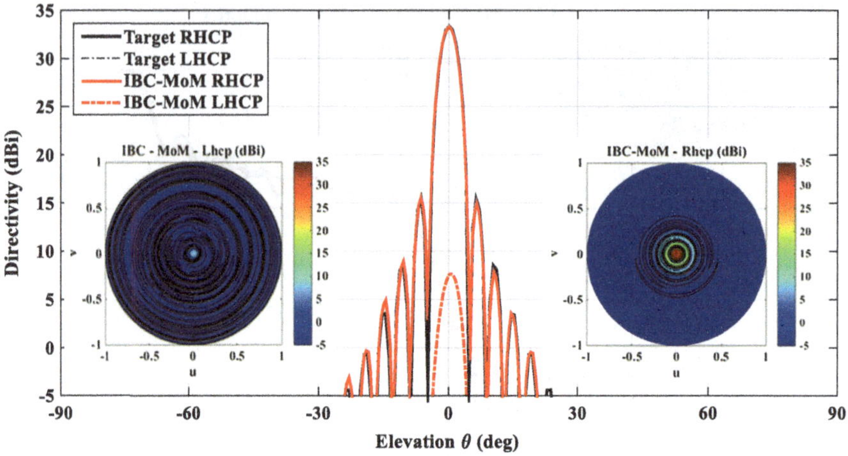

Fig. 9.19 Directive pattern from a MTS designed for a high tapering efficiency (85%). The results from the IBC-MoM (*red lines*) are compared with the target directivity patterns (*black lines*). Solid lines are relevant to RHCP components, *dotted-dashed lines* to the LHCP components. The *insets* show the radiation diagrams in the *u*-*v* spectral plane from the IBC-MoM

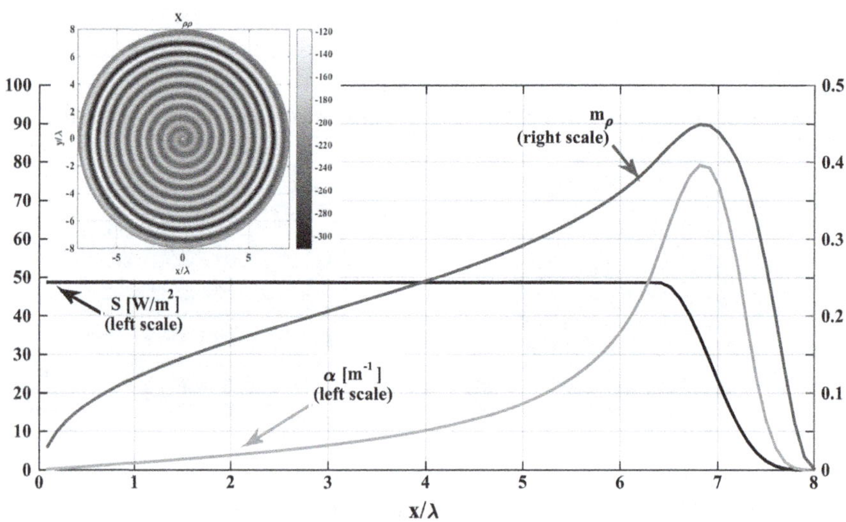

Fig. 9.20 The *inset* shows the $X_{\rho\rho}$ component of the modulated impedance for the highly efficient beam antenna. The main frame reports the following quantities taken along the *x*-cut of the MTS: *black line*, amplitude of the Poynting vector component *S* along the normal to the surface (*left* y-axis), *red line*, attenuation parameter of the SW on the MTS (*left* y-axis), modulation index (ρ-component) of the impedance surface

uniform the amplitude of the aperture distribution (described by S) up to a distance of 6.5 λ from the center of the surface.

9.8.3 Wide Band Antenna

Despite the fact that the LW radiation typically suffers of beam squinting with the frequency, a proper shaping of the BCs allows to overcome this limitation. The study for the bandwidth enlargement of modulated MTS antennas is, however, currently in progress and here we briefly sketch the concept of wideband modulated MTS antenna supporting it with a numerical example of a wideband, circularly polarized, broadside antenna. To widen the operational bandwidth of the MTS antenna we operate on the local periodicity of the impedance modulation, implementing a transition of the period's length across the aperture according to the operational frequencies. Such transition is a device to trade the available space with the operative bandwidth in a similar way as in spirals and log-periodic antennas. At the same time, we operate on the amplitude of the modulation to keep constant the peak directivity as the operational frequency change, namely reducing the effective area of the surface as the wavelength decreases. Figure 9.21 reports the frequency behavior of a synthesized broadside pointing, broadband aperture resulting from the IBC-MoM code in [37]. Figure 9.21 also shows some samples of directivity patterns at different frequencies over the wide band. The antenna has a radius of 16.7 cm on a substrate with $\varepsilon_r = 6.15$, thickness 0.635 mm. The low aperture efficiency is a trade off with the resulting wide bandwidth of about 50%.

9.9 Efficiency of Metasurface Antennas

Till now, we have discussed about the design process, starting from the definition of the continuous impedance BCs allowing to reproduce a desired aperture field. We have shown how the continuous impedance BCs are analyzed and implemented, and finally how the textured layout is verified through a global full-wave analysis. The effectiveness of the procedure has been discussed through numerical examples, and will be proved next by the experimental results. However, before going through the practical realization of MTS antennas, we spend a few comments on the efficiency of these antennas. The definition of the efficiency for MTS antennas involves several aspects related to the radiation mechanism [45]. When the input port of the feeder is fed by a power P_{in}, part of it is directly radiated in free space (P_{feed}), while the remaining part is delivered as SW (P_{sw}). The SW power is partly lost due to losses (P_{Ω}), partly radiated as LW (P_{lw}) and the remaining part reaches the

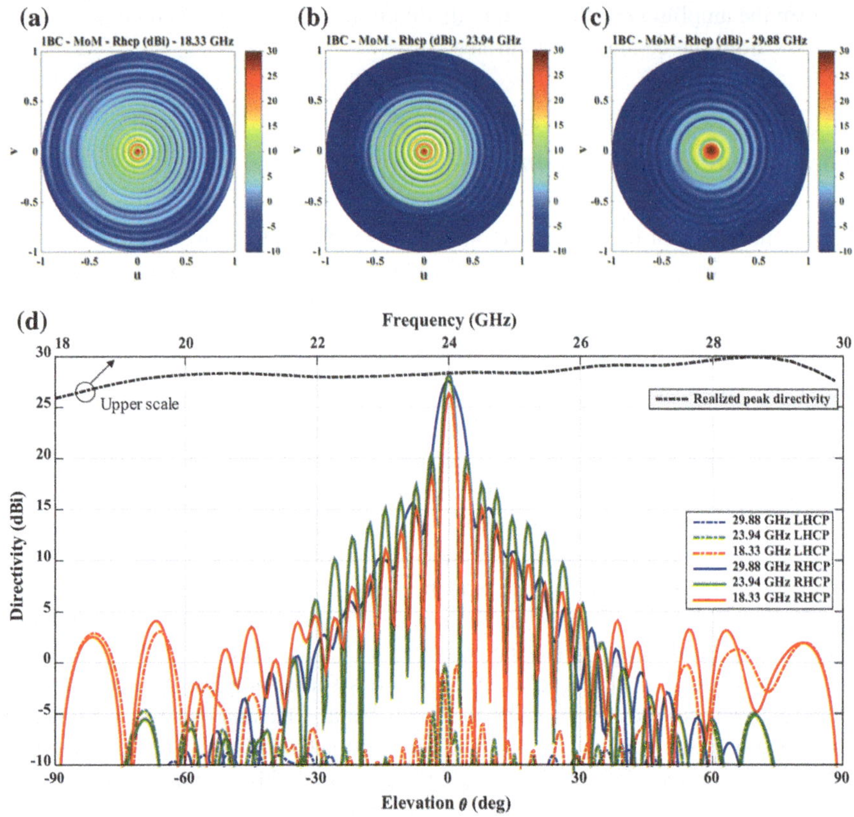

Fig. 9.21 Directive patterns for a broadband modulated MTS antenna obtained from a continuous impedance boundary conditions (IBC) MoM solver [35]. RHCP patterns are shown in the u-v spectral plane in at 18.33 GHz **(a)**, at 23.94 GHz **(b)** and at 29.88 GHz **(c)**. The relevant cuts along the $v = 0$ spectral line are in **(d)**. In the same frame, it is reported the peak directivity (*top scale*) which has a good stability within the Ka-band in the range 18–30 GHz

aperture rim giving rise to edge effects. All the efficiencies relevant to the introduced wave mechanism can be characterized more in detail as explained next.

9.9.1 *Feed Efficiency*

The feed efficiency ε_{feed} is defined as the ratio between the input power and the power delivered to the SW, i.e.,

$$\varepsilon_{feed} = P_{sw}/P_{in} = 1 - P_{feed}/P_{in} \tag{9.19}$$

Ideally, the feeder of a MTS antenna delivers all the input power to the SW, thus minimizing P_{feed}. The problem of the optimum excitation of surface waves for this kind of antennas is studied in details in [45] where it is shown that high values of ε_{feed} (around 90%) can be reached with simple SW launcher as a coaxially fed monopole on a circular patch, provided the size of the patch, the substrate and the average opaque impedance are properly chosen. More complex feeding solution can be sought for obtaining higher values of ε_{feed}, but the discussion on this topic is beyond the scope of this book and the reader is referred to [45–53] for further insights.

9.9.2 Ohmic Efficiency

The ohmic efficiency ε_Ω is defined assuming lossless metallic cladding and ground plane, and a lossy substrate. This is reasonable in the microwave regime where the dielectric losses dominate on the metal losses. The ohmic efficiency is defined as

$$\varepsilon_\Omega = \frac{P_{lw}/P_{sw}}{P_{lw}|_{P_\Omega=0}/P_{sw}} \tag{9.20}$$

where $P_{lw}|_{P_\Omega=0}$ is the LW power radiated in absence of losses.

9.9.3 Conversion and Tapering Efficiency

Conversion efficiency ε_{conv} represents the fraction of radiated LW power with respect to the SW power. In absence of losses, i.e., when $P_\Omega = 0$, the conversion efficiency is

$$\varepsilon_{conv} = P_{lw}|_{P_\Omega=0}\Big/P_{sw} \tag{9.21}$$

The tapering efficiency ε_{tap} is related to the directivity loss of a given aperture illumination with respect to a uniform distribution. For broadside beams, it is quantified as

$$\varepsilon_{tap} = \frac{\left|\int\int_A \mathbf{E}_t dA\right|^2}{A\int\int_A |\mathbf{E}_t|^2 dA} \tag{9.22}$$

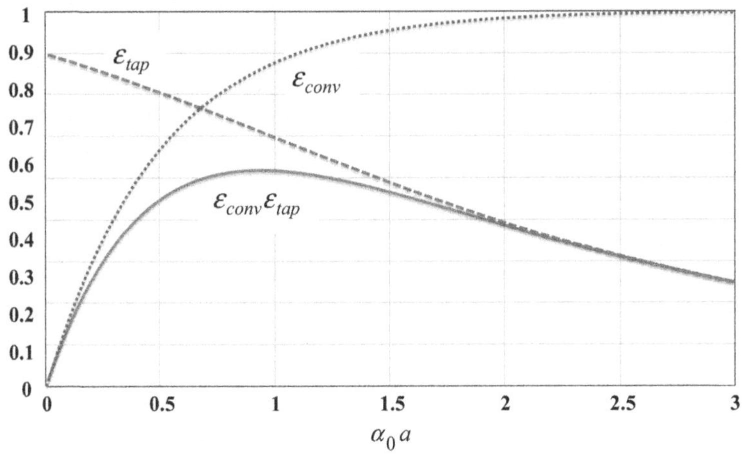

Fig. 9.22 Tapering and conversion efficiencies and their product as a function of $\alpha_0 a$ for modulation with uniform amplitude

The quantities in (9.21) and (9.22) are controlled by the attenuation parameter α, which, in turn, is controlled by the amplitude of the modulation [36]. When the amplitude of the modulation is uniform over the surface, the leakage parameter has a uniform value α_0, which depends on the depth of the modulation. Figure 9.22 shows the values assumed by ε_{conv} and ε_{tap} as a function of $\alpha_0 a$. Since the modulation is uniform, there is no control on the aperture distribution: the tapering efficiency is high only for small $\alpha_0 a$, where the conversion efficiency is small, and hence the radiated power is a little amount. The peak of efficiency is reached around $\alpha_0 a = 0.9$, where the product $\varepsilon_{tap}\varepsilon_{conv}$ is 58%. This limit can be overcome by shaping α, recurring to a nonuniform amplitude modulation, thus making the aperture distribution more efficient and at the same time increasing the amount of radiated power (Fig. 9.19).

9.9.4 Overall Efficiency

Figure 9.23 shows a plot of the overall efficiency for broadside, circularly polarized, MTS antennas with different sizes. The overall efficiency has been obtained and as the product of all the efficiencies introduced in the previous section. We have considered several lossy substrates and several values of opaque impedances that can be realized easily by a printed patch technology. It is seen that the overall efficiency can easily reach 75% by designs that make use of standard substrates, a nonuniform modulation and a simple feeder consisting of a vertical electric dipole with a metallic disk printed on the MTS (see top-left inset of Fig. 9.23). The values

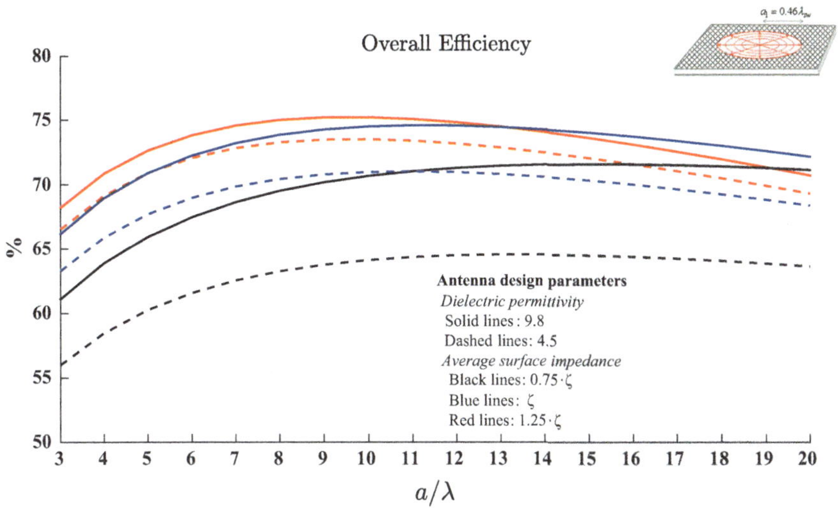

Fig. 9.23 Overall efficiency for MTS antennas of several sizes, radiating a broadside, circularly polarized beam. The substrate has a thickness corresponding to a phase change of 0.2 radians and a loss tangent of 0.002. The curves are obtained for several substrates and several values of the opaque impedance

shown in Fig. 9.23 are to be read as typical minimum overall efficiency figures for a good design. They can be exceeded with a more refined design and under specific conditions. On the other side, additional requirements on the antenna pattern may adversely affect the efficiency, like for any other antenna, e.g., a need to further reduce the side lobes level or the cross-polar radiation.

9.10 Practical Realization

Several prototypes have been built and measured during the recent years to prove the effectiveness of the concept of modulated MTS antennas. In this section we recap some of them. The earliest prototypes are all based on a modulation of the opaque reactance instead of the transparent reactance. On the other hand, the most recent ones, designed according to the theory presented in Sect. 9.4, are based on the modulation of the transparent reactance and they are currently under manufacturing or measurement.

Figure 9.24 shows the picture of a MTS antenna realized according to the theory presented in Sect. 9.4 for the synthesis of the transparent reactance. The substrate on which the patch texture has been printed, has a diameter of 24 λ at the operational frequency of 20 GHz, with $\varepsilon_r = 6.15$ and thickness 0.762 mm. The patch

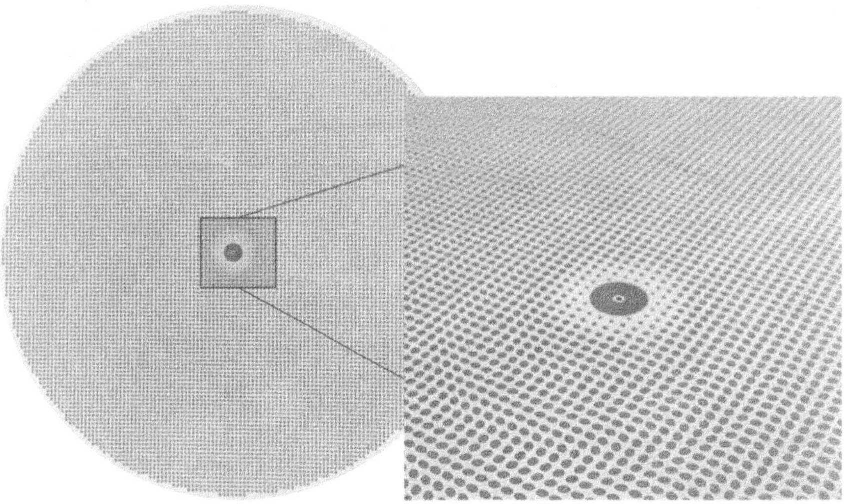

Fig. 9.24 Picture of the high gain modulated metasurface antenna prototype. The modulation of the transparent reactance is realized by a texture of elliptic patches on surface with diameter 24λ at the operational frequency of 20 GHz. The feeding system is embedded on the substrate ($\varepsilon_r = 6.15$, thickness 0.762 mm) and is constituted by a small vertical dipole, top loaded by a slotted circular patch

texture is composed of elliptical patches and it is shown more in detail in the right side of Fig. 9.24 together with the feeding system. This latter is a coaxially fed monopole, top-loaded with a patch slotted with an annular ring. Close to the feeding region, the patch sizes are progressively reduced to improve the input matching of the SW with the modulated impedance surface. The antenna radiates a circularly polarized directive beam with 35.5 dBi of gain tilted at 30° with respect to the normal to the surface. This prototype is still under measurement, and currently we can only provide the numerical prediction of the performance. Figure 9.25 compares the results from the flat optics theory (FO) of Sect. 9.3, with the ones coming from the continuous impedance MoM of Sect. 9.5 (IBC-MoM) and from the full-wave solver based on the fast multiple method (Sect. 9.7) of the textured layout (FMM-MoM). The three methods are in good agreement, except for a slight depointing of the peak resulting from the numerical simulation of the textured surface (inset of Fig. 9.25).

All the prototypes of modulated MTS antennas presented from here on are designed by shaping the opaque impedance rather than the transparent one. They represent the first realizations of such antennas and they were designed according to the theory described in [6]. One of the earliest prototypes of modulated MTS antenna is shown in Fig. 9.26. It is realized on a substrate with $\varepsilon_r = 3.66$, thickness 1.524 mm. The panel radius is 10.1 cm, corresponding to 5.7 λ at the operational frequency of 17 GHz. The MTS is formed by a dense texture of small square

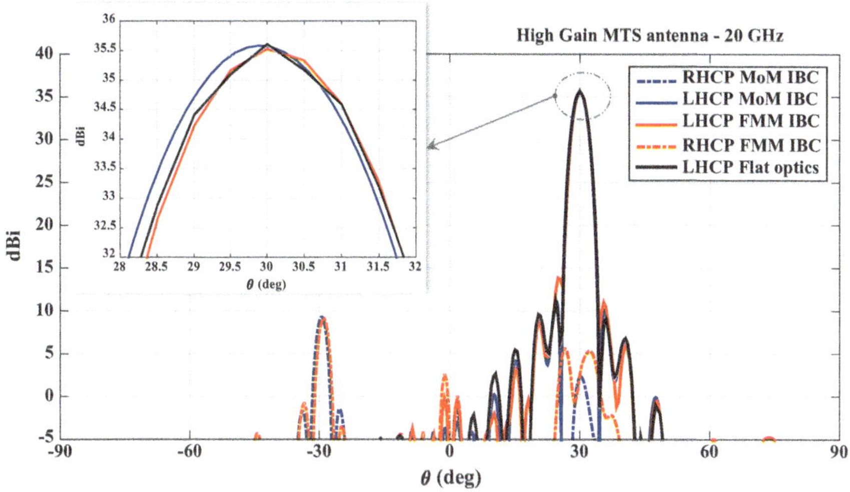

Fig. 9.25 Numerical prediction of the directivity of the antenna presented in Fig. 9.24. *Solid lines* are relevant to the copolar RHCP component, *dot-dashed lines* are relevant to the LHCP cross-polar components. *Blue lines*: pattern resulting from the IBC-MoM analysis of the continuous reactance. *Red lines*: pattern resulting from the FMM-MoM analysis of the textured surface. *Black line*, directivity diagram pattern resulting from the Flat-Optics model of the continuous impedance surface

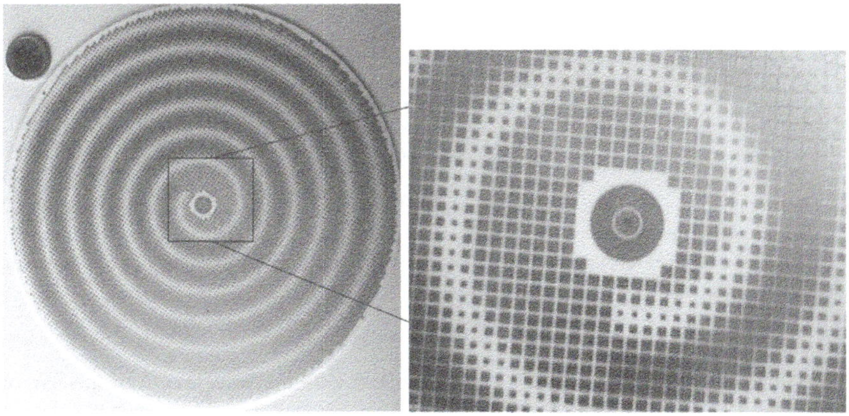

Fig. 9.26 Prototype of modulated MTS antenna realized on a substrate with $\varepsilon_r = 3.66$, thickness 1.524 mm and radius 10.1 cm. The MTS is formed by a dense texture of small *square patches* arranged in a *spiral shape*, resulting in an opaque isotropic impedance with sinusoidal modulation

patches arranged in a spiral shape, resulting in an opaque impedance with sinusoidal modulation. As the patches do not have geometrical asymmetries, the resulting modulation is isotropic, meaning that the TM nature of the SW is not significantly modified while propagating on the MTS. This latter has a radial

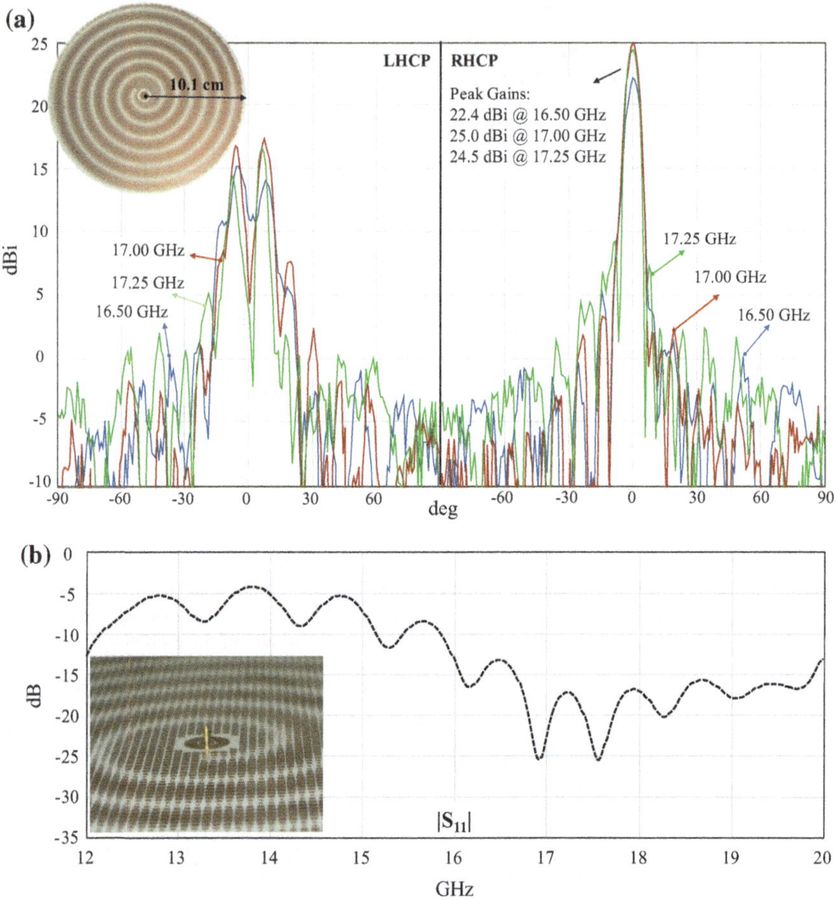

Fig. 9.27 Gain measurement for the modulated spiral MTS antenna. LCHP and RHCP components **(a)**. The antenna (*top-left* inset in **(a)** is designed at 17 GHz on a panel with radius 10.1 cm. The measured reflection coefficient is in **(b)** with a detail of the feeding vertical monopole (*bottom-left* inset in **(b)**)

periodicity $p_r = \lambda/\sqrt{1+(\bar{X}_{op}/\zeta)^2}$ synchronized with the wavelength of the SW on the average opaque impedance $\bar{X}_{op}$. Intuitively speaking, the spiral shape of the modulation makes any two orthogonal sectors radiating at the broadside with phase quadrature. Figure 9.27a shows the measured gain patterns at several frequencies around the designed working frequency of 17 GHz, respectively for the copolar (RHCP) and the cross (LHCP) components. As the implemented impedance surface is isotropic, the cross-polar levels are not controlled, differently from the prototype of Figs. 9.24 and 9.25. The antenna is fed through a small vertical monopole,

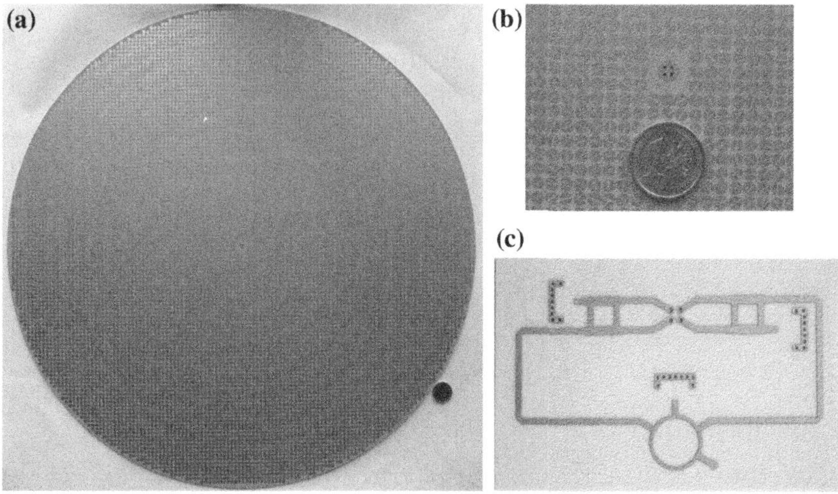

Fig. 9.28 Picture of the proof-of-concept MTS antenna radiating a CP isoflux beam realized for Earth observation missions on LEO satellites (**a**). Detail of the feeding system: the *circular patch* has the twofold role of adjusting the broadside gain and launching the SW on the modulated MTS (**b**). Picture of the feeding circuitry responsible for sequentially exciting four small vertical monopoles with initial phases of 0°, 90°, 180°, and 270° (**c**)

coaxially fed, and placed at the center of the device. The measured reflection coefficient is presented in Fig. 9.27b together with a detail of the feed in the bottom-left inset. Further details can be found in [6, 13].

A proof-of-concept MTS antenna radiating a CP isoflux beam realized for Earth observation missions on LEO satellites is shown in Fig. 9.28a. The antenna on the satellite platform shall radiate a homogenous power density toward the Earth surface, increasing the gain toward the angular regions where the path is longer, to compensate for the power loss. The realized antenna works in the X-band and the main frame of Fig. 9.29 shows the measured gain patterns both for LHCP and RHCP components around 8.6 GHz. The MTS is realized on a dielectric panel having radius of 27 cm with $\varepsilon_r = 9.8$, and thickness 1.575 mm and composed of about 12000 patches having a circular shape with a small cut in it. The small cut within the patch shape allows to realize an anisotropic impedance surface, therefore allowing for a full control of the aperture field polarization. The patch texture modulates the opaque impedance sinusoidally along the radial direction, without variation along the azimuthal direction. The radial periodicity of the modulation is such that it produces a conical radiation with peak at 60° off-axis, while the feeder (Fig. 9.28b) has the twofold role of adjusting the broadside gain and providing the SW on the modulated metasurface. The SW is excited with a rotational phase

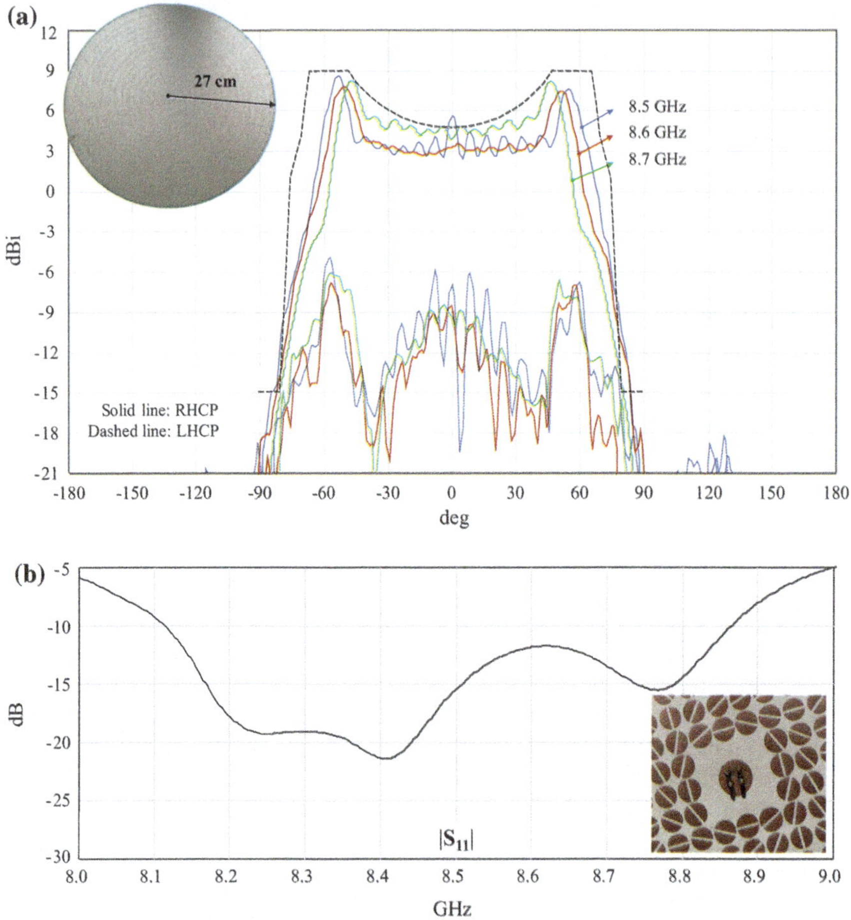

Fig. 9.29 Measured gain (RHCP vs. LHCP) of the isoflux metasurface antenna prototype **(a)** and measured reflection coefficient **(b)**

distribution, by four small vertical monopoles, sequentially fed with initial phases of 0°, 90°, 180° and 270° (Fig. 9.28c). The input reflection coefficient is presented in Fig. 9.29b together with a picture of the input feeding system in the bottom-right inset of the same figure. Further details about design, analysis, and realization can be found in [6, 8].

One of the most interesting feature of these antennas is that just by changing the impedance boundary conditions provided by the MTS, completely different performances can be obtained. This key feature of modulated MTS antenna allows a unique support for late electrical performance refinement and design re-usability which is very appealing for space applications. Figure 9.30 shows two broadside circularly polarized MTS antennas designed on the same dielectric panel used for

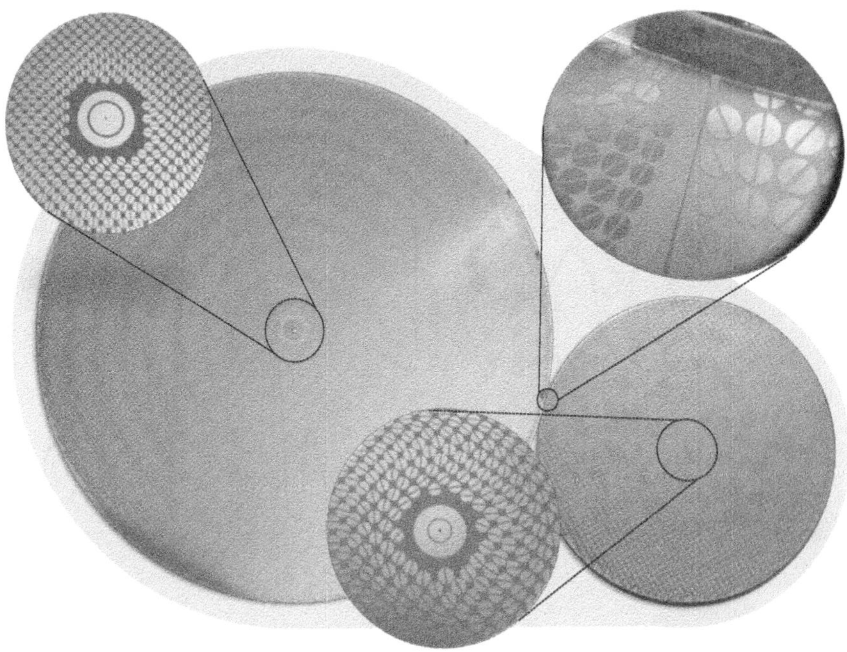

Fig. 9.30 Photos of the two realized prototypes. *Left side* Tx antenna. *Right side* Rx antenna, Details of the feeders and of the printed elements are shown in the *insets*

the isoflux antenna prototype. The two antennas can be set up on the same platform to get a TX-RX system with low coupling at the two ports. They are proof-of-concept devices in the X-band designed using typical requirements for deep space science and automated exploration missions.

The main frames of Fig. 9.31a and b present the peak-normalized radiation patterns relevant, respectively, to the Rx antenna and the Tx antenna. The Rx antenna exhibits a Gain larger than 21 dBi over a bandwidth of 4%, the Tx antenna has gain larger than 27 dBi over a bandwidth of 2.4%. It is to notice that, even without operating on the modulation depth profile, efficiencies of about 35% are obtained. Also, measurements prove good polarization purity at broadside being the CX component close to −15 dB below the copolar component level. The measured reflection coefficients is reported in Fig. 9.32. Further details can be found in [9].

The results shown in this section confirm the attractiveness of modulated MTS antennas, which preserve all its simplicity and handiness with different far field performances. All the reported prototypes were realized using standard PCB process, ensuring cost-effectiveness together with good performance of gain and polarization.

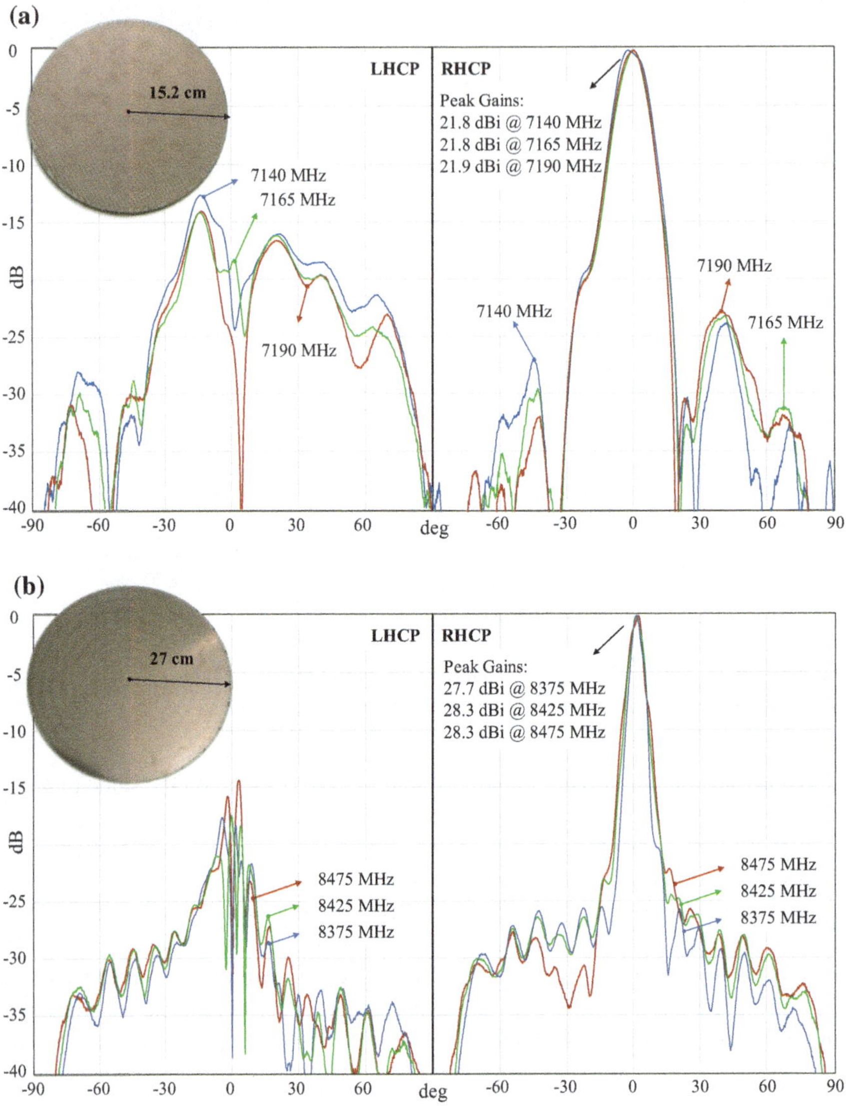

Fig. 9.31 Measured normalized far field patterns relevant to two X-band prototypes designed as proof-of-concept devices for deep space exploration missions. LHCP and RHCP component of the Rx antenna at three different frequencies **(a)**. LHCP and RHCP component of the Tx antenna at three different frequencies **(b)**. The *top-left insets* show the picture of both the antennas

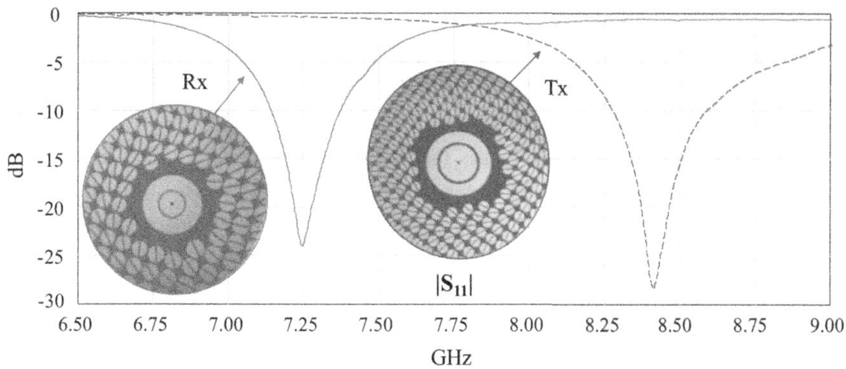

Fig. 9.32 Measured reflection coefficients relevant to two X-band prototypes designed as proof-of-concept devices for deep space exploration missions. Rx prototype (**a**) and Tx prototype (**b**). Both prototypes are fed by a vertical monopole, coaxially fed and top-loaded with a slotted patch to ensure the input matching. Pictures in the *insets* show the details of the feeding points

9.11 Discussion and Future Trends

Modulated metasurface antennas constitute a significant improvement from traditional antenna solutions. Rather than shaping a fixed boundary condition (mostly PEC) to achieve the desired radiation characteristics, they offer the revolutionary possibility to design the boundary condition itself to achieve the desired behavior, with an antenna having a predetermined global shape. All the designs shown are in fact obtained using antennas differing only in the metasurface layout and feeding details.

The full control of the aperture field and hence of the far-field properties (beam shape, pointing angle, polarization) is strictly related to the capability of accurately designing the MTS. Here, an effective synthesis method has been introduced which permits the control of amplitude, phase, and polarization of the aperture field by adjusting the boundary conditions imposed by the MTS. The complete procedure is described in [36]. The polarization control is based on the local value of the anisotropy, the phase is controlled by both the shape and the periodicity of the modulation, and the amplitude control is obtained by designing the local leakage attenuation distribution of the leaky wave. The synthesis is based on the *adiabatic* Floquet-wave expansion of currents and fields over the surface introduced in [35]. The main advantages of this process with respect to the ones already introduced are summarized below.

- Introduction of a systematic way to synthesize the aperture field with analytical formulas and to validate it in the systematic frame of Fig. 9.3.
- Development of new formulas based on the energy conservation, which allow for controlling the aperture field amplitude—not possible before—and therefore designing antennas with higher efficiency than before.

- Use of the transparent boundary conditions (in place of the opaque ones used by the authors in previous schemes) in a rigorous Green's Function scheme. This allows for accounting for space dispersion of the grounded slab in practical implementations.

Although the problem of the losses has not been treated here, small local losses in the dielectric and in the metal can be easily introduced in the formulation. In our experience, losses may have an impact when dealing with antennas with more than 40 dBi of gain.

We have also shown that despite the fact that modulated MTS antennas are leaky-wave antennas, it is possible to overcome the typical beam squinting with frequency by a proper shaping of the BCs imposed by the MTS. However, the enlargement of the pattern bandwidth is still under study and the issue requires a long discussion that will be the subject of a future paper. Concerning the bandwidth of MTS, in some recent works [54, 55] it has been proposed the use of cells with glide symmetry for realizing planar MTS lenses with ultra wide band.

Future research trends span from the design of conformal modulated MTS antennas to the design of a reliable dynamic impedance pattern. Conformal MTSs will allow to further reduce the encumbrance of these antennas by perfectly matching them to curved surfaces. For instance, one may use conformal MTSs matched to the fuselage of airplanes, reducing the impact of the antenna on the aerodynamics. Dynamic control of the impedance patterns is of importance for all the tracking applications, requiring to hold on a link between moving platforms. For instance, radar applications, data connection between satellite and ground station or Internet connections on board of trains. Also, MTS antennas with multiple operational frequencies, contoured beams and multi-beam designs are challenging and interesting topics. However, among all of them, the dynamic control of the MTS is the one that most attracts the attentions of the telecom-market. It is a research topic that appears ambitious and feasible at the same: in modulated MTS antennas, the excitation spectrum and the surface impedance pattern completely control the beam shape and polarization. Therefore, by dynamically controlling the impedance, virtually all the degrees of freedom of the radiated wave can be addressed, with conceptually very simple means and in a way that lends to attractive implementations. Possible solutions for reconfigurable dynamic metasurfaces can involve mechanical stretching of a proper dielectric material or electronic reconfigurations. For pencil beams antennas, a mechanically reconfigurable surface seems to be particularly suitable. At the opposite end of the spectrum of solutions an electronically reconfigurable surface could offer more potentialities in terms of flexibility of radiative characteristics. Materials having characteristics similar to liquid crystals at RF or based on arrays of MEMS are also good candidates for adaptive surfaces, as well as electronic devices like PiN diodes. The final goal is to realize a device able to radiate a wide range of arbitrarily shaped beams, circularly or linearly polarized, by simply modifying the surface pattern impedance, without changing the in-plane feeder.

References

1. C.L. Holloway, A. Dienstfrey, E.F. Kuester, J.F. O'Hara, A.K. Azad, A.J. Taylor, A discussion on the interpretation and characterization of metafilms/metasurfaces: the two dimensional equivalent of metamaterials. Metamaterials **3**, 100–112 (2009)
2. M.G. Silverinha, C.A. Fernandes, J.R. Costa, Electromagnetic charactrerization of textured surfaces formed by metallic pins. IEEE Trans. Antennas Propag. **56**(2), 405, 415 (2008)
3. C.L. Holloway, E.F. Kuester, J.A. Gordon, J. O'Hara, J. Booth, D.R. Smith, An overview of the theory and applications of metasurfaces: the two-dimensional equivalents of metamaterials. IEEE Antennas Propag. Mag. **54**(2), 10–35 (2012)
4. G. Minatti, F. Caminita, M. Casaletti, S. Maci, Spiral leaky-wave antennas based on modulated surface impedance. IEEE Trans. Antennas Propag. **59**(12), 4436–4444 (2011)
5. A.M. Patel, A. Grbic, A printed leaky-wave antenna based on a sinusoidally-modulated reactance surface. IEEE Trans. Antennas Propag. **59**(6), 2087–2096 (2011)
6. G. Minatti, M. Faenzi, E. Martini, F. Caminita, P. De Vita, D. Gonzalez-Ovejero, M. Sabbadini, S. Maci, Modulated metasurface antennas for space: synthesis, analysis and realizations. IEEE Trans. Antennas Propag. **63**(4), 1288–1300 (2015)
7. B.H. Fong, J.S. Colburn, J.J. Ottusch, J.L. Visher, D.F. Sievenpiper, Scalar and tensor holographic artificial impedance surfaces. IEEE Trans. Antennas Propag. **58**(10), 3212–3221 (2010)
8. G. Minatti, S. Maci, P. De Vita, A. Freni, M. Sabbadini, A circularly-polarized isoflux antenna based on anisotropic metasurface. IEEE Trans. Antennas Propag. **60**(11), 4998–5009 (2012)
9. M. Faenzi, F. Caminita, E. Martini, P. De Vita, G. Minatti, M. Sabbadini, S. Maci, Realization and measurement of broadside beam modulated metasurface antennas, in Antennas Wirel. Propag. Lett. IEEE **15**, 610–613, (2016)
10. S. Pandi, C.A. Balanis, C.R. Birtcher, Design of scalar impedance holographic metasurfaces for antenna beam formation with desired polarization. IEEE Trans. Antennas Propag. **63**(7), 3016–3024 (2015)
11. M. Casaletti, M. Śmierzchalski, M. Ettorre, R. Sauleau, N. Capet, Polarized beams using scalar metasurfaces. IEEE Trans. Antennas Propag. **64**(8), 3391–3400 (2016)
12. M. Sabbadini, G. Minatti, S. Maci, P. De Vita, Method for designing a modulated metasurface antenna structure. Patent WO 2015090351 A1, 25 June 2015
13. S. Maci, G. Minatti, M. Casaletti, M. Bosiljevac, Metasurfing: addressing waves on impenetrable metasurfaces. IEEE Antennas Wirel. Propag. Lett. **10**, 1499–1502 (2011)
14. E. Martini, S. Maci, Metasurface Transformation Theory, in *Transformation Electromagnetics and Metamaterials*, ed. by D.H. Werner, D.H. Know (Springer, London, 2013), pp. 83–116
15. C. Pfeiffer, A. Grbic, A printed, broadband Luneburg lens antenna. IEEE Trans. Antennas Propag. **58**(9), 3055–3059 (2010)
16. M. Bosiljevac, M. Casaletti, F. Caminita, Z. Sipus, S. Maci, Non-uniform metasurface Luneburg lens antenna design. IEEE Trans. Antennas Propag. **60**(9), 4065–4073 (2012)
17. C. Pfeiffer, A. Grbic, Metamaterial Huygens' surfaces: tailoring wave fronts with reflectionless sheets. Phys. Rev. Lett. **110**(19), 197401 (2013)
18. M. Selvanayagam, G. Eleftheriades, Discontinuous electromagnetic fields using orthogonal electric and magnetic currents for wavefront manipulation. Opt. Express **21**(12), 14409–14429 (2013)
19. N. Yu, P. Genevet, F. Aieta, M.A. Kats, R. Blanchard, G. Aoust, J.-P. Tetienne, Z. Gaburro, F. Capasso, Flat optics: controlling wavefronts with optical antenna metasurfaces. IEEE J. Sel. Topics Quantum Electron. **19**(3), 4700423–4700423 (2013)
20. P.S. Kildal, E. Alfonso, A. Valero-Nogueira, E. Rajo-Iglesias, Local metamaterial-based waveguides in gaps between parallel metal plates. IEEE Antennas Wirel. Propag. Lett. **8**, 84–87 (2009)

21. A.M. Patel, A. Grbic, Transformation electromagnetics devices based on printed-circuit tensor impedance surfaces. IEEE Trans. Microwave Theory Tech. **62**(5), 1102–1111 (2014)
22. R. Quarfoth, D. Sievenpiper, Surface wave scattering reduction using beam shifters. IEEE Antennas Wirel. Propag. Lett. **13**, 963,966 (2014)
23. A. Vakil, N. Engheta, Transformation optics using graphene. Science **332**, 1291–1294 (2011)
24. R. Yang, Y. Hao, An accurate control of the surface wave using transformation optics. Opt. Express **20**(9), 9341–9350 (2012)
25. R. Yang, W. Tang, T. Hao, Wideband beam-steerable flat reflectors via transformation optics. IEEE Antennas Wirel. Propag. Lett. **10**, 99–102 (2011)
26. W. Tang, C. Argyropoulos, E. Kallos, S. Wei, Y. Hao, Discrete coordinate transformation for designing all-dielectric flat antennas. IEEE Trans. Antennas Propag. **58**(12), 3795–3804 (2010)
27. M. Mencagli Jr., E. Martini, D. Gonzàlez-Ovejero, S. Maci, Metasurface transformation optics. J. Opt. **16**, 125106 (2014)
28. M. Mencagli Jr., E Martini, D Gonzàlez-Ovejero, S Maci Metasurfing by Transformation Electromagnetics. IEEE Antennas Wirel. Propag. **13**, 1767, 1770 (2014)
29. E. Martini, M. Mencagli, S. Maci, Metasurface transformation for Surface wave control. Phyl. Trans. R. Soc. A **A373**, 20140355 (2015)
30. M. Mencagli, E. Martini, S. Maci, Surface wave dispersion for anisotropic metasurfaces constituted by elliptical patches. IEEE Trans. Antennas Propag. **63**(7), 2992–3003 (2015)
31. A.M. Patel, A. Grbic, Effective surface impedance of a printed-circuit tensor impedance surface (PCTIS). IEEE Trans. Microwave Theory Tech. **61**(4), 1403–1413 (2013)
32. M. Mencagli Jr., E. Martini, S. Maci, Transition functions for closed form representation of metasurface reactance. IEEE Trans. Antennas Propag. **64**(1), 136–145 (2016)
33. M. Mencagli Jr., C. Della Giovampaola, S. Maci, A Closed-form representation of isofrequency dispersion curve and group velocity for surface waves supported by anisotropic and spatially dispersive metasurfaces, **64**(6), 2319–2327 (2016)
34. O.M Bucci, G Franceschetti, G. Mazzarella, G. Panariello, Intersection approach to array pattern synthesis. IEE Proc. H Microwaves Antennas Propag. **137**(6), 349–357 (1990)
35. G. Minatti, F. Caminita, E. Martini, S. Maci, Flat optics for leaky-waves on modulated metasurfaces: adiabatic floquet-wave analysis. IEEE Trans. Antennas Propag. **64**(9), 3896–3906 (2016)
36. G. Minatti, F. Caminita, E. Martini, M. Sabbadini, S. Maci, Synthesis of modulated-metasurface antennas with amplitude, phase and polarization control. IEEE Trans. Antennas Propag. **64**(9), 3907–3919 (2016)
37. D. Gonzalez-Ovejero, S. Maci, Gaussian ring basis functions for the analysis of modulated metasurface antennas. IEEE Trans. Antennas Propag. **63**(9), 3982–3993 (2015)
38. S. Maci, M. Caiazzo, A. Cucini, M. Casaletti, A pole-zero matching method for EBG surfaces composed of a dipole FSS printed on a grounded dielectric slab. IEEE Trans. Antennas Propag. **53**(1), 70–81 (2005)
39. M.A. Francavilla, E. Martini, S. Maci, G. Vecchi, On the numerical simulation of metasurfaces with impedance boundary condition integral equations. IEEE Trans. Antennas Propag. **63**(5), 2153–2161 (2015)
40. V. Rokhlin, Rapid solution of integral equations of scattering theory in two dimensions. J. Comput. Phys. **36**(2), 414–439 (1990)
41. C.R. Anderson, An implementation of the fast multipole method without multipole. SIAM J. Sci. Stat. Comput. **13**(4), 923–947 (1992)
42. M. Albani, A. Mazzinghi, A. Freni, Asymptotic approximation of mutual admittance involved in MoM analysis of RLSA antennas. IEEE Trans. Antennas Propag. **57**(4) (2009)
43. A. Tellechea, F. Caminita, E. Martini, I. Ederra, J.C. Iriarte, R. Gonzalo, S. Maci, Dual circularly-polarized broadside beam metasurface antenna. IEEE Trans. Antennas Propag. **64** (7) 2944–2953 (2016)
44. A. Chakraborty, B.N. Das, G.S. Sanyal, Determination of phase functions for a desired one-dimensional pattern. IEEE Trans. Antennas Propag. **29**(3), 502–506 (1981)

45. G. Minatti, E. Martini, S. Maci, Efficiency of metasurface antennas. IEEE Trans. Antennas Propag. **65**(4), 1532–1541, (2017)
46. A.L. Cullen, The excitation of plane surface waves. Proc. IEE—Part IV Inst. Monogr. **101**(7), 225–234 (1954)
47. A. Kay, F. Zucker, Efficiency of surface wave excitation, in *1958 IRE International Convention Record* (New York, NY, USA, 1955), pp. 1–5
48. A.L. Cullen, A note on the excitation of surface waves. Proc. IEE—Part C Monogr. **104**(6), 472–474 (1957)
49. D.A. Hill, J.R. Wait, Excitation of the Zenneck surface wave by a vertical aperture. Radio Sci. **13**(6), 969–977 (1978)
50. S. Mahmoud, Y.M.M. Antar, H. Hammad, A. Freundorfer, Theoretical considerations in the optimization of surface waves on a planar structure. IEEE Trans. Antennas Propag. **52**(8), 2057–2063 (2004)
51. H. Hammad, Y.M.M. Antar, A. Freundorfer, S. Mahmoud, Uniplanar CPW-fed slot launchers for efficient TM_0 surface wave excitation. IEEE Trans. Microwave Theory Tech. **51**(4), 1234–1240 (2003)
52. B. Friedman, W. Elwyn Williams, Excitation of surface waves. Proc. IEE—Part C Monogr. **105**(7), 252–258 (1958)
53. G. Tsandoulas, Excitation of a grounded dielectric slab by a horizontal dipole. IEEE Trans. Antennas Propag. **17**(2), 156–161 (1969)
54. M. Ebrahimpouri, E. Rajo-Iglesias, Z. Sipus, O. Quevedo-Teruel, Low-cost metasurface using glide symmetry for integrated waveguides, in *2016 10th European Conference on Antennas and Propagation (EuCAP)*, Davos (2016), pp. 1–2
55. O. Quevedo-Teruel, M. Ebrahimpouri, M. Ng Mou Kehn, Ultrawideband metasurface lenses based on off-shifted opposite layers. IEEE Antennas Wirel. Propag. Lett. **15**, 484–487 (2016)

Chapter 10
Terahertz Antennas and Feeds

Goutam Chattopadhyay, Maria Alonso-delPino, Nacer Chahat, David González-Ovejero, Choonsup Lee and Theodore Reck

Abstract Terahertz antennas present a different set of challenges to the antenna designer typically striving for very high performance while at the very limit of the chosen fabrication process. Many of the same design techniques used at lower frequencies are still applied, but fabrication constraints impose significant limitations on the type of structure that can be used, forcing the designer to consider unique fabrication processes or completely new antenna structures. Through advances in fabrication and computational techniques, the variety of terahertz antennas is growing. This chapter presents a range of antennas applied at these frequencies from 1.9 THz horn antennas to superconducting planar arrays. The chapter will cover different antenna technologies and feeds such as corrugated horn antennas, smooth-walled profiled horn antennas, multi-flare angle horn antennas, lens antennas, microlens leaky wave antennas, metasurface antennas, antenna arrays, off-chip antennas, and others. It will detail theory, simulation, fabrication techniques, and state-of-the-art antenna results in all these different technologies at millimeter and terahertz frequencies. The chapter will also provide details for terahertz antennas in the context of terahertz systems.

G. Chattopadhyay (✉) · M. Alonso-delPino · N. Chahat · D. González-Ovejero · C. Lee · T. Reck
NASA Jet Propulsion Laboratory, California Institute of Technology, Pasadena, CA, USA
e-mail: goutam.chattopadhyay@jpl.nasa.gov

D. González-Ovejero
e-mail: david.gonzalez-ovejero@univ-rennes1.fr

D. González-Ovejero
Institut d'Électronique et de Télécommunications de Rennes—UMR CNRS 6164, 263 Avenue du Général Leclerc, Rennes 35042, France

A. Boriskin and R. Sauleau (eds.), *Aperture Antennas for Millimeter and Sub-Millimeter Wave Applications*, Signals and Communication Technology,
DOI 10.1007/978-3-319-62773-1_10

10.1 Introduction

Terahertz antennas present a different set of challenges to the antenna designer typically striving for very high performance while at the very limit of the chosen fabrication process. Many of the same design techniques used at lower frequencies are still applied, but fabrication constraints impose significant limitations on the type of structure that can be used, forcing the designer to consider unique fabrication processes or completely new antenna structures. Through advances in fabrication and computational techniques, the variety of terahertz antennas is growing. This chapter presents a range of antennas applied at these frequencies from 1.9 THz horn antennas to superconducting planar arrays.

Horn antennas dominate terahertz systems because of the low losses and excellent performance. In particular, corrugated horn antennas have superior radiation characteristics, but are difficult to fabricate at terahertz frequencies, thus one of the most commonly used horns above 300 GHz is the diagonal horn antenna, despite its high cross-polarization. In addition to describing multimode horn designs such as the Pickett-Potter horn, Sect. 10.2 presents two approaches to overcome fabrication obstacles for achieving high gain and low cross-polarization horns. The first relies on the stacking of platelets to form a corrugated horn. The second uses extensive optimization to create profiled horns that has been shown to be manufacturable as high as 1.9 THz.

Lens-coupled antennas have been widely used for decades in Terahertz systems because they avoid the high tolerances of horns. But their application is limited because Terahertz systems are often waveguide-based system. Section 10.3 describes the leaky wave fed lens that directly couples to hollow-waveguide. In addition, the leaky wave system reduces the area required for illumination area, increasing pixel-spacing for array systems, and enabling a batch-fabrication process for forming the lenses.

Section 10.4 applies the emerging technique of modulated metasurface antennas, discussed in detail in Chap. 12, to Terahertz frequencies. These designs result in a planar antenna which is attractive for space-borne instruments because high gains can be achieved with less mass and volume than a comparable reflector system.

Planar Terahertz antennas are discussed in Sects. 10.5 and 10.6. To avoid lossy on-chip antennas when coupling to CMOS components, membrane-suspended antennas are demonstrated in Sect. 10.5 that use beam-lead technology to couple directly to the CMOS wire-bond pads. Finally, a slot array antenna is described that is used on Cosmic Microwave Background experiment that uses superconducting transmission lines for routing the large-format array.

10.2 Feed Horn Antennas

The most common use for horn antennas [1–4] at submillimeter-waves is to feed reflectors and/or quasioptical systems [5]. Reflector antennas and Gaussian beam telescopes produce pencil beams, and to be effective, their gain has to be as high as possible within an acceptable sidelobe envelope. Therefore, in order to increase the spillover and aperture efficiencies, feed horns are required to have equalized beam-widths in all planes and low sidelobe levels. In addition, the reflector is normally required to transmit and receive orthogonally polarized waves, so one has to guarantee a low level of the cross-polar radiation patterns. Finally, for instruments involving complex optics and/or focal plane arrays for multiple beam operation it is also convenient to have good coupling to the fundamental Gaussian mode. These are the reasons why, over the years, there has been a need in Terahertz instruments for horn antennas which provide symmetric patterns in all planes, phase center coincidence for the E- and H-planes, sidelobe suppression, low cross-polarized fields, and highly Gaussian beams.

Pyramidal horns have been widely used, with good success, in many applications. Such radiators, however, possess non-symmetric beamwidths and undesirable sidelobe levels, especially in the E-plane. Conical horns, operating in the dominant TE_{11} mode, have a tapered aperture distribution in the E-plane. Thus, they exhibit more symmetric E- and H-plane beamwidths and lower sidelobes than the pyramidal horns. Nevertheless, conical horns require appropriate transitions to make them compatible with a rectangular waveguide input. To overcome some of the limitations of pyramidal and conical horns, and further improve their attractive characteristics, horn structures with multiple-mode excitations have been suggested to enhance the beamwidth equalization in all planes, and to reduce both the sidelobe level and cross-polarized fields. By appropriately combining the fundamental mode in the waveguide with higher order modes, multimode horns allow one to tailor the field distribution at the horn aperture. This strategy can be used to generate highly symmetric beams with low sidelobe levels and low cross-polarized fields.

Diagonal horns are, for its simplicity, one of the most popular feeds at submillimeter-waves [6, 7]. Figure 10.1 shows the picture of a diagonal horn with a rectangular WR-1.5 waveguide input. Its flared section consists of a pyramid which, at the throat of the horn, intersects a rectangular waveguide with its wide and narrow walls parallel to the diagonals of the pyramid's cross section. Love [6] used a circular transition from a TE_{10} mode, but the direct transition described above provides good performance with a relatively simple fabrication process. Figure 10.2a and b shows cross sections through the block at various points along

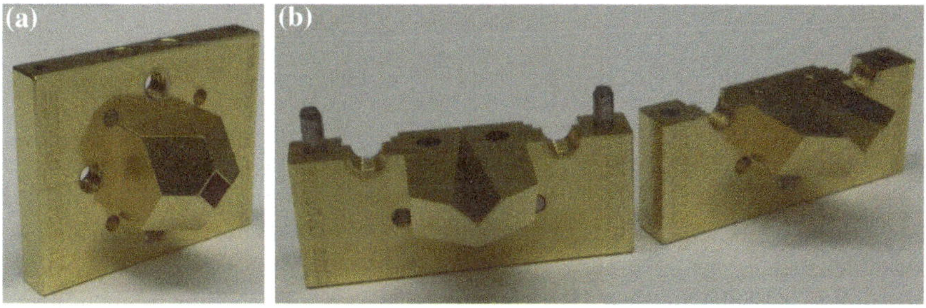

Fig. 10.1 Pictures of a split-block diagonal horn with WR-1.5 input from Virginia Diodes. **a** Assembled blocks including the horn and integrated flange. **b** *Upper* and *lower* blocks, where one can also observe the dowel pins used for alignment

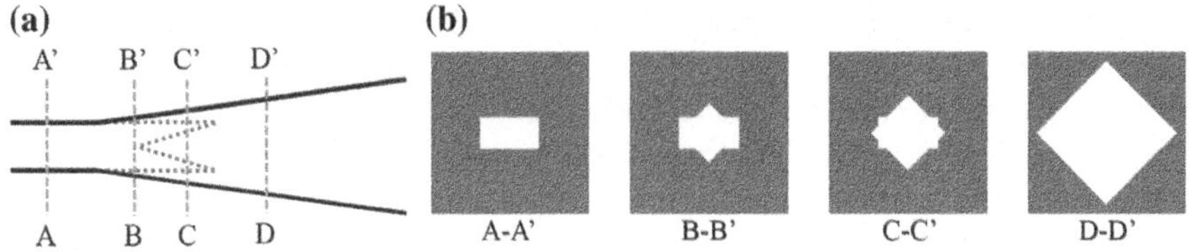

Fig. 10.2 **a** Longitudinal section of the transition from the rectangular waveguide to the diagonal horn's flared section. **b** Transverse cross sections through the block at various points along the transition

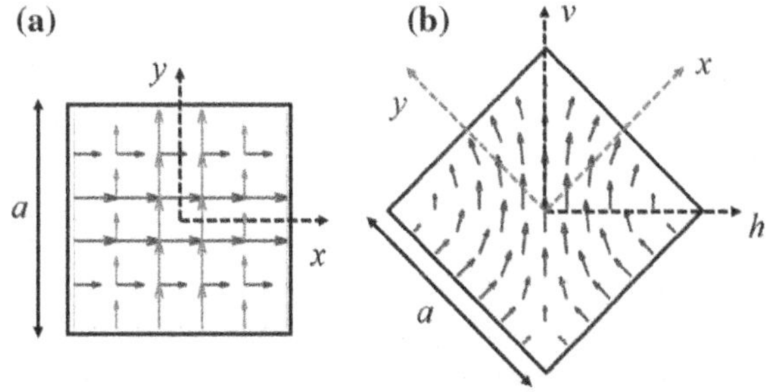

Fig. 10.3 **a** TE10 and TE01 modes coexisting in a square waveguide. **b** Result of combining the two modes in the diagonal horn's aperture

the transition. The field beyond the transition section consists of the superposition of two orthogonal TE_{10} modes, having power equally distributed between them, as shown in Fig. 10.3a. Bearing in mind the geometry in Fig. 10.3b, one can write the electric field at the aperture of the horn as

$$\mathbf{E}_A = E_0\left[\cos\left(\frac{\pi y}{a}\right)\hat{x} + \cos\left(\frac{\pi x}{a}\right)\hat{y}\right] \mathrm{e}^{-\mathrm{j}k\delta} \tag{10.1}$$

where k is the free space wavenumber, $|x| < a/2$, $|y| < a/2$, and $\delta = (x^2 + y^2)/2R_h$, with R_h being the slant length of the horn and a the side of the square aperture. Therefore, for small flare angles, the field structure within the horn aperture is

$$\begin{aligned} E_{\mathrm{v}} &= \mathbf{E}_A \cdot \hat{\mathbf{v}} = \frac{E_0}{\sqrt{2}}\left[\cos\left(\frac{\pi y}{a}\right) + \cos\left(\frac{\pi x}{a}\right)\right]\mathrm{e}^{-\mathrm{j}k\delta} \\ E_{\mathrm{h}} &= \mathbf{E}_A \cdot \hat{\mathbf{h}} = \frac{E_0}{\sqrt{2}}\left[\cos\left(\frac{\pi y}{a}\right) - \cos\left(\frac{\pi x}{a}\right)\right]\mathrm{e}^{-\mathrm{j}k\delta} \end{aligned} \tag{10.2}$$

such that the E-field vector is parallel to one of the diagonals (Fig. 10.3b). One can argue that, strictly speaking, the diagonal horn is not a multimode horn, given that it does not make use of higher order TE and TM modes. However, it presents the desirable attributes of the usual multimode horns, such as suppressed sidelobes and cross-polarized fields in the E- and H-planes and equal beamwidths in the principal planes. In addition, Johansson [7] showed by using a Gauss-Laguerre expansion of the aperture fields that the diagonal horn pattern couples to the fundamental Gaussian mode with an 84.30% efficiency when $w_A = 0.43a$, where w_A is the beam

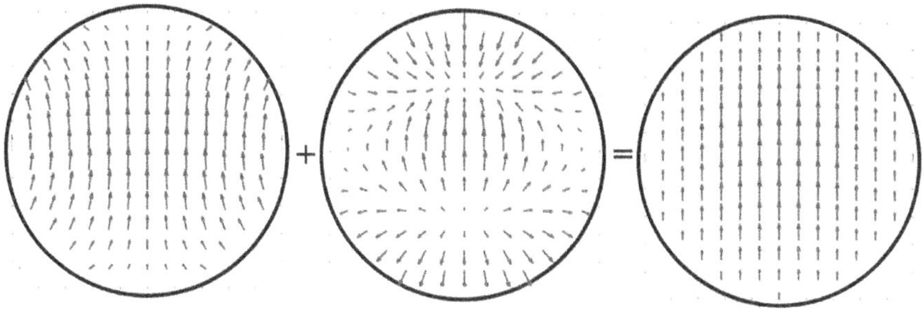

Fig. 10.4 E-field lines for a TE_{11} mode and TM_{11} and how a combination of both in a hybrid mode leads to a very symmetric beam with very pure polarization

radius at the horn's aperture. Nonetheless, all the attractive features above are accomplished at the expense of pairs of cross-polarized lobes in the intercardinal planes, making such a horn unattractive for applications where a high level of polarization purity is required. Indeed, the analysis in [7] predicts 10% of the power going into the cross-polarized component. Regarding their fabrication, diagonal horns present a geometry which is suited for being milled onto two symmetric separate pieces (split-blocks), which are then joined. The TE_{10} mode losses in the rectangular waveguide are low since the split occurs along the center of the broad walls of the waveguide. Diagonal horns are also easy to pack, which makes them attractive for focal plane imaging applications [7, 8].

A second option consists in using multimode axisymmetric horns. An appropriate combination of the circular TE_{11} and TM_{11} modes at the horn's aperture leads to far-fields with very good beam symmetry and pure linear polarization, as depicted in Fig. 10.4. Such a dual-mode horn, also known as Potter horn [9, 10], consists of a mode converter section, designed to excite the TM_{11} mode with a prescribed amplitude, and then a phasing section (the two modes have different cut-off frequencies) to ensure that both modes fulfill the correct relationship at the horn's aperture (see Fig. 10.5a). As a consequence of the required amplitude and phase balance, Potter horns are relatively narrowband, although they can meet the bandwidth requirements of some applications. Indeed, the Potter horn achieves complete beam-width equalization in all planes, complete phase center coincidence, and at least 30 dB sidelobe suppression in the electric plane over a somewhat narrowband.

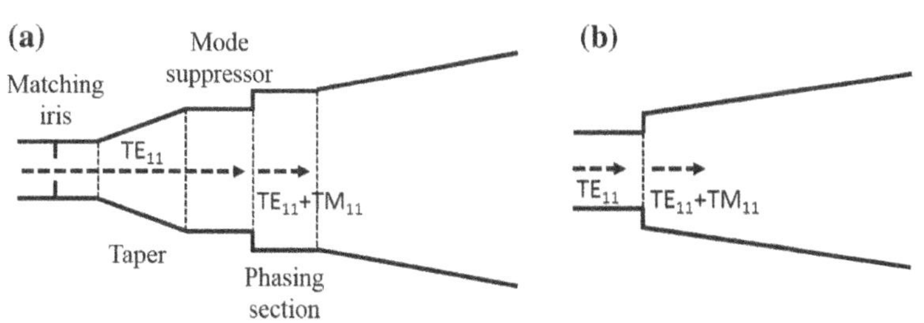

Fig. 10.5 **a** Cross-section of a Potter horn. **b** Cross-section of a Picket-Potter horn, the TM_{11} is generated at the step discontinuity

A similar strategy is used in Pickett-Potter horns [10, 11], which is a simplification of the Potter design. Essentially, they consist of a smooth-walled conical horn with a single step in the throat (Fig. 10.5b). An analysis analogous to the one carried out in [7] to determine the Gaussicity of the pyramidal horn was performed for dual-mode horns in [12]. The fraction of power coupled to the fundamental Gaussian mode is 96.3% when $w_A = 0.59a$, with a being the radius of the horn aperture. This efficiency compares well with the 98.1% of corrugated horns, normally taken as a reference. In addition, only 1.4% of the radiated power goes into cross-polarized fields. Another advantage of Potter and Pickett-Potter horns is that, as in the case of diagonal horns, they are easy to machine using split-blocks.

10.2.1 Corrugated Horns

In the previous section, we have described two different approaches to generate a hybrid mode at the horn aperture to obtain beam symmetry, low sidelobes, and highly pure linear polarization. Despite the improvement provided with respect to pyramidal and conical horns, the high level of cross-polarized fields that pyramidal horns present in the intercardinal planes, and the narrow bandwidth of Potter horns may hinder their use in some applications. The corrugated horn [2, 3, 13, 14], also known as scalar horn, is the feed of choice for high-performance systems in which the limitations of diagonal and Potter horns cannot be assumed.

The purpose of the corrugated surface is to provide the means to support the propagation of a hybrid HE_{11} mode, which results from the combination of the TE_{11} and TM_{11} modes. The geometry of a corrugated circular waveguide, including the definition of the different variables, is shown in Fig. 10.6 and the general expression of the HE_{11} mode reads:

$$
\begin{aligned}
E_x &= A\,J_0(k_\rho \rho) - B\frac{(X-Y)}{kr_1}J_2(k_\rho \rho)\cos(2\phi) \\
E_y &= B\frac{(X-Y)}{kr_1}J_2(k_\rho \rho)\sin(2\phi)
\end{aligned}
\tag{10.3}
$$

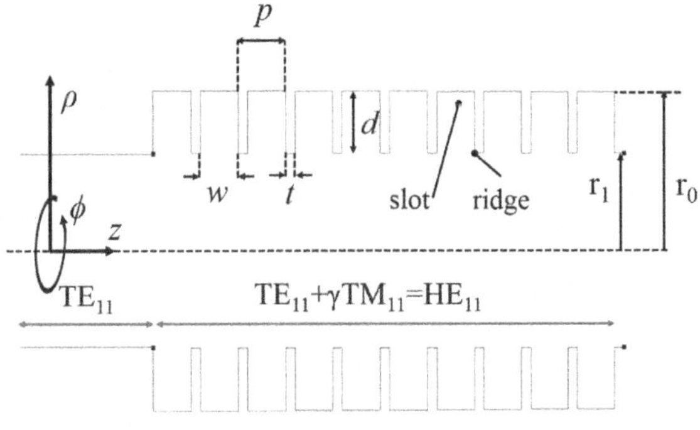

Fig. 10.6 Corrugated circular waveguide

where A and B are the amplitude coefficients, $J_n(k_\rho \rho)$ are the Bessel functions of the first kind and order n, and k and k_ρ are the free space and transverse -to-z wavenumbers. Finally, X and Y are the normalized reactance and admittance at $\rho = r_1$ defined as

$$X = -\mathrm{j}\frac{Z_\phi}{Z_0} = -\mathrm{j}\frac{E_\phi}{H_z}\sqrt{\frac{\epsilon_0}{\mu_0}}; \qquad Y = -\mathrm{j}\frac{Z_0}{Z_z} = \mathrm{j}\frac{H_\phi}{E_z}\sqrt{\frac{\mu_0}{\epsilon_0}} \tag{10.4}$$

One can easily realize from (10.3) that the fields are exclusively polarized in the x direction when $(X - Y) = 0$. Moreover, the E_x component will taper radially with ρ without any ϕ variation. The $(X - Y) = 0$ condition is usually referred to as balanced hybrid condition and it can be fulfilled either when $X = Y$ or when both X and Y are equal to 0. In a corrugated waveguide, provided that one has several slots per wavelength (preferably four or more), E_ϕ has to vanish at $\rho = r_1$, so X will be 0. In addition, for narrow corrugation ridges ($t < w/10$) and narrow slots ($w < \lambda/2$), the normalized admittance can be approximated as $Y = 1/\tan(2\pi d/\lambda)$ and, it will be thus 0 when $d = n\lambda/4$, with n being an odd integer. In other words, the short circuit is transformed to an open-circuit at $\rho = r_1$, which ensures the absence of axial currents resulting from H_ϕ and implies that $Y = 0$. In [14] and the references therein, a more refined set of formulas is provided in which d is modified by a correction factor resulting from a rigorous formulation of the surface reactance. The balanced hybrid condition will be fulfilled only at the frequencies in which $d = n\lambda/4$, so the absence of cross-polarization will be determined by the electrical length of d. Nevertheless, (10.3) shows that the cross-polar term decreases as kr_1 increases and the larger the diameter of the horn, the wider the bandwidth for a given cross-polar level. This is a general property of cylindrical hybrid mode feeds, where large apertures give an inherently better performance than feeds with a small aperture, although larger apertures will also lead to narrower beam (not suitable for prime feed focus).

We have succinctly explained in the paragraph above how the corrugated surface supports the propagation of the HE_{11} mode. However, the input of the corrugated horn will typically be the TE_{11} mode of a circular waveguide. Hence, preceding the flared section of the horn one has to introduce a mode converter to excite the hybrid mode. There are several schemes of mode converter, namely:

(a) variable depth slot mode converter [15, 16], which is the most commonly used mode converter, and provides up to 1.8:1 bandwidths (see Fig. 10.7a);
(b) ring-loaded slot mode converter [17], which provides up to 2.4:1 bandwidths (see Fig. 10.7b);
(c) and variable pitch-to-width slot mode converter [18], which can deliver up to 2.05:1 bandwidths, although is less commonly used (see Fig. 10.7c).

In practice, the number of slots (NMC) in a mode converter is between 5 and 7 for mode converters (*a*) and (*b*), and between 7 and 12 for the mode converter (*c*).

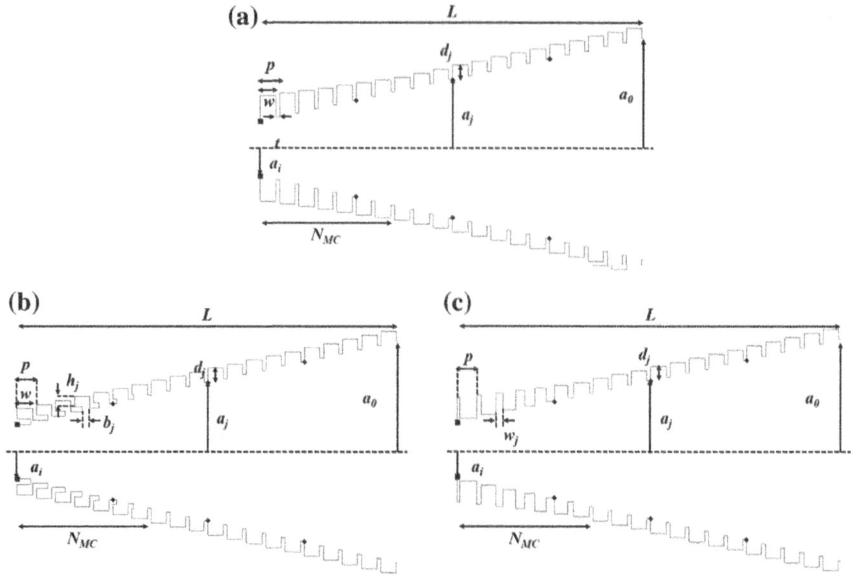

Fig. 10.7 Conical corrugated horn with: **a** variable depth slot mode converter, **b** ring-loaded-slot mode converter and **c** variable pitch-to-width slot mode converter

It is also important to note that corrugated horns are the classical solution for launching Gaussian beams. The optimum Gaussian coupling efficiency of a conical corrugated horn has been studied in [5] and it is equal to 98% when $w_a = 0.644\ a_0$, where w_a is the beam radius at the aperture and a_0 is the horn's aperture radius.

Even if conical corrugated horns (Fig. 10.7) are the most common option, it is possible to enhance the horn's performance by giving them a profile other than linear [19]. For instance, the authors of [20] proposed a Gaussian profiled horn antenna to improve its electromagnetic performance and/or optimize its geometry (reduce its length, weight, and/or output radius). Similarly, a horn with dual sinus-squared and exponential profile was presented in [21]. On the other hand, Maffei et al. [22] added a flared section at the end of a profiled horn, with the form heuristically developed to give the desired patterns and very low sidelobes. In general, the profile of the horn can be described using combination of exponential, sinusoidal, hyperbolic curves, etc., with control points that are optimized to obtain the desired far-field patterns using nonlinear least squares, minmax or genetic algorithms. Figure 10.8a and b show a corrugated horn with spline profile optimized to have a maximum coupling to the fundamental Gaussian mode (see the magenta line in Fig. 10.8d), while keeping a good match (S_{11} < 25 dB) across the 490–600 GHz band, as shown in Fig. 10.8c. On the other hand, the dual-profile (sine-squared/parallel and hyperbolic tangent/parallel) horns reported in [23] are obtained by exploiting the fact that the coupling from the HE mode in straight corrugated guide to a fundamental Gaussian mode in free space can be improved by

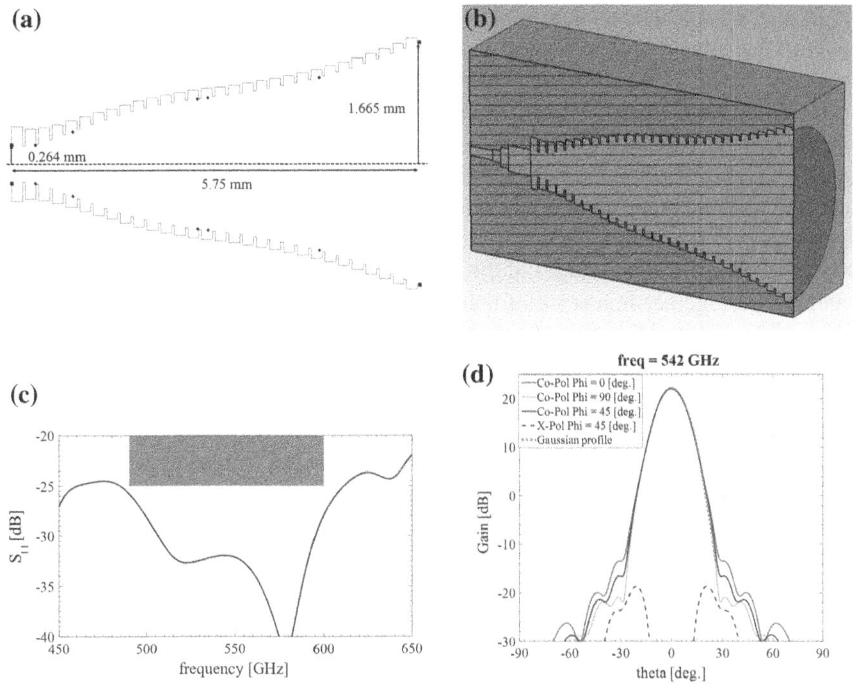

Fig. 10.8 **a** Spline-profiled corrugated horn. **b** Section of the horn including the rectangular waveguide to circular waveguide transition. **c** Magnitude of the reflection coefficient in dB. **d** Far-field radiation patterns at the central frequency

exciting higher order HE modes. This analysis, based on higher order HE modes, provides additional physical insight into the profile choice.

A crucial aspect of corrugated horns consists in how to fabricate the narrow ridges at Terahertz frequencies. The continuous evolution of computer numerical control (CNC) milling machines has made it possible to fabricate corrugated horns at millimeter and submillimeter frequencies by direct machining into two halves of a split-block, despite the difficulty of constructing azimuthal corrugations at short wavelengths. However, the length of narrow flare angle feed horns, and the depth of the matching grooves near the intersection of the circular waveguide, and the corrugated conical flared sections restrict the tool clearance available for direct machining operations. Hence, in some cases, a higher cost electroforming process will be required. The first step involves the high precision machining of the internal geometry of the circular waveguide and the corrugated conical section on an aluminum mandrel. The feed horn is then electroformed by electrochemical deposition of copper and nickel onto the mandrel surface until the desired wall thickness is achieved. Afterward, the mandrel is removed by a chemical etching process. Finishing includes flange soldering and a gold electroplate.

In modern instruments there is a trend toward increasing the number of detectors in the focal plane, which enables a significant increase in telescopes or imagers mapping speeds. Electroformed horns and machined split-blocks are not easy to adapt to the fabrication of focal planes with a very large pixel count. As an alternative, one can fabricate corrugated horns using platelets [24], which are thin metal sheets [25–27] or silicon wafers [28–30]. In order to create the corrugations for an array of horns, one has to etch two features in each sheet, namely, through holes of thickness p and radius a_j, and cylindrical steps of thickness w, and radius $a_j + d$, which serve to create the ridges of the corrugations (see Fig. 10.7). These sheets are sandwiched together in a stack of many layers either using screws or pins [25, 26, 30] or diffusion bonding them together to make a single piece [24]. With the design completed, and the photolithographic masks or machine programs created, many platelets can be reproduced accurately and economically. This technique is estimated to produce good results up to 1 THz.

Figure 10.9a shows the cross section of a platelet conical corrugated horn including a transition, manufactured using a Deep Reactive-Ion Etching (DRIE) process of silicon wafers. The first 20 platelets form a conical corrugated horn. An

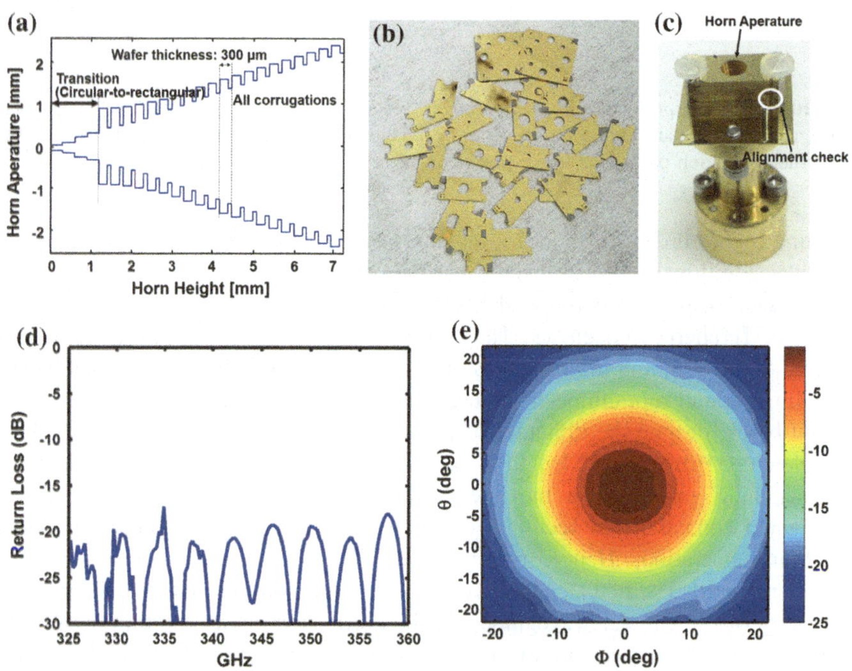

Fig. 10.9 **a** Detail of the 20 corrugations in the conical horn. **b** Different 300-µm-thick-silicon wafers before assembly. **c** Picture of the fabricated and assembled platelet antenna consisting of 24 300 µm-thick-silicon wafers. Apart from the silicon compression pins (not visible in the picture), two plastic screws and dowels pins were included for alignment. **d** Measured magnitude of the reflection coefficient in dB. **e** Measured co-polar radiation patter at 330 GHz, normalized to the maximum directivity value (21.2 dB)

additional 4 platelets form a circular-to-rectangular waveguide transition connected to the metal waveguide block at WR-2.2 band (330–500 GHz). Regarding micro-fabrication, it requires only two etch depths of 100 and 200 μm from the front- and back side of a 300 μm thick wafer, respectively. During the etching processes of 100/200 μm, screw holes, trenches for the alignment compression pins, and alignment marks all etched are simultaneously. A rectangular-to-circular waveguide transition is made on a separate 4-inch silicon wafer. All 24 silicon platelets are then coated with 2 μm thick gold using a metal sputtering system. Finally, the 24 gold-coated silicon platelets are precisely aligned and stacked together using a silicon compression pin technique [31].

The horn presents a return loss around 20 dB across the band (see Fig. 10.9c), better than 30 dB cross-polarization, and more than 20 dBi directivity over a 20% fractional bandwidth at 340 GHz. The measured radiation pattern also presents good beam symmetry, as shown in Fig. 10.9d.

To conclude this section, it is worth mentioning the importance that 3D printing techniques might have the future development of large focal plane arrays of corrugated horns. In [31, 32], the capability of fabricating W-band corrugated horns has been investigated.

Summarizing, corrugated horn antennas provide low cross-polarization, excellent beam-circularity, low sidelobe levels, nearly constant beam-width over a large frequency band, low loss, high Gaussian coupling efficiency, and low VSWR. However, at millimeter and submillimeter wave frequencies, the decreasing horn size results in increasing manufacturing difficulties, time, and cost. Therefore, corrugated horns are hardly compatible with multi-pixel arrays. We have discussed the design, fabrication, and assembly of platelet corrugated horns and focal plane arrays, which provide good performance and constitute a cost-effective alternative to electroformed corrugated horns. However, these techniques have their own limitations. In the case of platelet horns, the maximum wafer thickness has to be of the order of $\lambda/4$ to maintain the hybrid boundary condition on the horn's wall. Therefore, long horns (necessary for large f/D, which stands for the focal distance f over the reflector diameter D) will require stacking a large number of wafers, which complicates their assembly. Therefore, solutions such as multi-flared angle [34–36] and spline-profiled horns [37] have been extensively explored.

10.2.2 Multi-flare Angle

Multi-flare horns are a multi-mode horn consisting of a small number of sections (i.e., typically 2–3) with a linear profile (see Fig. 10.10). The simple interior profile of makes them much easier to fabricate than spline, diagonal, Potter, and corrugated horns because is enables a direct-drilling technique that avoids electroforming and split-block designs. As a result, multi-flare angle horns have been identified as promising solutions allowing fabricating large arrays out of a single metal block [34, 35].

The fabrication steps are straightforward:

(1) A machine tool is fabricated. Its cutting edges are manufactured with the shape of the optimized horn's interior.
(2) The horns are drilled out of a metal block one at a time.
(3) For a better surface accuracy, the input waveguide is machined after the horn steps are drilled.

This approach then shifts the challenge from fabrication to optimization, but using genetic algorithm optimization excellent designs have been achieved. Depending on the application, horn performance requirements are set in terms of gain, beam-width, sidelobe level, cross-polarization, beam-efficiency, and return loss. The feed needs to comply with f-number (*f/D*) for a given edge taper. The directivity D_0 of a horn can be determined for a given f-number using [5]:

$$D_0 = \frac{8\pi^2 w_{0feed}^2}{\lambda^2} \tag{10.5}$$

where $w_{0\text{feed}}$ is the beam waist radius of the feed given by

$$w_{0feed} = 0.216 \cdot [Te(dB)]^{0.5} \frac{f}{D} \lambda \tag{10.6}$$

where *f* is the focal length, and *D* is the reflector diameter.

It is necessary to optimize a horn accurately based on the aforementioned requirement over a given frequency range. We will illustrate the optimization of four multi-flare angle horns as summarized in Table 10.1.

Table 10.1 Optimized multi-flare angle horn

	Frequency band (GHz)	Bandwidth (%)	*F/D*	Edge taper (dB)	Directivity (dBi)
Horn #1	1700–2100	21	6	−10	31.2
Horn #2	118–183	43	2.1	−12	22.1
Horn #3	230–390	52	2.1	−12	22.1
Horn #4	650–690	6	2.97	−12	25.9

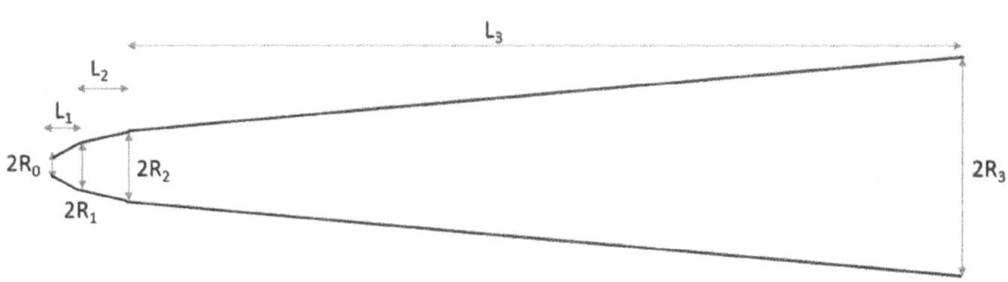

Fig. 10.10 Multi-flare angle horn with three sections

10.2.2.1 Optimization Method

The multi-flare horn antennas are optimized using an in-house BoR-FDTD (Body-of-Revolution FDTD) solver combined with real-valued and binary-coded genetic algorithm (GA) [38]. A detailed description of the BoR-FDTD technique can be found in [38–40]. The employed GA is associated to linear-combination selection, crossover, and mutation operator of the chromosomes [41]. The CAD tool is a combination between a time domain full-wave analysis solver (BoR-FDTD) and the global search algorithm (GA). This CAD tool is designed to find the global optimum complying with the desired specifications in radiation (i.e., directivity, cross-polarization, and beam circularity) and return loss over a specified frequency band. It is based on a time domain solver (FDTD) where one simulation can cover the entire frequency range. Indeed, this can be done using Fourier transform of the time domain signal. This makes the tool very useful where such large frequency band needs to be analyzed.

To reduce the computation time during the optimization process, the typical FDTD mesh size is $\Delta\rho = \Delta z = \lambda_0/20$, where λ_0 is the free space wavelength. To check the validity of the staircase approximation, the numerical result is checked with finer mesh ($\Delta\rho = \Delta z = \lambda_0/50$) after the optimization is completed.

The optimization accounts for the cross-polarization, the antenna directivity, and beam circularity. In addition, this optimization is effected for multiple frequency points ranging from the minimum frequency $f_{\min}$ to the maximum frequency $f_{\max}$. Hence, we chose a cost function that is minimized during the optimization process for horns having a given directivity D_0, low cross-polarization level (i.e., X-pol < −20 dB), high beam circularity, and a maximum length $L_{\max}$ (i.e., $L_1 + L_2 + L_3 \leq L_{\max}$). The cost function is given as:

$$Fitness = \sum_{f=f_{Min}}^{f_{Max}} \sum_{i=1}^{N} \sum_{j=1}^{M} W(f)\,\Delta(\theta_i, \phi_j) \tag{10.7}$$

where

$$\Delta(\theta_i, \phi_j) = \begin{cases} M_{Low} - G & (if\; G < M_{Low}) \\ 0 & (if\; M_{Low} \leq G \leq M_{Up}) \\ G - M_{Up} & (if\; G > M_{Up}) \end{cases} \tag{10.8}$$

$$W(f) = 1, \forall f: \quad f_{Min} \leq f \leq f_{Max} \tag{10.9}$$

In this relation, $G(\theta_i, \phi_j)$ is the horn antenna radiation pattern, and $M_{Low}(\theta_i, \phi_j)$ and $M_{Up}(\theta_i, \phi_j)$ are the desired masks of the pattern models. N and M denote the number of observation points in θ and φ directions of the 3D patterns. In this work N, M, θ_i and φ_i equal 361, 3, [0:0.25:90]° and [0:45:90]°, respectively. The function $\Delta(\theta_i, \phi_j)$ measures the differences between the horn radiation pattern $G(\theta_i, \phi_j)$ and the lower mask $MLow(\theta_i, \phi_j)$ and the upper mask $MUp(\theta_i, \phi_j)$ (10.8). The fitness function is precoded to include the weighting control *W(f)* for stressing on the optimizing frequency. In this case, all the frequencies have the same weight, and therefore the weight coefficients are assigned to 1 (10.9).

To control the cross-polarization level and the maximum length value of the horn antenna, we use the penalty technique as demonstrated in [38]:

- if the cross-polarization level > −20 dB is detected, the cost function will be double (*Fitness* = *Fitness* × 2),
- if the total length of the horn is bigger the maximum length $L_1 + L_2 + L_3 \geq L_{max}$, the cost function will be triple (*Fitness* = *Fitness* × 3).

The input waveguide radius R_0 is fixed and its value is determined to provide a good starting point for obtaining a good return loss, with TE_{11} well above the cut-off frequency and TM_{11} below the cut-off frequency at the design frequency. As such antennas are optimized by combining the TE_{11} mode with a TM_{11} mode, it is important to make sure no TM_{11} mode can propagate through the input port. Otherwise, a degenerative TM_{11} mode could affect the radiation pattern of the antenna. The value R_3 is approximated using a closed-form equation for a standard linear horn providing directivity D_0. The parameters R_1, R_2, L_1, L_2, and L_3 are then optimized.

10.2.2.2 1900 GHz Multi-flare Angle Horn Antenna

Design optimization

A feed horn with a f-number f/D of 6 is designed within the 1700–2100 GHz frequency band. An f/D of 6 translates into a gain of 31.2 dBi for an edge taper Te [−10 dB] (see Eqs. 10.5 and 10.6). The optimized horn has a gain of 31.2 dBi at 1900 GHz with a total length of 9.45 mm. Its cross-polarization is under −22 dB and its Gaussicity higher than 97% within the 1700–2100 GHz frequency range. The optimized dimensions are summarized in Table 10.2.

The radiation patterns obtained using the BoR-FDTD code are shown in Fig. 10.11 at 1700 GHz and 2100 GHz in the *E*-, *D*-, and *H*-plane. The cross-polarization level remains below −22 dB across the frequency range and a good beam-circularity is obtained.

Fabrication

The multi-flare angle horn antenna is machined out of a single metallic block of oxygen-free copper (Fig. 10.12) using an end-mill that is shaped in accordance with

Table 10.2 Optimized dimensions in micrometer for the 1.9THz multi-flare angle horn

	R_0	R_1	R_2	R_3	L_1	L_2	L_3
Optimized	74	189	261	1279	237	237	8976
Measured	83	185	253	1262	231	226	8867

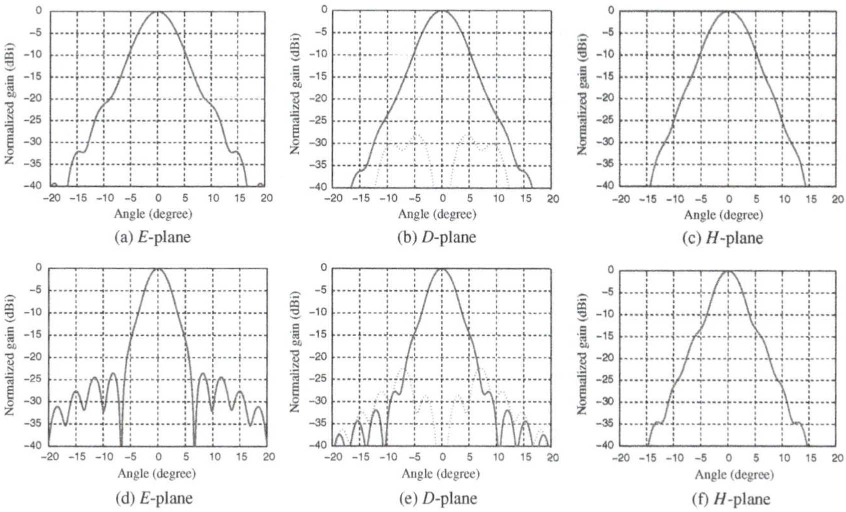

Fig. 10.11 Calculated radiation patterns with IETR BoR-FDTD at (*top*) 1700 GHz and (*bottom*) 2100 GHz

the multi-flare angle horn contour. The fabricated multi-flare angle horn antenna is shown in Fig. 10.12. Four horns were manufactured and one was cut in half for dimension inspection (Fig. 10.12). The designed and measured dimensions are compared in Table 10.2. The measured dimensions are all within 5% of the design except the throat radius. A better alignment of the input waveguide to the horn was obtained when the circular input waveguide was milled through the aperture of the horn, compared to attempting to mill from the backside. However, doing so requires an extremely high aspect ratio end-mill to reach down the horn. Hence, the tool tends to bow as it cuts and produced a wider throat.

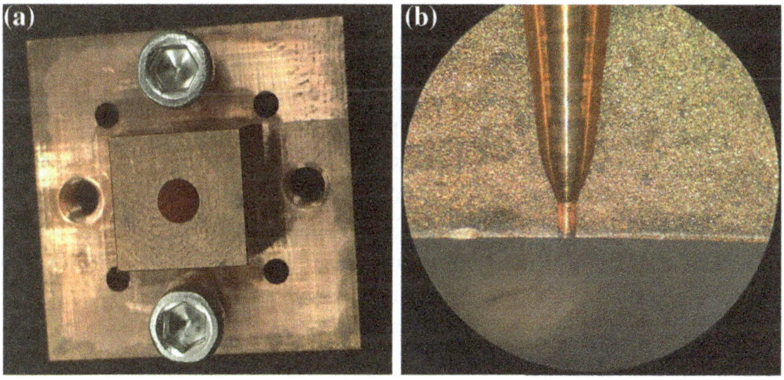

Fig. 10.12 Fabricated 1.9 THz multi-flare angle horn prototype. **a** Antenna prototype. **b** Cross-section of the antenna prototype

(a)

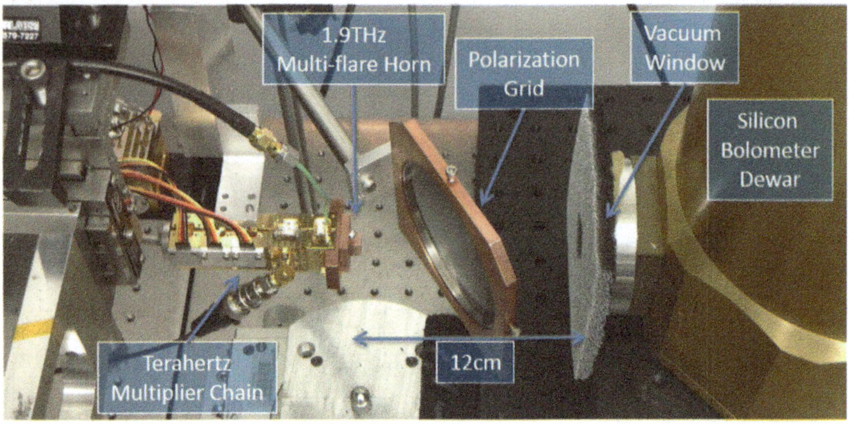

(b)

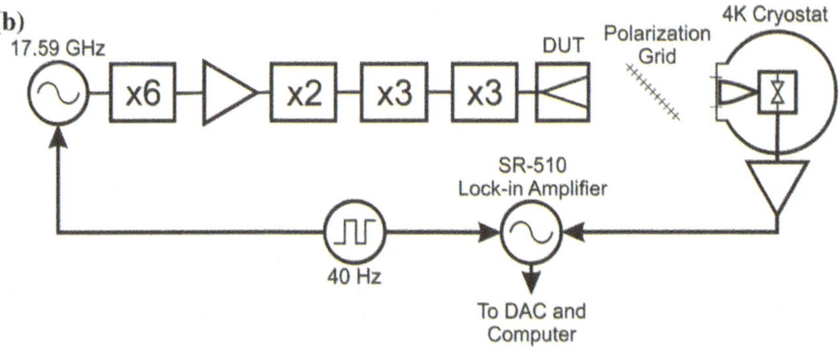

Fig. 10.13 Radiation pattern measurement set-up using an Infrared Labs HLR-5 bolometer. **a** Photograph and **b** block diagram of the measurement set-up

Measurement

The multi-flare angle horn is measured in transmission mode with an Infrared Labs HLR-5 bolometer (Fig. 10.13). A 1.9 THz multiplier chain consisting of a Millitech WR-10 AMC, a JPL GaAs WR-10 power amplifier chain, a JPL designed doubler, and two JPL triplers are assembled to drive the horn [35]. The output of the 1.9 THz tripler includes a rectangular-to-circular transition. The output power at the circular waveguide that feeds the horn is approximately 600 nW. The bolometer is a cryogenically cooled silicon bolometer with a responsivity of 220,000 V/W focused by a Winston cone with an *f*/*D* of 3.8. The bolometer is insensitive to polarization so a grating is used to filter either the co- or cross-polarization. The multiplier chain is mounted on a stepper-motor controlled mount that rotates in two axes around the approximate phase center of the horn. The synthesizer driving the multiplier chain is cycled at 50 Hz and a lock-in amplifier (SR-830) is used to detect the measured power. Finally, a computer is used to record the output signal from the lock-in with the stage's position.

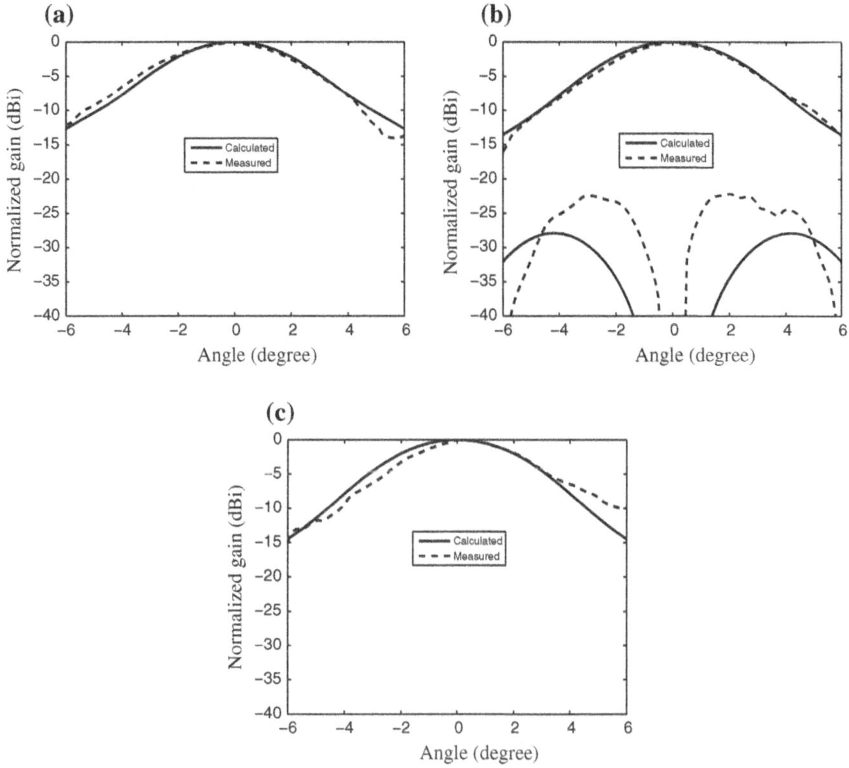

Fig. 10.14 Calculated and measured antenna radiation pattern in the **a** E-, **b** D-, and **c** H-planes

The radiation pattern is measured at 1.9 THz in the *E*-, *D*-, and *H*-planes. The calculated and measured results are shown in Fig. 10.14 are in good agreement. The directivity is calculated from the normalized intensity measured over a ±6-degree scan in both elevation and azimuth. The measured directivity of 31.7 dBi is very close to the design value of 31.6 dBi. To demonstrate the fabrication yield, the three fabricated horns are measured and only 0.1 dB variation in gain is observed. The measured cross-polarization remains below −20 dB. Using the Friis formula, the horn efficiency is assessed to be around 80 ± 15% [35].

10.2.2.3 TWICE Multi-flare Angle Horn Antenna

Three multi-flare angle horns are designed for the mission Tropospheric Water and Cloud ICE (TWICE) which consists of a radiometer integrated in a 6U CubeSat measuring water vapor and ice clouds in upper troposphere. First, the 183 GHz channel and 380 GHz channels are chosen to retrieve water vapor in the troposphere and upper troposphere/lower stratosphere (UTLS). To constrain the water

Table 10.3 Optimized dimensions in millimeter for TWICE's multi-flare angle horns

	R_0	R_1	R_2	R_3	L_1	L_2	L_3
Horn #2	0.900	1.995	2.518	5.500	6	8.911	11.608
Horn #3	0.450	0.884	1.211	2.746	4.339	4.777	6.141
Horn #4	0.230	0.496	0.613	2.004	0.492	1.346	6.506

vapor retrievals, 118 GHz channel measures tropospheric temperature profile using O_2 absorption line.

Horn #2 and #3 are optimized across broad frequency bands (i.e., bandwidth > 40%) and horn #4 is optimized across a narrow frequency band (i.e., bandwidth = 6%) (Table 10.1). The optimized dimensions for each antenna are summarized in Table 10.3. The optimized radiation pattern for each antenna is shown in Fig. 10.15 at the lower and upper band-edge frequencies. The multi-flare angle horn antennas clearly demonstrate low cross-polarization levels (i.e., <−19 dB), high beam circularity, and high Gaussicity.

As this specific mission uses single feed elements, the feed horns are machined by milling machines, operated under computer numerical control (CNC) and using a split-block package (Fig. 10.16). This allows to integrate the rectangular-to-circular waveguide transition in the same block.

Horn #4 was fabricated and measured. The calculated and measured radiation patterns at 670 GHz (Fig. 10.17) are in excellent agreement.

10.2.3 Smooth-Walled Horn

It is also possible to define the profile of our axisymmetric horn using cubic splines and then optimize the horn by shaping the splines. Such spline profile is typically described using a few nodes (two extreme nodes and five inner nodes, as shown in Fig. 10.18, in order to simplify the optimization process by limiting the number of parameters. The electromagnetic parameters of the horn are evaluated using modal matching techniques [3] or body-of-revolution solvers [38–41] at each optimization step. The initial value of the parameters $d_1 \ldots d_6$ and $L_1 \ldots L_6$ are introduced by the user, and they can be retrieved from the radius and length of a corrugated horn with the desired beam waist, bearing in mind that spline-profiled horns are, in general, longer. An alternative staring profile is given by a $\sin^{0.75}$ profile followed by a Gaussian section. Then, the parameters are optimized to provide the desire radiation pattern by minimizing a penalty function. The penalty function used is analogous to the one described in the previous section. Finally, it is important to note that some constraints must be applied to the radii to force the optimizer to create shapes that are physically possible to manufacture.

Figure 10.19a shows a block with a spline-profiled horn, whereas in Fig. 10.19b one can see the two split-blocks and, in the right-hand side block, one can also

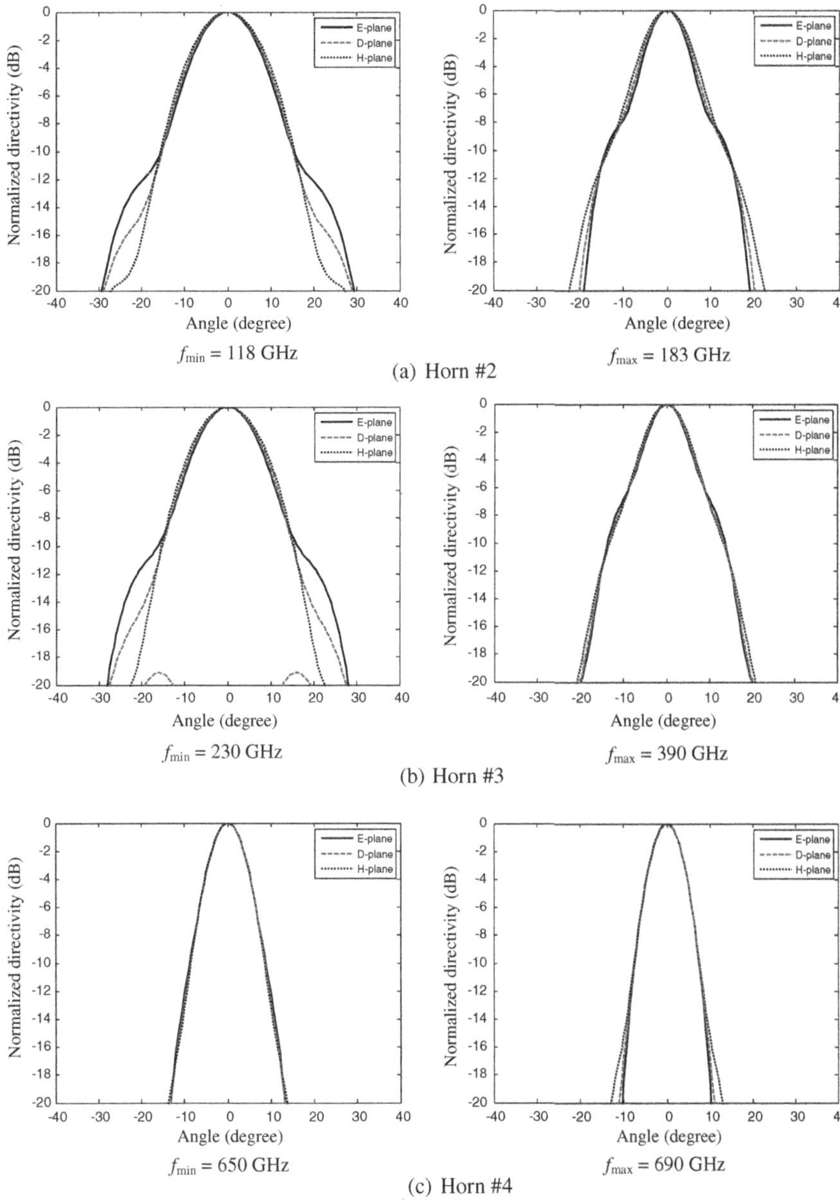

Fig. 10.15 Optimized radiation patterns of each horn

distinguish the dowel pins used for alignment. The horn has been designed for the 300–380 GHz band, where the simulated magnitude of the reflection coefficient is lower than −30 dB. Figure 10.19c shows the measured co-polar directivity pattern

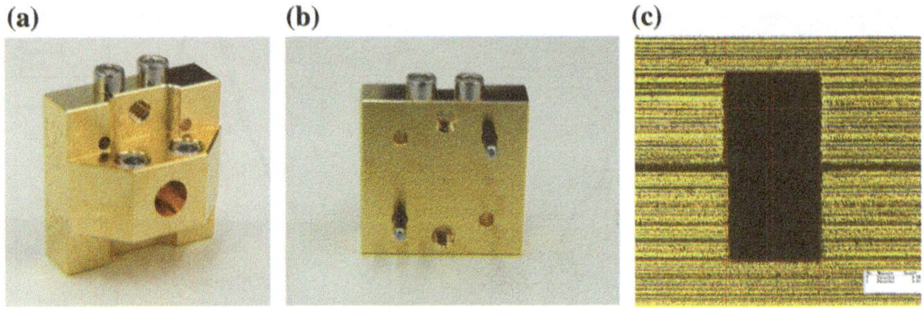

Fig. 10.16 Channel 3 multi-flare angle horn for TWICE

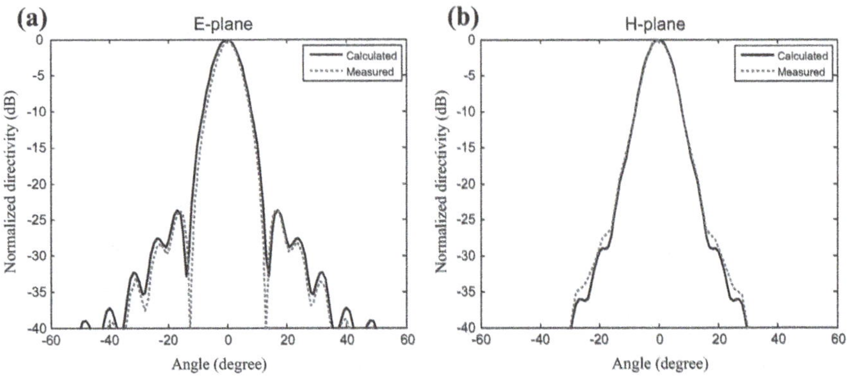

Fig. 10.17 Calculated and measured radiation pattern of the channel 3 multi-flare angle horn at 670 GHz

Fig. 10.18 Example of spline-profiled smooth-walled horn, including the variables used to define the cubic splines nodes

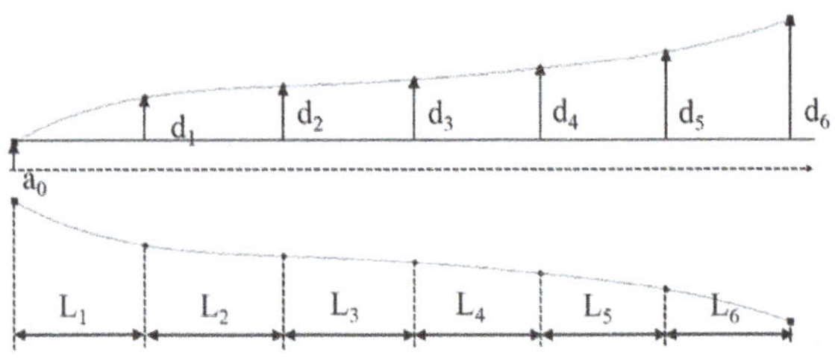

at the central frequency, while Fig. 10.19d shows the cross-polar one. The measured directivity is 24.25 dB. The ripples in the main lobe are due diffraction and standing wave effects in the measurement set-up. In this particular design, beam symmetry and cross-polar discrimination have been slightly sacrificed in order to obtain a good impedance matching over a larger frequency band.

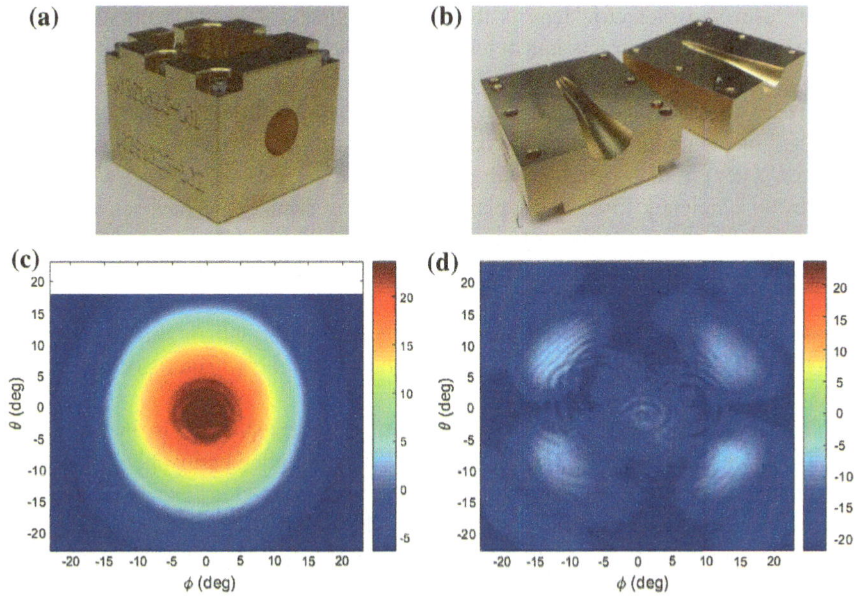

Fig. 10.19 **a** Block with a spline-profiled horn. **b** Split-blocks

10.3 Lens Antennas

10.3.1 Introduction and State of the Art

As it has been explained in the previous section, horn antenna can be fabricated either by split-block machining, drilling a metal block with custom drill tips or stacking layers of etched holes in silicon to form platelet horns. Although these techniques can provide good efficiencies at terahertz, higher directivities are difficult to achieve. The higher the directivity the longer the horn, which increases the conductor losses and the fabrication tolerances. Thus, lenses may be preferred at submillimeter-wave frequencies; they can be fabricated with silicon micromachining and be easily integrated with the front end.

Dielectric lens antennas are taken from the optical regime, replacing the waveguide horn for a planar printed antenna on a thick dielectric and a silicon lens [42]. The advantages of these hybrid antennas compared to waveguide systems are clear: low loss, easy integration, and low cost of manufacture, as the antennas can be fabricated using photolithographic techniques and the lenses using laser micromachining. They are commonly found in focal plane array configurations for imaging applications as in [43, 44], providing high efficiency over a narrow bandwidth with double slots antennas. They can also be used as standalone antennas using broadband printed antenna such as bow-ties, logarithmic spirals as photo-mixing antennas for spectroscopy applications as in [45–49].

The use of a dielectric lens is the most practical solution for obtaining high efficiency from a planar antenna at terahertz frequencies. When planar antennas are printed on dielectric substrates, they are prone to suffer from power loss due to the surface wave modes propagating within the substrate. These surface waves are confined inside the dielectric (they have a propagation constant larger compared to the free space) and absorb a certain percentage of the radiated power, which constitutes an efficiency loss. If the planar antenna is printed on the back of a high dielectric lens, the radiation will mostly go to the highest dielectric and by the use of a lens profile this radiation will be released into free space. This solution has been extensively used in the literature because of its simplicity and mechanical robustness compared to the use of substrate membranes, which should be in the order of $0.04\lambda_0$ or $0.01\lambda_0$ for a low loss operation of an elemental slot or a dipole antenna, respectively [43].

The dielectric lens design that has been most extensively is the extended hemispherical lens proposed by Rutledge [52]. The extended hemispherical shape is synthesized from an ellipsoidal shape where the antenna feed is placed at the second focus F of the ellipse with an eccentricity $e = 1/\sqrt{\varepsilon_r}$ in order to achieve maximum directivity and provide good coupling to an optical system. The rays emerging from the focus point above the critical angle will radiate outside the lens and will be perfectly collimated, while the rays that hit below will be reflected in different directions. The performance is constrained by three effects, all caused by the illumination of the lower part of the lens. The first is the dispersion effect caused by the multiple reflections that the rays undergo when illuminating the lower portion of the lens, even with the use of a matching layer (see Fig. 10.20a). The second is the phase error associated to the difference between an ideal elliptical and an extended hemispherical lens (see Fig. 10.20b). The last is the coupling loss to the optical system, i.e., how good the antenna is coupled to a Gaussian beam, defined as Gaussicity (see Fig. 10.20c). Thus, to improve the performance of these lens antennas the solution has been to employ directive feeds which focus the radiation on the upper part of the lens. That is why double slots, which avoid the illumination

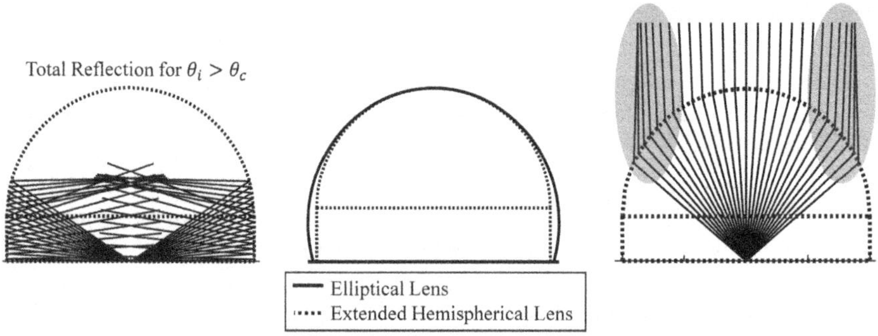

Fig. 10.20 Lens illumination scheme representing the reflection losses, loss in Gaussicity and phase error of extended hemispherical dielectric lenses

of these areas, are preferred as lens antenna feed compared to a simple dipole or slot [44].

A solution for this limitation was recently presented in [53, 54]. It consists of the combination of these simple printed antennas with a leaky wave cavity between the lens and the feed antenna. This section will be centered on the use of a leaky wave mechanism for a waveguide type of antenna feed. It is a work that was first started on [55] and has been continued on [56, 57]. The general design rules and the fabrication will be covered in this chapter, with a focus on the development and challenges for these dielectric lens waveguide antennas.

10.3.2 Silicon Microlens Antennas Fed by Leaky Wave Waveguide Feeds

From the guidelines of Jackson [58], Neto et al. proposed in [59] and then in [53, 54] the use of leaky wave cavities to increase the gain of planar printed antennas at terahertz frequencies. The cavity increases the directivity of the antenna feed that excites the dielectric lens in order to diminish the impact of the reflections at the dielectric air interface.

Instead of using resonant dielectric layer stratifications as in [60], the lens can be considered an infinite layer of dielectric. By leaving a gap between the antenna feed and the lens, the directivity of the feed increases. The highest directivity is achieved at broadside when the cavity thickness is $\lambda_0/2$, being λ_0 the wavelength in the air, which leads to a resonance at f_0. There are three modes propagating in the cavity: the dominant TE_1 and TM_1 modes that point toward broadside, and the non-desired TM_0 mode that radiates toward larger angles. The radiation of the TM_0 mode will not be intercepted by the lens, resulting in a spillover loss. The spillover loss of the lens is defined as the amount of power transmitted that is not caught by the aperture of the lens.

For the correct excitation of the leaky wave modes using a waveguide as a feeding structure, a membrane with a double arc slot iris is required. The pair of slots is excited in phase and is separated by a distance such that their contributions cancel each other. This double slot iris configuration cancels the spurious mode TM_0 mode generated inside the leaky wave cavity [61]. By adjusting the length and the width of the slot, the antenna impedance is matched to the TE_{10} mode of the waveguide. An example of a design is shown in Fig. 10.21b, where two double slots are tuned to operate 550 GHz on a 15% bandwidth for the dimensions specified. Dual-polarization can be achieved if another set another pair of arc slots are placed orthogonally from the original (see the inset of Fig. 10.21a) and excite the waveguide feed with two linear TE_{10} modes.

Using the dimensions of the provided example, we can examine the radiation patterns along the infinite silicon dielectric. An example of such patterns for the leaky wave feed designed at 550 GHz is shown in Fig. 10.22a and b. As it can be

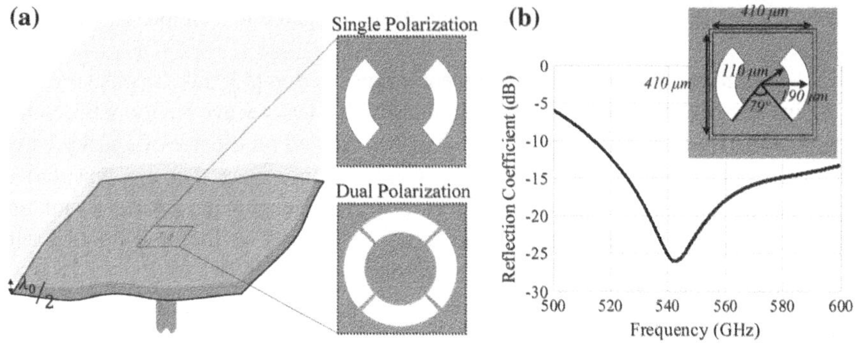

Fig. 10.21 **a** Scheme of the leaky wave waveguide feed and an inset of the iris for single polarization (*up*) and for double polarization (*down*). **b** Reflection coefficient example at 550 GHz for the dimensions shown in the inset. The air cavity has a thickness of 273 μm and the membrane is 15-μm thick

seen, the peak directivity in the broadside direction and the radiation patterns are very symmetric. This pattern makes it very suitable to feed a focusing element such as a lens with a moderate f-number (F/D). Due to the high directivity of the feed, the lens surface can be in the near or far-field of the antenna, depending on the lens aperture one needs to use. An example is shown in Fig. 10.22b, which illustrates that broader patterns are achieved in the case of the operation in the near-field compared to the far-field patterns. Thus, special care needs to be taken into account when optimizing the lens antenna.

The phase center for the leaky wave is not in the plane of the waveguide aperture, it is below the waveguide aperture. This is clearly perceived in the phase of the radiation patterns shown in Fig. 10.22 example, which have been calculated considering the reference plane is at the ground plane of the waveguide feed. This displacement of the phase center from the waveguide feed can be compensated by adjusting the height of the lens, as it will be shown next.

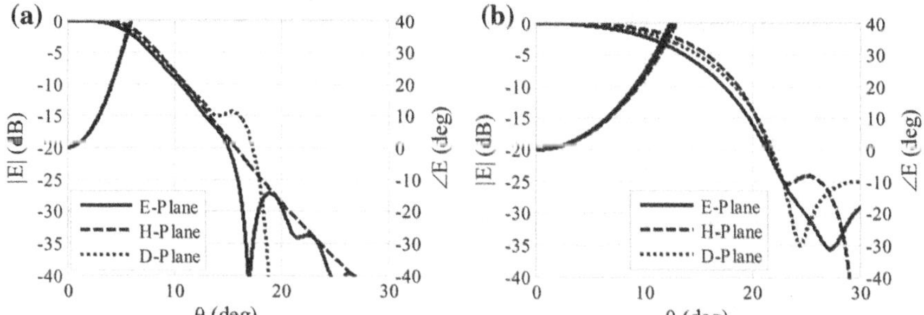

Fig. 10.22 Directivity (right axis) and phase (left axis) of the leaky wave waveguide feed inside the silicon, **a** at the far-field and **b** at the near-field distance of $4.5\lambda_0$

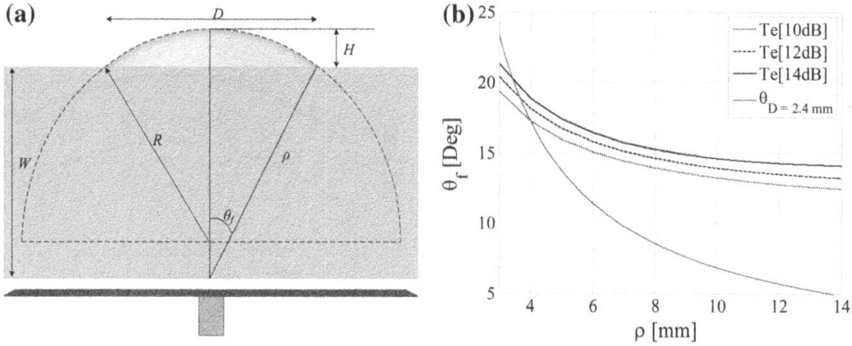

Fig. 10.23 **a** Sketch of the silicon microlens geometry. **b** Example that illustrates the relationship between the taper angle θf and the wafer thickness W at 550 GHz

The lens geometry, shown in Fig. 10.23a, is characterized by a silicon wafer of thickness W on top of the cavity and a silicon lens defined by a certain diameter D, edge angle θ_f, radius of the hemisphere R, and height of the lens H. All these parameters need to be optimized to achieve maximum gain and Gaussicity.

In order to do the optimization of the antenna and lens, a Physical Optical (PO) tool is essential to decrease the computational time that a full 3D electromagnetic simulator would need to compute and optimize the whole structure. For all the optimizations, a combination of a full 3D simulation and a PO tool has been employed. The full 3D simulator provides accuracy in the computation of the primary field of the feed (the leaky wave cavity, membrane and iris). This primary field is used to compute the distribution of the electric and magnetic fields across the spherical surface of the extended hemispherical lenses. The Fresnel transmission coefficients are used at the boundary to calculate the reflected and radiated field values. After the fields have been multiplied by the appropriate transmission coefficients, the equivalent and magnetic current densities are determined just outside the spherical surface. From these currents, the far-fields are then calculated by integrating over the lens surface. For the cases considered in this chapter, the lens is not placed in the reactive near-field region of the feed, but in the radiative region, and therefore the PO can still be performed.

The aim is to design the lens for optimal performance in terms of a gain and optical coupling to a Gaussian beam (Gaussicity) for a certain aperture diameter D. Thus, the lens needs to have at a certain height, H, and it needs to be placed at a certain distance from the feed W (see the scheme of Fig. 10.24a). The procedure to design the dielectric lens antenna for a certain diameter/aperture is the following:

- The waveguide, double slot, and air cavity are designed at the central frequency, in the example is 550 GHz. The double slot dimensions are optimized with a full-wave simulator for maximum radiation efficiency. The electric field components in the cavity plane are exported from the 3D simulator into a custom-made program that will perform the optimization of the lens geometry.

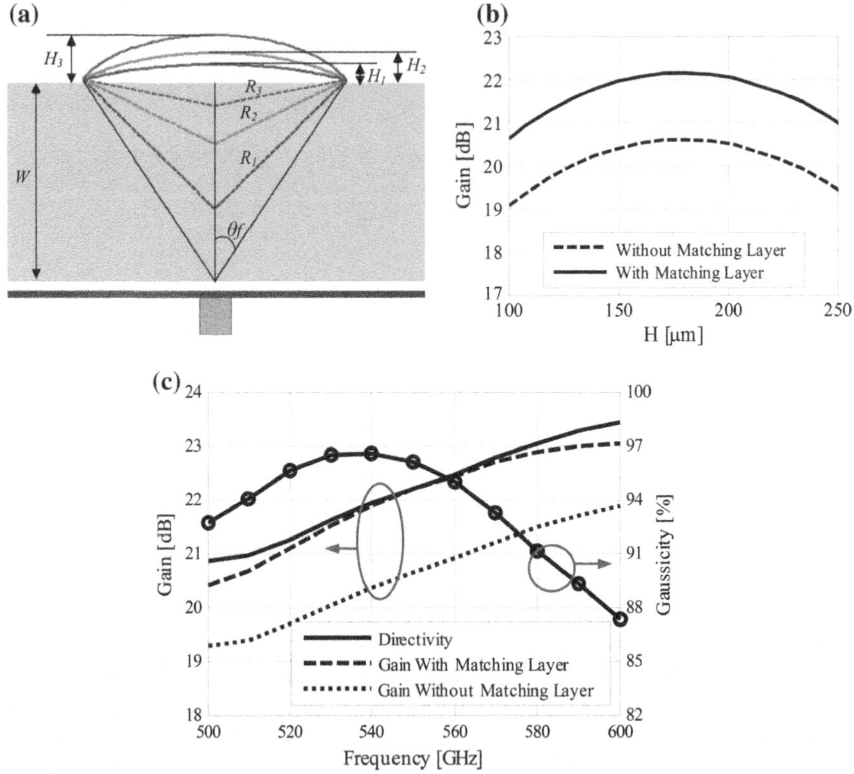

Fig. 10.24 **a** Sketch of the dependency between the lens height H and the radius of the extended hemispherical lens. **b** Gain with and without matching layer as a function of the lens height H. **c** Directivity and gain as a function of the frequency with and without matching layer (*left axis*), Gaussicity as a function of the frequency (*right axis*)

- The optimum distance W is determined based on the frequency, the feed antenna illumination and the diameter of the lens. First, the taper edge angle θ_f associated with the field level at the edge of the lens aperture is determined for certain ρ. For example, a taper of 12 dB, *Te*[12 dB], means that the field level at the edge of the lens diameter is 12 dB below the field level at the broadside direction of the lens. The field taper is chosen according to the tradeoff between the spillover loss and the taper efficiency. The taper efficiency expresses the uniformity of the field distribution on the aperture of the lens. The optimum taper is between 10 and 14 dB, similar to reflector antennas.

Thus, we will compute the field from the antenna feed inside the silicon for certain distances (in the near or far-field depending on the distance) and compute the taper edge angle θ_f for a taper field of *Te*[10 dB], *Te*[12 dB] and *Te*[14 dB]. The intersection of these values with the theoretical value of the taper for a fixed aperture $\theta_f = \sin^{-1}(D/2\rho)$ will give the optimum distance where the lens should be placed. An example is shown in Fig. 10.23b, where the dashed lines present the taper angle obtained from the computation and the solid lines present the theoretical value of the taper for a diameter of 2.4 mm. Thus, for a lens aperture of $D = 2.4$ mm and a taper of $Te[12\text{ dB}]$, the lens aperture should be placed at $W = \sqrt{\rho_{opt}^2 - (D/2)^2} = 3.8\,\text{mm}$. This computation is essential when the lens is placed in the near-field of the primary feed, as it can be noted in Fig. 10.23b, when we look at the dependency between the distance and the taper angle. When we move toward the far-field region this dependency fades, and the curves in Fig. 10.23b get flatten (θ_f becomes constant with ρ).

- Once the optimum distance W is set for a certain diameter, the optimum height of the lens H can be tuned to achieve maximum of directivity or taper efficiency. This tuning performs the optimization of the phase error associated to the phase center displacement and the phase error associated to the fact that the extended hemispherical lens is not quite an ellipse. In general, the larger the diameter the higher the optimum height. Following the previous example, for a lens of 2.4 mm at 550 GHz, the optimum lens height is $H = 181$ μm which provides a directivity of 22.2 dB (see Fig. 10.24b). The gain includes spillover losses and the reflection losses inside the lens with and without the use of a matching layer. The reflections due to the use of high dielectric substrates can strongly impact the performance of the antenna, causing a large power loss. As Fig. 10.24b shows, the total gain improves by 2 dB because of the use of a matching layer. The optimum matching layer is designed with a dielectric of $\varepsilon_m = \sqrt{\varepsilon_{Si}}$ and has a thickness of $\lambda_m/4 = \lambda_0/4\sqrt{\varepsilon_m}$, being λ_0 the wavelength at the desired central frequency. Since this antenna has a relatively narrowband, the use of one matching layer will be sufficient to cover the whole band. The frequency behavior of the resulting antenna is shown in Fig. 10.24c, where the directivity, the gain with and without matching layer, and the Gaussicity are presented as a function of the frequency.

10.3.3 Fabrication and Assembly

As it has been previously mentioned in the introduction, advanced fabrication processes and techniques are often required when working at terahertz frequencies, and it is one of the major constraints when designing these antennas. Silicon dielectric lens antennas have the advantage that the whole antenna can be fabricated using a silicon micromachining process, which provides excellent precision and accuracy, as well as a high level of integration with front-end transceivers.

For instance, the silicon lens fed by the leaky wave waveguide feed can be completely discredited among silicon wafers, processed with a DRIE silicon micromachining process and then stacked together. Specifically, it can be assembled in three parts: the first one is the leaky waveguide feed that contains waveguide, membrane and air-cavity; the second one are the blank silicon wafers required to achieve the correct thickness of the lens, *W*, and the third one is the lens surface. The following sections describe in detail the fabrication process for the two main parts.

10.3.4 Fabrication of the Leaky Wave Waveguide Feed

The three elements of the feed are the waveguide, the membrane with a double slot and the $\lambda_0/2$ air cavity. All three can be fabricated on the same double SOI wafer, using a three-step etch process.

The double SOI wafer sketch is presented in Fig. 10.25, and it contains two SiO_2 layers that act as etch stops to define the boundaries of the air cavity, the membrane with the iris and the waveguide length. For the leaky wave feed example shown in Fig. 10.21, designed at 550 GHz, the SOI wafer scheme consists of: 272 μm of silicon that provides the depth for the air cavity, a 250 nm layer of SiO_2 as the etch stop, a silicon layer 17 μm thick as the membrane for the iris, another 250 nm layer of SiO_2 as the etch stop and a final silicon layer of 995 μm for the rectangular waveguide.

The wafer can be patterned using conventional photolithographic techniques over the photoresist. The etching of the silicon can be performed using DRIE processes for deep selective silicon trenches which can achieve a selectivity of 300:1. The SiO_2 features can be etched using inducted coupled plasmas (ICP). Both DRIE and ICP allow the fabrication of clean and perfectly controlled silicon membranes.

After etching, both sides of the SOI wafer are sputtered with gold to metallize the waveguide and iris. A uniform coverage can be achieved by a good optimization of the sputtering process, even in structures that contain vertical walls with high aspect ratios.

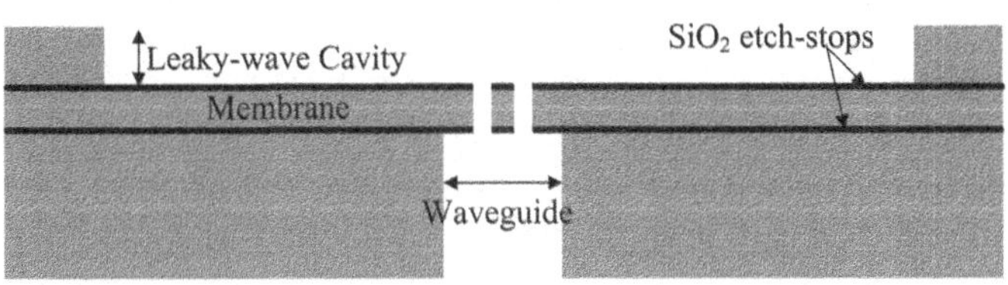

Fig. 10.25 Sketch of the etching steps in the SOI wafer for the leaky wave waveguide feed fabrication

10.3.5 Fabrication of the Shallow Silicon Lens

The fabrication of the shallow silicon lens consists of a single photolithographic process that can control the diameter, height and curvature of the lens, based on control of the shape of the deposited photoresist and the etching selectivity between the photoresist and the silicon.

The basic idea is to transfer the lens shape into a photoresist lens and transfer the shape onto the silicon wafer by a DRIE process as Fig. 10.26 shows. The process consists of four steps:

(1) Deposit a thick layer of photoresist and pattern it with the desired lens aperture using conventional photolithographic process.
(2) Reflow the photoresist pattern with heat in order to obtain the lens profile. By coating the photoresist above the glass transition temperature, the surface tension applied to the photoresist makes the photoresist to reflow in the resulting spherical shape of the lens.
(3) Etch the photoresist and silicon simultaneously with a DRIE process.
(4) The process finishes when a complete etching is achieved and the lens profile is transferred to the silicon wafer.

This process can control both the diameter and the curvature of silicon lens. The diameter of the lens depends on the thickness of the photoresist that can be coated and patterned. So far, the lens diameter of around 6.5 mm with a height of 400 μm has been successfully microfabricated as in [63]. Here, we will focus on a lens fabricated with a diameter of 3 mm that will follow the example at 550 GHz. The curvature of the shallow lens can be controlled by two process variables:

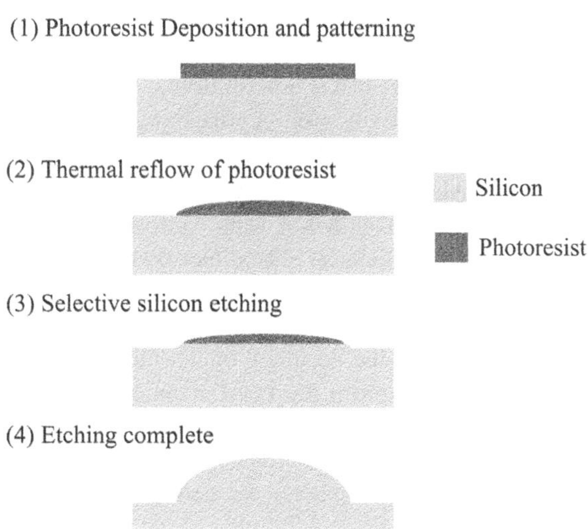

Fig. 10.26 Sketch of the silicon lens fabrication process using photolithographic techniques

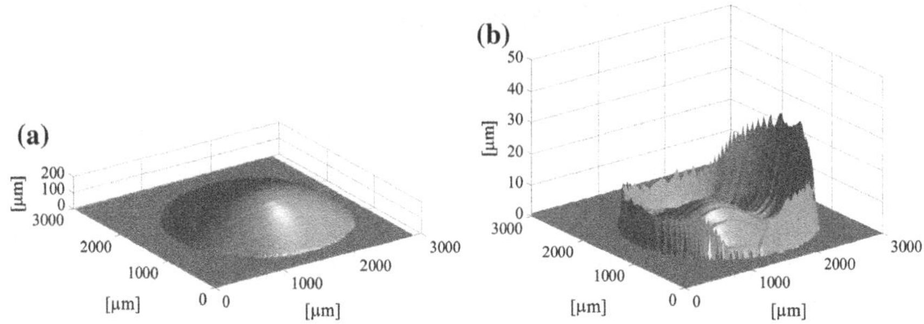

Fig. 10.27 **a** Resulting scanning of the fabricated lens profile. **b** Error Surface between the fabricated lens profile and a spherical surface

- Amount of photoresist: The pattern and the height of the photoresist applied onto the silicon wafer will determine the curvature of the resulting lens. The thicker the photoresist deposited, the thicker the lens can be.
- Etching selectivity applied to the photoresist and silicon: The curvature of the silicon lens can be controlled by the adjusting the selectivity between of the silicon and photoresist, which is optimized by controlling the gas ratio between the CF_4 and O_2, the bias power and RF power. For example, if the etching selectivity is 1:3, meaning 1 unit of photoresist etches three times the rate of the silicon, the curvature of the silicon lens will be three times higher than the curvature of the photoresist lens. In the lens shown in Fig. 10.27, the original height of photoresist was 210 μm but because the etching selectivity was 1:3, the height of the silicon lens was around 280 μm.

The surface roughness achieved with this process on the silicon lens is better than 0.5 μm which is more than sufficient for terahertz frequencies.

The use of a matching layer on dielectric lenses is essential to reduce the high reflection losses produced by the use of high permittivity substrates. At terahertz frequencies the material that is commonly used as a matching layer for silicon lenses is Parylene. It has a permittivity of 2.69 which is not ideal for silicon (the optimum permittivity for a matching layer for a silicon lens would be $\varepsilon_m = \sqrt{11.9} = 3.45$) but close enough to reduce considerably the losses by reflection. Also, Parylene polymer coatings can be applied using vapor deposition equipment, which allows a very conformal deposition with high control of the coating rate and thickness. The dimer (Parylene-C) is vaporized under vacuum and heat to form a dimeric gas. The gas is pyrolized to transform it into its monomeric form so it can be deposited as a transparent polymer film.

Characterizing the accuracy of the fabricated lens compared to an ideal hemispherical lens has been the key for understanding and optimizing the fabrication. The surface of the fabricated lens can be characterized and studied from the mapping performed by a surface profilometer. In order to study how the fabricated surface resembles a sphere, we can use the optimization toolbox in Matlab to

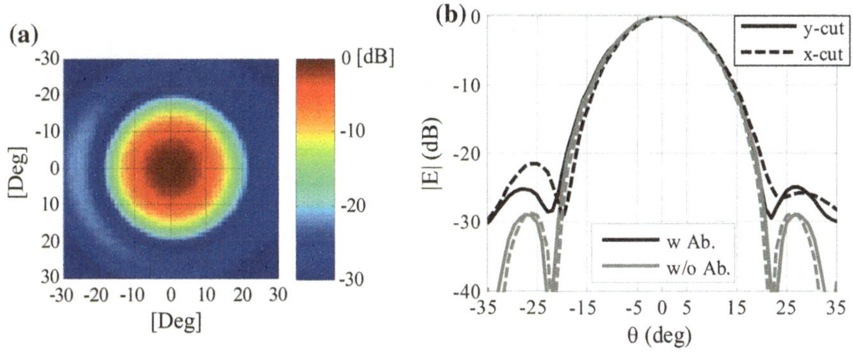

Fig. 10.28 **a** 2D and **b** 1D cuts of the radiation pattern of a Gaussian aperture with a field taper of 14 dB which contains the aberrations of the fabricated microlens at 550 GHz

compute the sphere dimensions. The sphere has a certain radius, R, aperture diameter, D, sphere height, H, that minimizes the error with the fabricated surface. As for example the fabricated lens shown in Fig. 10.27a, the optimum dimensions were $R = 3.90$ mm, $D = 2.93$ mm and $H = 285.6$ μm, and the standard deviation of the error was $\sigma = 18.36$ μm $= 0.11\ \lambda_{Si}$. From the error surface (see Fig. 10.27b) we can see that the highest error is contained in the edges of the lens. For instance, if 100 μm of the edges of the lens are not considered, the standard deviation goes down to $\sigma = 5.4$ μm $= 0.04\ \lambda_{Si}$. This adjustment in the illumination can be achieved by the tuning of the wafer thickness, W, which in this case will be designed to obtain a field taper of −14 dB at the border of this new effective lens aperture. Thus, by reducing the wafer thickness from the original $W = 3.8$ mm to $W = 3.21$ mm, we will reduce the effective diameter of the lens from 2.93 to 2.38 mm.

The effect of this aberrations can be understood as a phase distortion in the aperture field of the lens. Assuming a tapered Gaussian aperture of the diameter of the lens we can estimate what would be the effect of this phase error over the radiation pattern. For the error surface shown in Fig. 10.27b, the lens distortion would correspond to an increase of the secondary lobes in the radiation pattern and a tilt in the main direction of the main beam, as Fig. 10.28 shows.

10.3.6 Silicon Microlens Antenna Performance and Limitations

The key advantage of using this silicon dielectric lens antenna at terahertz frequencies is the fact that the lens antenna can be fabricated entirely using a silicon micromachining process. This means that from the simple processing of one wafer, over thousands of elements can be obtained. Figure 10.29, shows some examples of

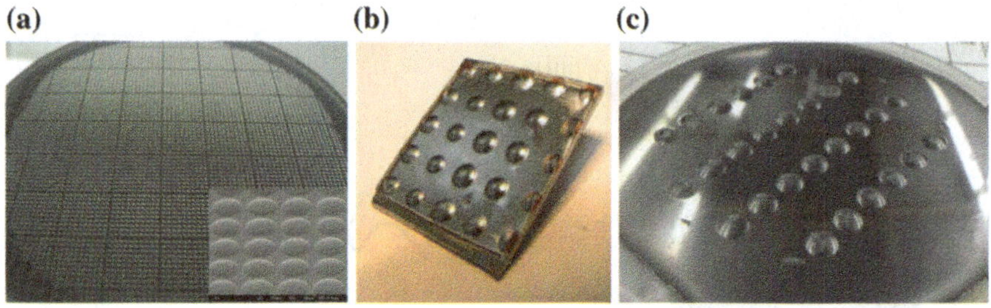

Fig. 10.29 Examples of lens arrays manufactured using silicon micromachining processes **a** One of the first 5 × 5 lens array of D = 2.5 mm, **b** 50 arrays of 10 × 10 lenses of D = 1.5 mm, **c** 4 linear arrays of 6 and 8 lenses of D = 6.4 mm

silicon lens arrays fabricated with this technology, with diameters of 1 mm, 3 mm, and 6.5 mm.

Lenses from the arrays (shown in Fig. 10.29) were scanned to check their accuracy [62]. The agreement to a spherical profile is good particularly for the small lenses (see Fig. 10.30). From the prototypes, several lens profiles have been studied with the technique presented before, and even individual antenna lens prototypes have been fabricated and measured. In this section, we will present one of the published prototypes in [56] to give an idea of how the assembly and measurements were performed.

Figure 10.31 shows the lens and the iris prototype at 550 GHz. Both pieces are assembled using metal pins that provide alignment and screws that provide the pressure to the wafers. The thickness between the lens and leaky wave is optimized for the illumination of an effective diameter of 2.39 mm and reduce the aberrations from the fabricated profile. To achieve the desired thickness, five silicon wafers are placed in between. These silicon wafers, as well as the lens wafers, need to be of high resistivity, i.e., of 10 kΩ cm, in order to lower the dielectric loss. In this prototype, the dielectric losses are calculated to be around 0.06 dB for the entire silicon stack. Moreover, double-sided polished wafers are employed in order to have good surface contact between all the wafers, avoiding air gaps.

Radiation pattern measurements were performed of the prototype at 550 GHz. Figure 10.32 presents the E and H-planes (solid and dashed lines, respectively) of the measured normalized radiation pattern at 530 GHz, 550 GHz and 590 GHz. The agreement with the simulations (shown in gray) is very good. Well-focused beams were achieved, and the difference with the simulations were primarily due to a small misalignment between the lens and the antenna feed and, as it was portrayed before, the aberrations of the shallow lens surface that would increase the secondary lobes and a tilt on the radiation pattern.

The results achieved so far show a great potential to use these dielectric lens antennas in the development of future focal plane arrays at terahertz frequencies. Using the leaky wave waveguide feed, we only need a small part of the surface of the lens, which reduces the reflection losses and phase errors that these types of

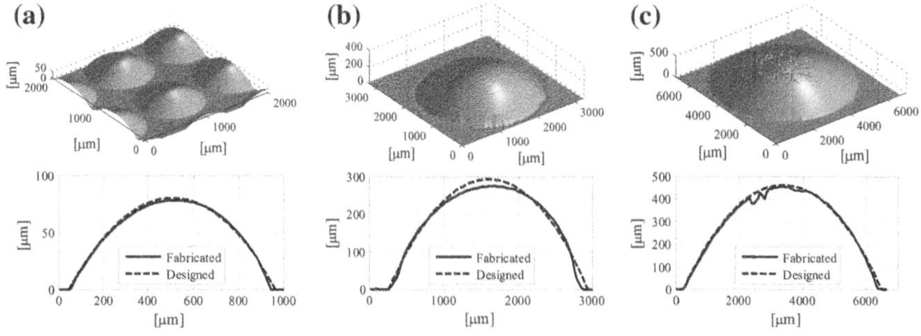

Fig. 10.30 Examples of fabricated lenses with diameters **a** 1.5 mm, **b** 2.5 mm and **c** 6.5 mm, and their comparison with a perfect hemispherical surface

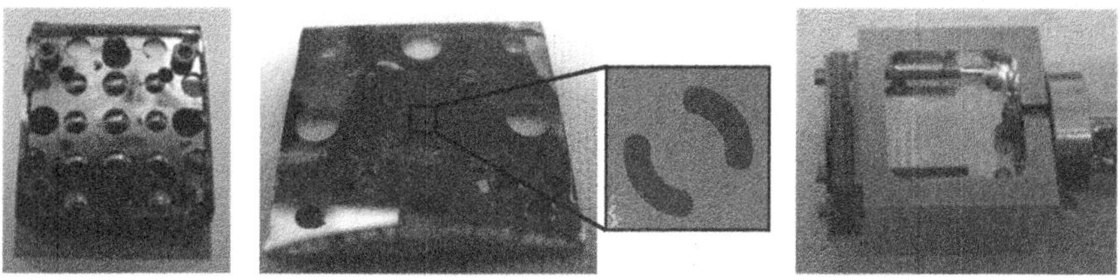

Fig. 10.31 Photograph of the fabricated prototype at 550 GHz

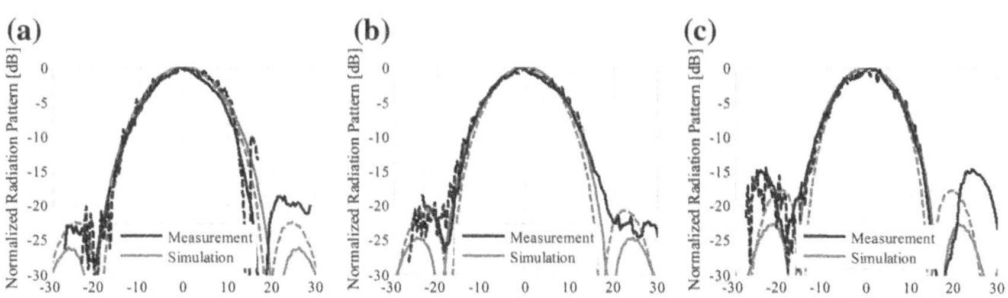

Fig. 10.32 Measurements of the radiation patterns at 530, 550 and 590 GHz. The *solid lines* correspond to the E-plane and *dashed line* to the H-Plane

lenses suffer. But most of all, it allows the use of photolithographic process when fabricating the lens. The fabrication of the lenses using photolithographic process reduces the cost, with the comparable performance achieved with other fabrication methods, such as laser micromachining. At the moment, it is still challenging to predict precisely some of the lens parameters in the fabrication, particularly the curvature, when we millimeter diameter silicon lenses are attempted. However, by characterizing accurately the fabricated profiles, we can partially compensate the aberrations and still obtain acceptable performances while continuing to keep improve the fabrication process of these lenses.

10.4 Terahertz Modulated Metasurface Antennas

The reader can find a comprehensive description of the analysis and design of modulated metasurface (MTS) antennas in Chap. 9. These antennas analyzed consist of planar apertures made of subwavelength patches printed on a grounded dielectric slab. This class of antenna is, thus, inherently flat and presents a low mass. Its most attractive feature is the capability of obtaining high- to very high gains with low-profile structures. Indeed, the aforementioned characteristics make modulated MTSs particularly appealing for space [64, 65]. Among their other advantages it is important to note their capability of conforming to curved surfaces [66], and providing a simple on-surface control of the aperture fields for beam shaping, pointing and scanning.

10.4.1 Sinusoidally Modulated Inductive Surfaces

Radiation by modulated MTS stems from the interaction of a surface wave (SW), which can be launched using a coaxial pin, and a modulated inductive plane, realized in the microwave range by means of printed subwavelength patches [64, 65]. The interaction between the SW and the periodic modulation makes the (−1) indexed Floquet mode enter the visible region so it becomes a radiating mode. To illustrate the concept, let us consider a scalar one-dimensional modulation at the $z = 0$ plane, invariant in the y direction, and with propagation along x. The sinusoidally modulated inductance can be written as

$$\mathrm{X}_s(x) = \overline{\mathrm{X}}\left(1 + M\sin\left(\frac{2\pi}{d}x\right)\right) \tag{10.10}$$

where $\bar{\mathrm{X}}$ is the average surface reactance, M is the modulation factor and d is the period of the modulation. This case was studied by Oliner and Hessel in [67]. The solution to the periodic problem consists in an infinite series of Floquet modes, the transverse-to-z wavenumber for the n-indexed mode is

$$k_{t,n} = \beta_{\mathrm{sw}} + \beta_{\Delta} - \mathrm{j}\alpha + 2\pi n/d \tag{10.11}$$

where β_{Δ} and α are perturbations in the phase and attenuation constants, respectively, which depend on $\bar{\mathrm{X}}$, M and d. In turn, β_{sw} stands for the unperturbed value of

k_t ($M = 0$), obtained by imposing the transverse resonance condition between $\mathrm{j}\bar{X}$ and the free space TM impedance

$$\mathrm{j}\overline{X} + Z_{0,\mathrm{TM}} = 0 \tag{10.12}$$

Given that we are looking for SWs $\beta_{\mathrm{sw}} > k$ and $Z_{0,\mathrm{TM}} = -\mathrm{j}\zeta\sqrt{\beta_{\mathrm{sw}}^2 - k^2}/k$, with k and ζ being the free space wavenumber and impedance, respectively. Solving for β_{sw} in (10.12) gives $\beta_{\mathrm{sw}} = k\sqrt{1 + (\overline{X}/\zeta)^2}$. When $|\Re\{k_{t,n}\}| < k$, the corresponding mode enters in the visible region of the spectrum, and it can be identified as a leaky wave solution. The $n = -1$ mode is the dominant leaky-mode, and it radiates in a direction given by

$$\beta_{\mathrm{sw}} + \beta_{\Delta} - 2\pi/d = k \sin\theta_0 \tag{10.13}$$

where θ_0 is the angle with respect to the z axis. Consequently, for a single beam directed at an angle θ_0, $\bar{X}$ and d can be related using $\beta_{\mathrm{sw}} = k\sqrt{1 + (\overline{X}/\zeta)^2}$ in (10.13). For instance, given a free space wavelength λ, one can generate a single forward beam using a period

$$d = \lambda/\left(\sqrt{1 + (\overline{X}/\zeta)^2} - \sin\theta_0\right) \tag{10.14}$$

when $\overline{X}/\zeta > \sqrt{4\sin\theta_0(1 + \sin\theta_0)}$. In the derivation of (10.14), the effect of β_{Δ} ($<<\beta_{\mathrm{sw}}$) has been neglected.

10.4.2 Constituent Elements for THz Metasurfaces

As mentioned in Sect. 10.4.1, one can implement modulated MTSs by artificial surfaces realized with subwavelength patches, which offer a good performance below 100 GHz. However, there are some aspects which hinder their use in terahertz (THz) antennas. First, the losses in the dielectric become, in most applications, a limiting factor, so all-metal structures are preferred. Second, it is challenging to combine a coaxial SW launcher with the rectangular waveguide (RW) output of solid-state frequency-multiplied continuous-wave sources, which are the most commonly used sources at THz frequencies [68]. One could, alternatively, use a graphene sheet transferred onto a back-metalized substrate to realize the modulated impedance plane [69]. Such modulation is achieved via graphene's field effect by applying adequate DC bias voltages to gating pads beneath the graphene sheet. However, the bias network renders the structure more complex, and graphene cannot be used to realize anisotropic inductance surfaces, which limits the degree of control on the aperture fields.

In the remainder of this section, we will show the suitability of silicon micromachined elements for modulated MTS antennas in the THz range. Among the existing techniques, deep reactive-ion etching (DRIE) is particularly well-adapted for micromachining subwavelength features at submillimeter wavelengths. Since it is based on etching, one may argue that it is challenging to maintain straight sidewalls and uniform depth across the wafer for each depth step. Nevertheless, these drawbacks can be overcome by extensive process development [33, 70] and a thorough design. After carrying out the desired number of etching steps, 2 μm of gold is deposited by sputtering to metallize the structure.

The proposed MTS consists in an array of metallic cylinders of elliptical cross section (see insets in Fig. 10.33, placed on a ground plane and arranged in a square lattice with subwavelength unit cell size [71, 72]. Such a structure, which resembles a Fakir's bed of nails, has been used in the past for synthesizing artificial surfaces of inductive nature [73], and can be easily fabricated using DRIE. The surface reactance corresponding to an elliptical cylinder is, in general, a tensor, which depends on the transverse wave vector $\underline{\beta}_{\rm sw} = \beta^x_{\rm sw}\hat{x} + \beta^y_{\rm sw}\hat{y}$ and relates the transverse electric and magnetic fields (evaluated at the upper interface) as $\vec{E}_t|_{z=0^+} = {\rm j}\underline{\underline{\rm X}}_S \cdot \hat{z} \times \mathbf{H}_t|_{z=0^+}$, where $\underline{\underline{\rm X}}_S = [[{\rm X}_{xx}{\rm X}_{yx}]^{\rm T}[{\rm X}_{xy}{\rm X}_{yy}]^{\rm T}]$ is defined in Cartesian coordinates, and $\hat{z}$ is the normal to the MTS plane.

The elliptical cylinder possesses two orthogonal symmetry axes, which principal directions can be identified, in the low frequency regime, with the principal axes of the dispersion ellipse. By dispersion ellipse we denote the curve that, for a fixed frequency, provides the $\beta^x_{\rm sw} - \beta^y_{\rm sw}$ pairs corresponding to waves propagating at such a frequency. The surface reactance tensor at a given frequency can be retrieved from its isofrequency dispersion ellipse by fitting, in the least square sense, the simulated data with [74]

$$\begin{aligned}&\left(1+{\rm X}_{xx}{\rm X}_{yy} - {\rm X}_{xy}{\rm X}_{yx}\right)kk_z + \left({\rm X}_{xy}+{\rm X}_{yx}\right)\beta^x_{\rm sw}\beta^y_{\rm sw}\\&\quad+\left[\left({\rm X}_{xx}+{\rm X}_{yy}\right)k^2 - {\rm X}_{xx}(\beta^y_{\rm sw})^2 - {\rm X}_{yy}(\beta^x_{\rm sw})^2\right] = 0\end{aligned} \tag{10.15}$$

where $k_z = \sqrt{(\beta^x_{\rm sw})^2 + (\beta^y_{\rm sw})^2 - k^2}$. Figure 10.33a shows the isofrequency dispersion ellipses for an elliptical cylinder with minor axis equal to 30 μm and axial ratio equal to 4, in a square unit cell with side a = 138.5 μm. This case presents at 300 GHz an artificial tensor surface with $\underline{\underline{\rm X}}_S = \zeta[[1.04, 0.46]^{\rm T}[0.46, 1.03]^{\rm T}]$. By changing the height, orientation, and axial ratio of the elliptical cylinders, each component of the tensor will undergo a different modulation. This feature is exploited to get the required degrees of freedom to control the aperture fields [65, 75].

In the limiting case of a circular cross section, for electrically small period, the isofrequency dispersion ellipse becomes a circumference and the surface reactance is scalar. Figure 10.33b shows the dispersion curves (obtained with the eigenmode solver of a commercial software) [76] of the fundamental TM-SW modes

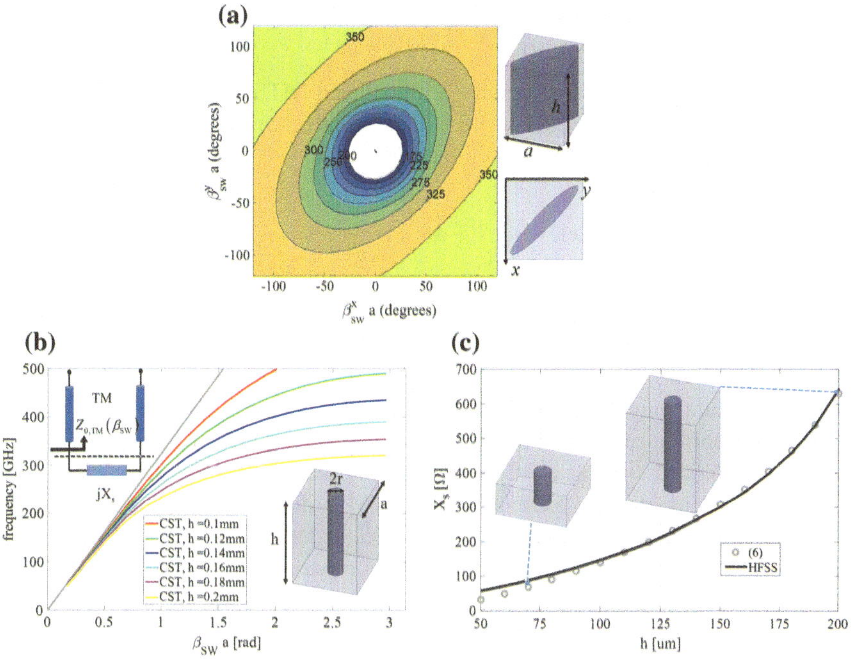

Fig. 10.33 **a** Isofrequency dispersion contours for the unit cell depicted in the insets. The unit cell's side is $a = 138.5$ μm, the height of the cylinder is $h = 180$ μm, and the elliptical cross-section has minor axis equal to 30 μm and axial ratio equal to 4. **b** Dispersion curves of the fundamental TM-SW modes propagating on a set of bed of nails. Each curve corresponds to an infinite array of cylinders of height h, radius $r = 17.5$ μm, and arranged on a square lattice with side $a = 147$ μm. Results obtained using the eigenmode solver in [76]. **c** Equivalent surface reactance of an infinite array of cylinders on a ground plane and arranged on a square lattice at 300 GHz, as a function of the height h of the cylinders. The radius of the cylinders is $r = 17.5$ μm, and the side of the unit cell is $a = 147$ μm. The solid line represents the values calculated solving (10.12) for $\bar{X}$, after using in (10.12) the β_{sw} obtained for each h with a full-wave eigenmode solver [76]

propagating on a set of bed of nails. Each curve corresponds to an infinite array of cylinders with a given height h, radius $r = 17.5$ μm, and arranged on a square lattice with side a. The period d has been computed at 300 GHz using (10.14) for an average impedance $\overline{X} = 0.7\zeta$ and a forward beam pointing at $\theta_0 = 5^o$, which for $N = 6$ implies $a = 147\,\mu$m. The surface reactances retrieved through (10.12) are shown in Fig. 10.33c. Unfortunately, long cylindrical rods are difficult to realize with DRIE while keeping a good precision and a simple process. Therefore, upon choosing $\bar{X}$ and N, one has to verify that the required heights satisfy $h/r \leq 10$ before proceeding with the design. Modulated surface reactances can be realized using the data in Fig. 10.32c under a local periodicity assumption.

10.4.3 Design Example

The design below consists in a spirally modulated MTS antenna. The spiral has been designed at 300 GHz and it provides a broadside pencil beam with circular polarization. The synthesized surface reactance is

$$X_s(x) = \overline{X}\left(1 + M \sin\left(\frac{2\pi}{d}\rho - \phi\right)\right) \tag{10.16}$$

where ρ and ϕ represent the position on the MTS plane in polar coordinates. The reader is referred to [65] for further details. In our case, $\bar{X} = 0.7\zeta$, $M = 0.65$, $N = 6$ and $a = 138.5\ \mu\text{m}$. The structure is shown in Fig. 10.34a, where each color represents different height of the metallic cylinders. Figure 10.34b shows a SEM picture of a portion of the bed of nails fabricated using DRIE. The structure is fed with an overmoded circular waveguide with only the TM_{01} mode propagating, this mode provides an efficient excitation of the TM surface wave on the MTS plane, while avoiding the use of a coaxial feed. The feeder (not shown here) is placed underneath the MTS and it transforms the TE_{10} mode in the input RW to the TM_{01} in the CW. The magnitude of the S_{11} in dB obtained with the aforementioned feeding structure is shown in Fig. 10.34c. The far-field gain patterns have been computed with HFSS [76] and they are shown in Fig. 10.34d for two orthogonal planes at 300 GHz. A similar performance has been verified for the patterns in the rest of the 290–315 GHz band.

10.5 CMOS Antennas

Complementary metal oxide semiconductor (CMOS) technology has recently made tremendous improvement in millimeter-wave capabilities because of the aggressive technology scaling during the past decade. As a demonstration of this rapid progress, full transceivers have been demonstrated in the millimeter-wave band [77, 78]. More recently, the NASA's Jet Propulsion Laboratory has extensively studied the applicability of system on chip communication, radar, and spectrometer terahertz solutions to the exploration of planets and moons in the outer solar system.

CMOS technology offers a pathway to improve the payload size, weight, and power of instruments and telecom systems on spacecraft directly through higher levels of integration and added calibration functionality. In Fig. 10.35, a mm-wave receiver of an existing radiometer/spectrometer pixel is compared to a receiver fully implemented in System on Chip (SoC) technology.

In the past few years, NASA's Jet Propulsion Laboratory in collaboration with UCLA, has been exploring the use of CMOS SoC in millimeter-wave and terahertz instruments [79–83]. The severe metal density rules imposed to the CMOS

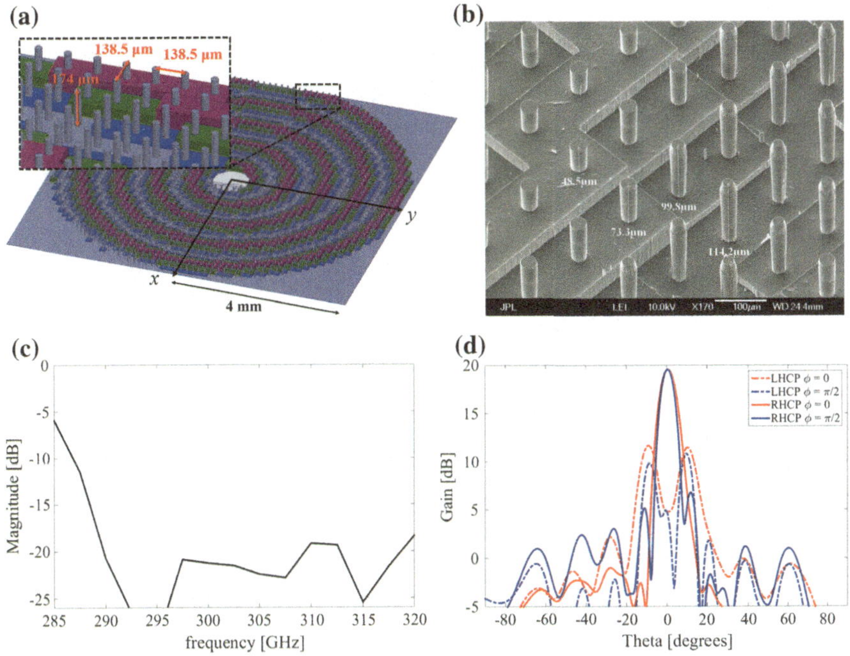

Fig. 10.34 **a** Simulated spiral modulated MTS antenna. **b** SEM picture of a bed of nails fabricated by means of DRIE. **c** Magnitude of the S11 in dB at the input RW port of the antenna. **d** Right-handed circular polarized (RHCP) and left-handed circular polarized (LHCP) gain patterns at 300 GHz plotted with *solid* and *dashed lines*, respectively, on two orthogonal cuts

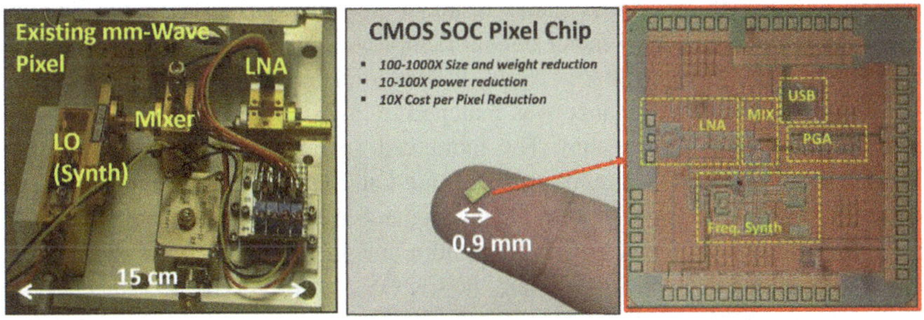

Fig. 10.35 Comparison of size between existing mm-wave receiver of an existing radiometer/spectrometer pixel and a receiver implemented in System on Chip (SoC) technology

fabrication process, making efficient on-chip antennas almost impossible to implement, can be overcome using off-chip antennas with beam-lead interconnections [79]. The beam-lead interconnection appears to be an excellent interconnection technique for CMOS circuits at terahertz frequencies that can be applied to

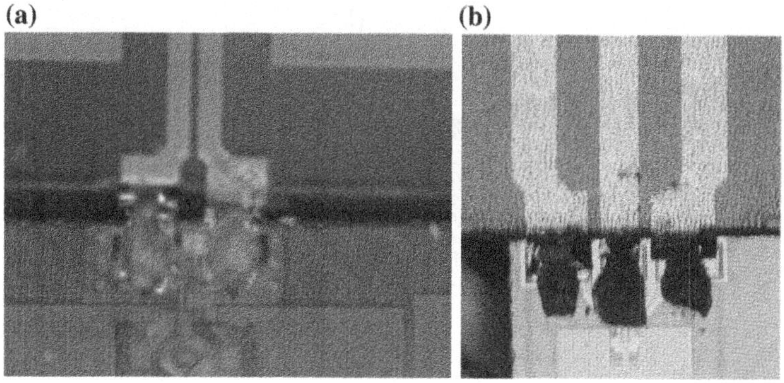

Fig. 10.36 Example of beam-lead interconnection for **a** ground-signal (GS) RFIC and **b** ground-signal-ground (GSG) pad RFIC

ground-signal (GS) pads (i.e., differential line) and GSG pads (CPW line) as shown in Fig. 10.36.

Several beam-lead antennas designed as part of CMOS SoCs are discussed here. The antenna shown in Fig. 10.37 consists of a micromachined silicon dipole antenna interconnected using the beam-lead technique. A dipole was selected as it naturally couples to the differential output of the CMOS device. The dipole is printed on a 1-μm thick Silicon Nitride membrane (Si_3N_4 with $\varepsilon_r = 7.5$) [79]. The hatched area in Fig. 10.37 is where the 300-μm-thick silicon wafer material is present below the Si_3N_4, while the light gray area surrounding the dipole is only made of the thin Si_3N_4 membrane. The cavity thickness was chosen to align the antenna and the chip at the same height. As the distance between the radiating element and the PCB ground plane is about 300 μm, it increases the antenna directivity. In addition, two metal bars are employed in order to reduce backward radiations toward the chip that is made of lossy dielectric.

The antenna input impedance was matched to $Z_{Ant} = Z_0^* = 29 + j38\ \Omega$, where Z_0 is the impedance of the chip. The fabricated antenna and its CMOS chip are shown in Fig. 10.38. The antenna was measured at 154 GHz (i.e., center frequency of the CMOS chip). This is an active measurement as the antenna is measured with its Tx CMOS chip. A standard pyramidal horn antenna was connected on one head of an ABmm measurement system and the CMOS chip prototype is connected to a DSP lock-in amplifier (Model SR830 from SRS, Sunnyvale, CA). Two rotating stages were employed to move the antenna in elevation and azimuth planes to scan the far-field. The measured and calculated radiation patterns are in good agreement as shown in Fig. 10.39. Measurement has shown that the antenna can reach up to 70% efficiency near the lossy CMOS transmitter chip. The overall transmitter operating at 154 GHz shows excellent performance as its effective isotropic radiated power (EIRP) and total radiated power (TRP) reach 20.6 dBm and 10.2 dBm, respectively [79].

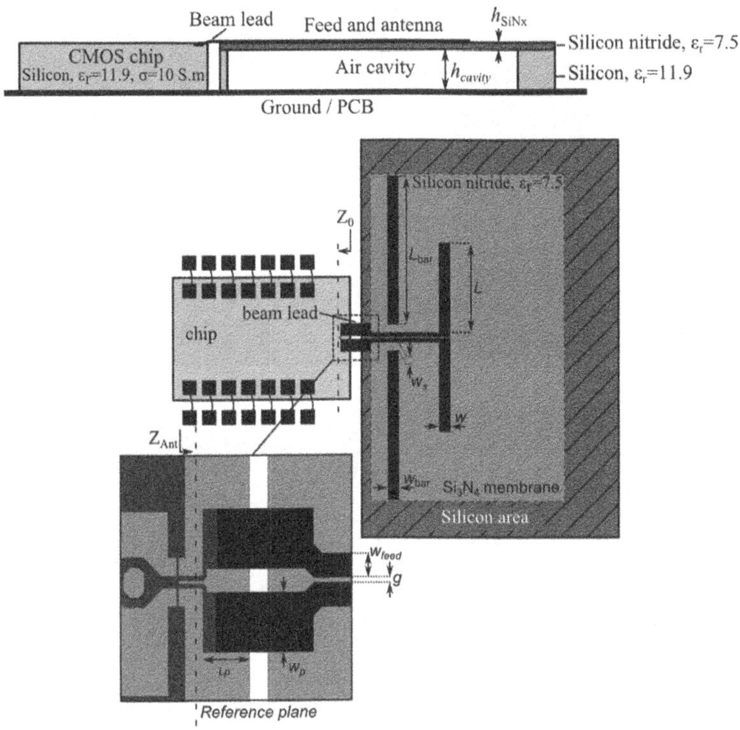

Fig. 10.37 Dipole antenna integrated with the Tx CMOS chip [79]. (*Top* cross section. *Bottom* top view

This antenna was also used as a part of a 65 nm CMOS 140 GHz, 27.3 dBm EIRP transmit array. This transmit-phased array offers beam steering capability [80]. It uses local phase-locked loop (PLL) reference generation system. Unlike traditional CMOS phased arrays, this enables the array to be formed over multiple chips while avoiding the challenges of distributing mm-wave signals between them. A 2 × 4 multi-chip array prototype was fabricated and tested using the beam-lead dipole membrane antenna.

The antenna shown in Fig. 10.40 was built on a high-resistivity silicon substrate for a full CMOS SoC navigation processor operating at 94 GHz. This navigation processor can be used to safely guide unmanned aerial vehicles (UAVs) and unmanned ground vehicles (UGVs) through cluttered and urban environments while consuming very little payload [83]. The navigation system defines multiple pathways using mm-wave base-stations called path generators and then uses a single CMOS SoC containing a receiver, ADC and an FFT processor to detect and navigate these pathways. The demonstrated confined pathway SoC (CP-SoC) occupies 5.4 mm^2 of silicon area in 65 nm technology, and consumes only 199 mW, making it suitable for lightweight payloads associated with UAVs and UGVs.

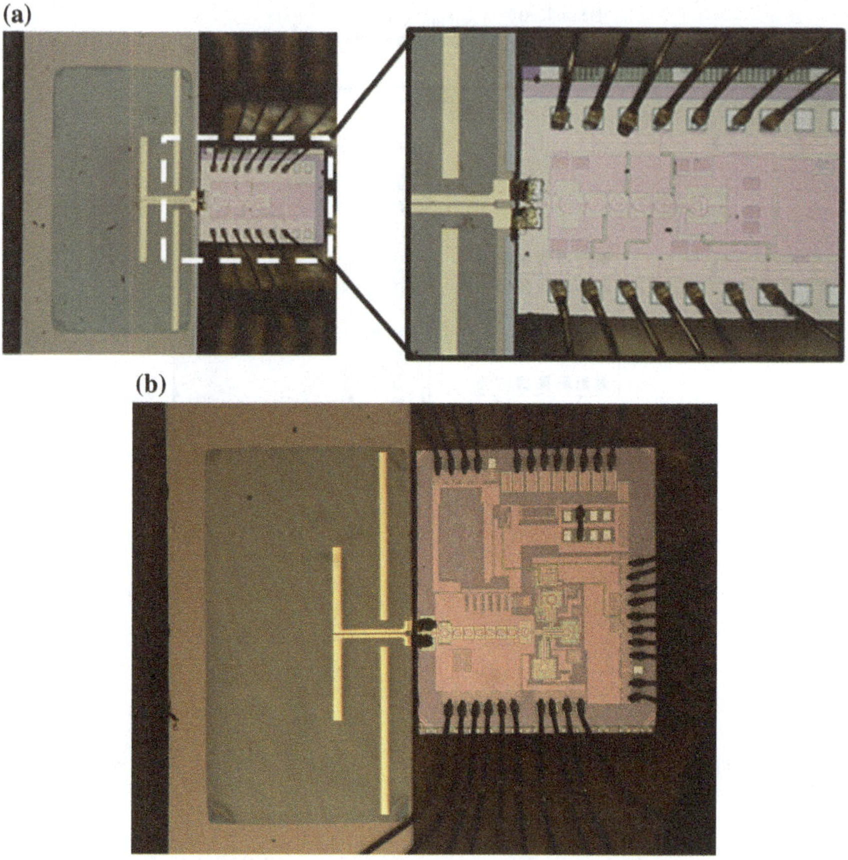

Fig. 10.38 **a** Die photo of a proposed SoC Transmitter operating at 150 GHz [79]. **b** Die photo of a 140 GHz transmit array for highly scalable multi-chip phase arrays [80]

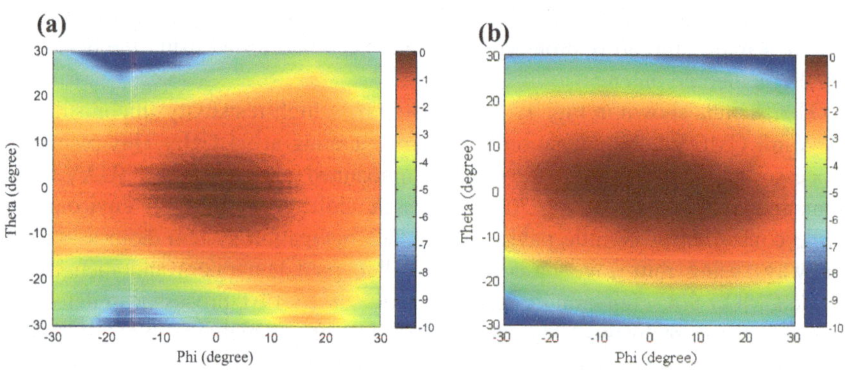

Fig. 10.39 **a** Measured and **b** simulated normalized 2D radiation patterns of the CMOS off-chip dipole antenna at 154 GHz (co-polarization)

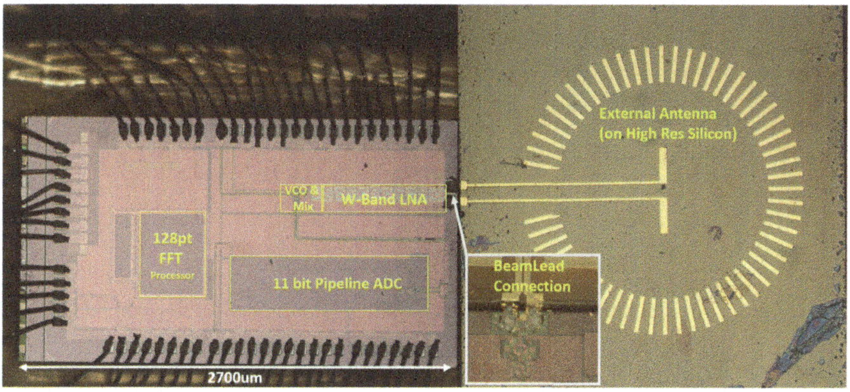

Fig. 10.40 Die photo of the proposed SoC navigation processor combined with the external antenna module (on a HiRes Si substrate) used to capture incoming radiation [83]

10.6 Slot Antennas in BICEP2

The measurement of the polarization of the cosmic microwave background (CMB) is the holy grail of present day cosmology. CMB contains the most valuable relic of the early universe [84]. The details of the structure and evolution of the universe are believed to be encoded in the anisotropy of the CMB. It is believed that the direct evidence of the inflationary epoch—the rapid expansion of the universe after the Big Bang—lies in the traces of gravitational waves in the CMB, which manifests in the temperature anisotropy and polarization of the CMB. If inflationary gravitational waves do indeed exist, they will be the oldest relic of the universe, created 500,000 years before the CMB was emitted. The measurement of the CMB polarization—the B-mode polarization in particular—will be the indirect proof of the existence of the inflationary gravitational waves and will give insight into the epoch of inflation.

CMB polarization detection instruments require large focal planes with thousands of detectors to have large throughput improving overall sensitivity of the instrument. To reduce stray radiation pickup, the field of view of these detectors needs to be restricted. The purpose of BICEP (i.e., Background Imaging of Cosmic Extragalactic Polarization) is to measure the polarization of the CMB radiation; in particular, to measure the B-mode of the CMB. In BICEP1, the first generation of the BICEP instruments, the collimation of the incoming beam is achieved using metal feed horns (Fig. 10.41a). Even though feed horns have excellent performance, their mass, size, and expense make them unsuitable for large arrays with hundreds and thousands of pixels.

A monolithic array of antenna-coupled detectors on a planar substrate, which can be fabricated with photolithographic techniques, would solve the aforementioned problem. However, most planar antenna designs use thin dielectric substrate to avoid excitation of surface waves, and they produce broad beam patterns. Most

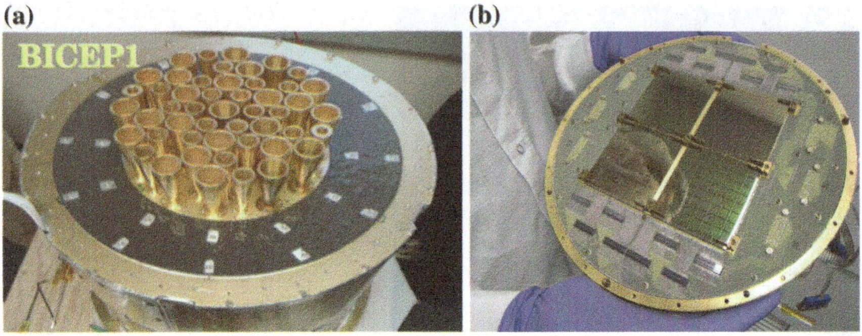

Fig. 10.41 **a** BICEP1 antenna using feed horns with different apertures. **b** BICEP2 antenna using planar dual-polarized phased arrays

often, they require substrate lenses or micromachined horns for efficient coupling to the telescope optics. As a significant improvement, the second generation instrument BICEP2 uses a novel dual-polarization planar slot antenna phased array which produces quite a narrow beam with no additional optical coupling elements. The output from the antenna array is two superconducting thin-film microstrip lines, one for each polarization, which can be efficiently coupled to bolometer detectors [6] to produce a single pixel in an imaging focal plane array. Figure 10.41 illustrates the BICEP1 and BICEP2 instruments.

The "T" shaped dual-polarized antenna structure chosen for this work is ideal for low cross-polarization performance. The T-slots were arrayed in a compact architecture to form the slot array antenna.

The antenna is fabricated on a high dielectric silicon substrate ($\varepsilon_r = 10.8$) with thickness between ¾ λ_g and λ_g. The slots are illuminated through substrate to take advantage of the stronger antenna response on the dielectric side. A quarter wavelength thick quartz, whose dielectric constant is optimum for silicon anti-reflection coating, is used on the silicon substrate to minimize reflection at the air–substrate interface.

Due to the thick dielectric substrate, excitation of the lossy grating lobes for the array is a concern. The antenna slots in each detector must be spaced to Nyquist sample the focal plane surface to avoid generating the grating lobes. The radiation pattern of each axis of an array is calculated from the N elements per linear dimension spaced at distance S as follows:

$$A(\theta) = \sum_{m=-(N-1)/2}^{(N-1)/2} e^{-j2\pi \frac{mS\sqrt{\varepsilon_r}}{\lambda_0}\sin(\theta)} = \frac{\sin(N\pi S\sqrt{\varepsilon_r}\sin(\theta)/\lambda_0)}{\sin(\pi S\sqrt{\varepsilon_r}\sin(\theta)/\lambda_0)}$$

where ε_r the relative permittivity of the surrounding medium and λ_0 is the free space wavelength.

The key features of the design are the slot length *L*, slot width *W*, and the slot separation *S*. Due to the relatively thick substrate with no coupling lenses, the parameter to optimize not to excite grating lobes within the frequency band of interest, is the slot separation distance S, which must follow:

$$S \leq \frac{\lambda_{0,\min}}{\sqrt{\varepsilon_r}}\left(1-\frac{1}{N}\right)$$

where $\lambda_{0,\min}$ is minimum wavelength of operation. With an operating bandwidth of 150–180 GHz (i.e., $\lambda_{0,\min} = 1.7$ mm), the spacing must satisfy $S \leq 460$ μm.

Another key design parameter is the length of the slots. Since the design is constrained by the slot separation distance *S*, the length of the slots that can used is also constrained. Ideally, for broadband performance, the slot antennas are used at their second resonance where the real part of the slot impedance is relatively constant and its value can be optimized between 20 and 60 Ω, when fed at the center of the slot. However, for the slots to operate in the second resonance, the length of the slots needs to be more than 0.3 $\lambda_{0.}$ The grating lobe constraint makes it impossible to operate the slots in their second resonance. Therefore, the slots for the BICEP2 antenna are operational in their first resonance. Unfortunately, the impedance bandwidth of a single slot operating in the first resonance is relatively narrow and the real part of the impedance is very high, in the 100–150 Ω range, depending on the specific design (caused by a current node at the slot center as shown in Fig. 10.42). The impedance can be optimized to somewhat lower value (50–60 Ω) in a "T" slot array configuration. Since the impedance of the detectors used for the BICEP2 instrument is relatively low, in the 20–25 Ω range, it needs a matching circuit to match the detector impedance with the antenna impedance when the slots are fed at the center. One way to overcome this is to feed the slots off-center. The real part of the impedance of a slot antenna operating at the first resonance decreases as the feed is moved away from the center, toward the edge of the slot. However, that breaks the symmetry of the slot array, leading to cross-polar coupling, and poor cross-polarization performance. The conflicting requirements of low value for the real part of the antenna at first resonance and low cross-polarization performance was addressed by feeding the slots at two points, toward the edge of the slots. This made the antenna a symmetric structure and at the same time providing low value for the real part of the antenna impedance (~22 Ω), as shown in Fig. 10.43c.

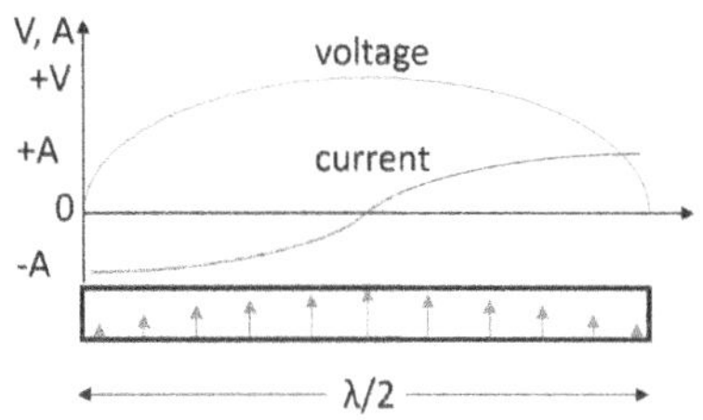

Fig. 10.42 Current and voltage distribution on a half-wavelength slot. The impedance Z = V/I is maximum in the center of the slot. The impedance reduces as the feeding point moves away from the center

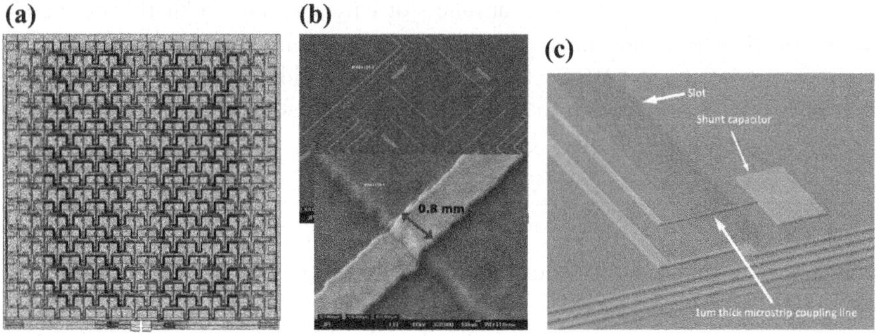

Fig. 10.43 **a** Layout of BICEP2 slot array. **b** SEM micrograph of the fabricated slot array showing the 1-μm-thick microstrip coupling line. **c** SEM micrograph of microstrip crossover and shunt capacitor at a sub-antenna slot

Provided that impedances between the microstrip line and slot are well matched, the coupling effectively transmutes the electric fields across the slot into fields. For the microstrip feed lines to match the antenna resistance, the thinnest lines that can be reproducibly fabricated is required (i.e., 1 μm as shown in Fig. 10.43b). The feed points also have a 5 Ω inductive reactance. It is tuned away using series capacitors that shunt current to ground (Fig. 10.43c). This inductive reactance is relative stable across the frequency and that is why a simple capacitor is enough.

The waves from the slots coherently sum in a corporate microstrip feed network. The network first combines incoming signals across rows and then sums the signals from each row in a column tree at the side of the detector. The signals sum in microstrip tee-junctions with proper impedance matching.

Each pixel of the phased array consists of 10 × 10 slot arrays for each polarization, 100 horizontal slots and 100 vertical slots, over a 10 mm × 10 mm area (see Fig. 10.44). The slots are excited inphase with superconducting microstrip transmission lines. There are two transition edge sensor (TES) detectors for each pixel, one for each polarization. The overall slot array is sized to match the *f*/2.2 camera optics for the BICEP2 instrument.

For characterization of the dual-polarization slot array, the TES outputs were read out by SQUIDs. To The detector system has an integrated microstrip bandpass filter to define the passband of the instrument. Two separate chips were fabricated and tested, one with the bandpass filter and the other without it. This was done to characterize both the filter and slot antenna designs independently. FTS response of the antenna with integrated bandpass filter and the measured co-polarized beam pattern of the antenna are shown in Fig. 10.45.

The slot array is currently integrated into the BICEP2 instrument and is collecting data every day in search of answers about the creation of the Universe, and all these are dictated by the performance of the dual-polarized "T" slot phased array antenna.

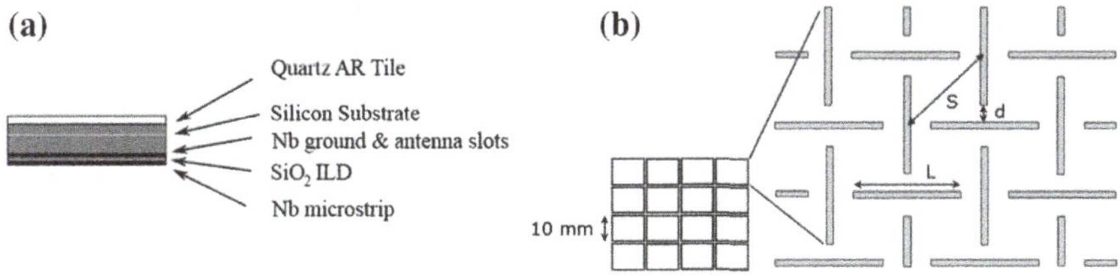

Fig. 10.44 Schematic diagram of the slot antenna array showing the slots for one pixel. **a** Cross-section of the pixel geometry. **b** Slots position for one pixel. Each pixel is 10 mm × 10 mm, and there are 10 × 10 pixels in the focal plane

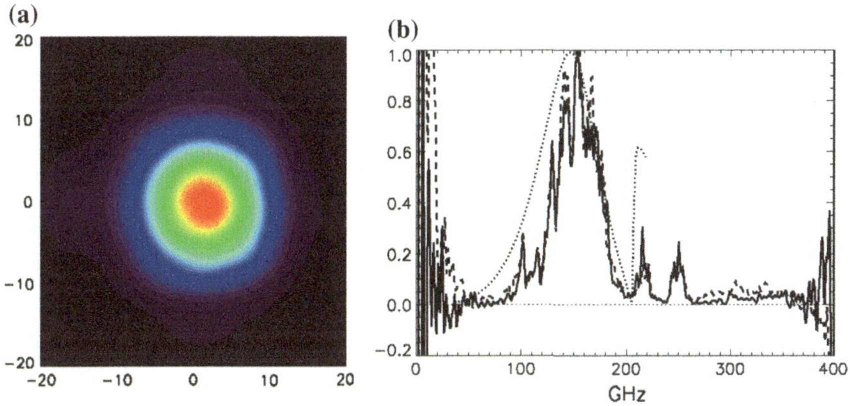

Fig. 10.45 **a** Measured co-polarized beam pattern. **b** FTS response of the 145 GHz slot antenna array

10.7 Conclusion

This chapter discusses the dominant types of antenna used at terahertz frequencies. For waveguide-based systems where polarization content is not a significant concern, diagonal horns are commercially available, simple to integrate into custom designs and have sufficient Gaussicity. In planar systems, a substrate integrated lens fed by a double slot or double-dipole the common choice because of the direct integration, high efficiencies, and relatively simple assembly required.

Because many applications at terahertz are for science instruments, higher performance is often demanded. Spline-profiled and multi-flare angled horns provide excellent performance while still being manufacturable with standard CNC milling processes. Corrugated horns are more challenging to produce, but have been demonstrated up to 340 GHz, with higher frequencies possible. The leaky wave coupled lens designs demonstrate a step forward in the lens-coupled designs by reducing the aperture size of the lens. This enables batch fabrication of the lenses and higher array densities. Finally, the slot array described in the final section shows a high gain, highly polarization sensitive design, enabled by superconducting transmission lines.

Acknowledgments Part of this work was supported by NASA Jet Propulsion Laboratory (JPL), CA. and "National Aeronautics and Space Administration" (NASA). Part of this work was also supported by a Marie Curie International Outgoing Fellowship within the 7th European Community Framework Programme.

References

1. A. Love, *Electromagnetic Horn Antennas* (IEEE Press, 1976)
2. P. Clarricoats, A. Olver, *Corrugated Horns for Microwave Antennas* (P. Peregrinus, 1984)
3. A.D. Olver, P.J. Clarricoats, A.A. Kishk, L. Shafai, *Microwave Horns and Feeds* (Institution of Engineering and Technology, 1994)
4. T.S. Bird, C. Granet, Profiled horns and feeds, in *Handbook of Reflector Antennas and Feed Systems: Volume 2—Feed Systems*, ed. by L. Shafai et al. (Artech House, 2013)
5. P.F. Goldsmith, *Quasioptical Systems: Gaussian Beam, Quasioptical Propagation, and Applications* (IEEE Press/Chapman & Hall, New York, 1998)
6. A.W. Love The diagonal horn antenna. Microw. J. **V**, 117–122 (1962)
7. J.F. Johansson, N.D. Whyborn, The diagonal horn as a sub-millimeter wave antenna. IEEE Trans. Microw. Theory Tech. **40**(5), 795–800 (1992)
8. T. Reck, C. Jung-Kubiak, J. Siles, C. Lee, R. Lin, G. Chattopadhyay, I. Mehdi, K. Cooper, A silicon micromachined eight-pixel transceiver array for submillimeter-wave radar. IEEE Trans. THz Sci. Technol. **5**(2), 197–206 (2015)
9. P.D. Potter, A new horn antenna with suppressed sidelobes and equal beamwidths. NASA Technical Documents, Technical report No. 32-354, 25 Feb 1963
10. P. Kittara, A. Jiralucksanawong, G. Yassin, S. Wangsuya, J. Leech, The design of potter horns for THz applications using a genetic algorithm. Int. J. Infrared Milli. Waves **28**(12), 1103–1114 (2007)
11. H.M. Pickett, J.C. Hardy, J. Farhoomand, Characterization of a dual-mode horn for submillimeter wavelengths. IEEE Trans. Microw. Theory Tech. **32**(8), 936–937 (1984)
12. J.F. Johansson, A Gauss-Laguerre analysis of the dual-mode ('Potter') horn, in *4th International Symposium on Space Terahertz Technology*, Apr 1993, p. 134
13. B.M. Thomas, Design of corrugated conical horns. IEEE Trans. Antenna Propag. **26**(2), 367–372 (1978)
14. C. Granet, G.L. James, Design of corrugated horns: a primer. IEEE Antennas Propag. Mag. **47**(2), 76–84 (2005)
15. G.L. James, Analysis and design of TE_{11}-to-HE_{11} corrugated cylindrical waveguide mode converters. IEEE Trans. Microw. Theory Tech. **29**(10), 1059–1066 (1981)
16. G. James, Design of wide-band compact corrugated horns. IEEE Trans. Antennas Propag. **32** (10), 1134–1138 (1984)
17. G.L. James, B.M. Thomas, TE_{11} to HE_{11} cylindrical waveguide mode converters using ring-loaded slots. IEEE Trans. Microw. Theory Tech. **30**(3), 278–285 (1982)
18. X. Zhang, Design of conical corrugated feed horns for wide-band high-frequency applications. IEEE Trans. Microw. Theory Tech. **41**(8), 1263–1274 (1993)
19. A.D. Olver, J. Xiang, Design of profiled corrugated horns. IEEE Trans. Antennas Propag. **36** (7), 936–940 (1988)
20. R. Gonzalo, J. Teniente, C. del Rio, Improved radiation pattern performance of Gaussian profiled horn antennas. IEEE Trans. Antennas Propag. **50**(11), 1505–1513 (2002)

21. G.G. Gentili, R. Nesti, G. Pelosi, V. Natale, Compact dual-profile corrugated circular waveguide horn. Electron. Lett. **36**(6), 486–487 (2000)
22. B. Maffei et al., Planck pre-launch status: HFI beam expectations from the optical optimisation of the focal plane. A & A **520**, A12 (2010)
23. J.E. McKay, D.A. Robertson, P.A.S. Cruickshank, R.I. Hunter, D.R. Bolton, R.J. Wylde, G. M. Smith, Compact wideband corrugated feedhorns with ultra-low sidelobes for very high performance antennas and quasi-optical systems. IEEE Trans. Antennas Propag. **61**(4), 1714–1721 (2013)
24. R.W. Haas, D. Brest, H. Mueggenburg, L. Lang, D. Heimlich, Fabrication and performance of MMW and SMMW platelet horn arrays. Int. J. Infrared Milli. Waves **14**(11), 2289–2294 (1993)
25. M. Kangas, K. Copsey, P. Lubin, A modular 100-GHz high-gain scalar corrugated nonbonded platelet antenna. IEEE Antennas Wirel. Propag. Lett. **4**, 89–92 (2005)
26. M.M. Kangas, M. Ansmann, B. Horgan, N. Lemaster, R. Leonardi, A. Levy, P. Lubin, J. Marvil, P. McCreary, T. Villela, A 31 pixel flared 100-GHz high-gain scalar corrugated nonbonded platelet antenna array. IEEE Antennas Wirel. Propag. Lett. **4**, 245–248 (2005)
27. L. Lucci, R. Nesti, G. Pelosi, S. Selleri, A stackable constant-width corrugated horn design for high-performance and low-cost feed arrays at millimeter wavelengths. IEEE Antennas Wirel. Propag. Lett. **11**, 1162–1165 (2012)
28. J.W. Britton, J.P. Nibarger, K.W. Yoon, J.A. Beall, D. Becker, H.-M. Cho, G.C. Hilton, J. Hubmayr, M.D. Niemack, K.D. Irwin, Corrugated silicon platelet feed horn array for CMB polarimetry at 150 GHz, in *Proceedings of SPIE*, vol. 7741, 2010, pp. 77 410T–77 410T-11
29. J.P. Nibarger, J.A. Beall, D. Becker, J. Britton, H.-M. Cho, A. Fox, G.C. Hilton, J. Hubmayr, D. Li, J. McMahon, M.D. Niemack, K.D. Irwin, J. Van Lanen, K.W. Yoon, An 84 pixel all-silicon corrugated feedhorn for CMB measurements. J. Low Temp. Phys. **167**(3), 522–527 (2012)
30. C. Lee, G. Chattopadhyay, E. Decrossas, A. Peralta, I. Mehdi, C.A. Leal-Sevillano, M.A. Pino, N. Llombart, Terahertz antenna arrays with silicon micromachined-based microlens antenna and corrugated horns, *International Workshop on Antenna Technology (iWAT)*, Mar 2015, pp. 70–73
31. E. Decrossas, T. Reck, C. Lee, C. Jung-Kubiak, I. Mehdi, G. Chattopadhyay, "Evaluation of 3d printing technology for corrugated horn antenna manufacturing," in IEEE Int. Symp Electromagn. Compat. **2**, (2016)
32. E. Decrossas, T. Reck, C. Lee, C. Jung-Kubiak, I. Mehdi, and G. Chattopadhyay, "Development of w-band horn antennas using 3d printing technologies," in IEEE Int. Symp. Antennas Propagation Soc. **2**, (2016)
33. T. Reck, C. Jung-Kubiak, J. Gill, G. Chattopadhyay, Measurement of silicon micromachined waveguide components at 500–750 GHz. IEEE Trans. THz Sci. Technol. **4**(1), 33–38 (2014)
34. J. Leech, B.K. Tan, G. Yassin, P. Kittara, S. Wangsuya, Experimental Investigation of a low-cost, high performance focal-plane horn array. IEEE Trans. Terahertz Sci. Technol. **2**(1), 61–70 (2012)
35. N. Chahat, T. Reck, C. Jung-Kubiak, T. Nguyen, R. Sauleau, G. Chattopadhyay, 1.9 THz multi-flare angle horn optimization for space instruments. IEEE Trans. Terahertz Sci. Technol. **5**(6), 914–921 (2015)
36. N. Chahat, R.E. Hodges, J. Sauder, M. Thomson, E. Peral, Y. Rahmat-Samii, CubeSat Deployable Ka-band mesh reflector antenna development for Earth science missions. IEEE Trans. Antennas Propag. **64**(6), 2083–2093 (2016)
37. C. Granet, G.L. James, R. Bolton, G. Moorey, A smooth-walled spline-profile horn as an alternative to the corrugated horn for wide band millimeter-wave applications. IEEE Trans. Antennas Propag. **52**(3), 848–854 (2004)
38. A. Rolland, M. Ettorre, M. Drissi, L. Le Coq, R. Sauleau, Optimization of reduced-size smooth-walled conical horns using BoR-FDTD and genetic algorithm. IEEE Trans. Antennas Propag. **58**(9), 3094–3100 (2010)
39. T.G. Jurgens, J.G. Blaschak, G.W. Saewert, Bodies of revolution, in *Computational Electrodynamics: The Finite-Difference Time-Domain Method*, 3rd edn., ch. 12, ed. by A. Taflove, S.C. Hagness (Artech House, Boston, MA, 2005), pp. 517–552

40. M. Celuch, W.K. Gwarek, Industrial design of axisymmetrical devices using a customized solver from RF to optical frequency bands. IEEE Microw. Mag. **9**, 150–159 (2008)
41. G. Godi, R. Sauleau, L. Le Coq, D. Thouroude, Design and optimization of three-dimensional integrated lens antennas with genetic algorithm. IEEE Trans. Antennas Propag. **55**(3), 770–775 (2007)
42. G.M. Rebeiz, D.P. Kasilingam, Y. Guo, P.A. Stimson, D.B. Rutledge, Monolithic millimeter-wave two-dimensional horn imaging arrays. IEEE Trans. Antennas Propag. **38** (9), 1473–1482 (1990)
43. G.M. Rebeiz, Millimeter-wave and terahertz integrated circuit antennas. Proc. IEEE **80**(11), 1748–1770 (1992)
44. D.F. Filipovic, S.S. Gearhart, G.M. Rebeiz, Double-slot antennas on extended hemispherical and elliptical silicon dielectric lenses. IEEE Trans. Microw. Theory Tech. **41**(10), 1738–1749 (1993)
45. T.H. Buttgenbach, An improved solution for integrated array optics in quasi-optical mm and submm receivers: the hybrid antenna. IEEE Trans. Microw. Theory Tech. **41**(10), 1750–1760 (1993)
46. R. Compton, R. McPhedran, Z. Popovic, G. Rebeiz, P. Tong, D. Rutledge, Bow-tie antennas on a dielectric half-space: theory and experiment. IEEE Trans. Antennas Propag. **35**(6), 622–631 (1987)
47. R. DuHamel, D. Isbell, Broadband logarithmically periodic antenna structures, *1958 IRE International Convention Record*, New York, NY, USA, 1957, pp. 119–128
48. J. Dyson, The equiangular spiral antenna. IRE Trans. Antennas Propag. **7**(2), 181–187 (1959)
49. T.H. Buttgenbach, R.E. Miller, M.J. Wengler, D.M. Watson, T.G. Phillips, A broad-band low-noise SIS receiver for submillimeter astronomy. IEEE Trans. Microw. Theory Tech. **36** (12), 1720–1726 (1988)
50. D.F. Filipovic, G.P. Gauthier, S. Raman, G.M. Rebeiz, Off-axis properties of silicon and quartz dielectric lens antennas. IEEE Trans. Antennas Propag. **45**(5), 760–766 (1997)
51. G. Rebeiz, W. Regehr, D. Rutledge, R. Savage, N. Luhmann, Submillimeter-wave antennas an thin membranes, in *Antennas and Propagation Society International Symposium*, 1987, pp. 1194–1197
52. D. Rutledge, M. Muha, Imaging antenna arrays. IEEE Trans. Antennas Propag. **30**(4), 535–540 (1982)
53. N. Llombart, A. Neto, THz time-domain sensing: the antenna dispersion problem and a possible solution. IEEE Trans. THz Sci. Technol. **2**(4), 416–423 (2012)
54. A. Neto, N. Llombart, J.J.A. Baselmans, A. Baryshev, S.J.C. Yates, Demonstration of the leaky lens antenna at submillimeter wavelengths. IEEE Trans. THz Sci. Technol. **4**(1), 26–32 (2014)
55. N. Llombart, G. Chattopadhyay, A. Skalare, I. Mehdi, Novel terahertz antenna based on a silicon lens fed by a leaky wave enhanced waveguide. IEEE Trans. Antennas Propag. **59**(6), 2160–2168 (2011)
56. N. Llombart et al., Silicon micromachined lens antenna for THz integrated heterodyne arrays. IEEE Trans. THz Sci. Technol. **3**(5), 515–523 (2013)
57. M. Alonso-DelPino et al., Design guidelines for a terahertz silicon micro-lens antenna. IEEE Antennas Wirel. Propag. Lett. **12**, 84–87 (2013)
58. D.R. Jackson, A.A. Oliner, A leaky-wave analysis of the high-gain printed antenna configuration. IEEE Trans. Antennas Propag. **36**(7), 905–910 (1988)
59. A. Neto, UWB, non dispersive radiation from the planarly fed leaky lens antenna—Part 1: Theory and design. IEEE Trans. Antennas Propag. **58**(7), 2238–2247 (2010)
60. N. Llombart, A. Neto, G. Gerini, M. Bonnedal, P. De Maagt, Impact of mutual coupling in leaky wave enhanced imaging arrays. IEEE Trans. Antennas Propag. **56**(4), 1201–1206 (2008)

61. M. Qiu, G.V. Eleftheriades, M. Hickey, A reduced surface-wave twin arc-slot antenna element on electrically thick substrates, in *Proceedings of IEEE AP-S Society International Symposium*, Boston, MA, USA, 2001, vol. 3, pp. 268–271
62. C. Lee, G. Chattopadhyay, K. Cooper, I. Mehdi, Curvature control of silicon microlens for THz dielectric antenna, in *37th International Conference on Infrared Milli. THz Waves (IRMMW-THz)*, Wollongong, NSW, 2012, pp. 1–2
63. C. Lee, G. Chattopadhyay, M. Alonso-DelPino, N. Llombart, "6.4 mm Diameter silicon micromachined lens for THz dielectric antenna," *39th Int. Conf. Infrared Millimeter, and THz waves (IRMMW-THz)*, Tucson, AZ, 2014, pp. 1–1
64. B. Fong, J. Colburn, J. Ottusch, J. Visher, D. Sievenpiper, Scalar and tensor holographic artificial impedance surfaces. IEEE Trans. Antennas Propag. **58**(10), 3212–3221 (2010)
65. G. Minatti, M. Faenzi, E. Martini, F. Caminita, P. De Vita, D. González-Ovejero, M. Sabbadini, S. Maci, Modulated metasurface antennas for space: synthesis, analysis and realizations. IEEE Trans. Antennas Propag. **63**(4), 1288–1300 (2015)
66. D. Sievenpiper, J. Colburn, B. Fong, J. Ottusch, J. Visher, Holographic artificial impedance surfaces for conformal antennas, in *Proceedings of IEEE AP-S Society International Symposium*, vol. 1B, 2005, vol. 1B, pp. 256–259
67. A. Oliner, A. Hessel, Guided waves on sinusoidally-modulated reactance surfaces. IRE Trans. Antennas Propag. **7**(5), 201–208 (1959)
68. G. Chattopadhyay, Technology, capabilities, and performance of low power terahertz sources. IEEE Trans. THz Sci. Technol. **1**(1), 33–53 (2011)
69. M. Esquius-Morote, J. Gomez-Diaz, J. Perruisseau-Carrier, Sinusoidally modulated graphene leaky-wave antenna for electronic beamscanning at THz. IEEE Trans. THz Sci. Technol. **4**(1), 116–122 (2014)
70. C. Jung, B. Thomas, C. Lee, A. Peralta, J. Gill, K. Cooper, G. Chattopadhyay, E. Schlecht, R. Lin, I. Mehdi, Compact submillimeter-wave receivers made with semiconductor nanofabrication technologies, in *Proceedings of IEEE International Microwave Symposium (IMS)*, June 2011, pp. 1–4
71. D. González-Ovejero, T.J. Reck, C.D. Jung-Kubiak, M. Alonso-DelPino, G. Chattopadhyay, Silicon micromachined modulated metasurface antennas in the terahertz range, in *Proceedings of 10th European Conference on Antennas and Propagation*, 2016, pp. 1–4
72. D. González-Ovejero, T.J. Reck, C.D. Jung-Kubiak, M. Alonso-DelPino, G. Chattopadhyay, A class of silicon micromachined metasurface for the design of high-gain terahertz antennas, in *Proceedings of IEEE AP-S Society International Symposium*, 2016, pp. 1–4
73. R. King, D.V. Thiel, K. Park, The synthesis of surface reactance using an artificial dielectric. IEEE Trans. Antennas Propag. **31**(3), 471–476 (1983)
74. H. Bilow, Guided waves on a planar tensor impedance surface. IEEE Trans. Antennas Propag. **51**(10), 2788–2792 (2003)
75. D. González-Ovejero, S. Maci, Gaussian ring basis functions for the analysis of modulated metasurface antennas. IEEE Trans. Antennas Propag. **63**(9), 3982–3993 (2015)
76. ANSYS Inc., HFSS, Version 15, Pittsburgh, PA, 2015
77. A. Tang et al., A CMOS D-band transmitter with IF envelope feed-forward pre-distortion and injection locked frequency tripling synthesizer. IEEE Trans. Microw. Theory Tech. **60**(12), 4129–4137 (2012)
78. A. Tang et al., A low overhead self-healing embedded system for ensuring high performance yield and long-term sustainability of a 60 GHz 4 Gb/s Radio-on-a-Chip, in *IEEE International Solid-State Circuits Conference Digest of Technical Papers*, Feb 2012, pp. 316–318
79. N. Chahat, A. Tang, C. Lee, R. Sauleau, G. Chattopadhyay, Efficient CMOS systems with beam–lead interconnects for space instruments. IEEE Trans. Terahertz Sci. Technol. **5**(4), 637–644 (2015)
80. A. Tang, N. Chahat et al., A 65 nm CMOS 140 GHz 27.3 dBm EIRP transmit array with membrane antenna for highly scalable multi-chip phase arrays, in *Proc. IEEE International Microwave Symposium (IMS)*, Tampa, FL, 2014

81. A. Tang, N. Chahat, CMOS mm-wave transceiver techniques beyond 50 GHz, in *Proceedings of IEEE Global Conference* on *Signal and Information Processing (GlobalSIP)*, Austin, TX, 2013, pp. 715–718
82. A. Tang et al., CMOS (Sub)-mm-wave system-on-chip for exploration of deep space and outer planetary systems, in *Proceedings of IEEE Custom Integrated Circuits Conference*, San Jose, CA, 2014
83. A. Tang et al., A 95 GHz centimeter scale precision confined pathway system-on-chip navigation processor for autonomous vehicles in 65 nm CMOS, in *Proceedings of IEEE International Microwave Symposium (IMS)*, Phoenix, AZ, 2015, pp. 1–3
84. R.R. Cladwell, M. Kamionkowski, Echoes from the Big Bang. Sci. Am. 38–43 (2001)

Chapter 11
Transformation Optics Applied to Antennas and Focusing Systems

Rhiannon C. Mitchell-Thomas and Oscar Quevedo-Teruel

Abstract This chapter describes the way in which transformation optics can be used to design lens antennas. It begins with a brief introduction to the analytical method and describes the difficulties faced by practical applications, in terms of anisotropic material requirements and bandwidth restrictions. Two further variants of transformation optics are provided to circumvent these issues, which are the quasiconformal transformation technique, and non-Euclidean transformations for surface devices. The use of both these methods results in material property requirements that can be easily implemented with dielectrics. These two techniques are thoroughly described, with the benefits and drawbacks they present detailed. The discussion is supported throughout with a number of application examples in each section, which include planar lenses, bespoke lenses, curved surface wave lenses, and a surface wave cloak. Examples are chosen to illustrate the variety of ways in which transformation optics can be applied to antenna systems, and are all supplemented by simulation or measurement data.

11.1 Introduction to Transformation Optics

The advent of transformation optics [1, 2] has provided electromagnetic engineers and physicists with a very powerful tool. This mathematical technique allows a user to predefine the propagation characteristics of electromagnetic waves, and implement this wave behavior in a device. This is possible by employing complex

R.C. Mitchell-Thomas
Department of Physics and Astronomy, University of Exeter, Physics building, Stocker road, Exeter EX4 4QL, United Kingdom
e-mail: r.c.mitchell-thomas@exeter.ac.uk

O. Quevedo-Teruel (✉)
School of Electrical Engineering (EES), Electromagnetic Engineering (ETK), KTH Royal Institute of Technology, Osquldas väg 6, fifth floor, SE-100 44 Stockholm, Sweden
e-mail: oscarqt@kth.se
URL: http://www.etk.ee.kth.se/personal/oscarqt/

A. Boriskin and R. Sauleau (eds.), *Aperture Antennas for Millimeter and Sub-Millimeter Wave Applications*, Signals and Communication Technology,
DOI 10.1007/978-3-319-62773-1_11

material properties to direct waves, which are generally spatially dependent and anisotropic. It can be used to enable waves to perfectly transmit through sub-wavelength apertures [3], to allow waves to avoid a certain region of space [4], or for waves to concentrate to a particular point in space [5], illustrating the versatility of the technique.

Over the last 20 years as the research on this topic has developed and intensified, the areas in which the method has been used have broadened significantly, and range from theoretical concepts such as black hole analogies [6], to the applied end of the spectrum including improvements in solar cell technology [7]. The versatility and opportunities posed by this technique may appear endless; however, as the field has progressed, the understanding of the limitations and difficulties has also grown. Efforts to circumvent these problems have resulted in a number of variants of transformation optics being developed, two of which are outlined here.

This chapter focusses on the application of transformation optics to a range of lens antenna systems. This has been a particularly successful area where the use of transformation optics has provided solutions to difficult engineering problems such as completely removing the reflection caused by waveguide bends [8] creating isotropic radiators [9] or designing retroreflectors that do not need infinite refractive indices [10]. The challenges posed when designing antennas for the mm and sub-mm range using technology commonly used for the GHz range, such as waveguides, reflectors, or phased arrays is that they need extremely high precision, which increases the cost of manufacture considerably. Therefore, different routes need to be sought, and this chapter proposes the use of transformation options to overcome these issues.

Here, a description of how transformation optics can be used to design a range of lenses which are appropriate for increasing the directivity of high-frequency antennas will be given. These are alternatives to those designed using ray tracing, for example, the Fresnel lens, the hyperbolic lens, or the Luneburg which either suffer from bandwidth restrictions or complex geometry. Overcoming these issues mean that they can be employed in transport and automotive communication systems, where weight and volume are important issues, or radioastronomy where ultrawide bandwidths are required. In addition, surface wave devices where curvature plays a crucial role will be shown. For example, curvature can be employed so that dielectric is either not necessary, or lower permittivities can be used, thus reducing the losses, which is highly desirable for aerospace applications.

11.2 Analytical Transformation

In its purest and original form, the transformation optics technique relies on an exact transformation of coordinates. It was pioneered by Prof. Pendry and Prof. Leonhardt in 2006 [1, 2], when two seminal papers were published in the same Science journal issue. The process defines the modification of constituent material parameters required, so that the electromagnetic behavior remains rigorously

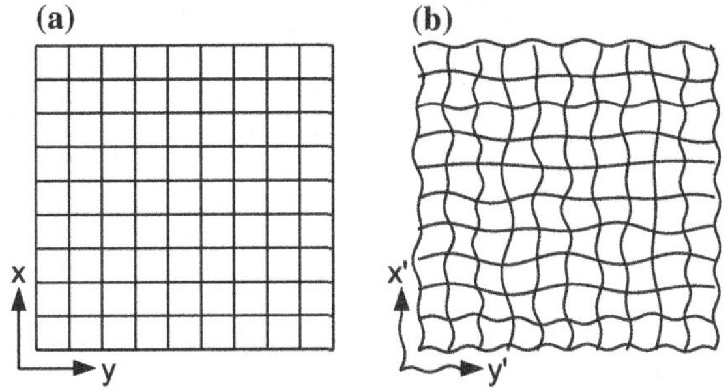

Fig. 11.1 Illustration of a Cartesian coordinate system (**a**), and an example of a transformed coordinate system (**b**)

invariant in both the original and new coordinate systems. An example of two coordinate systems is given in Fig. 11.1, where in the original Cartesian space, labeled (x, y), the coordinate lines can be interpreted as rays propagating in straight lines through a vacuum. The transformed system, labeled by (x′, y′), now exhibits ray propagation that follows curved paths.

In order to achieve this equivalent behavior, the nature of the chosen coordinate transformation is subsumed into the material parameters, namely the permeability and permittivity tensors, which are of the same form and depend on spatial parameters. Mathematically, this transformation of the permittivity and permeability is calculated as follows:

$$\varepsilon' = \frac{J\varepsilon J^T}{|J|}, \mu' = \frac{J\mu J^T}{|J|}, \tag{11.1}$$

where J represents the Jacobian transformation tensor that connects the two coordinate systems; ε and μ represent the permittivity and permeability tensors in the original space; and ε' and μ' are the same tensors in the transformed space. The Jacobian transformation tensor takes the form

$$J = \begin{vmatrix} \frac{\partial x'}{\partial x} & \frac{\partial x'}{\partial y} & \frac{\partial x'}{\partial z} \\ \frac{\partial y'}{\partial x} & \frac{\partial y'}{\partial y} & \frac{\partial y'}{\partial z} \\ \frac{\partial z'}{\partial x} & \frac{\partial z'}{\partial y} & \frac{\partial z'}{\partial z} \end{vmatrix}. \tag{11.2}$$

This type of analytical transformation has been employed for a number of antenna applications [11], such as the redesign of traditional devices such as polarization rotators [12], wave collimators and beam benders [13], or reduction of the size of conventional lenses [14, 15]. However, one of the most simple transformations that can be employed is the dilation of one of the axes of a Cartesian coordinate system [15], where $x' = x$, $y' = y$, and $z' = \eta z$, and will be used here as an example. Figure 11.2 displays this transformation in two dimensions, where the vertical axis can be either x or y.

If we consider the original space to be a vacuum, then $\varepsilon = \mu = 1$, and therefore it is simple to deduce that the forms of the transformed permeability and permittivity

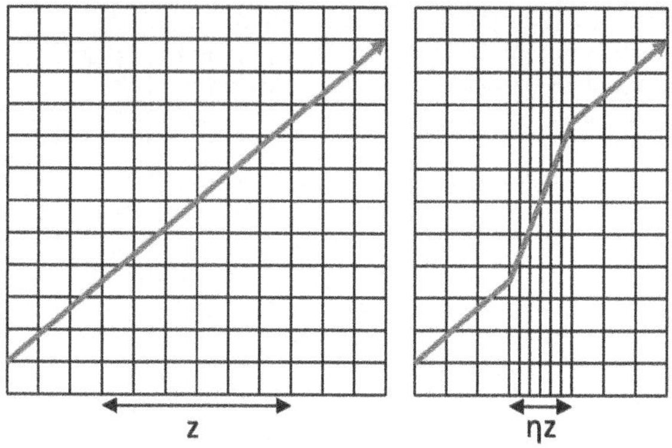

Fig. 11.2 A representation of a region of space before, a, and after, b, a compression in the z-direction. The *arrow* represents a ray propagating through the space

that are required to reproduce the same effect as a dilation of one of the coordinate axes are

$$\varepsilon' = \mu' = \begin{vmatrix} \frac{1}{\eta} & 0 & 0 \\ 0 & \frac{1}{\eta} & 0 \\ 0 & 0 & \eta \end{vmatrix}. \tag{11.3}$$

These material properties now exhibit uniaxial anisotropy, where two of the elements are lower than unity, which can only be achieved with metamaterials, and therefore introduces the narrowband behavior generally associated with them. One area that this type of analytical transformation has found application is in the domain of metasurfaces. In this case, advantage can be taken off the inherently 2D nature, so that a surface impedance tensor can be employed, which has a reduced number of elements that need to be controlled. For certain cases, the required anisotropy can be achieved, details of which are covered in Chap. 10 of this book devoted to metasurface antennas. There is a rich literature in this area, and much progress has been made in the theoretical description of such anisotropic metasurfaces, which is detailed in a number of publications [16–22]. This method has successfully been used to design beam shifters [19, 22] and beam splitters [17, 20] for surface waves, in order to reduce the scattering [22]. While this is a thriving area for transformation optics, the current chapter will focus on methods which do not rely on anisotropy, and can be simplified to isotropic material requirements.

For most of the bulk lens antenna systems designed using this technique, the extreme anisotropy required is very difficult to achieve, particularly when combined with spatial variation in three directions. This means that relatively few have been experimentally validated, and those that have are only valid for a given polarization. The following section will present a route that succeeds to escape these limiting factors.

11.3 Quasi-conformal Transformation

The ability to determine the path of electromagnetic waves is a concept that holds a great deal of promise for the design of electromagnetic devices. However, the main limitation of the analytical transformation method is the necessity to fabricate anisotropic materials. Even for very simple transformations, such as the dilation of one of the Cartesian coordinates given above, it is necessary to implement anisotropy in both the permeability and permittivity tensors. For more complex transformations, these material tensors can require non-zero off-diagonal elements and some of the tensor elements can reach infinity or zero. While the advances of metamaterials are promising, and some degree of anisotropy can be realized, there are still strong limitations to what is achievable. Additionally, the resonant nature of these metamaterials implies that high losses would also be present. These difficulties still remain a barrier to the deployment of transformation optics.

One method developed to overcome these problems is the quasi-conformal transformation technique [23]. A conformal transformation is one which retains the angles and shape of the original coordinate grid. In this case, the material tensors would be spatially dependant but, crucially, are isotropic. In the majority of problems, it is not possible to analytically find the exact conformal transformation required, but the quasi-conformal technique uses numerical methods to approximate a conformal transformation. An example of a quasi-conformal grid is given in Fig. 11.3, where (a) is the original Cartesian coordinate system, and (b) is the approximation of a conformal map within a new domain.

Once again, this transformation is effected by a change of the material parameters to play the role of the coordinate transformation. To ease the issues with implementation, only the permittivity is altered, and value of the permittivity for each individual mesh element is given by the following equation:

$$\varepsilon' = \frac{\varepsilon}{\det(J)}. \tag{11.4}$$

As it can be assumed that the transformation is conformal, and therefore the angles and shape of each individual element of the mesh are conserved after the

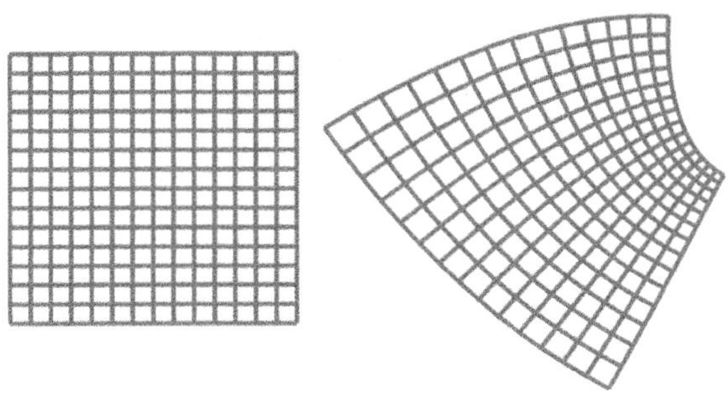

Fig. 11.3 An example of a quasi-conformal grid on a square domain, and a distorted domain

distortion, then the permittivity simply reduces to a ratio of the change of the area of the cell, given by [24]

$$\varepsilon' = \frac{\Delta x \Delta y}{\Delta x' \Delta y'}, \tag{11.5}$$

where ε' is the new permittivity value for each cell of the coordinate systems. This technique negates the use of anisotropic materials, but it also introduces its own problems. After a quasi-conformal transformation is effected, there are usually regions in the domain which have grown in volume. This means that the index at that position will reduce as can be appreciated from Eq. (11.5), and therefore values less than unity exist in the permittivity map. For this reason, care must be taken when employing this technique, and two considerations must be taken into account. The first is an appropriate coordinate system that must always be chosen to ensure that the areas of the sub-unity permittivity areas are minimized, and the second is to ensure that the mesh cells are not warped to an extent that the approximation of isotropy is invalid [25].

The quasi-conformal transformation technique has been applied to antenna systems in a number of ways [25–28], and four examples will be given here. The first is the possibility to alter the shape of existing lenses, such as a hyperbolic lens [27]. In this redesign of the lens, the focusing effect caused by the curved shape of the homogenous dielectric lens can be reproduced by a flat lens with a graded index after applying the quasi-conformal transformation technique. Figure 11.4a shows the shape of the original lens, plus the quasi-conformal map where the color gives an indication of the area of each mesh element. Figure 11.4b displays a photograph

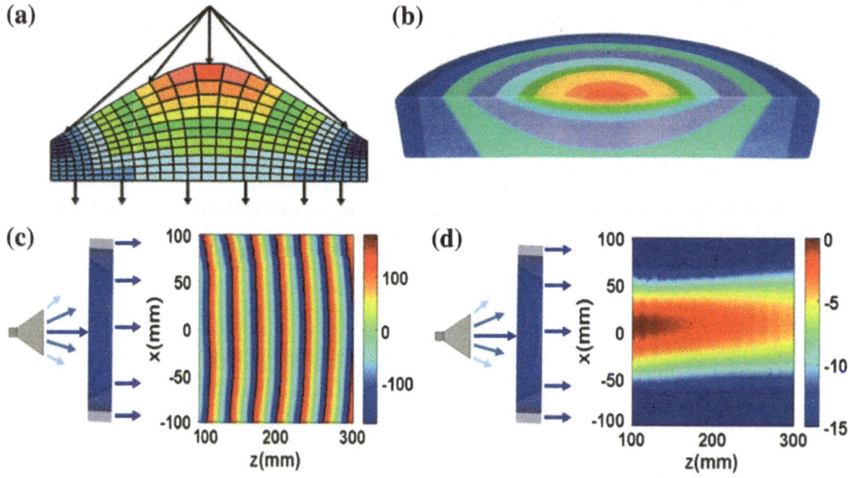

Fig. 11.4 **a** A hyperbolic lens complete with a quasi-conformal coordinate grid, where the colored regions relate to the area of each mesh cell, and the *arrows* emit from the focal point. **b** A photograph of the cross section of the graded index lens, where each color corresponds to a different permittivity of the lens. **c** and **d** show experimental measurements of the lens with the phase illustrated in (**c**) and the collimated beam amplitude shown in (**d**). Extracted from [27]

of a half lens, and now the color refers to the values of permittivity and shows the internal discretisation of the lens. Figure 11.4c, d shows the measured phase and magnitude of a beam that has been emitted from a horn antenna and collimated by the lens. It can be appreciated that this redesigned planar lens has served to increase the directivity of the horn.

Another way in which the quasi-conformal technique can be applied is to the redesign of standard graded index lenses so that their shapes are more appropriate for a specific application. The Luneburg lens in one such lens has a spherically symmetric refractive index profile which converts a focal point on the surface to a plane wave on the opposite side. It allows steering of the beam to be achieved simply by changing the position of the feed source, while the lens itself remains stationary. However, one of the main drawbacks is that the focal plane is a hemisphere, and therefore the mechanical actuation required to position the feed is not ideal. It is possible to implement a quasi-conformal transformation so that the feed surface becomes a flat plane, and the beam-steering can be achieved much more simply [25, 29]. This example is shown in Fig. 11.5, where (a) is the

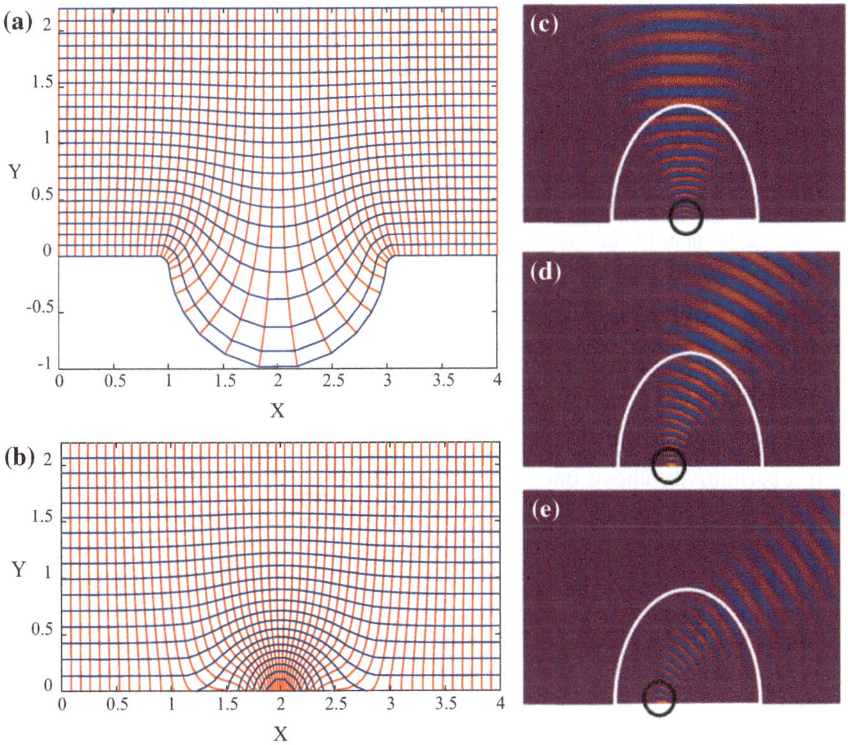

Fig. 11.5 Transformation of a Luneburg lens to provide a flat feed plane. **a** The quasi-conformal map showing the hemispherical Luneburg feed plane. X′, Y′ plotted on the X, Y coordinates. **b** The inverse map, X, Y plotted on X′, Y′ coordinates. c–e Full-wave simulations of the transformed lens, with three different beam angles

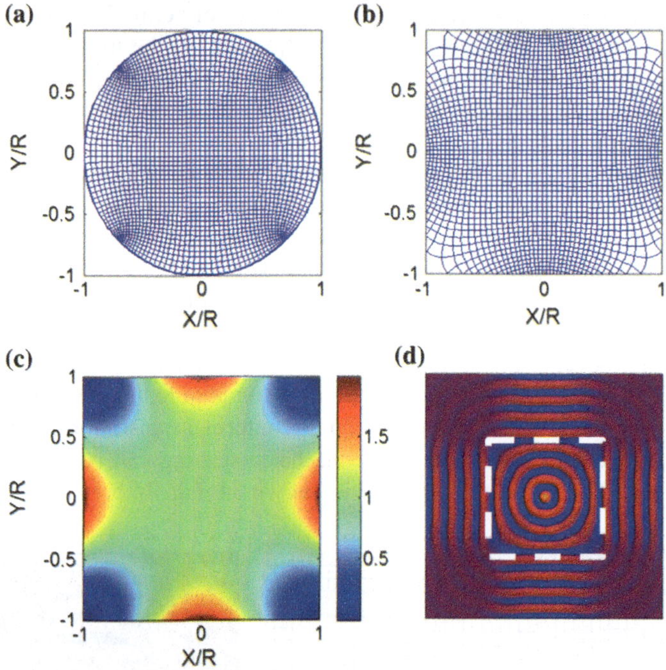

Fig. 11.6 Mapping of a circular wavefront to a four-beam lens. **a** A quasi-conformal map of the domain terminated by a circular wavefront. **b** The inverse map of the transformed square grid. **c** The permittivity map calculated from the mesh element areas. **d** A full-wave simulation of a point source positioned at the center of the lens showing the four planar beams

quasi-conformal map of a domain encompassing the Luneburg lens along with a region of the surrounding free space, and (b) shows the inverse map. Often in these designs, some of the cells experience an increase in the area, which would result in a reduction of the permittivity. If this falls below unity, these regions can be approximated to free space. In this example, the area of the lens that is implemented with a permittivity above one is bounded by the white curve in (c–e). Full-wave simulations are also shown for three different feed positions (illustrated by the black circle), each producing a different beam angle.

The quasi-conformal technique can also be used to design lenses from scratch, rather than to alter an existing one, using the phase-front shape of a source as the original domain, and transforming it to desired phase-front shape, which defines the new domain. This can be seen in Fig. 11.6 where (a) is a quasi-conformal coordinate map on a circle, where the latter represents the emitted phase front from a point source; and (b) is the inverse map on the transformed domain, which has the shape of a square. The result of this transformation is to create a lens which has four directive beams, due to the planar phase fronts [30]. The permittivity map required is given in (c) where, as explained earlier, some of the regions are above, and some below unity. The full-wave simulation shown in (d) includes the entire,

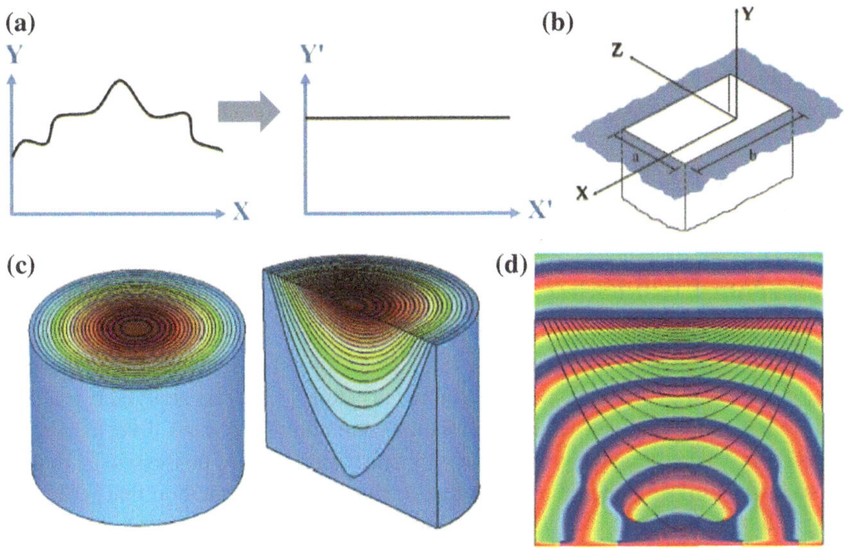

Fig. 11.7 A bespoke lens designed for a waveguide aperture. **a** The *curved line* in the *left* plot represents an arbitrary shaped wavefront, and the transformation to the *flat line* in *right* plot represents a planar wavefront. **b** A simple waveguide aperture. **c** The graded index lens showing a cross section of the layers of various permittivities. **d** A full-wave simulation showing the efficient transformation of the phase fronts from the aperture at the *bottom* of the lens, to a planar output at the *top* of the lens. Extracted from [31]

un-approximated permittivity map, bounded by the white dashed line, and shows that the four beams are evident emitted from each side of the lens.

The final example is to design a tailored lens for a particular feed type. Rather than using a standard lens, which is designed using geometrical optics for a perfect point focus, it is possible to employ this method to use the radiation characteristics of the feed to design a bespoke graded index lens [31]. This is done by taking a line of constant phase that has been emitted from the feed, which can be of arbitrary shape, an example of which is shown in Fig. 11.7a. This curve then defines the upper boundary of the domain to be transformed. The other three sides of the domain are chosen appropriately so that the conformality of the map is maximized. This is then transformed to a domain in which the upper boundary is planar, also illustrated in Fig. 11.7a. Here, a rectangular waveguide aperture is used as an antenna feed, as shown in (b). The complete lens design is shown in (c), where the colors indicate the regions of different values of permittivity and (d) is a full-wave simulation showing that the phase of the wave emitted from the waveguide aperture has been effectively transformed to a plane wave output through the use of the bespoke lens.

From the range of examples given here, it can be appreciated that this technique is very versatile for lens antenna design. The challenge of spatially varying the dielectric properties within these lenses is not to be underestimated. Additionally, the loss presented by materials with a high permittivity is also a necessary consideration. However, many advances have been made in the fabrication of graded index devices [27], and together with the design method described here, a real opportunity for a step-change in lens design is possible.

The next section will depart from the volumetric lens designs given above, and describe a method for designing a whole range of surface wave lens antennas.

11.4 Non-Euclidean Transformation

A non-Euclidean transformation, as the name suggests, deals with the transformation between geometries which exhibit curvature [32]. The analytical transformation optics method is a very powerful tool and allows the most diverse range of geometries to be transformed. However, one restriction it possesses is that all geometries being considered must be flat, or Euclidean. The method that will be set out in this section will overcome this limitation, and allow the same propagation characteristics to be displayed on two two-dimensional surfaces where at least one of these surfaces is curved [33]. This additional degree of freedom, the inclusion of curvature in the geometries being transformed, will be shown to allow more flexibility in device design [34]. This type of transformation is most useful for surface wave applications because all the geometries being considered are necessarily two-dimensional to allow for curvature to be employed. Additionally, as this takes a geometrical optics approach, all the devices must be significantly larger than the wavelength of operation for the approximations to hold. The resultant material requirements are always isotropic and have values of refractive index above unity, and are therefore very simple to implement using either dielectric layers or metasurfaces [35].

Three different applications of this method will be described in this section, including surface wave cloaking [34, 35], lenses on curved surfaces [36], and using curvature to remove the requirement of singular refractive indices in some lenses [10]. We will begin with the design of a surface wave cloak, which has the simplest mathematical formulation.

In order to hide an obstruction on a surface so that a surface wave would not be scattered by it, it is possible to guide the wave up and over the object with the use of a curved portion of the surface. However, although the scattering from the object has now been removed, the curvature will leave a signature on the wave because of the difference in optical path lengths experienced by different parts of the wave. We want our curved surface to behave as if it were a flat homogenous surface free from obstruction, and this is made possible if we add a radially dependant refractive index profile to the curved surface. Consider the two surfaces represented in Fig. 11.8. The bottom-flat surface has a refractive index of one and it is the propagation of a wave on this surface that we want to emulate on the curved surface. The upper curved surface, bounded by a radius, a, and with a maximum height, b, has an arbitrary rotationally symmetric shape. The refractive index, $n(\theta)$,

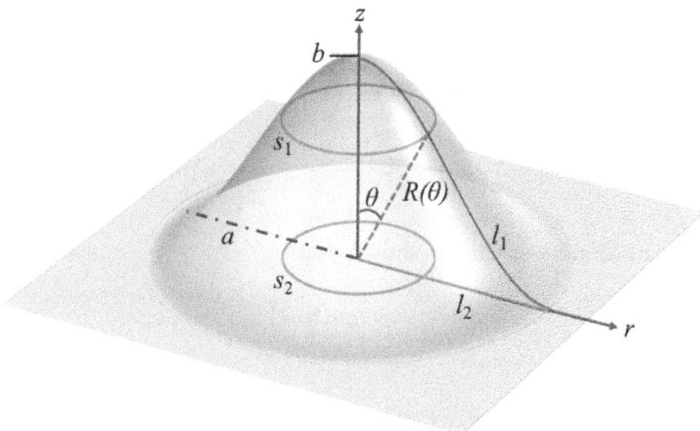

Fig. 11.8 Diagram of the two surfaces under consideration. The curved surface has two ray paths labeled as s_1 and l_1, and the flat surface has two ray paths labeled s_2 and l_2. Extracted from [33]

that we have to cover the curved surface with so that it appears flat can be calculated if we consider the ray paths labeled as s_1, s_2, l_1, and l_2.

The optical path lengths of l_1 and l_2, when equated, give [34]

$$n(\theta)\sqrt{R(\theta)^2 + R'(\theta)^2}d\theta = dr, \tag{11.6}$$

and the optical path lengths of the two circular paths, as s_1 and s_2, can be given by

$$2\pi R(\theta)sin(\theta)n(\theta) = 2\pi r. \tag{11.7}$$

Dividing Eq. (11.6) by $d\theta$ gives

$$\frac{dr}{d\theta} = n(\theta)\sqrt{R(\theta)^2 + R'(\theta)^2}. \tag{11.8}$$

If we take the derivative with respect to θ of Eq. (11.2), we find

$$\frac{dr}{d\theta} = \frac{d}{d\theta}(R(\theta)sin(\theta)n(\theta)) \tag{11.9}$$

$$\frac{dr}{d\theta} = R'(\theta)sin(\theta)n(\theta) + R(\theta)cos(\theta)n(\theta) + R(\theta)sin(\theta)n'(\theta). \tag{11.10}$$

Equating Eqs. (11.8) and (11.10) gives

$$\frac{n'(\theta)}{n(\theta)} = \frac{R'(\theta)sin(\theta)n(\theta) + R(\theta)cos(\theta)n(\theta) + R(\theta)sin(\theta)}{\sqrt{R(\theta)^2 + R'(\theta)^2}}. \tag{11.11}$$

This equation can then be solved either analytically or numerically to find $n(\theta)$. One example that can be analytically calculated is that of a hemispherical bump. In this case, R is no longer dependent upon θ, and the solution is found to be

$$\frac{n'(\theta)}{n(\theta)} = \frac{1 - cos(\theta)}{sin(\theta)}. \tag{11.12}$$

And using an identity it can be shown that

$$n(\theta) = \frac{tan(\theta/2)}{sin(\theta)}. \tag{11.13}$$

Equation (11.13) gives the exact refractive index profile that will make a hemispherical surface appear as if it were flat to rays propagating across it. This is shown in Fig. 11.9, where a parallel plate waveguide that follows the shape of the curvature is employed to illustrate the operation. In Fig. 11.9a, the hemispherical-shaped waveguide is filled with a homogeneous dielectric with a refractive index of two, and it can be seen that the curvature changes the behavior of

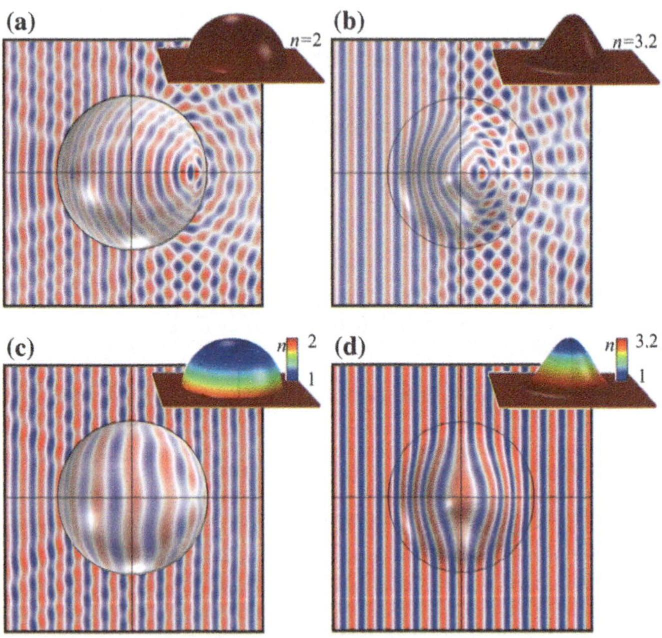

Fig. 11.9 Simulations performed using planar waveguides following the curvature of the surface being considered, with a plane wave source incident from the *left*. **a**, **c**—A hemispherical waveguide filled with; a homogeneous dielectric (**a**) and a rotationally symmetric profile (**c**). **b**, **d**—a cosine-shaped waveguide filled with; a homogeneous dielectric (**b**) and a rotationally symmetric profile (**d**). Extracted from [34]

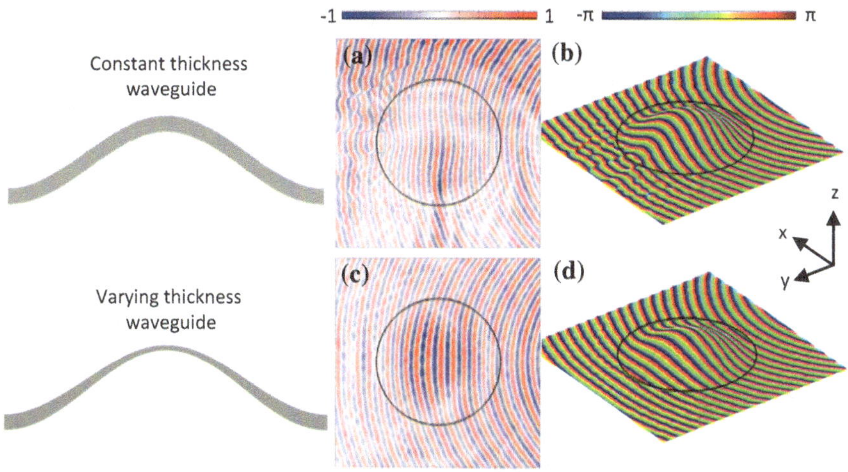

Fig. 11.10 Experimental data of the electric field 0.5 mm above the curved dielectric waveguide surface. **a**, **b**—A constant thickness waveguide, **c**, **d**—A varying thickness waveguide with the graded index properties of the cloak. **a**, **c**—Electric field component E_z, **b**, **d**—Phase data. The figures on the *left* show the cross section of the two waveguides. Extracted from [35]

the waves, creating an imperfect focal point. When the hemispherical guide is filled with the angle-dependent refractive index profile given in Eq. (11.13), the planar wavefronts that are incident from the left are reformed on the far side of the bump, demonstrating that the guide appears as if the curved portion were absent (Fig. 11.9c). However, this choice of guide shape creates reflections at the interface between the flat and curved regions of the guide. This can be overcome if a cosine-type function is instead chosen, as is the case in Fig. 11.9 b, d. Once again it can be seen that the curvature in the guide distorts the shape of the planar wavefronts incident from the left (Fig. 11.9b), but when the guide is filled with the appropriate refractive index profile, the planar nature of the wave is perfectly restored after propagating over the curved guide, so that it would not be possible to detect a signature from the curved surface on the wave.

One option to implement the required graded index surfaces shown in Fig. 11.9 for the purpose of creating a surface wave cloak is to employ a varying thickness dielectric waveguide. This waveguide has a metallic sheet on the underside of the curved guide and is open at the upper surface. In order to calculate the mode index of a dielectric waveguide of a given thickness, either numerical simulations or an analytical solution can be used. For a given frequency, this mode index will alter depending on the thickness of the guide, and by slowly varying the thickness it is possible to create a graded mode index medium. One such example is shown in Fig. 11.10, where a cosine-shaped obstacle is chosen, and both a graded mode index cloak and a homogeneous mode index comparison are shown. The surface wave on both samples was imaged using a motorized 3-stage scanner to obtain the field distribution 0.5 mm above the upper surface of the dielectric waveguide. It can be seen in the measurement data shown in Fig. 11.10 that the constant thickness waveguide, which exhibits a homogeneous mode index, causes a disturbance of the circular wave propagating across it (Fig. 11.10a, b). However, in Fig. 11.10c, d, the circular nature of the wavefronts has been restored by employing the appropriate variation in guide thickness to achieve the mode index profile required to remove the influence of the curvature, and thus act as a surface wave cloak.

This type of surface wave cloak has advantages over other designs that employ the analytical transformation method due to the fact that the material requirements are isotropic and therefore it is very simple to fabricate. In addition, the cloaks designed here are also very thin (the order of $\lambda/6$) and possess omnidirectional operation in the plane, as opposed to those designed using the quasi-conformal transformation method [37].

However, the design of surface wave cloaks is not the only use of this technique. Another application is to redesign rotationally symmetric lenses, for example, the Luneburg lens, to operate effectively when applied to a curved surface [36]. As it has been demonstrated, there is an equivalence between curvature and refractive index profiles for rotationally symmetric surfaces. It has previously been shown that it is possible to design lenses using this principle. One well-known example of this is the Maxwell fish-eye lens, which emulates the propagation of rays on the surface of a hemisphere using a flat surface with a refractive index profile. Because of the curvature, the rays emitted from a point on the circumference will diverge until they meet the mid-point, before converging once again to a point on the opposite side of the circumference. Although these rays are not changing speed, the curvature serves to direct the rays on a particular path. This exact same behavior is exhibited by the index profile given by the Maxwell fish-eye lens. Once again, a point source positioned on the circumference of the lens emits rays that diverge; however, this time the rays are being directed by the change in speed created by the index profile. Just like the propagation on the hemisphere, the rays then converge at a point on the opposite side of the lens circumference.

The Luneburg lens has been widely used in antenna design since the 60s. This lens focusses an incident plane wave to a point on the opposite side of the circumference, and vice versa, using a rotationally symmetric index profile to guide the rays. Its index profile has the form $n_L = \sqrt{2 - (r/R_0)^2}$, where R_0 is the lens radius. The index profile also has an equivalent homogenous surface which can be used to remove the requirement of a graded index medium. This provides an appealing opportunity to create a surface wave Luneburg with a single-mode index surface. However, the problem with this surface is that when it is applied to a flat surface, there is a 90-degree transition between the curved surface and the flat. This, once again, causes unwanted reflections and scattering of the surface wave. To circumvent this problem, it is possible to calculate the profile of the Luneburg lens that is appropriate for other surfaces, for example, a cosine-type function that smoothly attaches to the flat place. In order to calculate this, Eqs. (11.6) and (11.7) must include an index profile $n_1(r)$ on the flat surface, resulting in the following equation:

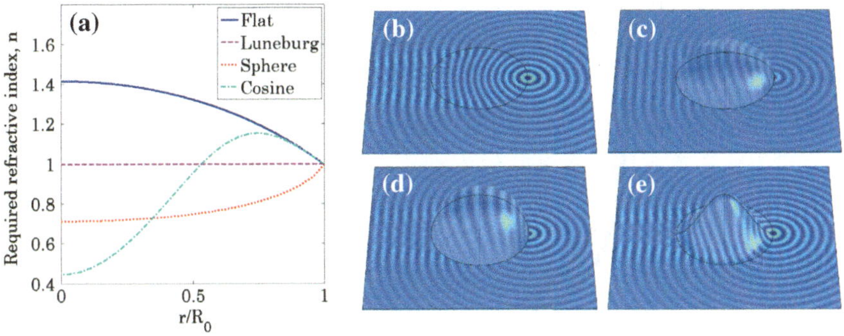

Fig. 11.11 **a**—Required refractive index profiles for a Luneburg lens on each of the four labeled surfaces. **b–e**—Electric field distribution in parallel plate curved waveguides, each filled with one of the refractive index profiles given in (**a**), which are **b**—Flat, **c**—Luneburg shape, **d**—Hemispherical shape, and **e**—Cosine shape. Extracted from [36]

$$\frac{n'(\theta)}{n(\theta)} = \frac{\left(1 + r\frac{n_1'(r)}{n_1(r)}\right)\sqrt{R(\theta)^2 + R'(\theta)^2} - R'(\theta)sin(\theta) - R(\theta)cos(\theta)}{R(\theta)sin(\theta)}. \tag{11.14}$$

Using this equation, if the flat Luneburg profile given above is substituted into $n_1(r)$, it is possible to calculate the index $n(\theta)$ that will create the same behavior for any surface represented by the function $R(\theta)$. In Fig. 11.11, four different surface shapes are chosen, and the corresponding index profile for the surface to function as a Luneburg lens is given in Fig. 11.11a. Figure 11.11b–e shows the electric field on the surface of a curved planar waveguide, each illustrating effective transformation of the point source positioned at the circumference of each device, to a plane wave propagating to the left on the far side of the lenses.

In addition to the Luneburg and the Maxwell fish-eye lenses mentioned above, there is a whole family of graded index lenses that exhibit interesting behavior [33]. Some examples of which are the Eaton lens, which redirects incident rays back in the direction that they originated from, the monopole lens that allows a point source of rays to only emit in one direction, and the invisible lens that can only be detected by a temporal delay of the rays incident on it. The ray tracing for these three remarkable lenses can be seen in the inset of Fig. 11.12. What these lenses have in common is that at the center of the lens the refractive index goes to infinity. This is a major barrier to their implementation, as it is impossible to create a dielectric medium with a singular index, and it is also practically impossible to fabricate such a steeply increasing graded index profile. However, using Eq. (11.14), it is possible to solve for $R(\theta)$ to find the equivalent surfaces [36].

All of the equivalent surfaces of singular lenses exhibit a tip which has the same influence on the ray trajectories as the infinity in the refractive index. This simple addition, and sculpting of the surface, means that lenses that were formerly impossible to implement for practical applications are now achievable. In addition,

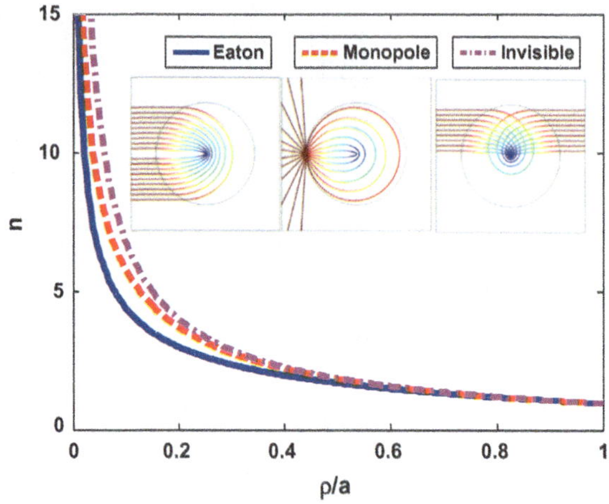

Fig. 11.12 The index profiles of three lenses with a singularity at the *center*. Inset—Ray tracing demonstrating how each of the lenses performs, where the color of the ray trajectory is the speed of the ray at that position

it is also possible to combine the use of finite, and low contrast index profiles with curvature to allow for more appropriate surface shapes to be employed, provided that the tip in the surface remains. One other method for removing the singularities is called transmutation [38, 39]. However, these designs require highly anisotropic materials, so once again are very difficult to implement, in contrast to the technique used here. The limitation common to the two techniques is when the index profile reaches infinity at a rate equal to or faster than $\frac{1}{r^2}$, it is no longer possible to use either transformation method to redesign the lens. This is the condition for a black hole, where all rays incident on the profile are unable to escape the device, and spiral into the center [40].

An example is given in Fig. 11.13, which is of an Eaton lens. This design requires the modest index contrast of $\sim$1.65 (shown in Fig. 11.13c), and the surface smoothly transitions to a flat plane so that a surface wave incident upon it will not suffer any reflections. It can be seen in Fig. 11.13a that the beam incident at the bottom left of the lens is rotated around the tip and emitted in the same direction it was incident from, creating a retroreflecting effect.

11.5 Conclusions

This chapter has introduced the concept of transformation optics and explained how it can be used to design a range of different devices, particularly in the field of lenses for antenna engineering. In the mm and sub-mm ranges, these lenses are of specific interest because the volumes required are no longer prohibitive and the

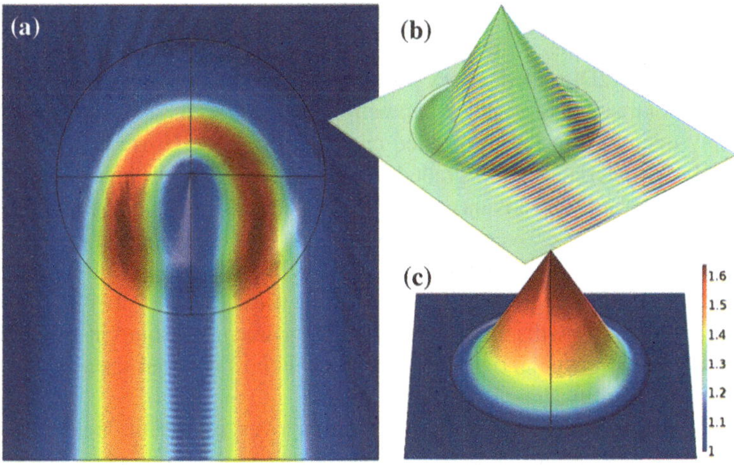

Fig. 11.13 The Eaton lens utilizing a hybrid shaped surface and a finite refractive index profile. *Left*—Electric field magnitude of a beam incident from *bottom left* being rotated and exiting *bottom right*. *Top right*—Instantaneous electric field viewed from the side. *Bottom right*—Refractive index profile. Extracted from [10]

technologies traditionally used at lower frequency need very high manufacturing tolerances, making them expensive to fabricate. We have described in detail three different variants of the method, namely the analytical technique, quasi-conformal transformation optics, and non-Euclidean transformations. Each has its own merits and limitations, all of which were outlined here. The most general formulation is the analytical technique; however, even simple devices designed using this technique require anisotropic materials. While it is possible to fabricate some degrees of anisotropy using metamaterials, more often than not approximations are required leading to single polarization devices and/or a decrease in the accuracy of the device. The quasi-conformal transformation method overcomes this problem, resulting in purely isotropic devices. It allows the redesign of common electromagnetic devices to make them planar in shape, while retaining the behavior of the original. It can also be employed to create bespoke lenses, designed to alter the radiation properties of specific antenna feed systems, for example, to improve the directivity, which is of particular importance for submm-wave antennas. For this method, care must be taken to choose appropriate coordinate system to ensure the approximations hold. The final technique allows non-Euclidean coordinate systems to be considered. This adds a degree of freedom that is not possible with the previous two techniques, however, and is only applicable for two-dimensional systems. For this reason, it is invaluable in the design of curved surface wave devices, and new types of surface wave cloak. It allows for the redesign of lenses that were previously impossible to fabricate, meaning that retro-reflectors and beam steerers can be created using either air-filled, or modest ranges of graded dielectric-filled curved waveguides.

There are still numerous challenges involved in the implementation of the aforementioned devices. Metasurfaces have proven to be instrumental in the development of 2D-graded index devices, in that they are able to achieve high-index contrast by varying the geometric parameters, can attain anisotropy, and are simply manufactured using printed circuit technology. However, the ability to fabricate bulk materials with a graded index that varies in all three dimensions is less well established. The advances in 3D printing technology bring much hope that this issue may soon be overcome. Research on printable low-loss and variable permittivity materials is well underway, which will provide a straightforward route to fabricating complex graded dielectric geometries. Coupled to the printing of conductive inks, this will also offer a less labor-intensive solution to the manufacture of graded metamaterial structures, with exotic material properties such as anisotropy or chirality.

In summary, the transformation optics tool is a very powerful one and has great potential to be utilized in the design of electromagnetic lenses, particularly in the mm-wave region where the size of the lenses is not prohibitive. It is hoped that the methods described here provide the reader with the essentials to apply these techniques to their own antenna problems, and allow advantage to be taken from the opportunities posed by the use of transformation optics.

References

1. U. Leonhardt, Optical conformal mapping. Science **312**(5781), 1777–1780 (2006)
2. J.B. Pendry, D. Schurig, D.R. Smith, Controlling electromagnetic fields. Science **312**, 1780 (2006)
3. W. Tang, Y. Hao, F. Medina, Broadband extraordinary transmission in a single sub-wavelength aperture. Opt. Express **18**(16), 16946 (2010)
4. D. Schurig, J.J. Mock, B.J. Justice, S.A. Cummer, J.B. Pendry, A.F. Starr, D.R. Smith, Metamaterial electromagnetic cloak at microwave frequencies. Science **314**, 977 (2006)
5. M. Rahm, D. Schurig, D.A. Roberts, S.A. Cummer, D.R. Smith, J.B. Pendry, Design of electromagnetic cloaks and concentrators using form-invariant coordinate transformations of Maxwell's equations. Photon. Nanostruct. Fundam. Appl. **6**, 87 (2008)
6. H. Chen, R.-X. Miao, M. Li, Transformation optics that mimics the system outside a Schwarzschild black hole. Opt. Express **18**(14), 15183 (2010)
7. M.F. Schumann, S. Wiesendanger, J.C. Goldschmidt, B. Bläsi, K. Bittkau, U.W. Paetzold, A. Sprafke, R.B. Wehrspohn, C. Rockstuhl, M. Wegener, Cloaked contact grids on solar cells by coordinate transformations: designs and prototypes. Optica **2**, 850–853 (2015)
8. D.A. Roberts, M. Rahm, J.B. Pendry, D.R. Smith, Transformation-optical design of sharp waveguide bends and corners. Appl. Phys. Lett. **93**, 251111 (2008)
9. P.-H. Tichit, S.N. Burokur, X. Wu, D. Germain, A. de Lustrac, Novel antenna concepts via coordinate transformation. Adv. Electromagn. **2**, 25 (2013)
10. S.A.R. Horsley, I.R. Hooper, R.C. Mitchell-Thomas, O. Quevedo-Teruel, Removing singular refractive indices with sculpted surfaces. Sci. Rep. **4**, 4876 (2014)
11. D. Kwon, D.H. Werner, Transformation electromagnetics: an overview of the theory and applications. IEEE Antennas Propag. Mag. **52**, 24 (2010)
12. D. Kwon, D.H. Werner, Polarization splitter and polarization rotator designs based on transformation optics. Opt. Express **16**(23), 18731 (2008)

13. D. Kwon, D.H. Werner, Transformation optical designs for wave collimators, flat lenses and right-angle bends. New J. Phys. **10**, 115023 (2008)
14. A. Demetriadou, Y. Hao, Slim luneburg lens for antenna applications. Opt. Express **19**(21), 19925 (2011)
15. D.A. Roberts, N. Kundtz, D.R. Smith, Optical lens compression via transformation optics. Opt. Express **17**(19), 16535 (2009)
16. E. Martini, S. Maci, Metasurface transformation theory, in *Transformation Electromagnetics and Metamaterials*, ed. by D.H. Werner, D.-H. Kwon (Springer, London, 2014), pp. 83–116
17. E. Martini, M. Mencagli, D. González-Ovejero, S. Maci, Flat Optics for Surface Waves. IEEE Trans. Antennas Propag. **64**, 155–166 (2016)
18. E. Martini, M. Mencagli, S. Maci, Metasurface transformation for surface wave control. Phil. Trans. R. Soc. A **373**, 20140355 (2015)
19. M. Mencagli, E. Martini, D. González-Ovejero, S. Maci, Metasurfing by transformation electromagnetics. IEEE Antenn. Wirel. Propag. Lett. **13**, 1767–1770 (2014)
20. A. Patel, A. Grbic, Transformation electromagnetics devices based on printed-circuit tensor impedance surfaces. IEEE Trans. Microw. Theory Tech. **62**, 1102–1111 (2014)
21. A. Patel, A. Grbic, The effects of spatial dispersion on power flow along a printed-circuit tensor impedance surface. IEEE Trans. Antennas Propag. **62**, 1464–1469 (2014)
22. R. Quarfoth, D. Sienvenpiper, Surface wave scattering reduction using beam shifters. IEEE Antenn. Wirel. Propag. Lett. **13**, 963–966 (2014)
23. J. Li, J.B. Pendry, Hiding under the carpet: a new strategy for cloaking. Phys. Rev. Lett. **101**, 203901 (2008)
24. W. Tang, C. Argyropoulos, E. Kallos, W. Song, Y. Hao, Discrete coordinate transformation for designing all-dielectric flat antennas. IEEE Trans. Antennas Propag. **58**(12), 3795–3804 (2010)
25. N. Kundtz, D.R. Smith, Extreme-angle broadband metamaterial lens. Nat. Mater. **9**(2), 129 (2010)
26. D.H. Kwon, Quasi-conformal transformation optics lenses for conformal arrays. IEEE Antennas Wirel. Propag. Lett. **11**, 1125–1128 (2012)
27. O. Quevedo-Teruel, W. Tang, R.C. Mitchell-Thomas, A. Dyke, H. Dyke, L. Zhang, S. Haq, Y. Hao, Transformation optics for antennas: why limit the bandwidth with metamaterials? Sci. Rep. **3**, 1903 (2013)
28. R. Yang, W. Tang, Y. Hao, Wideband beam-steerable flat reflectors via transformation optics. IEEE Antennas Wirel. Propag. Lett. **10**, 1290 (2011)
29. H.F. Ma, T.J. Cui, Three-dimensional broadband and broad-angle transformation-optics lens. Nat. Commun. **1**, 124 (2010)
30. Q. Wu, Z.H. Jiang, O. Quevedo-Teruel, J.P. Turpin, W. Tang, Y. Hao, D.H. Werner, Transformation optics inspired multibeam lens antennas for broadband directive radiation. IEEE Trans. Antennas Propag. **61**(12), 5910–5922 (2013)
31. M. Ebrahimpouri, O. Quevedo-Teruel, Bespoke Lenses Based on Quasi Conformal Transformation Optics Technique. IEEE Trans. Antennas Propag. **65**(5), 2256–2264 (2017)
32. U. Leonhardt, T.G. Philbin, *Geometry and Light: The Science of Invisibility* (Dover, New York, 2010)
33. M. Sarbort, T. Tyc, Spherical media and geodesic lenses in geometrical optics. J. Opt. **14**, 075705 (2012)
34. R.C. Mitchell-Thomas, T.M. McManus, O. Quevedo-Teruel, S.A.R. Horsley, Y. Hao, Perfect surface wave cloaks. Phys. Rev. Lett. **111**, 213901 (2013)
35. R.C. Mitchell-Thomas, O. Quevedo-Teruel, J.R. Sambles, A.P. Hibbins, Omnidirectional surface wave cloak using an isotropic homogeneous dielectric coating. Sci. Rep. **6**, 30984 (2016)
36. R.C. Mitchell-Thomas, O. Quevedo-Teruel, T.M. McManus, S.A.R. Horsley, Y. Hao, Lenses on curved surfaces. Opt. Lett. **39**, 3551–3554 (2014)
37. P.A. Huidobro, M.L. Nesterov, L. Martin-Moreno, F.J. Garcia-Vidal, Transformation optics for plasmonics. Nano Lett. **10**, 1985–1990 (2010)

38. Y.G. Ma, C.K. Ong, T. Tyc, U. Leonhardt, An omnidirectional retroreflector based on the transmutation of dielectric singularities. Nat. Mater. **8**(8), 639 (2009)
39. T. Tyc, U. Leonhardt, Transmutation of singularities in optical instruments. New J. Phys. **10**, 115038 (2008)
40. E.E. Narimanov, A.V. Kildishev, Optical black hole: broadband omnidirectional light absorber. Appl. Phys. Lett. **95**(4), 041106 (2009)

Part II
Antenna Measurements

Chapter 12
Antenna Measurements at Millimeter and Submillimeter Wavelengths

Antti V. Räisänen, Juha Ala-Laurinaho, Thomas Crowe, Sergiy Pivnenko, Manuel Sierra Castañer and Ville Viikari

Abstract Testing of electrically large, high-gain antennas as well as that of small integrated antennas at millimeter and submillimeter wavelengths is extremely challenging. Basically, there are three methods for measuring radiation properties of an antenna: the far-field method, the near-field method, and the compact antenna test range (CATR). In case of large antennas, the classical far-field method has two major obstacles at mm and submm wavelengths: impractically large measurement distance and high atmospheric loss. The planar near-field scanning method has been used up to 1 THz. However, the applied near-field methods often give useful information only on the main beam and its vicinity, because the field-sampling is typically very sparse. Reflector-based and hologram-based compact antenna test range (CATR) measurements have been demonstrated up to 500 GHz and 650 GHz, respectively. In the case of small integrated antennas, various techniques for on-wafer measurements have been developed. This chapter discusses the theory, techniques and limitations of the various test methods—the far-field method, planar near-field scanning and CATR as well as on-wafer measurements. Also, antenna pattern correction techniques are discussed.

A.V. Räisänen (✉) · J. Ala-Laurinaho · V. Viikari
School of Electrical Engineering, Department of Electronics and Nanoengineering, Aalto University, Espoo, Finland
e-mail: antti.raisanen@aalto.fi

T. Crowe
Virginia Diodes, Inc., Charlottesville, VA, USA

S. Pivnenko
Antenna Systems Solutions S.L., Santander, Spain

M.S. Castañer
Dpto. Señales, Sistemas Y Radiocomunicaciones, ETSI Telecomunicación, Universidad Politécnica de Madrid (UPM), Madrid, Spain

A. Boriskin and R. Sauleau (eds.), *Aperture Antennas for Millimeter and Sub-Millimeter Wave Applications*, Signals and Communication Technology,
DOI 10.1007/978-3-319-62773-1_12

12.1 Introduction

12.1.1 General Overview of Antenna Measurements at mm and submm Wavelengths

Testing of electrically large, high-gain satellite antennas as well as that of small integrated antennas at millimeter and submillimeter wavelengths is extremely challenging. Basically, there are three methods for measuring radiation properties of an antenna: the far-field method, the near-field method, and the compact antenna test range (CATR), e.g. [1]. The natural choice for antenna measurements is the far-field method. However, in the case of electrically large antennas, the classical far-field method has two major obstacles. The separation between the antenna under test (AUT) and the source antenna must be so large that the spherical wave front radiated by the source antenna can be approximated as a plane wave. Typically the required far-field distance is $d = 2D^2/\lambda$, where D is the diameter of the AUT and λ is the wavelength. For example, the required separation is 9.8 km for a 1.5-m antenna at 650 GHz. A larger separation may be needed when very low sidelobe levels are measured. At mm- and submm wavelengths, the large attenuation and the distortions to the signal due to the atmosphere are the major problems. Variations of temperature and humidity in time and space cause distortions to the amplitude and phase of the signal. However, the far-field method is readily applicable for mm-wave antennas with gain less than 20 dB, because in this case the far-field measurement can be done in a controlled environment such as an anechoic chamber.

Near-field range measurements can be done indoors in a relatively small space. The temperature and humidity can be controlled and thus the distortions due to the atmosphere become less severe. The amplitude and the phase of the AUT near-field are sampled with a probe antenna and the far-field radiation characteristics are calculated using numerical methods. The sampling of the near-field can be accomplished over a plane, a sphere, or a cylinder. The sampling area has to be comprehensive so that all the significant energy transmitted by the AUT can be captured. The sampling has to be dense enough for satisfying the Nyquist sampling criterion, i.e., the distance between two sampling points has to be smaller than one half of the wavelength, i.e., $<\lambda/2$. For example, 1.8 m × 1.8 m rectangular sampling grid, which covers a 1.5-m antenna, includes about 60 million sampling points, if full sampling is carried out at 650 GHz. The sampling rate has to be extremely high to keep the acquisition time reasonable. The requirements for the stability of the measurement system are very stringent. Sparser sampling grid can be used, if only the main beam region and the beam alignment are of interest, e.g. [2]. Major error sources in the near-field scanning are the phase errors in flexing and bending cables, inaccurate positioning of the probe, and the reflections between the probe and the AUT.

In a compact antenna test range (CATR), the spherical wave front is transformed into a plane wave with a collimating element, which can be a reflector, e.g. [3], a lens, e.g. [4, 5], or a hologram, e.g. [6–8]. Also, an array of antennas may be used to produce a plane wave at lower frequencies. CATRs enable the antenna measurements indoors in plane wave conditions, i.e., the radiation pattern can be directly measured by rotating the AUT. The main beam of a high-gain antenna can be measured without rotating the AUT by scanning the CATR feed transversally. The AUT is placed into the quiet-zone (QZ), which is the extent of the volume where the wave front meets the requirements set on the plane wave. Typically, the maximum amplitude and phase deviations have to be less than ±0.5 dB and ±5°, respectively. Most of the CATRs are based on reflectors. The major problem is the very stringent surface accuracy requirement at high frequencies. Usually the root mean square (rms) surface error should be less than 0.01 λ, e.g., 4.6 μm at 650 GHz. The surface accuracy requirements for lenses are not as stringent as for reflectors, since the lens is a transmission type of an element. In practice, the major problem is to find suitable material for lenses. If dielectric constant is small, like for foam, the material is usually not homogeneous. The problem is more severe at high frequencies.

12.1.2 Specific Aspects Concerning the Instrumentation at mm and submm Wavelengths

As is clear from the descriptions above, the high test frequency introduces challenges for the RF instrumentation in any antenna measurement. The minimum requirement is to have a frequency-stable transmitter with sufficient output power and a sensitive receiver at the given mm-wave or THz frequency. However, the frequency range from 0.1 to 10 THz is often called also as the terahertz gap, because technologies for generating and detecting this radiation are less developed than those at microwave frequencies or at infrared. Generation of THz waves with microwave means may take place by using vacuum tube generators such as backward wave oscillators (BWO) or by multiplying semiconductor oscillator frequency from microwave or low millimeter wave frequency by a Schottky diode multiplier, see e.g. [9]. On the other hand, generation of THz radiation with photonics means may take place using optically pumped gas lasers or photomixers pumped with two infrared semiconductor lasers [10]. However, power of the THz signal from a photomixer is typically only microwatts and therefore not really usable for antenna measurements.

Amplitude (i.e., scalar) measurements may suffice in some applications, and the phase may also be retrieved by using holographic means as discussed in Sect. 12.5 (Chapter 13 is fully devoted to holographic principles). In such measurements a sensitive power meter or detector may be used as a receiver. A spectrum analyzer is an excellent incoherent receiver (when equipped with external down-conversion units). In most antenna measurements, however, a vector, i.e., both amplitude and phase, measurement is desired. Today, there are commercially available excellent vector network analyzer s (VNA) with extension units up to 1 THz based on frequency multiplication and harmonic mixing.

12.2 Millimeter- and Submillimeter-Wave Instrumentation

12.2.1 Background and Conceptual Review of VNA Frequency Extenders

The testing of antenna systems is greatly facilitated by the availability of high quality test and measurement equipment, such as the vector network analyzers (VNA) from leading equipment manufacturers such as Keysight Technologies [11], Anritsu [12] and Rohde & Schwarz [13]. Present VNA models operate to frequencies as high as about 70 GHz and are exceptionally useful tools for general laboratory measurements; including measurements of passive devices, amplifiers, and frequency conversion devices (such as mixers), as well as antenna testing. However, measurements above ~70 GHz require the use of external frequency extender modules, and these devices are the primary subject of this section.

A block diagram for a generic VNA extender module is shown in Fig. 12.1. The module requires two inputs from the VNA: the RF signal and the local oscillator (LO) power required to pump the mixers. The VNA must supply these signals with the appropriate power levels and frequency range, while maintaining the necessary frequency offset so that the IF signals generated by the module's mixers are generated at a precisely controlled frequency. Also, the VNA signals should have low phase noise. This is particularly important in the case of very high frequency extension modules, where the high multiplication factor degrades the phase noise. At measurement frequencies above 500 GHz, the VNA's synthesizer phase noise can be a limiting factor in the quality of the measurements.

The extender module consists of the transmitter and the measurement and reference receivers. The transmitter uses a cascade of amplifiers and frequency multipliers to increase the signal frequency to the desired band. The exact configuration of this amplifier-multiplier—chain is selected by the manufacturer based on a variety of considerations; including their technological capabilities, performance goals and cost. For example, a multiplier with a high harmonic factor is simplest to achieve and has the lowest cost, but a cascade of multipliers with lower harmonic factor may lead to greater transmit power and a more pure test signal. Other features of the transmitter may include a power attenuator and an isolator. The attenuator is useful for tests that require less power delivered to the device under test (DUT), for example measurements of the gain and linearity of an amplifier. The isolator is almost always beneficial in that it improves the source match of the system.

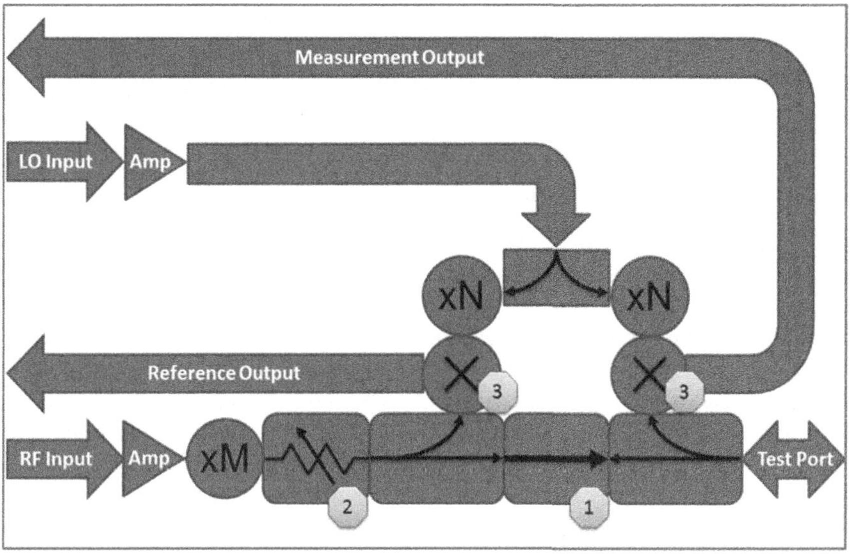

Fig. 12.1 A generic frequency extender module. The system consists of one signal transmitter, two receivers (signal and reference) and directional couplers. (*1*—Isolator, *2*—variable attenuator, *3*—frequency down-conversion mixers)

However, isolators are difficult to achieve at frequencies above ~220 GHz, and their insertion loss can be significant.

The two receiver systems are essentially identical in design, but serve different functions. The reference receiver samples the outgoing power. The intermediate frequency (IF) output from this mixer is coupled back to the VNA and is used to determine the magnitude and phase of the outgoing wave. The measurement receiver samples the power that enters the module from the test port. This can be power from the module's own transmitter that is reflected by the device under test, or transmitted power from a second module that is attached to a different port of the DUT. The IF output from the measurement receiver is coupled to the VNA and is used to determine the magnitude and phase of the incoming wave. A comparison of the measured IF to the reference IF allows the compensation of drift in the magnitude and phase of the RF test signal. As with the transmitter, the exact configuration of the receiver system is selected by the module manufacturer. The critical choices of when to split the LO signal and the design of the amplifiers, multipliers, and mixers have a very significant impact on the sensitivity, stability and manufacturing cost of the receivers.

The final major component of the module is the directional coupling system. The couplers send samples of the outgoing and incoming waves to the mixers. The simple model of the extender module assumes that each mixer receives only the intended signal, but leakage of other signals through the couplers complicates the signal analysis and degrades system performance. The design and manufacture of the couplers is thus a critical aspect of the development of the module, and each manufacturer carefully balances the trade-offs between different aspects of system performance, manufacturing complexity and cost.

12.2.2 Considerations for Frequency Extenders Well Above 110 GHz

The availability of electronic equipment for use in the upper part of the mm-wave band and into the submillimeter band has grown significantly in the past decade. This has been made possible by the development of improved simulation tools for the linear and nonlinear circuit elements, the advancement of microelectronic fabrication technologies and the development of advanced "terahertz" circuits with higher levels of integration. Together, these advancements have enabled the development of frequency multipliers and mixers that operate across full waveguide bands with excellent performance to frequencies above 1 THz. Most importantly, these components can now be reliably manufactured at an acceptable cost.

As an example, Fig. 12.2 shows a VDI integrated diode circuit that operates above 1 THz [14]. The entire chip is designed and simulated using modern CAD tools; including the diodes, fingers, pads, frequency filters, and waveguide probes. The wafer is then fabricated and singulated into individual circuits in the cleanroom facility. For a size reference, the thickness of the substrate is about five micrometers. This gives an indication of the extremely small feature sizes that are required for terahertz components and the quality of the fabrication technology. To complete the component, the split-block waveguide housing must be co-designed with the terahertz IC, and then precision manufactured. Finally, the IC is placed and secured within the housing and additional low-frequency components, including bond wires, are placed. Upon final assembly of the block, the component is RF tested and qualified units integrated into the frequency extender module.

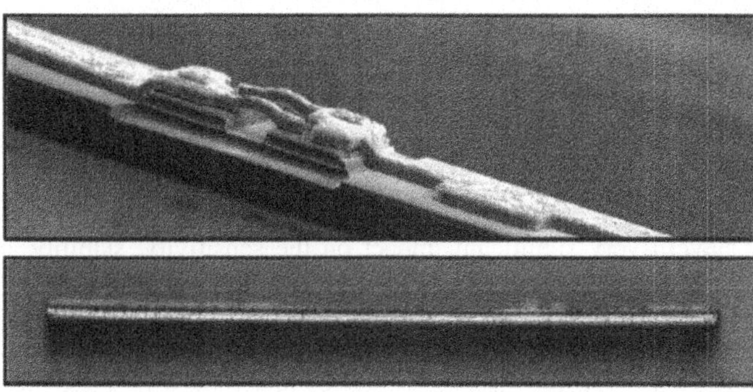

Fig. 12.2 Two SEM images of a THz integrated circuit focusing on the anti-parallel diode pair. The circuit substrate is ~5 μm thick. This gives an indication of the extremely small feature sizes that are required for terahertz components and the quality of the fabrication technology

A VNA extender module to 1 THz requires of about ten multiplier and mixer stages, each of which must be fully optimized for performance across the correct waveguide band using the exact power level that is available from the preceding component in the chain. Also, these components, and the transmitter and receiver systems as a whole, must be developed so that they are stable and do not generate significant power at unwanted frequencies. Thus, the design of very high frequency extender modules requires not only the development of individual components covering the frequency bands of interest, but also the development of a system architecture that yields exceptional performance across the full waveguide band without spurious signals or oscillations.

12.2.3 Extender Module Configuration, Considerations for Antenna Test

A complete VNA extender module is depicted in Fig. 12.1. Two of these modules are required for full S-parameter characterization of a two-port DUT. These modules connect to a VNA system that supplies the required RF and LO signals, and processes the four IF outputs from the modules. Additionally, a switch in the VNA system is used to alternately direct the RF test signal from one extender module to the other, since only one transmitter should be active at any time. This is commonly achieved with a four-port VNA or the use of some sort of additional signal controller. The LO distribution is somewhat simpler. Since all four mixers are intended to be active at all times, it is necessary only to split the power between the modules. However, it should be considered that any relative drift in the phase of the LO signal delivered to the two modules will create measurement drift that cannot easily be calibrated or otherwise compensated.

The chart in Fig. 12.3 shows various configurations of the VNA extender modules. For antenna measurements the full two-port solution is not required, and a simpler configuration will not only reduce costs, but can also improve performance. For the typical case of an antenna range only S_{21} is required and the "Through" configuration (TxRef-Rx) reduces the cost by nearly a factor of two. This is because

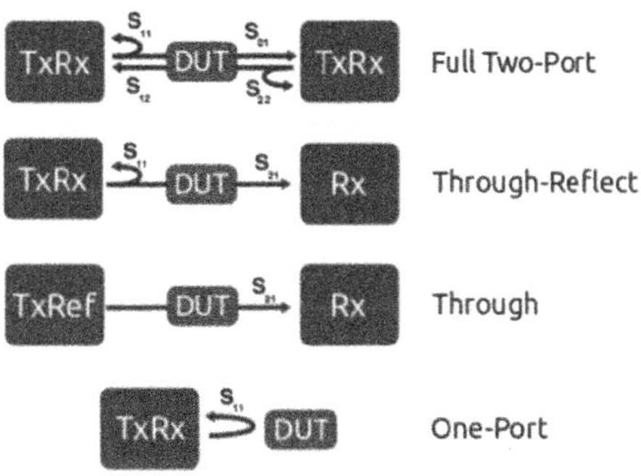

Fig. 12.3 Various configurations of VNA modules and the supported S-parameter measurements

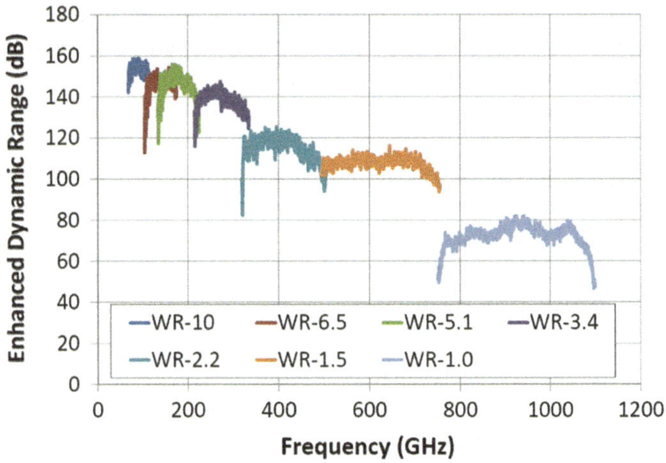

Fig. 12.4 The measured "enhanced dynamic range" of various VDI VNA extender modules for antenna test systems [14]

the TxRef includes only the transmitter and reference receiver, and the Rx consists of only a measurement receiver. Also, one of the two coupler systems is eliminated.

This simplification also allows performance improvements for certain applications. Considering first the Rx module, the elimination of the coupler and other components allows all of the power that passes through the DUT to be coupled to the mixer. This can increase the measurement receiver sensitivity, referred to the input of the module, by orders of magnitude. Also, the simplification of the transmitter module (TxRef) can allow a re-optimization of the internal components to achieve greater test port power. This improvement is not as great, but can still be significant.

Figure 12.4 shows a graph of the "enhanced dynamic range" of VDI extender modules that are optimized for antenna test systems in the 75–1100 GHz frequency range. The data shown represents the ratio of the test port power of the transmitter to the noise power of the receiver module referred to its input port in a 10 Hz measurement band. This would equal the system dynamic range provided the power does not saturate either the module's mixer or the low-frequency mixers in the VNA, which is unfortunately not the case for a low loss DUT. However, in the case of a lossy DUT, such as an antenna test range, the power actually reaching the Rx module may be low enough to prevent saturation. For example, in the WR-10 band the enhanced dynamic range is about 150 dB. However, practical considerations, including the saturation of the extenders' mixers and the dynamic range limits of the VNA itself, normally prevent the actual instantaneous dynamic range from exceeding about 125 dB. But, if this module configuration is used in an antenna range with 30 dB minimum range loss, then there will be no saturation effects and the remaining dynamic range for measurements, after the 30 dB loss is overcome, will be 120 dB. This represents about a 30 dB improvement over the use of two full

Fig. 12.5 A size comparison of the original modules (11″ x 5″ x 3″) with the new mini-modules (8.5″ x 3″ x 1.5″). The volume is reduced by 75% and the weight by more than 50%

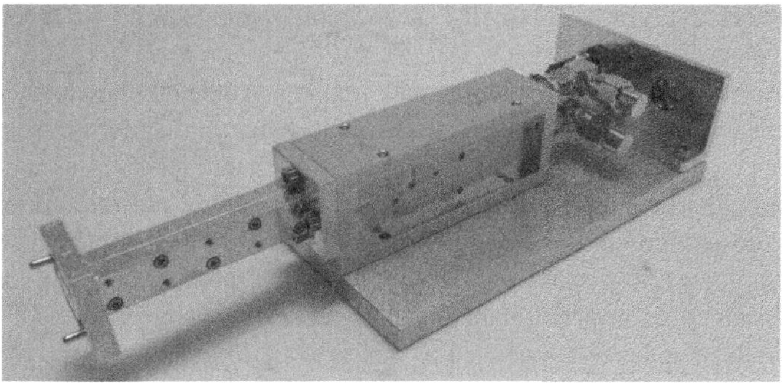

Fig. 12.6 A complete Rx module for the WR-10 waveguide band, including a two inch test port and a mounting plate; both are removable. The Rx module measures 2.5″ x 1″ x 0.75″. The LO and IF connectors are shown, but the DC power connector is not visible

TxRx modules and this increased dynamic range can be used to measure deep nulls in the antenna pattern and/or to greatly increase the measurement speed by increasing the resolution bandwidth.

A final benefit of this simplified configuration is the ability to greatly reduce the size and weight of the Rx module. Recently VDI introduced new "mini-modules" with reduced size (Fig. 12.5). This introduction has begun with the lower frequency modules, WR-15 through WR-6.5, and will soon extend to higher bands. This improvement is made possible through the development of new internal components and greater levels of system integration. However, for the Rx modules, it is possible to achieve an even smaller package. A WR-10 "micro" receiver module is shown in Fig. 12.6. This is a complete receiver system with LO input and IF output suitable for connection directly to the VNA. This small, lightweight package is ideal for use in antenna test systems, especially in the case of on-wafer pattern measurements where the available space is very limited. An additional benefit of this integrated module is the elimination of several waveguide flange connections, which are a primary source of unwanted signal leakage.

12.2.4 Future Perspectives

The advancement of VNA extender technology is made possible through the continued development of advanced terahertz components such as frequency mixers, multipliers, and amplifiers. Also, the development of VNAs that are better optimized for high frequency extenders can have a significant impact. In May of 2015 VDI demonstrated a TxRx-Rx extender system that yielded about 70 dB of dynamic range (10 Hz) in the 1.1–1.5 THz range [14]. However, this level of performance required the use of external, ultra-low-noise synthesizers. Given the demand for higher frequency measurements, it is expected that all major VNA manufacturers will strive to improve the phase noise performance of their top-level vector network analyzers. This will lead to improvements in all extender modules operating above 500 GHz.

Modern vector network analyzers and frequency extender modules allow full two-port measurements with excellent dynamic range and stability. The extender modules are now being optimized for specific applications, such as antenna test. This includes the development of more compact modules and the optimization of the system to overcome range losses while maintaining excellent dynamic range. These technologies now enable high quality antenna measurements up to 1.1 THz, and further advances in performance and frequency range are expected.

12.3 Near-Field Measurements

12.3.1 Planar Near-Field Scanning

Aperture antennas at mm-wave bands usually have very large electrical size, with the characteristic aperture dimensions ranging from several tens to several hundreds of wavelengths. For these kinds of antennas, due to their very large directivity and thus primary radiation in one direction orthogonal to the aperture, specifically planar near-field (PNF) technique has found wide application.

In the PNF technique, the field is measured by a small probe on a plane in front of the aperture antenna on a regular grid of points. The most often used measurement grid is a rectilinear grid, in which the probe is moved by two linear orthogonal slides along two coordinates usually called x and y (Fig. 12.7a). Other types of regular grids are also used, for example, plane-polar grid, in which the probe is moved along a linear slide, while the antenna is rotated along the vertical

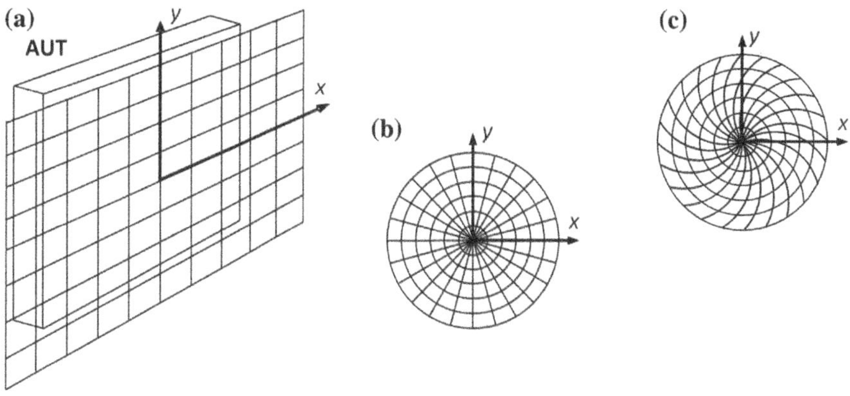

Fig. 12.7 Examples of the scanning grids in planar near-field technique: **a** rectilinear grid, **b** plane-polar grid, and **c** bipolar grid

axis (Fig. 12.7b), and bipolar grid, in which both the probe and the antenna are rotated (12.7c).

For the rectilinear grid, a rigorous solution exists for the expansion of the measured field over the plane waves, while for the other two grids, such a solution is not available and thus the measured field must first be interpolated into the rectilinear grid, which introduces interpolation errors. On the other hand, flexing cables following the probe moving on the rectilinear grid and mechanical uncertainties of the slides are among the largest error sources at very high frequencies, while the errors of the rotary joints used in the polar grids and mechanical uncertainties of the rotating stages introduce smaller effects and are more systematic thus allowing their more effective compensation.

12.3.2 Plane-Wave Expansion and Far Field

By means of the equivalence principle [15], the far-field of an antenna can uniquely be determined by means of suitable equivalent currents on an imaginary surface in front of the antenna; in the case of the PNF measurements, this surface is an imaginary flat aperture parallel to the antenna aperture, Fig. 12.7a. Thus, instead of considering the real antenna, we consider an aperture located for convenience in $z = 0$ plane of the coordinate system on which the tangential components of the electric field $\mathbf{E}_t = E_x\hat{x} + E_y\hat{y}$ are assumed known.

The principle of near-field antenna measurements is based on expansion of the fields measured on the chosen surface close to the antenna into a set of free space modes, which represent a solution of Maxwell's equations in the chosen coordinate system [16].

In a source-free and lossless medium, the electric field satisfies the following equations [15]

$$\nabla \times \nabla \times \mathbf{E} - \omega^2 \mu\varepsilon \mathbf{E} = 0; \tag{12.3.1}$$

$$\nabla \cdot \mathbf{E} = 0. \tag{12.3.2}$$

Using a vector identity $\nabla \times \nabla \times \mathbf{E} = \nabla\nabla \cdot \mathbf{E} - \nabla^2 \mathbf{E}$ and substituting $\omega^2 \mu\varepsilon = k_0^2$ the first equation becomes

$$\nabla^2 \mathbf{E} + k_0^2 \mathbf{E} = 0. \tag{12.3.3}$$

In Cartesian coordinates (x, y, z) these two equations are expressed as

$$\left(\frac{\partial^2}{\partial x^2} + \frac{\partial^2}{\partial y^2} + \frac{\partial^2}{\partial z^2} + k_0^2\right)\mathbf{E}(x, y, z) = 0; \tag{12.3.4}$$

$$\frac{\partial E_x(x, y, z)}{\partial x} + \frac{\partial E_y(x, y, z)}{\partial y} + \frac{\partial E_z(x, y, z)}{\partial z} = 0. \tag{12.3.5}$$

The Fourier transform of a two-dimensional function $u(x, y)$ (with exp $(j\omega t)$ time notation) is

$$U(k_x, k_y) = \int\int_{-\infty}^{\infty} u(x, y)\, e^{j(k_x x + k_y y)}\, dxdy \tag{12.3.6}$$

with the corresponding inverse Fourier transform

$$u(x, y) = \frac{1}{4\pi^2}\int\int_{-\infty}^{\infty} U(k_x, k_y) e^{-j(k_x x + k_y y)}\, dk_x dk_y. \tag{12.3.7}$$

Applying the Fourier transform to the Eqs. (12.3.4) and (12.3.5) with respect to x and y we obtain

$$\left(\partial^2/\partial z^2 + \left(k_0^2 - k_x^2 - k_y^2\right)\right)\mathbf{E}(k_x, k_y, z) = 0; \tag{12.3.8}$$

$$k_x E_x(k_x, k_y, z) + k_y E_y(k_x, k_y, z) + j\partial E_z(k_x, k_y, z)/\partial z = 0, \tag{12.3.9}$$

where $\mathbf{E}(k_x, k_y, z)$ is the Fourier transform of the electric field with respect to x and y.

Now we let $k_0^2 - k_x^2 - k_y^2 = k_z^2$ and note that (12.3.8) has only derivative with respect to z and thus the problem of solving a partial differential equation has been reduced to solving an ordinary differential equation. The general solution to this equation is

$$\mathbf{E}(k_x, k_y, z) = \mathbf{f}_1(k_x, k_y)e^{-jk_z z} + \mathbf{f}_2(k_x, k_y)e^{jk_z z}, \tag{12.3.10}$$

where $\mathbf{f}_1(k_x, k_y)$ and $\mathbf{f}_2(k_x, k_y)$ are unknown constants with respect to z. The two terms in (12.3.10) represent plane waves propagating in positive and negative directions, respectively, with respect to z. Since in our formulation the antenna under test (AUT) is located in the hemisphere $z < 0$ and we consider only waves propagating outwards from the AUT, in the positive z-direction, we shall keep only the first term $\mathbf{f}(k_x, k_y)e^{-jk_z z}$ omitting the subscript for $\mathbf{f}$.

Substituting the solution into (12.3.9) we obtain

$$k_x f_x + k_y f_y + k_z f_z = \mathbf{k} \cdot \mathbf{f} = 0, \tag{12.3.11}$$

which shows that only two components of the vector $\mathbf{f}$ are independent.

Using the inverse Fourier transform, the electric field can then be found

$$\begin{aligned} \mathbf{E}(x, y, z) &= \frac{1}{4\pi^2} \int\int_{-\infty}^{\infty} \mathbf{f}(k_x, k_y) e^{-jk_z z} e^{-j(k_x x + k_y y)}\, dk_x dk_y \\ &= \frac{1}{4\pi^2} \int\int_{-\infty}^{\infty} \mathbf{f}(k_x, k_y) e^{-j\mathbf{k}\cdot\mathbf{r}}\, dk_x dk_y. \end{aligned} \tag{12.3.12}$$

The unknown function $\mathbf{f}(k_x, k_y)$ is found by noting that for $z = 0$ the solution must be equal to the assumed known tangential electric field $\mathbf{E}_t$ in the aperture, thus

$$\mathbf{f}_t(k_x, k_y) = \int\int_{-\infty}^{\infty} \mathbf{E}_t(x, y, 0) e^{j(k_x x + k_y y)}\, dx dy. \tag{12.3.13}$$

Using (12.3.13) to find the tangential components $\mathbf{f}_t$ of the plane-wave spectrum and then evaluating (12.3.12), the electric field can be found everywhere in the hemisphere $z > 0$.

In the far-field of the AUT, the integral in (12.3.12) can be asymptotically evaluated by the method of stationary phase [15, 17]

$$\mathbf{E}(r) \sim \frac{jk_0}{2\pi} \cos\theta \frac{e^{-jk_0 r}}{r} \mathbf{f}(k_0 \sin\theta \cos\phi, k_0 \sin\theta \sin\phi), \tag{12.3.14}$$

where $k_x = k_0 \sin\theta\cos\phi$, $k_y = k_0 \sin\theta \sin\phi$ and θ, ϕ are the spherical coordinate angles. Equation (12.3.14) expressed in terms of the spherical components, with $\hat{\theta}$ and $\hat{\phi}$ being the spherical unit vectors, becomes

$$\mathbf{E}(r) \sim \frac{jk_0}{2\pi} \frac{e^{-jk_0 r}}{r} \left[(f_x \cos\phi + f_y \sin\phi)\hat{\theta} + \cos\theta (f_y \cos\phi - f_x \sin\phi)\hat{\phi} \right]. \tag{12.3.15}$$

12.3.3 Scattering Matrix Approach of Antenna-Antenna Interactions, Transmission Formula

The assumption regarding the known, somehow measured, tangential aperture field is not valid in practical cases, since the aperture field cannot be measured using an ideal probe—Hertzian dipole of zero physical extent. All practically used probes have finite dimensions, usually on the order of a wavelength, for example, a commonly used rectangular open-ended waveguide probe has the aperture size of some 0.375 λ by 0.75 λ at the center frequency. Due to these finite dimensions, the received signal at the probe terminals is not proportional to the field in a point, but represents a weighted value of the incident field over the probe aperture. Thus the receiving properties of the probe must be properly compensated in order to find the correct characteristics of the AUT. In order to get the relation between the received signal at the probe terminals depending on the probe location in front of the AUT, and the provided signal at the AUT terminals, we shall use the scattering matrix formulation for the AUT and probe.

An antenna can be represented by a multi-port network with one-feed port—the feeding line with a single propagating mode, for example, the TEM mode in a coaxial waveguide, and with an infinite number of the radiation ports representing the plane waves radiated, received, and scattered by the antenna. The relation between the ports can be expressed through the scattering matrix, which contains the reflection coefficient Γ_{AUT}, the receiving coefficients $\boldsymbol{R}_s$, the transmitting coefficients $\boldsymbol{T}_s$, and the scattering coefficients $\boldsymbol{S}_{s,\sigma}$, see Fig. 12.8.

It is noted that at a given frequency, the reflection coefficient is a complex number, while the other coefficients represent continuous complex functions reflecting the property of continuous spectra of the plane waves. The indices s and σ are used to count the two orthogonal components of the plane waves—in our formulation here, these components will be parallel to the axes x and y of the antenna coordinate system, see Fig. 12.9.

The complex amplitude of the outgoing wave b_0 at the antenna feed port can be written as

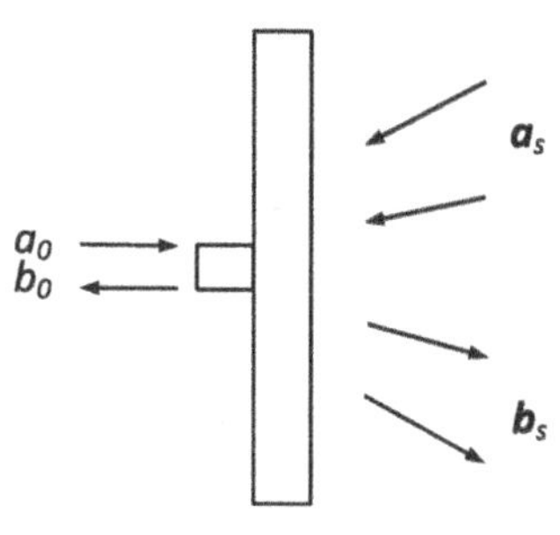

$$\begin{bmatrix} b_o \\ \boldsymbol{b}_s \end{bmatrix} = \begin{bmatrix} \Gamma_{AUT} & \boldsymbol{R}_s \\ \boldsymbol{T}_s & \boldsymbol{S}_{s,\sigma} \end{bmatrix} \begin{bmatrix} a_o \\ \boldsymbol{a}_s \end{bmatrix}$$

Fig. 12.8 Scattering matrix of an antenna

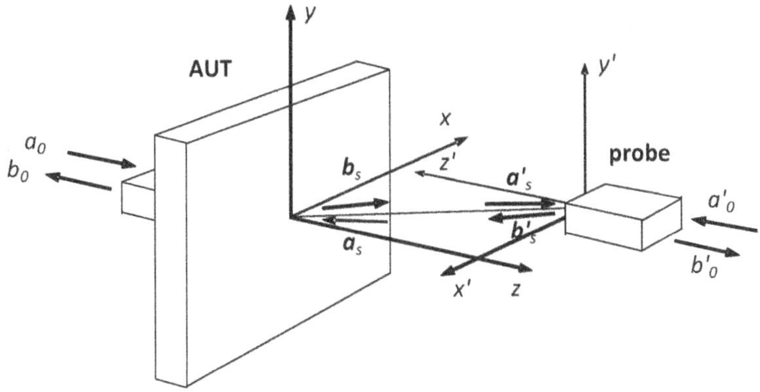

Fig. 12.9 Antenna and probe coordinate systems

$$b_0 = \Gamma_{AUT} a_0 + \frac{1}{4\pi^2} \int \int_{-\infty}^{\infty} \boldsymbol{R}_s(k_x, k_y) \boldsymbol{a}_s(k_x, k_y)\, dk_x dk_y \tag{12.3.16}$$

where a_0 is the complex amplitude of the incoming wave at the antenna feed port and $\boldsymbol{a}_s$ are the complex amplitudes of the incoming plane waves at the radiation ports. Similarly, the complex amplitudes of the outgoing waves $\boldsymbol{b}_s$ at the antenna radiation ports can be written as

$$\boldsymbol{b}_s(k_x, k_y) = \boldsymbol{T}_s(k_x, k_y) a_0 + \frac{1}{4\pi^2} \int \int_{-\infty}^{\infty} \sum_{\sigma=1}^{2} \boldsymbol{S}_{s,\sigma}(k_x, k_y; \kappa_x, \kappa_y) \boldsymbol{a}_\sigma(\kappa_x, \kappa_y)\, d\kappa_x d\kappa_y \tag{12.3.17}$$

In the absence of the incoming plane waves, which is a usual assumption in near-field antenna measurements, the second term in (12.3.17) is 0.

Similar formulation is applicable for the probe, in its own coordinate system with the z'-axis pointing outward from the probe aperture, and using the primed symbols, to distinguish from the AUT: the complex amplitude of the outgoing wave b'_0 at the probe port is written similar to (12.3.16). Assuming for simplicity a matched load at the probe port, the first term is equal to 0. For the coordinate system with the z-axis pointing inwards to the probe aperture, the following relations are valid, for rotation around the y'-axis: $\hat{x}' = -\hat{x}$, $\hat{y}' = \hat{y}$, $\hat{z}' = -\hat{z}$, and thus $k'_x = -k_x$, $k'_y = k_y$. We also need to use a relation between the probe receiving and transmitting coefficients, which according to the reciprocity relation can be expressed as $\boldsymbol{R}'_s(k'_x, k'_y) \propto \boldsymbol{T}'_s(-k'_x, -k'_y) = \boldsymbol{T}'_s(k_x, -k_y)$.

The probe is moved on a plane in front of the AUT at a fixed distance z_0, but with arbitrary x' and y', and such that the probe z'-axis is parallel to the AUT z-axis

(Fig. 12.9). The relation between the incoming waves in the probe coordinate system and the outgoing waves in the AUT coordinate system is written as

$$\boldsymbol{a}_s'\left(k_x', k_y'\right) = \boldsymbol{b}_s\left(k_x, k_y\right)e^{-j\mathbf{k}\mathbf{r}_0} = \boldsymbol{T}_s\left(k_x, k_y\right)a_0 e^{-j\mathbf{k}\mathbf{r}_0}, \tag{12.3.18}$$

where $e^{-j\mathbf{k}\mathbf{r}_0} = e^{-j\left(k_x x + k_y y + k_z z_0\right)}$.

Finally, the relation between the signal received at the probe output and the signal at the AUT input depending on the probe position in the AUT coordinate system is written as

$$b_0'(\mathbf{r}_0) = \frac{a_0}{4\pi^2}\int\int_{-\infty}^{\infty} \boldsymbol{T}_s'\left(k_x, -k_y\right)\boldsymbol{T}_s\left(k_x, k_y\right)e^{-jk_z z_0}e^{-j\left(k_x x + k_y y\right)}\, dk_x dk_y. \tag{12.3.19}$$

This is recognized as a two-dimensional inverse Fourier transform over the product of the transmission coefficients of the AUT and the probe in their own coordinate systems. By applying the Fourier transform, the product of the AUT and probe coefficients is found as

$$\boldsymbol{T}_s'\left(k_x, -k_y\right)\boldsymbol{T}_s\left(k_x, k_y\right) = B\left(k_x, k_y\right) = \frac{e^{jk_z z_0}}{a_0}\int\int_{-\infty}^{\infty} b_0'(\mathbf{r}_0)e^{j\left(k_x x + k_y y\right)}\, dxdy \tag{12.3.20}$$

This represents one equation with the two unknowns $\boldsymbol{T}_s\left(k_x, k_y\right) = T_x\left(k_x, k_y\right)\hat{\mathbf{x}} + T_y\left(k_x, k_y\right)\hat{\mathbf{y}}$. Another equation is obtained by performing a second measurement with the probe rotated orthogonally to the first orientation or using a dual-polarized probe and the signals from the two orthogonal ports. The two components of the AUT transmission coefficients are then resolved from the two equations as follows, written in the matrix form:

$$\begin{bmatrix} T_x \\ T_y \end{bmatrix} = \frac{1}{T_{xy}'}\begin{bmatrix} T_{y,2}' & -T_{y,1}' \\ -T_{x,2}' & T_{x,1}' \end{bmatrix}\begin{bmatrix} B_1 \\ B_2 \end{bmatrix}. \tag{12.3.21}$$

The indices 1 and 2 indicate the two probe orientations and $T_{xy}' = T_{x,1}'T_{y,2}' - T_{x,2}'T_{y,1}'$. It is noted that in this derivation, for the probe orientation 1, with its port aligned along the x-axis, $T_{x,1}' \gg T_{y,1}'$, while for the probe orientation 2, with its port aligned along the y-axis, $T_{x,2}' \ll T_{y,2}'$.

12.3.4 Discretization and Sampling Intervals

In practice, the probe signal is not measured continuously, but in a series of discrete points both in x and in y directions with the intervals Δx and Δy. The integral in (12.3.20) is then replaced by a double summation

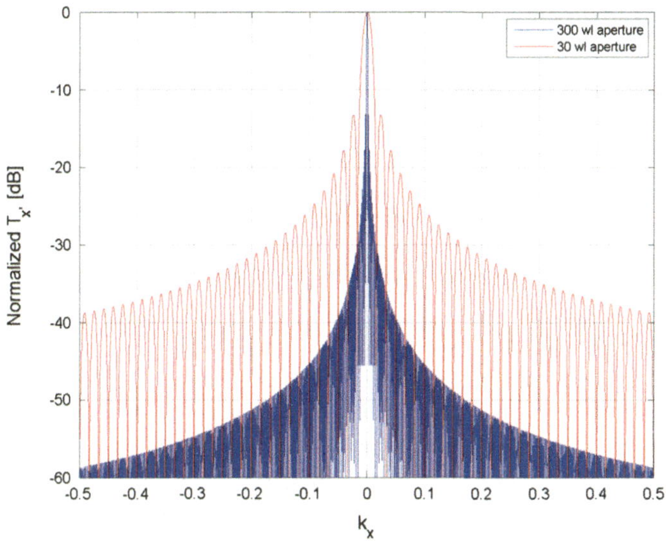

Fig. 12.10 Normalized plane-wave spectra of uniform aperture antennas

$$\int\int_{-\infty}^{\infty} b_0'(\mathbf{r}_0)e^{j\left(k_x x + k_y y\right)}\,dxdy = \Delta x \Delta y \sum\nolimits_{n,m} b_0'(x_n, y_m, z_0)e^{j\left(k_x x_n + k_y y_m\right)}. \quad (12.3.22)$$

The values of the sampling intervals are found recalling that the spectrum of the plane waves becomes evanescent for $k_x, k_y > k_0 = 2\pi/\lambda$. It is then found that with the sampling intervals $\Delta x = \Delta y \leq \lambda/2$ the whole propagating part of the plane-wave spectrum is obtained.

In some cases, especially at the very high frequencies with the AUT aperture being electrically very large, for example, hundreds of wavelengths, the plane-wave spectrum falls down very rapidly, reaching negligibly small values already at parts of k_0. This is illustrated in Fig. 12.10, where the normalized simulated plane-wave spectra of two uniform aperture antennas are shown. It is seen that for the 300 λ antenna the spectrum drops below -50 dB around $k_x = 0.17$ and thus, provided that errors at the level of -50 dB are acceptable, the sampling intervals can be increased 6 times from $\lambda/2$ to 3λ thus significantly decreasing the number of the measured points and the total measurement time. On the other hand, using the sampling intervals larger than $\lambda/2$ will only provide the spectrum in the proportionally smaller part of the k-domain and thus smaller angular sector, according to the relation $\Delta x = \Delta y = \lambda/(2\sin(\theta_{max}))$, and it also results in aliasing of the calculated spectrum thus increasing the uncertainty of the calculated far-field [18].

12.3.5 Measurement Errors and Uncertainties

One inherent feature of the PNF technique is the truncation of the measurement plane at some finite dimensions, either due to the finite size of the scanner or driven by the condition of minimizing the measurement time. Similar to the effects of increasing the sampling interval, this truncation limits the obtained angular sector and introduces truncation errors into the calculated field. The valid angle for the calculated field is found from simple geometrical relations of the measurement setup as follows $\theta_v = \tan^{-1}((L - D)/2z_0)$, where L is the size of the measurement plane in a given direction, for example, along the x-axis, D is the AUT aperture size in the same direction, and z_0 is the distance between the antenna aperture and the measurement plane—the measurement distance.

Among other errors and uncertainties, the following must be noted due to their increasing importance at very high frequencies. Since the probe is moved along two linear paths, the RF cable connected to the probe changes the shape and thus its electrical characteristics are slightly varied. This results in induced variations on the received signal and uncertainties of the calculated far-field.

To mitigate the phase errors caused by bending cables, a pilot signal based phase correction method has been developed [19]. A pilot signal, slightly at different frequency than the measurement signal, is injected in the bending measurement cable. A filter is implemented at the end of the cable inducing a reflection at the pilot-signal frequency while letting the measurement signal pass through. The phase of the pilot signal transmitted twice through the bending cable is observed and any changes in the phase indicate also changes at the measurement frequency. A correction can be applied taking into account the frequency difference. The correction method is applied at 310 GHz and at 650 GHz for quiet-zone field verification in hologram compact antenna test ranges [7, 8]. Harmonic mixers were used as receivers and microwave frequency LO signals were transmitted through flexing cables. Phase changes at microwave frequencies were translated to the phase corrections at submillimeter wave test frequencies.

Another important uncertainty factor is nonideal planarity of the measurement plane. Variations in the probe z position not exceeding $\lambda/100$ rms are usually required; this results in equivalent phase variations of $\pm 3.6°$, which is considered acceptable. However, for example, at 650 GHz this converts to the planarity requirement of ± 4.6 µm, which is not that easy to achieve. In order to ensure accurate planar movement of the probe, granite reference plates are used for measurements at 550 GHz [20]. Tracking laser interferometers can be used to measure the planarity of the near-field scanner. The measured planarity data can be used either for phase correction of the near-field data or for active correction of the probe movement during the data acquisition using an additional translation stage in z-direction.

The difficulties in accurate phase measurements have led to the development of phaseless, amplitude only measurements where the phase is retrieved with numerical methods [21–23]. However, these methods require measurements at two different distances or use of reference antennas.

Other uncertainties of the PNF technique, listed here in no particular order, are receiver nonlinearity, thermal noise, drift, leakage and cross-talk, multiple reflections between the antenna and probe and reflections from the surroundings, probe x- and y-position uncertainties, etc. The detailed investigations of the PNF uncertainties can be found in [24, 25].

12.4 Compact Antenna Test Range (CATR)

12.4.1 CATR: Applications and Specifications

CATRs have been widely used for antenna measurements since 1968, when the first range was designed at Georgia Tech in the United States. Basically, a compact antenna test range [26] is a hardware setup that transforms the spherical wave generated by a feeder in a plane wave, to simulate the far-field conditions. Although there are different setups, the most common are based on reflectors. This chapter examines the application of reflectors, lens, and holograms.

CATRs are used for two main applications: antenna measurements and radar cross-section (RCS) measurements. In the first case, the feeder generates a continuous wave, and after being reflected or modulated by the intermediate hardware, a plane wave impinges in the volume where the antenna under test (AUT) is placed. This volume is called quiet-zone. Therefore, the measurements can be performed as in far-field systems [1]. Figure 12.11a shows the VAST12 antenna measured in the facilities of Universidad Politécnica de Madrid. In the case of RCS measurement, the signal (pulsed or continuous wave) impinges in the target, and it is reflected. The same feeder (in a monostatic configuration), or another horn (in a quasi-monostatic configuration) receives the field, and this field is compared to the one reflected by the reference target (sphere, disk, or rectangular plate). Figure 12.11b shows the measurement of a target in the facilities of INTA, Spanish National Research Centre for Aerospace techniques.

The maximum dimension of AUT and frequency band depends on the Quiet-Zone, the volume, where the AUT is located, in which the illuminating field amplitude and phase differ from those of a plane wave by less than a pre-established amount. Obviously, the field in the quiet-zone is not perfectly a plane wave; but a taper and a ripple disturb this field. The taper is the amplitude variation in the quiet-zone border induced by the feed pattern and path lengths to the reflector surface. Typical reflector edge illumination is −1.5 to −2 dB and typical taper specifications are from −0.5 to −1 dB. The ripple includes the field variations produced by edge reflector diffractions and reflections from the room walls: typical

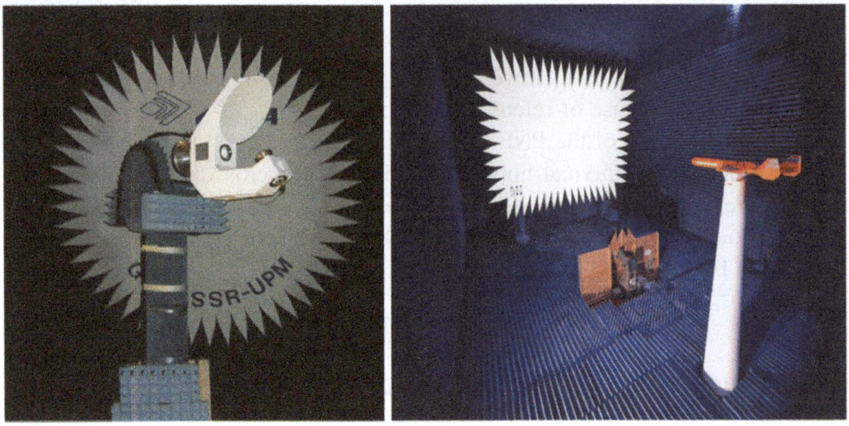

Fig. 12.11 Compact antenna test ranges: **a** LEHA-UPM (Universidad Politécnica de Madrid) for antenna measurements and **b** INTA (Spain) for radar cross-section measurements

Fig. 12.12 Sources of the ripple and taper in a CATR

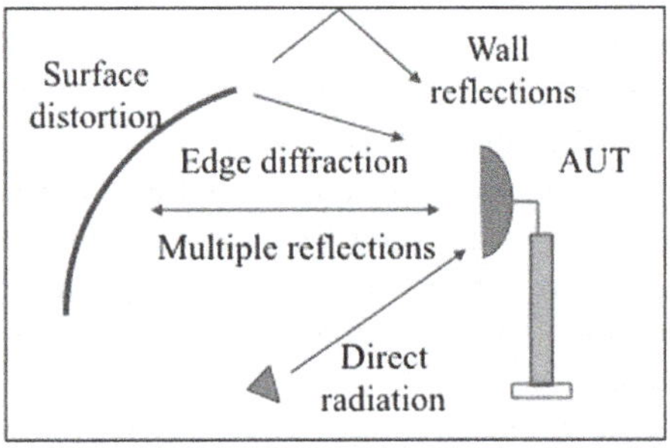

specifications for the ripple are ±0.5 dB in amplitude and ±5° in phase. This taper and this ripple create some errors and uncertainties in the measurements and are different for each AUT (depending on the size) and for each frequency bands. The sources of ripple and taper are shown in Fig. 12.12, and can be summarized in:

- Surface inaccuracy, important for higher frequencies as millimeter waves.
- Edge diffraction, important for lower frequencies.
- Direct feed radiation to quiet-zone.
- Reflections from chamber walls.
- Multiple reflections between AUT and CATR feed.

All these sources of errors create spurious waves those are added with the desired wave (the one reflected in the parabolic), and disturb the plane wave. In general, the minimum frequency is limited by the diffraction associated with the electrical size of the reflectors. For a 30-λ reflector (serrations included) the ripple is of 2 dB peak to peak; therefore a 10 m reflector is required for measuring at 1 GHz. The maximum frequency is limited by the phase errors associated with the accuracy of the reflectors surfaces. It is required a λ/100 accuracy at the maximum desired frequency.

The last important factor to be considered is the cross-polar level. This level depends on the system used and on the feed cross-polar quality. Single offset reflectors intrinsically introduce cross-polarization in the antisymmetric plane and are not adequate for antenna measurements, although the cross-polar level can be reduced using reflectors with a large focal distance to diameter ratio (F/D). Also, dual reflector systems can be designed, using Mizuguchi–Dragone condition [27, 28] to cancel this cross-polarization.

12.4.2 Absorbers and Anechoic Chambers

CATRs are located in anechoic chambers. These chambers are closed areas, shielded, and covered by electromagnetic absorbing material, which simulates free space propagation conditions, due to the absorption of the radiation absorbing material (RAM) [29]. These anechoic chambers are not only used for antenna or RCS measurements, but also for anechoic chambers are used for measurements of radiated emissions, radiated susceptibility or electromagnetic compatibility in general. The main advantages of using anechoic chambers in antenna measurements are the all weather conditions operation, the controlled environment (for instance in temperature or cleanness), the security aspects, and the absence of interferences. The main drawback is the large maximum dimension leading to a large, expensive special space.

Anechoic chambers are covered by electromagnetic absorbers. These absorbers are fabricated with a material that simulates free space conditions. This material is fabricated from polystyrene with a dielectric constant close to one in order to avoid reflections, filled with carbon graphite material to absorb the impinging signal. Usual absorbing materials have pyramidal shape in order to avoid the reflections and they should be long enough in terms of wavelengths. Therefore, for low frequencies these pieces of absorbers can be very long and heavy. The signal should impinge close to normal incidence to maintain good absorption properties; therefore the chambers cannot be very long and narrow. Table 12.1 presents the specifications of one of the providers of absorbing materials (AEMI) [30]. In this case they are the classical pyramidal ones. It can be observed the reflectivity in terms of the frequency and the height of the material: it can be observed that around 9 wavelengths are required for getting 50 dB of reflectivity.

There is also another kind of absorbing materials: in the walls of the CATR, wedge materials are placed: these materials guide the signal to the back wall of the chamber, and there, the pyramidal ones absorb the signal. For millimeter wavelengths, convoluted materials are used instead of pyramidal ones. For covering

Table 12.1 Specifications of pyramidal absorbers (AEMI-MVG) at RF and microwave frequencies

		AEP-4	AEP-8	AEP-18	AEP-36	AEP-60	AEP-96
Geometry	Normal (N)/ Twisted (T)	N	N	N	N/T	N	T
Height	cm	10.2	20.3	45.7	91.4	152.4	243.8
Pyramids/block		144	64	16	4	1	1
Weight	kg/pc.	1.1	2	4.1	8.4	13.6	27.2
Power	W/m^2	775	775	775	775	775	775
Absorption (dB) at normal incidence at	30 MHz				5	8	15
	125 MHz				15	30	35
	500 MHz			30	37	40	45
	1.0 GHz		30	40	42	47	50
	3.0 GHz	30	37	45	50	50	50
	10.0 GHz	40	50	50	50	50	50
	18.0 GHz	50	50	50	50	50	50

Table 12.2 Maxima of monostatic reflectivities of absorbers at 310 GHz [31]

Absorber	Reflectivity, max (dB) H-pol	Reflectivity, max (dB) V-pol	Transmission (dB) V-pol
FIRAM-500			−24
Wedges horizontal	−30	−26	
Wedges vertical	−31	−30	
TERASORB-500			−33
Wedges horizontal	−29	−41	
Wedges vertical	−41	−30	
TK THz RAM	−51	−56	−50
Eccosorb VFX-NRL-2	−50	−55	−50
Carpet #1	−40	−40	−17
Carpet #2	−40	−40	−25
Carpet #3	−21	−20	−5

masts and towers, laminated materials are used, while for the paths, there are special pieces as walkway ones. During the last years, rubberized materials have been fabricated to improve the cleanness and durability of the material. Also, there are some providers that offer weatherproof absorbers for outside measurements. Table 12.2 shows measurement results of some commercial absorbers at 310 GHz.

12.4.3 Reflector Compact Antenna Test Ranges

Among the different kinds of compact antenna test ranges, reflectors based are the most common. There are different designs, but in a first division, there are single or dual reflectors. The single offset reflectors, also known as virtual vertex systems, are the simplest ones. They use a large focal distance to diameter (F/D), 0.8–1.2, and low offset angle, in order to improve the cross-polar performance and to equalize the amplitude tapering. In this case, the focus of the parabolic reflector coincides with the phase center of the feeder. The main advantages are that they are much less expensive than dual reflector systems and easier to design and align. To reduce the scattering produced by the edge of the reflector, two main techniques are used: serrated and rolled edge reflectors. Figure 12.13 shows both designs (from ORBIT/FR [30]). The serrations control the direction of the diffracted signals, sending them in the direction of the peak. The rolled edge controls the direction of the reflected signal on the edge of the reflector. In both cases, these designs make a better quiet-zone. In a first approach the diameter of the quiet-zone is 35% of the diameter of the reflector. The quiet-zone is defined by the taper (typically 1 dB in amplitude and ±2° in phase), ripple (typically ±0.5 dB in amplitude and ±5° in phase), and cross-polar level (typically −30 dB for single offset reflectors). The minimum frequency depends on the dimension of the reflector, and typically the diameter is around 30 wavelengths.

If the cross-polar level has to be improved, dual reflectors should be designed. The easiest system is the dual cylindrical reflector configuration, or Vokurka system [32]. It uses two cylinder parabolic reflectors with single curvature: the subreflector with the horizontal axis and the main reflector with the vertical axis. This system collimates in elevation and azimuth the spherical wave from a point source, with a moderate reflector cost with high accuracy (single curvature reflectors). This system

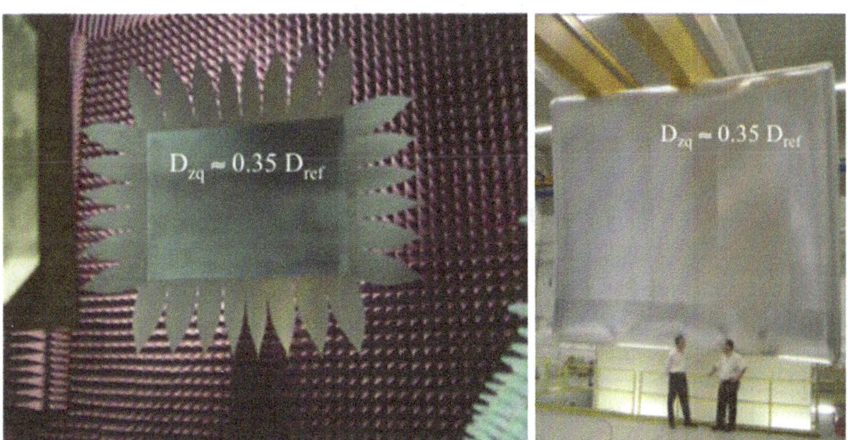

Fig. 12.13 Serrated and rolled edge reflectors (ORBIT/FR systems)

Fig. 12.14 ESA/ESTEC 100000 cleanroom compensated compact range in the HERTZ facility

gets good linear cross-polarization performance in V and H planes (although bad performance in diagonal planes ±45°). If a better cross-polar is desired, compensated compact ranges are designed, using Cassegrain or Gregorian configurations. Cassegrain configuration is composed by a hyperboloid, as subreflector, and a paraboloid as main reflector. Figure 12.14 shows the ESA/ESTEC [33] Compensated Compact Range using a Cassegrain configuration. This CATR shares the room with a near-field antenna measurement range, both planar and spherical. Another design for improving the cross-polar is the Gregorian configuration, that is, an ellipsoid and a paraboloid. As in the case of the Cassegrain configuration, one of the foci of the ellipsoid is located in the phase center of the feeder, and the other coincides with the focus of the parabolic reflector.

12.4.4 Lens and Hologram CATR

As mentioned in earlier section, reflectors are most commonly used as the CATR collimating elements. However, the stringent surface accuracy requirements, and therefore, difficulties in accurate manufacturing and high costs have lead a way to the research of the use of other collimating elements such as lenses or holograms for millimeter and submillimeter wavelengths. Figure 12.15 shows schematics of CATRs based on a dielectric lens and a transmission type of hologram.

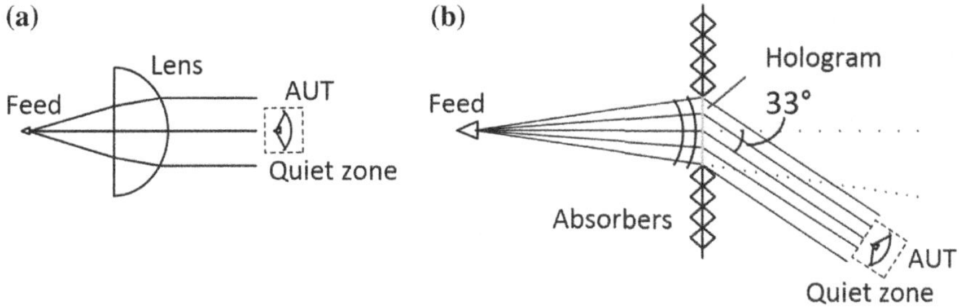

Fig. 12.15 Schematic of CATRs based on **a** a lens and **b** a transmission type hologram

Lens CATR

The first lens-based CATR was presented in the end of the 1970s [34]. Research on mm-wave lens CATR have been reported, e.g., in [5] where a CATR based on a Teflon lens operating at 110 GHz is presented and in [4] where measurements with a 50 cm lens to facilitate the designing of a 30 cm reflector antenna at 94 GHz are presented. The lens CATR could include several lenses, e.g., the first one shapes the illumination of the second collimating lens. The achievable quiet-zone diameter is about 70…75% of the lens diameter [5, 35].

The surface accuracy requirement for the CATR lenses is not as stringent as for reflectors because of the transmission type of operation. The accuracy requirement set for a single CATR reflector is multiplied by a factor of $\sqrt{2}/\left(\sqrt{\varepsilon_r}-1\right)$ [5] for similar performance. This factor takes into account the two surfaces of the lens. The factor encourages the use of low-permittivity lens. However, low-permittivity materials are often foam-like materials the homogeneity of which is not easy to control. Also, the lens thickness increases with the decreasing permittivity of the lens material. For example, when $\varepsilon_r = 1.1$ and the *F*/*D* ratio is 3 the thickness of the lens is of the same order than the diameter [35]. Also, an integrated lens antenna could be used, i.e., range feed is in direct contact with the lens. However, also then the lens becomes large compared to achievable quiet-zone size. The dielectric constant of the lens affects also the reflection from the lens surface: the larger ε_r is the more of the power radiated by the feed is reflected from the lens surface. Matching layers can be used to reduce the reflections.

The edge diffraction has less effect on the performance of the lens CATR than on that of the reflector CATR [36]. However, for minimizing the edge diffraction, serrated metal plates may be used with the lens [5] or the surfaces of the lens can be shaped, analogous to rolled edges in reflectors, in order to achieve an appropriate aperture distribution.

Hologram CATR

The use of computer-generated holograms as the collimating element of a CATR was introduced in the beginning of the 1990s [37]. The hologram-based CATR is a potential antenna measurement method especially for millimeter and submillimeter wavelengths. Figure 12.15b shows the principle of the CATR based on a hologram in a transmission type of operation. The planar hologram structure is an interference pattern of the CATR feed antenna and the desired planar wave front. The hologram modulates the spherical wave front radiated by a feed horn in such a way that a planar wave is formed. The transmission-type amplitude hologram is a binarized interference pattern on a thin metal sheet supported by a plastic substrate. The hologram either lets the field go through (slot) or blocks the field (metal). The slots are tapered towards the edges in order to reduce edge diffraction. The hologram is often designed in such a way that the plane wave leaves the hologram at an angle of 33° with respect to the hologram normal. Thus, the diffraction modes, which are propagating to the hologram normal direction, do not disturb the quiet-zone field [38].

The hologram CATR may also be based on a reflection-type amplitude hologram, or reflection- or transmission-type of phase holograms [39, 40]. The phase holograms are manufactured by milling a groove structure to a dielectric (transmission type) or metal (reflection type) plate. In transmission type of operation the surface accuracy requirements are more relaxed compared to the reflection type of operation. In particular, for the phase holograms, the difference is similar to that between the reflectors and lenses discussed in the previous section.

So far, transmission type amplitude holograms are used for antenna measurements [8, 41]. The hologram is assembled into a frame, which ensures sufficient planarity for the hologram. The accuracy requirement for the hologram pattern is about 0.01–0.02 λ, which is of the same order as the surface accuracy for the reflector. However, the manufacturing of the two-dimensional hologram is easier than the manufacturing of the three-dimensional reflector surface, thus the manufacturing costs are reduced substantially. Manufacturing techniques allow production of large holograms up to 3-m diameter with accurate enough pattern resolution and accuracy up to THz operation. Figure 12.16 shows an artistic view of a hologram CATR used for 1.5-m reflector antenna tests at 650 GHz [8].

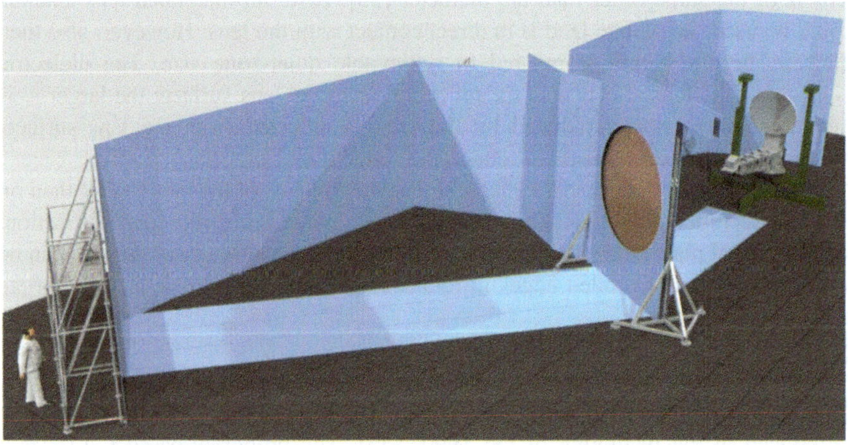

Fig. 12.16 Artistic view of hologram CATR used for 1.5-m test antenna measurements [8]. Some absorbing material is removed for clarity

The relation between the hologram and the quiet-zone sizes depends on the required size of the quiet-zone and frequency in a similar way as in the case of reflector CATRs. The proportion of the tapering area decreases when the size of the hologram increases at the given frequency. Therefore the ratio of the quiet-zone and hologram sizes is larger for larger holograms. For example, at 119 GHz for the 550 mm hologram the quiet-zone diameter is about 43% of the hologram diameter while for the 2.4 m × 2.0 m hologram the quiet-zone diameter is about 70% of the smaller dimension [6, 42].

A CATR based on a reflector or a lens operates equally well with all polarisations and frequencies (the edge diffraction and the surface accuracy give limitations). However, the operation of a hologram depends fairly strongly on the frequency and polarization, because it is a diffractive element. In principle, a hologram can have an octave bandwidth [43] if the feed location is moved while changing the operation frequency. However, the substrate thickness of mechanically reasonable substrate may reduce the bandwidth of the hologram CATR.

12.4.5 Antenna Measurements in CATR

Antenna measurements in a CATR are carried in a similar way than in a conventional far-field antenna test range. The AUT is placed in the quiet-zone region where the planar wave front is created. Antenna is rotated with an antenna positioner and the radiation pattern is recorded.

Feed scanning, i.e., movement of the CATR feed can be used for radiation pattern measurements in small angular region. The plane wave steers with respect to the feed offset from the focal point of the collimating element. The applicable angular region depends on the quiet-zone field requirements. See also Sect. 12.4.7 for feed scanning antenna pattern comparison [44].

For the antenna gain measurements, a gain comparison method is well applicable. The measured amplitude level of the AUT is compared to the amplitude level of a known reference antenna. In principle, the gain of the reference should be of the same order as the AUT gain in order to obtain accurate results. It is recommended, that the quiet-zone of the CATR is verified by probing the field with a small probe antenna, e.g., an open-ended waveguide or a small horn antenna. If the probe gain is known, the gain of the AUT can be estimated by comparison against the probe antenna gain. For electrically large antennas the gain difference may be large, e.g., 50–60 dB which leads to stringent linearity requirement of the measurement system or use of accurate attenuators.

12.4.6 RCS Measurement in CATR

CATR can be also used for radar cross-section measurement. In this case, the signal is reflected by the target and received by one probe. The received signal is compared with the one received using a reference target. For this kind of measurements, it is important to separate the effect of the target from the effect of the surrounding elements. In this case, the effect of the pylon is really important. Figure 12.11b shows a knife pylon, designed to reduce the reflection. Other pylon designs are cylindrical polystyrene ones: in this case the dielectric constant is close to 1, to reduce the reflected wave. However, other techniques are used to reduce the effect of the reflections in the surroundings of the target. One of them is the "empty chamber" method. The measurement is repeated with target and without target, and the complex difference between both measurements is calculated. The reflectivity of the target can be calculated with the following equation:

$$\sqrt{\sigma_{\text{target}}} = \sqrt{\sigma_{reference}} \cdot \left|\frac{E_{\text{target}}}{E_{reference}}\right| = \sqrt{\sigma_{reference}} \cdot \left|\frac{E_{meastarget} - E_{emptychamber}}{E_{measreference} - E_{emptychamber}}\right| \tag{12.4.1}$$

For measuring RCS, pulsed systems are usually used, in order to isolate the transmitted ray from the received wave. However, these systems are expensive, and not always available. Continuous wave transmitters can be used if a measurement in a wideband is performed and time filtering is used. The received signal is transformed to time domain, a time filter is applied, and it is transformed again to frequency domain. In this case, monostatic configurations do not provide enough isolation between transmitter and receiver, and quasi-monostatic configurations (with two different probes) are used.

12.4.7 Antenna Pattern Correction and Range Evaluation Techniques

The quality of the quiet-zone (QZ) field limits the measurement accuracy in a CATR. The QZ field is ideally a plane wave, but in practice it is distorted by many unwanted effects. Common sources of distortions include (1) edge diffraction of the collimating element, (2) surface errors of the collimating element, (3) direct leakage from the range feed to the QZ, (4) multiple reflections between the AUT and the range, and (5) range reflections. Figure 12.17 illustrates these error sources. The level of the distortion signals compared with the level of the desired plane wave should be much lower than the measured side lobe level of the AUT. Otherwise distorting signals may cause a significant error the measured results.

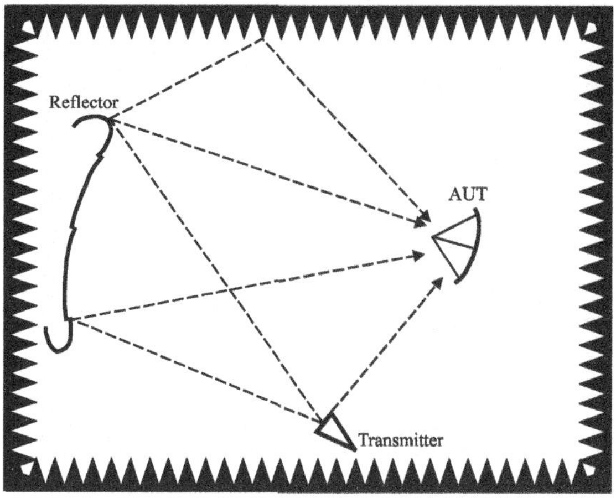

Fig. 12.17 Common error sources in a CATR include edge diffraction, surface errors of the collimating element, direct leakage from the range feed to the QZ, range reflections, and multiple reflections

The presence of distorting signals in CATRs raises many questions, such as:

- How accurately can a given antenna be tested in the CATR?
- How could the CATR be improved by identifying and eliminating distortion sources?
- As distortion signals are always present, is it possible to reduce computationally their effect?

Many methods have been developed to address the above questions. Some of the methods have been primarily developed for correcting measured antenna pattern in a distorted quiet-zone field, while others have been developed for identifying distortion sources in the range. Most of the methods, however, are suitable for both purposes.

Antenna pattern correction is always based on some additional information. For instance, the antenna can be measured multiple times in different locations in the CATR or the measurements can be performed at different frequencies. Based on the additional information used, the methods can be categorized into four groups:

1. Methods utilizing the time or frequency response of the CATR
2. Methods utilizing the spatial response of the CATR
3. Methods utilizing the probed quiet-zone field of the CATR
4. Methods utilizing information on the physical size of the AUT.

The following reviews antenna pattern correction techniques in each group. Some of the reviewed techniques may not have been developed for CATRs, but are reviewed here because of their educational aspects or because they are potentially feasible for CATRs. Finally, this section introduces methods that are solely meant for identifying distortion sources in a CATR.

Time and frequency techniques

Distorting signals often follow other than the wanted signal path from the range feed to the AUT. In such a case, the distortion and desired signals delay differently during

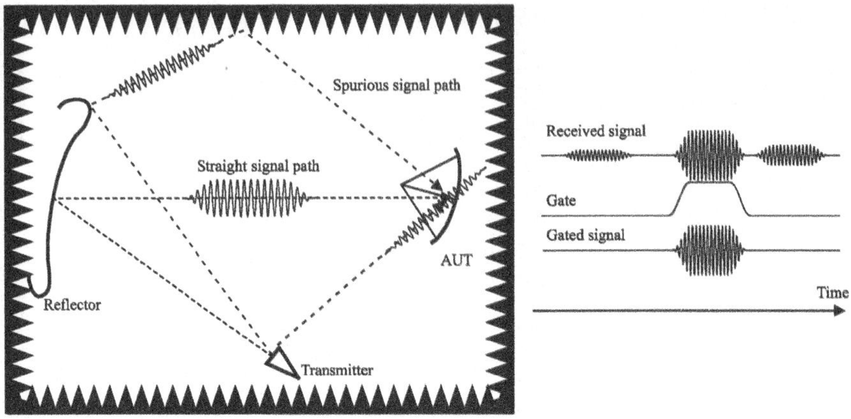

Fig. 12.18 Spurious signals that arrive after a different time-delay than the desired signal, are gated out in the time gating method

their propagation through the CATR. Time and frequency techniques utilize the time or frequency response of the CATR for eliminating the undesired signal contributions from the measured response. Figure 12.18 illustrates a situation where there are the desired signal and distortion signals in a CATR. When the range feed transmits a pulse, the receiver records three differently time-delayed and attenuated copies of it. The desired signal can be picked out from other signals in the time domain.

Multipath signals can be separated from each other when the pulse length is smaller than the path length difference. The usable pulse length and thus also the correction efficiency is limited by the bandwidth of the equipment or the AUT. For instance, if the relative bandwidth is 10%, multipath signals whose path length differs by 10% or more from that of the desired signal can be eliminated.

Measurements can be performed either in time or frequency domain, because signals in both domains contain the same information. Measurements are often performed with a network analyzer offering the possibility to sweep the frequency. This possibility also makes the implementation of time and frequency methods convenient: No additional equipment is needed and the measurement time does not considerably increase.

There are various time and frequency domain techniques differing in how the corrected pattern is obtained from the measured time or frequency domain response. Simple Fourier transform is often used to obtain the time domain response from the frequency response. An implementation of a Fourier transform based time gating at microwave frequencies is presented in [45]. Instead of the Fourier transform, other methods have been used to obtain multipath components, such as the Matrix-pencil [46] and oversampled Gabor transform [47]. These methods are potentially more accurate over the Fourier transform.

Depending on the measurement equipment, it is not always possible to preserve phase coherence across the whole frequency band. In such a case, Fourier

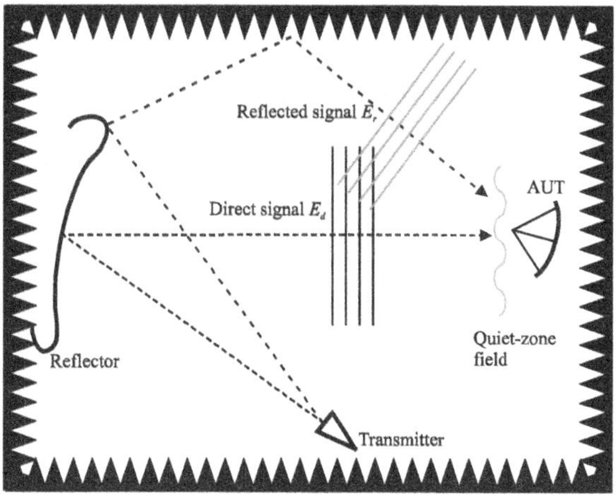

Fig. 12.19 A compact antenna test range with a direct signal and a reflected signal

transformation cannot be used. A frequency shift technique capable of correcting the antenna pattern from noncoherent frequency response is described and demonstrated at 310 GHz in [48].

Spatial methods

Spatial methods utilize the spatial response of the test range. That means that the AUT is measured several times at different locations in the test range. Let us consider the test range shown in Fig. 12.19, where there is the desired signal and one reflected signal arriving at the AUT. In each measurement point, the AUT receives a signal that is a vector sum of the desired and reflected signal. The phase of each vector depends on the spatial location of the AUT. For example, if the AUT is moved transversally to the propagation direction of the direct signal, the phase of direct signal remains unaltered, whereas the phase of reflected signal changes. This way signals arriving from different directions can be separated from each other.

The antenna pattern comparison (APC) method was first used for determining the reflectivity level of the chamber [49], but the method can also be used for antenna pattern correction. In this method, the antenna pattern of the AUT is measured several times at different locations in the range. The variations between the measured patterns are related to the reflectivity level of the range and the corrected pattern can be obtained as an average of the measured patterns.

Instead of simple averaging of the measured antenna patterns, the corrected pattern can be obtained in other ways. In the virtual array method the antenna pattern is measured twice. During the first measurement, the AUT is kept in place, whereas it is spatially displaced as a function of rotation angle during the second measurement. The displacement is selected such that when the two patterns are summed, the desired signals sum in phase and the signals arriving through the main beam of the antenna sum in the opposite phase cancelling each other.

Somewhat similar approach can also be used even if the AUT is not displaced as a function of the antenna rotation angle. The measurements performed at different locations form a synthetic antenna array at each rotation angle of the AUT, provided that the measurements are coherent in each position. If the precise locations of the AUT are known, it is possible to synthesize an array that has a high gain towards the desired signal, and high attenuation towards possible spurious signals. Examples of such methods have been presented in [50–52].

The aforementioned spatial methods necessitate the possibility to move the AUT. This may be challenging in particular if the AUT is heavy and if it needs to be moved with a sub-wavelength precision. It has been shown in [44], that instead of potentially heavy AUT, the range feed can be moved to obtain a phase difference between the direct and reflected signals. The method has been demonstrated at 310 GHz in a hologram-based CATR.

Techniques utilizing the measured Quiet-Zone field

If the quiet-zone field differs from the plane wave field, the measured antenna pattern is erroneous. However, if the quiet-zone field is precisely known, it is, at least in principle, possible to solve the true antenna pattern from the measured antenna pattern and the quiet-zone field. There are two ways of measuring the quiet-zone field. The quiet-zone field can either be probed with a near-field scanner or be found out by measuring a precisely known reference antenna in it. A correction method utilizing measured quiet-zone field is presented in [53].

Somewhat similar method was used in a hologram-based CATR at 650 GHz [8]. Here the quiet-zone field was probed and it was calculated how the simulated antenna pattern should look if it was measured in that quiet-zone field. This result was then compared with the measured antenna pattern. Although the measured antenna pattern was not corrected with the method, it was found out that some unexpected features in the measured antenna pattern were due to quiet-zone field instead of fabrication errors of the AUT.

Methods utilizing information on the size of the AUT

Antenna diagnostics refers to an operation in which the near-field or surface currents of the AUT are solved from its measured far-field pattern. Near-field or surface currents are solved for diagnostic purposes, i.e., to identify possible malfunctions or manufacturing errors in the antenna structure. This approach offers a possibility for antenna pattern correction as well. For instance, the near-field may reveal that part of radiation comes outside the physical volume of the antenna. In such a case, the field outside the physical volume of the AUT can be filtered out. An example of such a method is presented in [54].

Other range diagnostic techniques

A quiet-zone field of a CATR is often probed in order to verify that the field meets certain specifications. In addition to range verification, the probed field can be used for identifying spurious signals in the range. After identifying the signal sources, the range can be modified to remove them.

Various methods for locating the spurious signals have been presented. Many of the methods are reviewed in [55]. One way is to calculate the plane-wave spectrum of the quiet-zone field. Large contributions outside the desired signals indicate a potential spurious signal. The plane-wave spectrum has been calculated with a Fourier transform [7] and superresolution techniques [56, 57]. These methods provide information on the direction of possible spurious signal sources.

Plane-wave spectrum gives correct information on the direction of spurious signal sources when the signal source is in the far-field of the probed area. In CATRs, however, the spurious signal sources are typically in the near-field of the probed area and, therefore, they do not appear as a peak in the plane-wave spectrum. To overcome this challenge, a near-field focusing method is proposed for locating the sources of spurious signals [55]. In this method, the phase of the probed field is adjusted to focus at a certain point in the image domain. After repeating this operation for every point in the image domain, spurious signal sources should appear as high-intensity points in the image.

The aforementioned methods can also be utilized in time domain provided that quiet-zone field is probed in time or frequency domain. Time dimension gives additional information for example on the distance of the source of the spurious signal.

12.5 Special Topics

12.5.1 Holographic Principles Used in Antenna Measurements at mm and submm Wavelengths

Holographic imaging is discussed in detail in Chap. 13 "Holographic imaging approach". It is highly recommended that the reader studies Chap. 13 before going into details of this subsection.

Microwave holography has been under intensive research since the mid-1960s, and has been applied, e.g., for antenna design and antenna measurements since 1970s. Holographic principles may be applied directly into the antenna metrology for phase retrieval when using a reference antenna, or in producing plane wave conditions for the measurement. In indirect holography, scalar measurements are sufficient, and the use of rather expensive vector measurement equipment described in Sect. 12.2 can be avoided.

In the following, we briefly discuss how holographic principles have been used in antenna metrology in the millimeter wave and THz bands. As we will see, microwave holography has provided many advantages in antenna metrology at millimeter and submillimeter wavelengths. In measurement of large radio astronomy telescopes, it has become a universal method in determining the reflector surface accuracy. In many other measurements it may bring more accuracy and better insight to the unideal antenna radiation performance. A computer-generated radio hologram is a prospective alternative to large reflectors as a collimating element in CATR at THz frequencies (see Sect. 12.4.4).

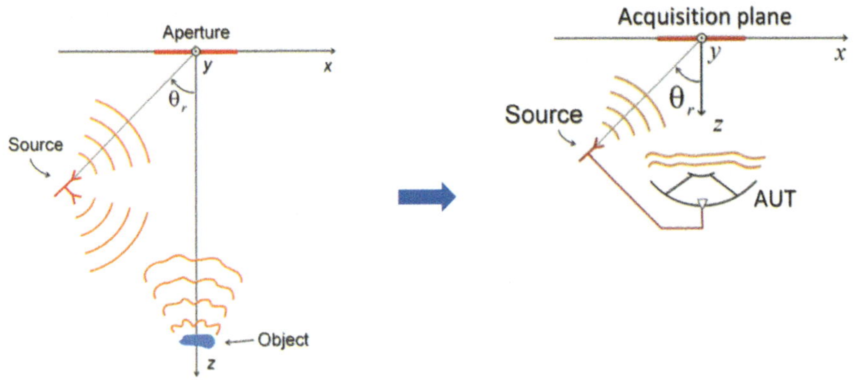

Fig. 12.20 Setup for indirect off-axis holographic imaging in order to be able to detect the interference pattern of the object field (field scattered by the object when illuminated by the source field) and the reference field (lhs), and the applied in antenna measurement (rhs)

Fig. 12.21 Hologram formation geometry (after [58])

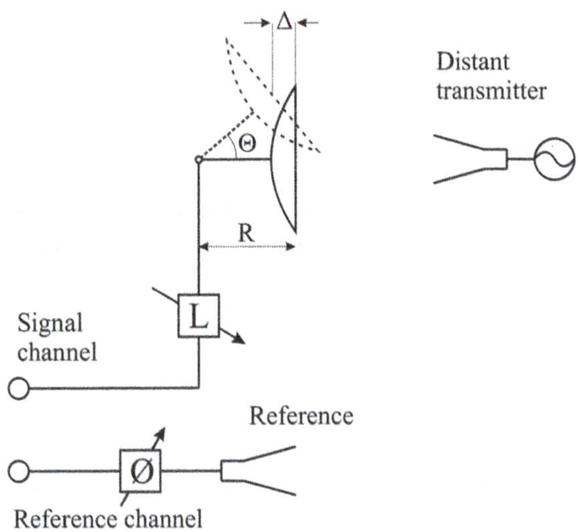

The schematic of indirect holographic imaging of Fig. 13.8 is repeated in Fig. 12.20 (lhs) and then applied to indirect off-axis holographic antenna measurement in Fig. 12.20 (rhs).

Holographic principles used in determining the surface accuracy of large reflectors
Microwave holographic metrology of large reflector antennas was introduced by Bennett et al. in 1970s [58], see Fig. 12.21. Since then many major millimeter wave and THz radio telescopes for radio astronomy have undergone a holographic measurement during the reflector panel adjustment or as a final check of the reflector surface quality, see, e.g., [59] (see Fig. 12.22), [60, 61].

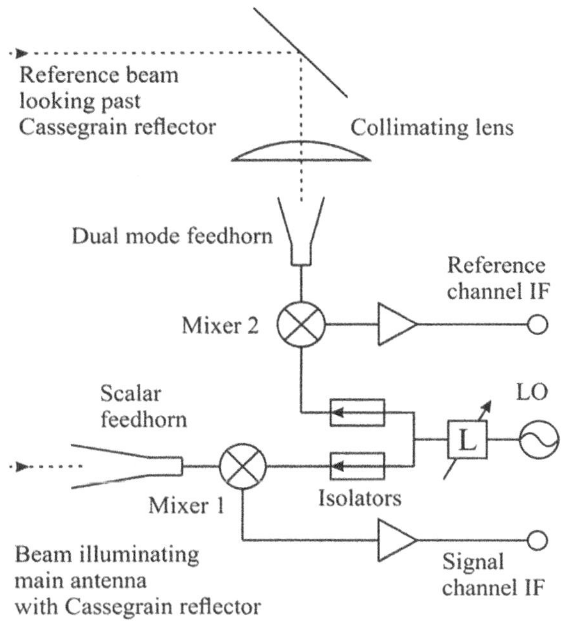

Fig. 12.22 Feed system for holographic measurements of the 13.7-m telescope of University of Massachusetts at 38 GHz. The MIT Lincoln Laboratory LES-8 geostationary satellite was used as the source (after [59])

The antenna under test (AUT), typically a large telescope antenna, is pointed to a far-field source: a radio astronomical source (maser) or a satellite equipped with a mm-wave transmitter, or to a far-away terrestrial source. Also a synchronous reference receiver with a lower-gain antenna is pointed to the same source. Then the AUT scans the source using its azimuth-elevation rotating mechanism. The phase difference pattern of the two signals (from AUT and reference receivers) over the aperture reveals possible deviations of the AUT reflector surface from the ideal parabolic shape.

Holography for phase retrieval in near-field antenna measurements

A method to determine the AUT aperture field distribution without phase measurement, but with a synchronous reference transmitter, was first studied by Napier and Bates in early 1970s [62].

This holographic method for phase retrieval in planar near-field antenna measurements at THz frequencies was theoretically studied by Junkin et al. [23, 63]. A small and coherent reference antenna is placed next to the antenna under test (AUT) and the interference pattern, i.e., hologram, of the ensemble of the AUT and reference antenna is recorded. The phase of the AUT can be retrieved from the hologram by numerical means when the radiation characteristics of the reference antenna are known. Recently Arboleya et al. have demonstrated a method for broadband antenna characterization from planar phaseless acquisitions [64]. The method is based on off-axis indirect holography and use of a reference antenna as the methods above. However, a frequency sweep is performed at every acquisition point which allows retrieval of the phase at each frequency point and at each spatial point independently. This allows less dense sampling than required in the conventional off-axis holography.

Holograms used to create plane wave in CATR

If a hologram is synthesized by computational methods; we call it a computer-generated hologram [65]. Such a hologram is determined by computing the image pattern that will change an incident beam into a desired beam. Usually such holograms are designed by assuming that the image pattern modulates either only the amplitude or the phase of the incident field known as amplitude or phase holograms. In order to facilitate the production of a hologram, binary coding is used: the amplitude or phase is described using only ones and zeros. For example, a binary amplitude hologram fully transmits the incident field at some positions, but blocks it completely at some other positions.

Such a computer-generated binary hologram is the basis for the holograms that can be used in antenna testing for creating a plane wave field needed in measurements. If the hologram is printed as a metal pattern on a thin dielectric substrate, the resulting hologram can be used as a collimating element in a compact antenna test range (CATR) instead of a lens or a reflector.

At Aalto University, use of radio frequency holograms in CATRs has been extensively studied. The holograms are planar structures, which is advantageous with respect to the manufacturing. Demonstrations have been carried out at the frequency range of 40–650 GHz. The major part of the experiments is carried out using transmission type binary amplitude holograms, which consist of an etched metal structure on a supporting dielectric film. Large test objects, such as the Odin space telescope 1.1-m antenna and the 1.5-m prototype reflector antenna by EADS Astrium have been tested at 119 GHz and at 320 GHz and 650 GHz, respectively, in a hologram CATR [6–8, 41, 42]. The challenges in the use of holograms are the rather narrow operation band ($\sim 10\%$) and the polarization limitations, especially with amplitude holograms. Very recently, also a wideband hologram-based CATR has been demonstrated [43].

From the above discussion we can conclude that microwave holography has provided many advantages in antenna metrology at millimeter and submillimeter wavelengths. In measurement of large radio astronomy telescopes, it has become a universal method in determining the reflector surface accuracy. In many other measurements it may bring more accuracy and better insight to the unideal antenna radiation performance. A computer-generated radio hologram is a prospective alternative to large reflectors as a collimating element in CATR at THz frequencies.

12.5.2 Measurement of Small Integrated Antennas

Advancement of integrated circuit technologies has enabled the use of even higher frequencies in wireless communication and other applications. At the same time, the antennas having size proportional to the wavelength are getting smaller. The aim is to integrate the antennas on an IC chip possibly measuring only a few mm.

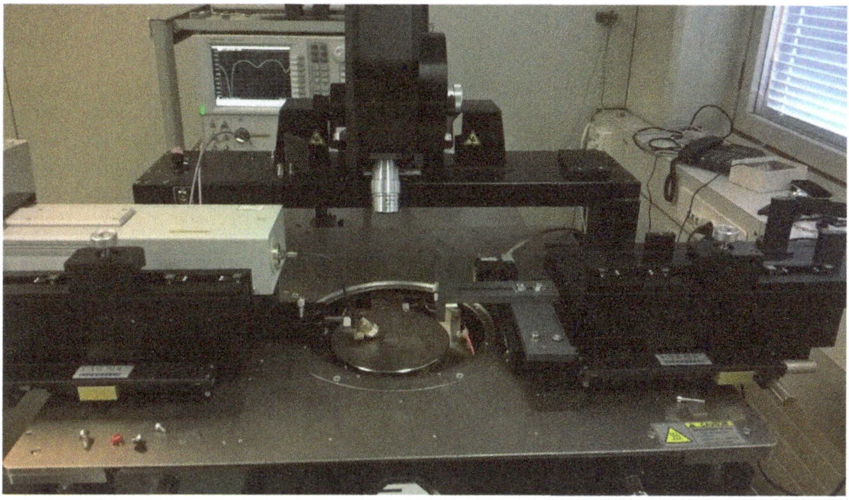

Fig. 12.23 A probe station and typical space around it. A microstrip antenna (AUT) at the center of chuck connected to a VNA with a ground-signal-ground (GSG) probe

The measurement techniques presented previously, far-field, near-field, and compact antenna test ranges, may encounter difficulties when measuring integrated antennas or antennas which can only be accessed with an on-wafer probe. There are often a lot of scattering objects in the vicinity of the AUT when it is measured on a probe station. See Fig. 12.23 for a typical probe station for on-wafer measurements.

Also, the rotation of the antenna is impractical because of the probe heads that are fixed on the probe station. Therefore, in far-field measurements, the measuring probe antenna is moved along a surface of a sphere centered at the AUT phase center. The movement of the probe is usually produced using pivoted arms as shown in Fig. 12.24. The free rotation of the probe antenna around the AUT may be limited by other blocking objects like microscopes and supporting structures. Therefore, custom-made mounting fixtures are required for the antennas in order to enable rotation of the measuring antenna around the AUT. Also, the polarization of the probe antenna should be correct. Far-field measurements of integrated antennas are reported, e.g. in [66–70].

Often, the compact antenna test ranges are not practical for measuring small integrated antennas because there is no possibility to rotate the AUT and, therefore, the collimating element with its feed should be rotated around the AUT.

Near-field technique for antenna characterization on a probe station has been demonstrated by Klein et al. [71]. The near-field technique is sensitive to phase measurement errors because the far-field properties of the AUT are solved from the complex near-field. The phase errors in the near-field result in an amplitude error in the antenna pattern. During the probe movement, the bending RF cables easily change their electrical length resulting in phase errors. The phase errors may be corrected in some extent using error correction techniques, e.g., pilot signal based

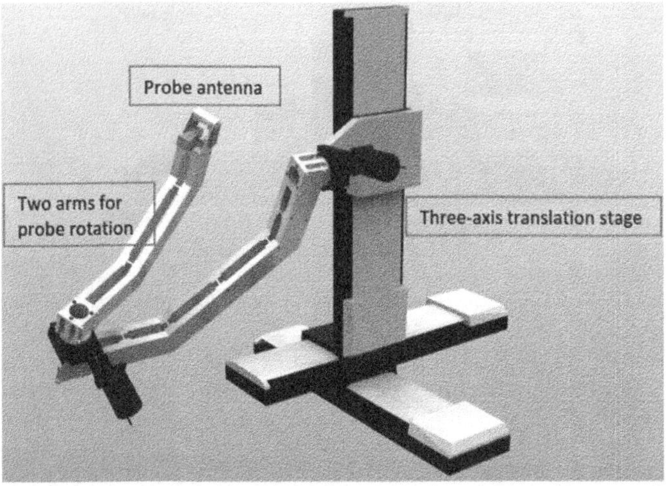

Fig. 12.24 Typical mechanical assembly for movement of the probe on a spherical surface

method described in Sect. 12.3.5. Accurate mechanical assembly of the near-field scanner may be difficult to accomplish because of the restricted space around the probe station.

The above-mentioned techniques require a probe antenna which may be difficult to facilitate on a probe station environment due to size limitations. Calazans et al. have proposed a technique where there is no need for a probe antenna [72]. A metal wire is moved in front of the AUT and the antenna radiation pattern is solved from the measured reflection coefficients. This method enables to solve the antenna pattern on the plane normal to the wire.

Another antenna measurement method avoiding the use of a probe antenna has been proposed in Aalto University [73, 74]. In this method, the reflection coefficient of the antenna is measured when there is a reflective object in the vicinity of the radiating AUT. The measurement is repeated with many known reflective objects. The field propagation from the AUT to the reflective object and after the reflection back to the AUT is modeled with Fourier techniques for each object using an initial guess for the antenna aperture field. Iterative methods are then used to find such an antenna aperture field that best explains the modeled propagation and measured reflection coefficients. The radiation characteristics can be calculated from the retrieved antenna aperture field. Figure 12.25 shows the concept of the measurement method. A flat metal strip is used as the reflective object and measurements are carried out with different strip positions. The setup can be arranged in a small volume and, because it avoids the movement of the probe antenna, the problems with flexing cables are also avoided. Therefore, the method offers some benefits for testing antennas on probe stations.

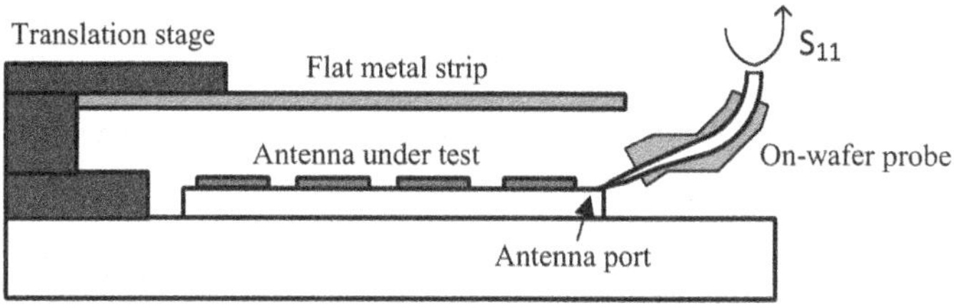

Fig. 12.25 Potential measurement setup for characterizing an antenna on a probe station. The translation stage is used to alter the position of the reflective flat metal strip

The method was first demonstrated with a pyramidal horn antenna [73] and then, in a probe station environment, using a planar antenna array as an AUT [74]. The limitations of the method are the limited angular range, at maximum the measurement range is the half hemisphere, and multiple reflections which may need to be mitigated by time gating or other methods. Also, the inversion problem is complex and the associated iterative numerical methods have to be selected carefully in order to ensure reliable convergence.

References

1. G.E. Evans, *Antenna Measurement Techniques* (Artech House Inc. 1990)
2. D. Slater, P. Stek, R. Cofield, R. Dengler, J. Hardy, R. Jarnot, R. Swindlehurst, A large aperture 650 GHz near-field measurement system for the earth observing system microwave limb sounder, in *Proceedings of 23rd AMTA*, Oct. 2001, Denver, USA, pp. 468–473
3. J. Hartmann, J. Habersack, H.-J. Steiner, J. Lemanczyk, P. de Maagt, Calibration and verification measurements in compensated compact ranges up to 500 GHz, in *Proceedings of 23rd AMTA*, Denver, USA, Oct 2001, pp. 377–382
4. W. Menzel, B. Huder, Compact range for millimeter-wave frequencies using a dielectric lens. Electron. Lett. **20**(19), 768–769 (1984)
5. T. Hirvonen, J. Tuovinen, A. Räisänen, Lens-type compact antenna test range at mm-waves, in *21st European Microwave Conference Proceedings* (Microwave Exhibitions and Publishers Ltd., Kent, England, 1991), pp. 1079–1083
6. J. Ala-Laurinaho, T. Hirvonen, P. Piironen, A. Lehto, J. Tuovinen, A.V. Räisänen, U. Frisk, Measurement of the Odin telescope at 119 GHz with a hologram type CATR. IEEE Trans. Antennas Propag. **49**(9), 1264–1270 (2001)
7. A. Lönnqvist, T. Koskinen, J. Häkli, J. Säily, J. Ala-Laurinaho, J. Mallat, V. Viikari, J. Tuovinen, A.V. Räisänen, Hologram-based compact range for submillimeter-wave antenna testing. IEEE Trans. Antennas Propag. **53**(10), 3151–3159 (2005)
8. A. Karttunen, J. Ala-Laurinaho, M. Vaaja, T. Koskinen, J. Häkli, A. Lönnqvist, J. Mallat, A. Tamminen, V. Viikari, A.V. Räisänen, Antenna tests with a hologram-based CATR at 650 GHz. IEEE Trans. Antennas Propag. **57**(3), 711–720 (2009)
9. A.V. Räisänen, Frequency multipliers for millimeter and submillimeter wavelengths. Proc. IEEE **80**(11), 1842–1852 (1992)
10. G. Carpintero, L.E. García Muñoz, H.L. Hartnagel, S. Preu, A.V. Räisänen (eds.) *Semiconductor Terahertz Technology: Devices and Systems at Room Temperature Operation* (Wiley, 2015)

11. Keysight Technologies, http://www.keysight.com/en/pc-1000000457%3Aepsg%3Apgr/network-analyzer?nid=-536902444.0&cc=US&lc=eng
12. Anritsu, http://www.anritsu.com/en-us/products-solutions/products/ms4640b-series.aspx
13. Rohde & Schwarz. https://www.rohde-schwarz.com/en/products/test-measurement/network-analyzers/pg_overview_64043.html
14. Virginia Diodes Inc. http://vadiodes.com/index.php/en/about-vdi
15. R.E. Collin, *Antennas and Radiowave Propagation* (McGraw-Hill, 1985)
16. A.D. Yaghjian, An overview of near-field antenna measurements. IEEE Trans. Antennas Propag. **34**(1), 30–45 (1986)
17. C.A. Balanis, *Antenna Theory: Analysis and Design*, 3rd edn. (Wiley, 2005)
18. M.H. Francis, Aperture-sampling requirements. IEEE Antennas Propag. Mag. **39**(5), 76–81 (1997)
19. J. Säily, P. Eskelinen, A.V. Räisänen, Pilot signal-based real-time measurement and correction of phase errors caused by microwave cable flexing in planar near-field tests. IEEE Trans. Antennas Propag. **51**(2), 195–200 (2003)
20. D. Slater, "A 550 GHz near-field antenna measurement system for the NASA Submillimeter Wave Astronomy Satellite," in *Antenna Measurement Techniques Association Conference*, October 3–7, 1994
21. R.G. Yaccarino, Y. Rahmat-Samii, A comparison of conventional and phaseless planar near-field antenna measurements: the effect of probe position errors, in *Proc. of IEEE Int. Conference on Phased Array Systems and Technology*, Dana Point, CA, USA, 21–25 May 2000, pp. 525–528
22. T. Isernia, G. Leone, R. Pierri, Radiation pattern evaluation from near-field intensities on planes. IEEE Trans. Antennas Propag. **44**(5), 701–710 (1996)
23. G. Junkin, T. Huang, J.C. Bennett, Holographic testing of terahertz antennas. IEEE Trans. Antennas Propag. **48**(3), 409–417 (2000)
24. A.C. Newell, Error analysis techniques for planar near-field measurements. IEEE Trans. Antennas Propag. **36**(6), 754–768 (1988)
25. *IEEE Recommended Practice for Near-Field Antenna Measurements*, IEEE Std 1720-2012 (IEEE, New York, 2012)
26. *IEEE Standards Test Procedures for Antennas*, IEEE Std. 149-1979 (Wiley, 1979)
27. Y. Mizuguchi, M. Akagawa, H. Yokoi, Offset gregorian antenna. Electron. Commun. Jpn. **61-B**(3), 58–66 (1978)
28. C. Dragone, Offset antennas with perfect pattern symmetry and polarization discrimination. Bell Syst. Tech. J. **57**(7), 2663–2684 (1978)
29. L.H. Hemming, *Electromagnetic Anechoic Chambers: A Fundamental Design and Specification Guide* (IEEE Press and John Wiley Interscience, 2002)
30. Microwave Vision Group, http://www.microwavevision.com/
31. A. Lönnqvist, A. Tamminen, J. Mallat, A.V. Räisänen, Monostatic reflectivity measurement of radar absorbing materials at 310 GHz. IEEE Trans. Microw. Theory Tech. **54**(9), 3486–3491 (2006)
32. V.J. Vokurka, New compact range with cylindrical reflectors and high efficiency factor, in *Proceedings of Electronica 76 Conference*, Munich, Germany (1976)
33. European Space Agency, http://www.esa.int/Our_Activities/Space_Engineering_Technology/Antenna_Test_Facilities_and_Electro-Magnetic_Compatibility_Laboratories/
34. A.D. Olver, A.A. Saleeb, Lens-type compact antenna range. Electron. Lett. **15**(14), 409–410 (1979)
35. A.A. Saleeb, Theory and design of lens-type compact antenna ranges, p. 279, Doctoral Thesis, Queen Mary & Westfield College, University of London, 1982
36. A.D. Olver, Compact antenna test ranges, in *Proceedings of the Seventh International Conference on Antennas and Propagation (ICAP)*, York, UK, (1991), pp. 99–108
37. J. Tuovinen, A. Vasara, A. Räisänen, A new type of compact antenna test range, in *Proceedings of the 22nd European Microwave Conference*, Espoo, Finland, vol. 1, (1992), pp. 503–508

38. A. Vasara, J. Turunen, A. Friberg, Realization of nondiffracting beams with computer-generated holograms. J. Opt. Soc. Am. A **6**, 1748–1754 (1989)
39. J. Meltaus, J. Salo, E. Noponen, M.M. Salomaa, V. Viikari, A. Lönnqvist, T. Koskinen, J. Säily, J. Häkli, J. Ala-Laurinaho, J. Mallat, A.V. Räisänen, Millimeter-wave beam shaping using holograms. IEEE Trans. Microw. Theory Tech. **51**(4), 1274–1280 (2003)
40. A. Tamminen, A. Karttunen, M. Vaaja, E. Noponen, J. Ala-Laurinaho, J. Mallat, A.V. Räisänen, Reflection-type phase hologram for beam shaping: Experimental results at 310 GHz, in *Proceedings of the 30th ESA Antenna Workshop*, Noordwijk, The Netherlands, 27–30 May 2008, pp. 470–473
41. J. Häkli, T. Koskinen, A. Lönnqvist, J. Säily, V. Viikari, J. Mallat, J. Ala-Laurinaho, J. Tuovinen, A.V. Räisänen, Testing of a 1.5 m reflector antenna at 322 GHz in a CATR based on a hologram. IEEE Trans. Antennas Propag. **53**(10), 3142–3150 (2005)
42. T. Hirvonen, J. Ala-Laurinaho, J. Tuovinen, A.V. Räisänen, A compact antenna test range based on a hologram. IEEE Trans. Antennas Propag. **45**(8), 1270–1276 (1997)
43. Z. Li, J. Ala-Laurinaho, Z. Du, A.V. Räisänen, Realization of wideband hologram Compact Antenna Test Range by linearly adjusting the feed location. IEEE Trans. Antennas Propag. **62**(11), 5628–5633 (2014)
44. V. Viikari, J. Häkli, J. Ala-Laurinaho, J. Mallat, A.V. Räisänen, A feed scanning based APC technique for compact antenna test ranges. IEEE Trans. Antennas Propag. **53**(10), 3160–3165 (2005)
45. A.M. Predoehl, W.L. Stutzman, Implementation and results of a time-domain gating system for a far-field range, in *Proceedings of the 19th Annual Meeting & Symposium of the Antenna Measurement Techniques Association (AMTA)*, Boston, MA, USA, 17–21 Nov 1997, pp. 8–12
46. B. Fourestie, Z. Altman, J. Wiart, A. Azoulay, On the use of the matrix-pencil method to correlate measurements at different test sites. IEEE Trans. Antennas Propag. **47**(10), 1569–1573 (1999)
47. B. Fourestie, Z. Altman, Gabor schemes for analyzing antenna measurements. IEEE Trans. Antennas Propag. **49**(9), 1245–1253 (2001)
48. V. Viikari, J. Mallat, J. Ala-Laurinaho, J. Häkli, A.V. Räisänen, A frequency shift technique for pattern correction in hologram-based CATRs. IEEE Trans. Antennas Propag. **54**(10), 2963–2968 (2006)
49. J. Appel-Hansen, Reflectivity level of radio anechoic chambers. IEEE Trans. Antennas Propag. **AP-21**(4), 490–498 (1973)
50. M.D. Migliore, Filtering environmental reflections in far-field antenna measurement in semi-anechoic chambers by an adaptive pattern strategy. IEEE Trans. Antennas Propag. **52** (4), 1112–1115 (2004)
51. V. Viikari, V.-M. Kolmonen, J. Salo, A.V. Räisänen, Antenna pattern correction technique based on an adaptive array algorithm. IEEE Trans. Antennas Propag. **55**(8), 2194–2199 (2007)
52. V. Viikari, A.V. Räisänen, Antenna pattern correction technique based on signal to interference ratio optimization. IEEE Antennas Wirel. Propag. Lett. **6**, 267–270 (2007)
53. P.L. Garcia Muller, J.L. Cano, R. Torres, A deconvolution method for correcting antenna measurement errors in compact antenna test ranges, in *Proceedings of the 17th Annual Antenna Measurement Techniques Association (AMTA) Meeting and Symposium*, USA, 1995, pp. 509–514
54. G. Hindman, A.C. Newell, Mathematical absorber reflection suppression (mars) for anechoic chamber evaluation and improvement, in *Proceeings of the 30th Annual Antenna Measurement Techniques Association (AMTA) Meeting and Symposium*, USA, 2008
55. I.J. Gupta, Stray signal source location in far-field antenna/RCS ranges. IEEE Antennas Propag. Mag. **46**(3), 20–29 (2009)
56. T.P. Delfeld, F.C. Delfeld, Use of the MUSIC algorithm in the analysis of compact range field probe data, in *Proceedings of the 11th Annual Antenna Measurement Techniques Association (AMTA) Meeting and Symposium*, Monterey, CA, USA, 9–13 Oct 1989, pp. 6-3–6-9

57. I.J. Gupta, W.D. Burnside, Imaging the compact range probe data, in *Proceedings of the 12th Annual Antenna Measurement Techniques Association (AMTA) Meeting and Symposium*, Philadelphia, PA, USA, 1990, pp. 14-13–14-17
58. J.C. Bennett, A.P. Anderson, P.A. Mclnnes, A.J.T. Whittaker, Microwave holographic metrology of large reflector antenna. IEEE Trans. Antennas Propag. **AP-24**(3), 295–303 (1976)
59. P.F. Goldsmith, N.R. Erickson, Holographic measurements, in Kraus, J.D (ed.) *Chapter 6 "Radio-telescope antennas" in the 2nd edition of Radio Astronomy* (Cygnus-Quasar Books, 1986), pp. 6-63–6-64
60. J.M. Payne, Millimeter and submillimeter wavelength radio astronomy. Proc. IEEE **77**(7), 993–1016 (1989)
61. J.W.M. Baars, R. Lucas, J.G. Mangum, J.A. Lopez-Perez, Near-field radio holography of large reflector antennas. IEEE Antennas Propag. Mag. **49**(5), 1–14 (2007)
62. P.J. Napier, R.H.T. Bates, Antenna aperture distributions from holographic-type radiation pattern measurement. Proc. IEEE. **120**(1), 30–34 (1973)
63. G. Junkin, Phase shifting holography for THz near-field/far-field prediction. Prog. Electromagnet. Res. Lett. **44**, 15–21 (2014)
64. A. Arboleya, J. Laviada, J. Ala-Laurinaho, Y. Álvarez, F. Las-Heras, A.V. Räisänen, Phaseless characterization of broadband antennas. IEEE Trans. Antennas Propag. **64**(2), 484–495 (2016)
65. W.-H. Lee, Computer-generated holograms: techniques and applications, in E. Wolf (ed.) *Progress in Optics XVI* (Elsevier, Amsterdam, 1978) pp. 121–231
66. L. Roy, M. Li, S. Labonte, N.R.S. Simons, Measurement techniques for integrated-circuit slot antennas. IEEE Trans. Instrum. Meas. **46**(4), 1000–1004 (1997)
67. R.N. Simmons, Novel on-wafer radiation pattern measurement technique for MEMS actuator based reconfigurable patch antennas, in *Proceedings of the 24th Annual Antenna Measurement Techniques Association Meeting and Symposium*, Cleveland, Ohio, 3–8 Nov 2002
68. T. Zwick, C. Baks, U.R. Pfeiffer, D. Liu, B.P. Gaucher, Probe based MMW antenna measurement setup, in *Proceedings of the IEEE AP-S Symposium 2004*, Monterey, CA, USA, vol. 1, 20–25 June 2004, pp. 747–750
69. S. Ranvier, M. Kyrö, C. Icheln, C. Luxey, R. Staraj, P. Vainikainen, Compact 3-D on-wafer radiation pattern measurement system for 60 GHz antennas. Microw. Opt. Technol. Lett. **51** (2), 319–324 (2009)
70. K. Mohammadpour-Aghdam, S. Brebel, A. Enayati, R. Faraji-Dana, G.A.E. Vandenbosch, W. Deraedt, RF probe influence study in millimeter-wave antenna pattern measurements. Int. J. RF Microw. Comput. Aided Eng. **21**(4), 413–420 (2011)
71. B. Klein, M. Jenning, P. Seiler, K. Wolf, D. Plettemeier, Verification and demonstration up to 67 GHz of an on-chip antenna pattern measurement setup, in *Proceedings of IEEE Conference on Antenna Measurements and Applications (CAMA)*, Antibes Juan-les-Pins, France, 16–19 Nov 2014
72. T. Calazans, H.D. Griffiths, A. L. Cullen, D.E.N. Davies, R. Benjamin, Antenna radiation pattern measurement using a near-field wire scattering technique. IEEE Proc. Microw. Antennas Propag. **145**(3), 263–267 (1998)
73. Z. Du, V. Viikari, J. Ala-Laurinaho, A. Tamminen, A.V. Räisänen, Antenna pattern retrieval from reflection coefficient measurements with reflective loads. Prog. Electromagn. Res. **148**, 15–22 (2014)
74. V. Viikari, Z. Du, V. Semkin, J. Ala-Laurinaho, A.V. Räisänen, Reflection coefficient method for characterizing antennas on probe stations, in *Proceedings of the 9th European Conference on Antennas and Propagation*, Lisbon, Portugal, 12–17 Apr 2015

Chapter 13
Holographic Imaging Approach

Christian Schildbach and Lorenz-Peter Schmidt

Abstract This chapter provides an overview over common categories of microwave holography systems. Terms like resolution or speckle are defined and design rules are presented. Performance and cost of the various microwave holography systems are discussed as well as their applications. A special focus is devoted to indirect holographic systems at millimeter wave frequencies. A setup using arrays of planar antenna coupled zero bias Schottky diodes as detectors is presented in detail.

13.1 Introduction

Before the performance of microwave holographic systems and their specific applications are discussed, at first the term holography is clarified. The word holography stems from the Greek words "holos" meaning the "whole" and "gráphein" which means to write or to draw. The expression signifies that by holography, one can record the complete wave front of scalar waves, such that these wave fronts can be reconstructed later. The process is usually called a two-step process with the two steps: first the recording of the hologram, and second the reconstruction of the wave front [1]. Therefore, the recorded hologram must contain both magnitude and phase information completely describing the wave front. This already implies the existence of a distinct frequency, the wave oscillates at. In short,

C. Schildbach (✉)
Am Schillerplatz 4/1, 71522 Backnang, Germany
e-mail: christian.schildbach@fau.de

L.-P. Schmidt
Schleifweg 33, 91093 Heßdorf, Germany
e-mail: lorenz-peter.schmidt@fau.de

C. Schildbach · L.-P. Schmidt
Lehrstuhl für Hochfrequenztechnik/Institute of Microwaves and Photonics,
Friedrich-Alexander-Universität Erlangen Nürnberg/Friedrich-Alexander-University
Erlangen Nuremberg, Cauerstr 9, 91058 Erlangen, Germany

A. Boriskin and R. Sauleau (eds.), *Aperture Antennas for Millimeter and Sub-Millimeter Wave Applications*, Signals and Communication Technology,
DOI 10.1007/978-3-319-62773-1_13

coherence is required. Holography can be applied to many different kinds of waves, including sound waves, particle waves, and with scalar approximation also to electromagnetic waves and others. In the following chapter, the focus lies on one specific part of the electromagnetic spectrum, namely the microwave holography. In the beginning, holography was not intended for use in the microwave regime, but its inventor Dennis Gabor searched for a way to improve the resolution of electron microscopes [2]. The challenge in the recording step is that usually intensities are recorded rather than complex phasors. Gabor realized in 1948 that to overcome this difficulty a phase reference is needed for comparison, which he called "coherent background". The superposition of both the wave front and the coherent background is called a hologram. Although he never used holography for its original purpose, Gabor was able to demonstrate the functionality in the optical domain. Because in those days, computers were not powerful enough to process or even store the amount of data contained in holograms, the reconstruction had to be done by a coherent optical setup. Another problem in both steps was the lack of powerful coherent sources. Gabor used a high pressure mercury lamp yielding a coherence length of about 0.1 mm, which was sufficient for about 200 fringes.

Leith and Upatnieks were the first to use holography for microwave imaging [3, 4]. They found out that the reconstruction of side-looking airborne radar images can be viewed as the reconstruction of a hologram. Although the data was recorded in a different range of wavelengths, they still stored it on a photographic film and processed it with a coherent optical setup [5]. So not only optical, but also microwave holography benefited from the invention of the laser with its superior power and coherence properties compared to any other source available. A lot of improvement has since been made to microwave holography with the most important one being the computer-based reconstruction. Performance of nowadays computers makes storage and reconstruction even available in real time.

Today, microwave holography has found a widespread field of applications. In the field of remote sensing, the special frequency range of microwaves offers some advantages. In the microwave region, wavelengths have dimensions reaching from about 1 mm to several centimeters. Structures in these dimensions can be produced with high precision. On the other hand, the lower limit of the resolution of imaging systems is more or less given by the wavelength. So the microwave region is a good compromise between resolution and complex technological processes. Setups in the microwave region often allow to distinguish the two orthogonal components of polarization, giving access to the complete polarimetric information about a target. Material properties differ from the easy to observe optical range, sometimes yielding important additional information. Microwave radiation is nonionizing and thus not hazardous. This allows for use in diagnostic medicine and security applications by visualizing the interior of nonconducting motives. Today's Synthetic Aperture Radar systems use holography for geophysical exploration. Other areas of application are nondestructive testing, production quality control or automotive radar, just to name a few. New microwave holography applications still emerge.

13.2 Resolution

13.2.1 Definition

One of the most important figures of merit of an imaging system is its resolution. In principal, it is defined as the minimum distance between two small targets such that those two targets can still be recognized as distinct. Since this definition is rather vague, several criteria have emerged that give an estimate for the resolving power of a system. They all use the point spread function (PSF), the image of a point like object. Because imaging systems are approximately linear space invariant systems, the PSF is comparable to an impulse response of the system. For linearity, two arbitrary targets imaged together have to produce an image whose brightness is the sum of the two images of the single targets. This is usually true for imaging systems. For space invariance, the PSF may not depend on the location of the target. This is only approximately true for most imaging systems that cover only a limited solid angle of the scenery.

The produced image is then the convolution of the real scenery with the PSF. There are various shapes of PSFs that have been investigated. The earliest one is the sinc^2 function, which was discussed by Rayleigh [6]. He claimed that two sinc^2 functions of equal width and magnitude can be resolved if the maximum of the first sinc^2 is situated at the first null of the second one as depicted in Fig. 13.1. Differently shaped square integrable PSFs like, e.g., Gaussian, Lorentzian or Voigt profiles may not have any null. Jones et al. compare the most common methods to calculate the width of a function [7]. They recommend the use of the equivalent width, the ratio of the area under the profile function, and its peak height.

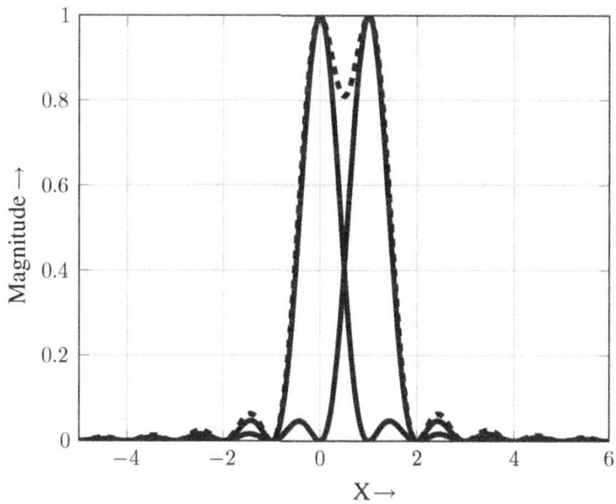

Fig. 13.1 Two sinc^2 functions displaced according to the Rayleigh criterion and their sum. The valley in the middle indicates that both peaks can clearly be resolved

Another method to derive the resolution of any given PSF is to examine the equivalent width of its spectrum. In the case of the sinc^2 function, the spectrum is a triangular function with an equivalent width of

$$\Delta k = \frac{2\pi}{\delta}, \tag{13.1}$$

where δ is the distance between the sinc's maximum and its first null. The other way round, the resolution can be calculated from the spectral width.

$$\delta = \frac{2\pi}{\Delta k} \tag{13.2}$$

13.2.2 Lateral Resolution Limited by Diffraction

In order to determine the spectral width of the PSF, a closer look at the imaging system has to be taken. Therefore, the propagation of the electromagnetic waves from the point source to the aperture is considered as well as the reconstruction of the image. The propagation of nonplane waves in free space is covered by diffraction theory, first discussed by Sommerfeld [8, 9]. Such waves originate from a radiator or a scattering object. There are many different theories on diffraction, getting simpler the more approximations are made. In the following, the approximations that will be used later are summarized. First, electromagnetic waves in the microwave region are treated as scalar waves. Silver [10] and Goodman [3] show that this approximation can give very good results if the fields are measured sufficiently far away from the diffracting object and if the diffracting object details are large compared to the wavelength. The corresponding wave equation then is:

$$\nabla^2 E + k_0^2 E = 0 \tag{13.3}$$

with k_0 representing the wavenumber in free space and E is the scalar complex field strength. The second approximation used in the following is called the Fresnel Approximation. Here, a spherical wave front of radius r is approximated by a second order Taylor polynomial. The resulting paraboloid wave front describes the wave over a limited region around the expansion point contained in a rectangle of size $\Delta x \cdot \Delta y$. For this approximation to be valid, the distance z from the center of the spherical wave front must be large compared to the distance from the axis of the paraboloid, as indicated in (13.4).

$$z \gg \frac{\Delta x^2 + \Delta y^2}{2\lambda} \tag{13.4}$$

As (13.3) is a linear homogeneous differential equation, all possible solutions can be constructed as superposition of a set of fundamental solutions. In the following, the two most important solutions are discussed.

First, there is the spherical wave originating from a point source, given by (13.5)

$$E_{\odot} = \frac{1}{r} \mathrm{e}^{-\mathrm{j}k_0 r}, \tag{13.5}$$

where r is the distance from the point source. Its significance for the resolution has already been discussed. In fact, every source distribution can be interpreted as a density of point sources, and therefore spherical waves with different source points are a set of fundamental solutions to the wave equation. Due to its symmetry, the spherical wave homogeneously propagates in each direction of the far-field.

Second, the plane wave solution to the differential equation only propagates in a single direction. This direction is given by the vector $\vec{k} = \left(k_x, k_y, k_z\right)^{\mathrm{T}}$. Then the field distribution

$$E_{|} = \mathrm{e}^{-\mathrm{j}\vec{k}^{\mathrm{T}} \cdot \vec{r}} \tag{13.6}$$

with the coordinate vector $\vec{r} = (x, y, z)$ solves the wave equation if $\left|\vec{k}\right| = k_0$. Evaluating the field along any straight line whose direction is given by the unit vector $\vec{e}_{\mathrm{u}}$ always yields a linearly changing phase. The corresponding wavenumber is k_{u}.

$$k_{\mathrm{u}} = \vec{k}^{\mathrm{T}} \cdot \vec{e}_{\mathrm{u}} \tag{13.7}$$

This implies that k_x, k_y and k_z are the wavenumbers observed, when evaluating the field along lines parallel to the coordinate axes. As the norm of the vector $\vec{k}$ is fixed, two components are sufficient to determine the third one.

$$k_z = \sqrt{\left(k_0^2 - k_x^2 - k_y^2\right)} \tag{13.8}$$

Plane waves with different directions are also a set of fundamental solutions to the wave equation. Hence, each possible field distribution can be described by a superposition of plane waves. Their use offers some advantages compared to spherical waves. For example, the calculation of the field value in some given points of interest in a Cartesian coordinate system is easier. In contrast to spherical waves, the magnitude of plane waves does not change, so only the phase has to be adjusted. Also this change in phase depends linearly on the change in coordinates, whereas the phase of a spherical wave changes with its distance from the source point. In particular, when considering a multitude of source points and points of field evaluation, the distance for each combination of points has to be calculated separately. For plane waves on the other hand, the calculation becomes extraordinarily simple if the field is evaluated in points, arranged in a cubic array. For each plane wave, only the phase in one point of interest and the change to the neighboring point in each direction have to be known. Going n steps further in an arbitrary direction, the phase has to be turned n times the amount for one step.

As stated above, each field distribution can be described as a superposition of fundamental solutions. When decomposing a field in plane waves, the result is called the angular spectrum of plane waves or k-space representation. If the field is known in each point of a plane, the decomposition becomes a two-dimensional Fourier transform. For instance, the angular spectrum of a spherical wave is uniformly distributed in all directions.

$$E_{\odot} = \int\int \exp\left\{-\mathrm{j}\left(k_x x + k_y y + \sqrt{\left(k_0^2 - k_x^2 - k_y^2\right)z}\right)\right\}\mathrm{d}k_x \mathrm{d}k_y \tag{13.9}$$

Reference [11] proves (13.9) to be accurate by using the method of stationary phase.

Using this and reconstruction methods, the spectral width of the PSF for an example setup as shown in Fig. 13.2 can be estimated. A rectangular aperture of width L_x and height L_y receives the signal from a point source. The source point is situated at a distance z_0 in front of the aperture. Reconstruction algorithms give an estimate $\hat{E}$ of the field E in front of the aperture based on the measured fields. The

Fig. 13.2 Imaging setup

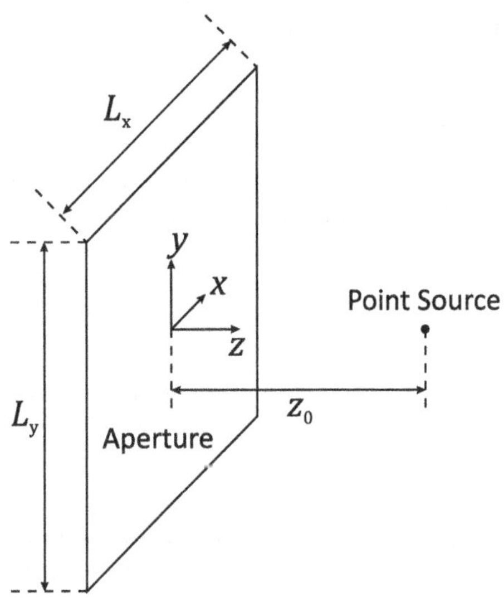

simplest way to reproduce the fields is called backpropagation or Back Projection (BP). The measured field in the aperture is decomposed into plane waves by Fourier analysis. The propagation of plane waves is easy to invert by just turning back the phase according to the direction and distance as indicated in (13.10).

$$\widehat{E} = FT_{2D}^{-1}\left\{FT_{2D}\{E(x,y,z=0)\}\cdot e^{j\sqrt{\left(k_0^2-k_x^2-k_y^2\right)}z}\right\} \tag{13.10}$$

Therefore, the spectral width of the PSF is identical to the spectral response of the measured field in the aperture to a point source. The same reconstruction can also be regarded in the space domain. Here, $\hat{E}$ at each position x, y, z is calculated as the correlation of the measured field in the aperture with the field of a spherical wave centered at x, y, z. There exist many different reconstruction algorithms with different implementations, but the basic idea always is the above described BP. A detailed analysis of the different algorithms cannot be given here.

Often, the received spectrum of plane waves is approximated by geometrical optics. The maximum angle of incidence is determined by a straight line going from the source point to the edge of the aperture. Actually, the angle between the straight line and the normal vector of the aperture gives the local spatial frequency (the analog of the instantaneous frequency) at the corresponding point in the aperture. As Mandel points out, the spectrum of the Fourier decomposition does not have a one-to-one correspondence to local spatial frequencies [12]. He also illustrates that the Fourier spectrum can be calculated by analyzing the instantaneous frequency of the function divided into small intervals, if the amplitude is smooth enough. The spectrum can then be approximated by the complex superposition of the contributions of the small intervals. To show this, the method of stationary phase can be applied. To confirm the applicability, the spectral content of the received signal in the aperture is analyzed by Fourier decomposition in the following. Using the Fresnel Approximation, the field at an arbitrary point in the aperture can be calculated:

$$E(x,y) \approx \frac{1}{z_0 + \frac{x^2+y^2}{2z_0}} e^{-jk_0 z_0} e^{-jk_0\frac{x^2+y^2}{2z_0}} \operatorname{rect}\left(\frac{x}{L_x}\right)\operatorname{rect}\left(\frac{y}{L_y}\right) \tag{13.11}$$

Also, the amplitude term only varies slowly and thus its influence can be neglected compared to the phase term.

$$E(x,y) \approx e^{-jk_0\left(z_0 + \frac{x^2+y^2}{2z_0}\right)} \operatorname{rect}\left(\frac{x}{L_x}\right)\operatorname{rect}\left(\frac{y}{L_y}\right) \tag{13.12}$$

$$\Rightarrow \Phi(k_x, k_y) = \int_{-\frac{L_x}{2}}^{\frac{L_x}{2}} \int_{-\frac{L_y}{2}}^{\frac{L_y}{2}} \mathrm{e}^{-\mathrm{j}k_0\left(z_0 + \frac{x^2+y^2}{2z_0}\right)} \mathrm{e}^{-\mathrm{j}k_x x} \mathrm{e}^{-\mathrm{j}k_y y} \mathrm{d}x\mathrm{d}y$$

$$= -\mathrm{j}\frac{\pi z_0}{2k_0} e^{j\left(-k_0 + \frac{\left(k_x^2+k_y^2\right)}{4k_0}\right)z_0}$$

$$\cdot \left[\mathrm{erf}\left(\sqrt{\frac{\mathrm{j}k_0}{2z_0}}\frac{L_x}{2} + k_x\sqrt{\frac{\mathrm{j}z_0}{2k_0}}\right) - \mathrm{erf}\left(-\sqrt{\frac{\mathrm{j}k_0}{2z_0}}\frac{L_x}{2} + k_x\sqrt{\frac{\mathrm{j}z_0}{2k_0}}\right)\right]$$

$$\cdot \left[\mathrm{erf}\left(\sqrt{\frac{\mathrm{j}k_0}{2z_0}}\frac{L_y}{2} + k_y\sqrt{\frac{\mathrm{j}z_0}{2k_0}}\right) - \mathrm{erf}\left(-\sqrt{\frac{\mathrm{j}k_0}{2z_0}}\frac{L_y}{2} + k_y\sqrt{\frac{\mathrm{j}z_0}{2k_0}}\right)\right], \tag{13.13}$$

where erf denotes the Gauss error function. Apart from the scaling factor and phase, there are two distinct terms for the x- and y-dependence, respectively. The error function changes its value rapidly in a region where the argument is close to zero and is almost constant elsewhere. This is also true for arguments with a phase of $\pi/4$. So if $k_0 L_i^2 \gg z_0$ for $i = x, y$, the error function can be approximated by the sign function.

$$\Phi(x, \mathrm{y}) \approx \frac{\pi z_0}{k_0} e^{j\left(-k_0 + \frac{\left(k_x^2+k_y^2\right)}{4k_0}\right)z_0} \mathrm{rect}\left(\frac{z_0 k_x}{L_x k_0}\right) \mathrm{rect}\left(\frac{z_0 k_y}{L_y k_0}\right) \tag{13.14}$$

For illustration, Fig. 13.3 shows the magnitude of the x-dependence of the k-space support in case that $z_0 = 0,175\,\mathrm{m}$, $L_x = 0,1\,\mathrm{m}$ and the frequency is $f = 300\,\mathrm{GHz}$. Under these conditions, the estimation using the local spatial

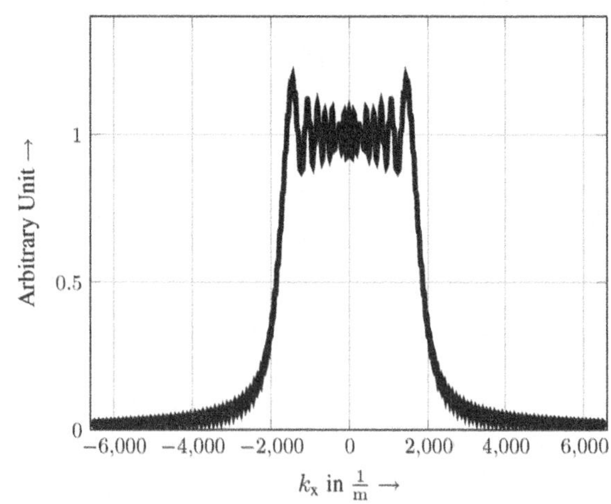

Fig. 13.3 k_x-dependence of k-space support of an aperture with $z_0 = 0,175\,\mathrm{m}$, $L_x = 0,1\,\mathrm{m}$ and $f = 300\,\mathrm{GHz}$

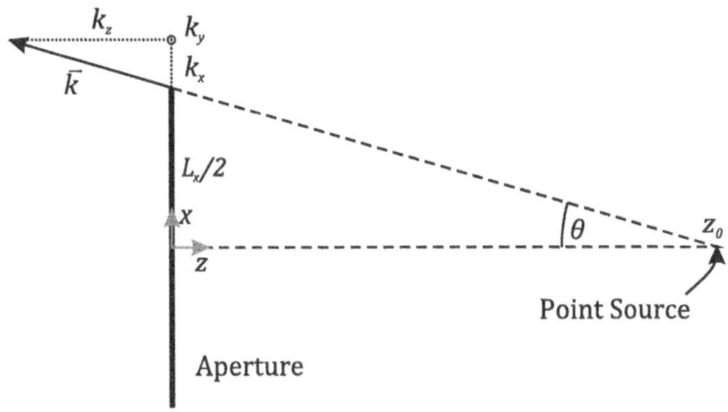

Fig. 13.4 Maximum local spacial frequency introduced to the aperture by a point source at a distance of z_0 in front of a rectangular aperture

frequencies predicts the width of the k-space support to be between $\pm 1800\,1/\text{m}$, agreeing very well with the results shown in Fig. 13.3.

In the above, it has been shown that the resolution in direction $i = x, y$ due to the PSF of a receiving square aperture is approximately given by (13.15), with c_0 being the speed of light.

$$\delta_i \approx \frac{c_0 z_0}{f L_i} \tag{13.15}$$

Accepting the approach using local spatial frequencies, the maximum local frequency in direction i can be calculated by (13.16) as illustrated in Fig. 13.4.

$$k_{i,\max} \approx \frac{k_0}{\sqrt{4\left(\frac{z_0}{L_i}\right)^2 + 1}} \tag{13.16}$$

$$\Rightarrow \delta_i \approx \frac{c_0}{2f}\sqrt{4\left(\frac{z_0}{L_i}\right)^2 + 1} \tag{13.17}$$

The total image spectrum may be much wider, if a multitude of sources in different directions is considered. For an object of size $L_{x,\text{Obj}} \cdot L_{y,\text{Obj}}$ for instance, (13.18) holds.

$$k_{i,\text{Obj},\max} \approx \frac{k_0}{\sqrt{4\left(\frac{z_0}{L_i + L_{i,\text{Obj}}}\right)^2 + 1}} \tag{13.18}$$

In some cases, it is also possible to use a second aperture for illumination. Such systems are called multistatic. Due to reciprocity, the same rules apply to the transmitting aperture as to the receiving one. They also create a PSF independently. The total PSF of the system is then the product of the PSFs of both the transmitting and the receiving aperture. Of course, the transmitting and receiving aperture may be close to each other or even overlapping.

13.2.3 Range Resolution

Also the resolution in range direction is determined by (13.2). The bandwidth has got the strongest impact on Δk_z. But even with a single frequency, the range resolution is finite due to the different angles of incidence on the aperture. So as comes out of Fig. 13.4, the k-space support and the according resolution are given by (13.19), (13.20) and (13.21).

$$k_{z,\max} = \frac{2\pi f_{\max}}{c_0} \tag{13.19}$$

$$k_{z,\min} \approx \frac{2\pi f_{\min}}{c_0} \frac{1}{\sqrt{1+4\left(\frac{z_0}{L_x}\right)^2+4\left(\frac{z_0}{L_y}\right)^2}} \tag{13.20}$$

$$\Rightarrow \delta_z \approx \frac{c_0}{2\left[\Delta f + f_{\min}\left(1-\frac{1}{\sqrt{1+4\left(\frac{z_0}{L_x}\right)^2+4\left(\frac{z_0}{L_y}\right)^2}}\right)\right]} \tag{13.21}$$

Moreover, the resolution in range direction is improved in the multistatic case.

13.3 Array Design

In the preceding section, a simple aperture has been assumed. In microwave holography, the field in the entire aperture has to be measured in magnitude and phase. This is usually accomplished by an array of small antenna elements, filling the aperture. That means that the field is not measured homogeneously everywhere in the aperture but it is rather sampled in discrete points of space. Just like the spatial sampling, the characteristic of each single element influences the performance of the imaging system. In the following, both effects will be examined.

As the single antenna elements are usually about the size of a wavelength or smaller, they are small compared to the distance to the object. So the object is in the far-field of the single element. Thus, the directivity of the single element to a certain direction yields an extra factor when receiving plane waves from this direction. Therefore, it is desirable to have an antenna element that covers the complete scenery as homogeneously as possible. On the other hand, elements in close proximity to each other can couple considerably, particularly if they are omnidirectional radiators. This must either be prevented by careful antenna design of the single element or calibrated and corrected later. Both topics will be covered later in this section.

Mathematically, equidistant sampling can be represented by multiplication with the two-dimensional Ш-function, an equidistant rectangular grid of Dirac δ-impulses. In the k-space, this corresponds to a convolution with the corresponding two-dimensional Ш-function. Hence the obtained spectrum is repeated. The relation between the sample distance d and the repetition in the spectrum k_r is given by (13.22).

$$k_r = \frac{2\pi}{d} \tag{13.22}$$

For an alias free sampling, $\frac{k_r}{2}$ has to be larger than any spatial frequency content in the aperture. Obviously, the maximum spatial frequency that can be obtained is k_0, so sampling with a distance of

$$d_{\min} = \frac{\lambda}{2} \tag{13.23}$$

is always small enough. But of course even larger sampling distances suffice if large angles of incidence $\varphi_{\max}$ can be excluded. Then the required sampling distance is:

$$d_{\max} = \frac{\lambda}{2\sin(\varphi_{\max})} \tag{13.24}$$

But the total number of antenna elements can be reduced even further. The reconstruction of a field measured with an antenna array is often compared to Digital Beam Forming (DBF). In fact, the only difference is that for DBF, the Fraunhofer far-field of the array is optimized, while in holography the focus is still in the Fresnel zone. Just like in the above, the performance of DBF depends on the k-space support of the array. Therefore, the far-field beam width of the array is also a measure for the resolution of the imaging system, although they are not the same thing. For beamforming applications, sparse arrays have been widely studied [13–19]. The array factor AF of a general plane array in the aperture of Fig. 13.2 is given by (13.25).

$$AF = \sum_{n=1}^{N} A_n e^{jk_0 \sin(\vartheta)(x_n \cos(\varphi) + y_n \sin(\varphi))} \tag{13.25}$$

Here, x_n and y_n are the coordinates of the n th element, which is weighted with the complex amplitude A_n. These complex A_n can be used either to produce or to receive beams from various directions. There are many text books on antennas like [20] that provide formulas for the case of antenna elements on a regular lattice. Obviously, (13.25) has the form of a Fourier series of the amplitude distribution A_n with the fundamental spatial frequencies

$$k_x = -k_0 \sin(\vartheta) \cos(\varphi) \tag{13.26}$$

$$k_y = -k_0 \sin(\vartheta) \sin(\varphi). \tag{13.27}$$

This shows the close relationship between beamforming and imaging. With a limited amount of elements and limited space of the aperture, the choice of the weights A_n is always a trade-off between a small beam width and a low grating lobe level [21]. Similarly, in imaging the trade-off occurs between resolution and side lobe level of the PSF. For sparse arrays, one simply has to set the weights of the left out antennas to zero to obtain the array factor of the remaining array. This of course worsens at either resolution or side lobe level. Those side lobes are one cause of clutter. Clutter is a disturbance in the image that appears random to the viewer and may have various causes. Further causes of clutter are noise, multiple reflections that cannot be reconstructed, approximations made for the reconstruction algorithm, imperfect calibration or coupling between the antenna elements. In [22], Gupta et al. show how the performance of an array is impaired by inter-element coupling. Decrease in side lobe level can be achieved by multiplication with taper functions as described, for example, in [23] or following specific thinning strategies [13, 16–18]. If clutter is dominated by noise, it might be beneficial to optimize the weights for maximum signal-to-noise ratio *SNR* by using the *SNR* of a single receiver as weight of the respective signal. Assuming equal noise level for each receiver, the *SNR* is proportional to the signal power itself. A drawback can be seen in the fact that the resulting resolution depends on the imaged scenery. If the clutter is dominated by other processes, an increased resolution may be attainable at the cost of only very little increase in clutter. According methods have been proposed very early. References [24] and [25], for example, increase the effective aperture area by extrapolating the measured signals to the outside of the real aperture. The extrapolated values are calculated depending on the measurement results within the aperture. Therefore, such methods can also be expressed as weighting for the individual receivers within the aperture. The other way round, setting the A_n of missing antenna elements to zero can be interpreted as interpolating the missing signals by zero. Due to Parseval's theorem, this will produce the spectrum containing the least amount of energy still explaining the measured data. It is on the other hand possible to minimize just the part of the spectrum according to the directions from where no radiation is expected. Solving this problem is covered by

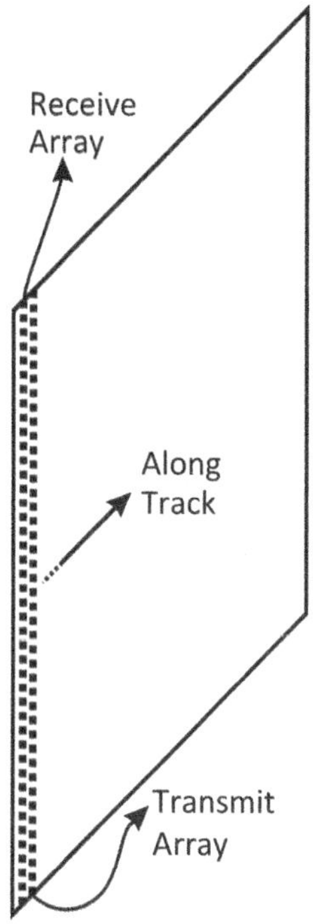

Fig. 13.5 Principle of the synthetic aperture radar (SAR): linear transmit and receive arrays are swept across an aperture

compressed sensing [26]. Sparse arrays are especially useful in the case of independent transmit and receive arrays [27, 28]. As it has been shown earlier, the far-field characteristics of both arrays multiply. By placing the zeros of the receive far-field characteristic at the angles of the grating lobes of the transmit array and vice versa, the total side lobe level of the system can be reduced dramatically. The combination of both arrays can also be viewed as an effective aperture distribution, calculated as the convolution of the field distributions of both arrays. A special form of multistatic imaging using sparse arrays is Synthetic Aperture Radar (SAR). Linear transmit and receive arrays are swept along a linear trajectory, spanning an area called synthetic aperture, as shown in Fig. 13.5 [29–31]. The effective aperture of this setup is the convolution of the transmit and receive arrays and the positions in along track direction. The length of the effective aperture is bounded by the beamwidth of the single elements. For comparison, supposing a nonsparse/fully occupied array, the effective aperture would be the transmit array convoluted with the receive array. Each one would be the respective linear array convoluted with the along track positions. So the effective aperture would contain the along track positions twice. This is not the case, because the measured radiometric information does not contain all combinations of the transmit and receive positions [32].

13.4 Categories

Holographic imaging setups can be split up into various categories. This section gives an overview over the most important categories of holography and discusses their properties. A structure of the described categories is depicted in Fig. 13.6. The imaging process relies on the knowledge of the fields in the aperture in magnitude and phase. One of the most important differentiation criteria between systems is the way how they gain the phase information. In order to detect the phase of a signal, the time information can be provided by a reference signal. So the detected quantity is the relative phase between the measured field and the reference signal. Such a phase difference is a measurable quantity. Direct and indirect holography use different ways to provide the reference signal to the detectors. The characteristics of direct and indirect holography will be the focus of this section, followed by further useful categories.

13.4.1 Direct Holography

In direct holography, the illumination and the reference signal are generated independently and in general at different frequencies. In the following, the frequency used for imaging will be called f_{RF} and the frequency of the reference signal will be called f_{LO}. Both the transmitted and received signals are measured in magnitude and phase using the same reference [33]. For the measurement of the transmitted illumination, usually a small fraction is coupled out from the generator, such that the remaining power can be used as illumination. The RF signals are measured by mixing them with the reference signal in a heterodyne mixer. The resulting signal will be at the intermediate frequency f_{IF}.

$$f_{IF} = |f_{RF} - f_{LO}| \tag{13.28}$$

In a heterodyne mixer, the RF and LO currents both flow through a component with a nonlinear characteristic. Thereby the currents create voltages at frequencies

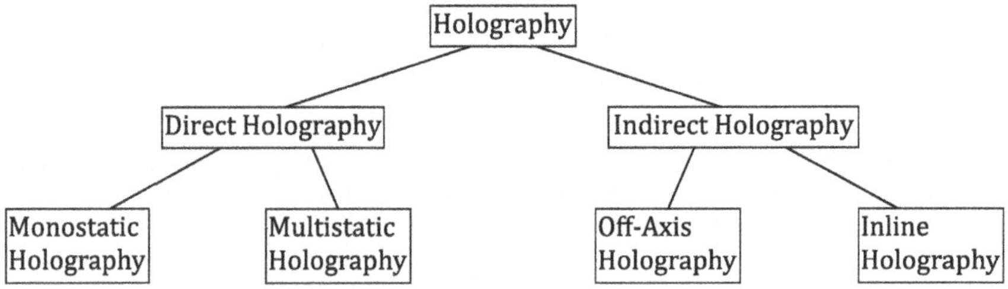

Fig. 13.6 Overview over described categories of holography

that are all linear combinations of f_{RF} and f_{LO}. A filter suppresses all frequency components except f_{IF}. The IF signal is then converted to digital information using an I/Q demodulator. This way both inphase and quadrature components of the signals are known and consequently magnitude and phase. The advantage of directly measuring the phase difference between transmit and receive elements comes at the cost of high system complexity. Because all these components are already combined in a vector network analyzer (VNA), these instruments are commonly used for imaging applications. Unfortunately, VNAs only come with a few ports and hence antennas are often swept over the aperture.

If an antenna array is used, the signal from each antenna has to be measured. Because the reference signal will experience different electrical lengths and attenuations when traveling to the respective antenna elements, calibration is obligatory. As stated earlier, calibration can also compensate for mutual coupling (between two transmit or receive antenna elements, respectively) and cross-coupling (from a transmit antenna element directly to a receive antenna element) [22]. When calibrating a system, objects with known properties, so-called standards, are measured. From the differences to the obtained measurement values, the systematic measurement errors can be modeled and later corrected. As the receivers translate field strength linearly to measured data, a linear error model is sufficient to compensate for magnitude and phase errors. The transmission between the antennas can be modeled using a signal flow graph. In Fig. 13.7, all transmission paths that involve at most one coupling term from transmit antenna j to receive antenna i are depicted. All paths including two or more coupling terms and loops with coupling edges are neglected, because coupling constants are usually very small. Paths connecting the transmit antenna i to the receive antenna j directly via the object are called Γ_i^j. Cross-coupling between the same antennas is called Ψ_i^j. Mutual coupling between two transmit antennas j and k is described by T_k^j and between the two receive antennas i and l is described by K_i^l. As a consequence of reciprocity, the mutual coupling coefficients show the following symmetry:

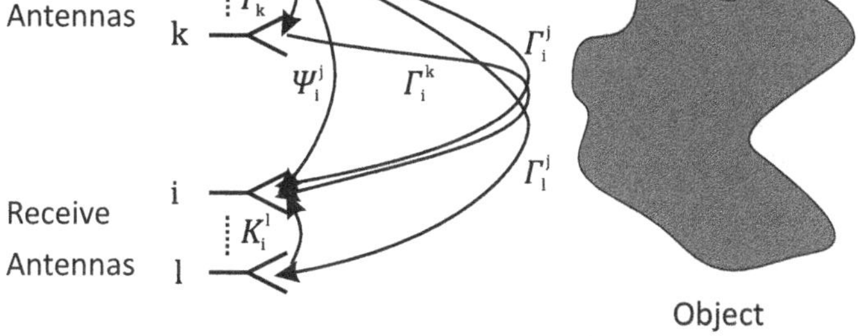

Fig. 13.7 Error terms for calibration: cross-coupling and mutual coupling are shown

$$T_k^j = T_j^k \tag{13.29}$$

$$K_i^l = K_l^i \tag{13.30}$$

To make it short a vector notation is adapted. $\boldsymbol{T}^j$ is the column vector of mutual coupling coefficients between the transmit antenna j and all other transmit antennas, and $\boldsymbol{\Gamma}^j$ is the column vector containing the imaging transmission paths from transmit antenna j to all receive antennas. Similarly, $\boldsymbol{K}_i$ and $\boldsymbol{\Gamma}_i$ are the row vector of mutual coupling coefficients between the receive antenna i and all other receive antennas and the row vector containing the imaging transmission paths from each transmit antenna to the receive antenna j, respectively. Adding up all relevant paths and considering possible differences of the reference signal at the respective mixers due to different transmission line length C_i^j, the measured signal M_i^j is given by (13.31).

$$M_i^j = C_i^j \left(\Gamma_i^j + \Psi_i^j + \boldsymbol{\Gamma}_i \boldsymbol{T}^j + \boldsymbol{K}_i \boldsymbol{\Gamma}^j \right) \tag{13.31}$$

One suitable standard is an absorbing half space. In this case, all Γ_i^j are zero. The corresponding measurement results ${}_aM_i^j$ then only contain C_i^j and the cross-coupling term Ψ_i^j.

$${}_aM_i^j = C_i^j \Psi_i^j \tag{13.32}$$

With this result, (13.31) reduces to (13.33) and further to (13.34).

$$M_i^j - {}_aM_i^j = C_i^j \left(\Gamma_i^j + \boldsymbol{\Gamma}_i \boldsymbol{T}^j + \boldsymbol{K}_i \boldsymbol{\Gamma}^j \right) \tag{13.33}$$

$$\Rightarrow M_i^j - {}_aM_i^j = \begin{pmatrix} \Gamma_i^j & \boldsymbol{\Gamma}_i & \boldsymbol{\Gamma}^{j\mathrm{T}} \end{pmatrix} \begin{pmatrix} C_i^j \\ C_i^j \ \boldsymbol{T}^j \\ C_i^j \boldsymbol{K}_i^{\mathrm{T}} \end{pmatrix} \tag{13.34}$$

Noise occurring in this absorber measurement will disturb any further measurement. It is therefore recommended to use a low IF bandwidth and accept longer measurement time for this one measurement. Also the return loss of the absorbing material is critical.

As the column vector in (13.34) still contains unknown variables, further calibration measurements are needed. Further standards may be implemented like reflecting sheets and spheres at different distances, because it is rather easy to calculate the expected fields. Each measurement yields one line generating a system of linear equations. Suppose that mutual coupling is considered between n_T transmit antennas and n_R receive antennas, respectively, then $\boldsymbol{T}^j \in \mathbb{C}^{n_T - 1}$ and $\boldsymbol{K}_i \in \mathbb{C}^{n_R - 1}$. Thus, there are $n_T + n_R - 1$ unknown variables yielding a minimum for the rank and hence for the number of linear equations. Clever spacing of the standards can prevent exact linear dependence of two equations, but because of the vast amount of combinations of transmit and receive antennas at different frequencies, usually some will occur, where the linear system shows a bad condition number. Each additional calibration measurement improves this condition number and unknown coupling constants are computed using Gaussian least squares.

It follows from the above that the required effort for the calibration of a system for direct holography grows quickly with the number of antenna elements. Fortunately, the majority of the coupling constants are usually small enough to be neglected. That is because coupling coefficients between two antenna elements decay quickly when one antenna element leaves the reactive nearfield of the other. As a rule of thumb, a coupling attenuation above 35 dB will cause a phase error smaller than 1°. In some cases, even all mutual coupling coefficients can be ignored. Then (13.34) becomes a simple single equation and the solution is given by (13.35).

$$C_i^j = \frac{\Gamma_i^j}{M_i^j - {}_a M_i^j} \tag{13.35}$$

Averaging this result for several standards again will increase accuracy.

The specific applications for direct holography are widespread. Personal security scanners in airports represent a quickly growing market. At the moment, four different commercial products are available, all using frequencies between 25 GHz and 100 GHz. In contrast to X-ray based systems, the nonionizing nature of these millimeter waves prevents damage of biological cells. In fact, millimeter waves hardly penetrate human tissue, while textiles are mostly transparent in this domain. This allows for detection of threats by concealed objects and weapons. As spaceborne SAR systems improve, they increasingly replace airborne systems. Still airborne systems are more flexible for surveillance of a nearby region, because satellite orbits cannot be changed at will, but the orbits of other objects have to be taken into account. For global operation on the other hand, spaceborne systems are advantageous. Nowadays spaceborne SAR systems in the X-band achieve resolvable distances below 1 m. While resolution performances of the two kinds of SAR systems are about to level up, the swath width observed by satellites is much larger, extending up to more than 100 km. Apart from military reconnaissance, many different areas of science and industries profit from such systems. In hydrology, the surface coverage with ice, snow and water can be surveilled. Continuously mapping earth's surface incorporating interferometric features yields a deeper insight to tectonics or volcanism. From those abilities, also disaster and crisis management benefit, as well as archaeology and the search for natural resources. Another geological application of direct holography is the ground penetrating radar, using frequencies from 10 MHz to 1 GHz. Microwave mammography is still in development. The systems use frequencies around 1 GHz. The required resolution can only be achieved because of the high permittivity of the tissue, greatly increasing the wavenumber.

13.4.2 Indirect Holography

In contrast to direct holography, in indirect holography the phase reference is not independent of the illumination, but one coherent source provides both. Although it is less popular, indirect holography features some general advantages. First, the phase of the illuminating source does not have to be measured, because its phase is inherently identical to the reference. Second, the architecture of a single receiver can be much simpler. A square law detector for power detection is sufficient. Third, if mutual coupling between the receive antennas can be neglected, calibration can be omitted. That is because there is only one transmit antenna, and phase difference between reference and object wave needs not to be characterized.

Indirect holography setups further divide into off-axis and inline holography. In the following, those two types are distinguished. Thereafter, receiver design will be discussed.

13.4.3 Off-Axis Holography

Off-axis holographic setups include a free space reference wave that is also projected to the receiving aperture, but from a steeper angle of incidence Θ_{Ref}. The situation is illustrated by Fig. 13.8. A well-collimated reference beam only has a small k-space bandwidth around a center k_{Ref} determined by Θ_{Ref}, whereas the object wave will exhibit a larger spatial bandwidth Δk_{Obj}.

$$k_{\mathrm{Ref}} = k_0 \sin(\Theta_{\mathrm{Ref}}) \tag{13.36}$$

Without loss of generality, let the axis of the reference beam be in the xz-plane as indicated in Fig. 13.8. Then E_{Ref} is described by

$$E_{\mathrm{Ref}}(x, y) = E_{\mathrm{Ref},0}(x, y)\mathrm{e}^{-\mathrm{j}k_{\mathrm{Ref}}x}, \tag{13.37}$$

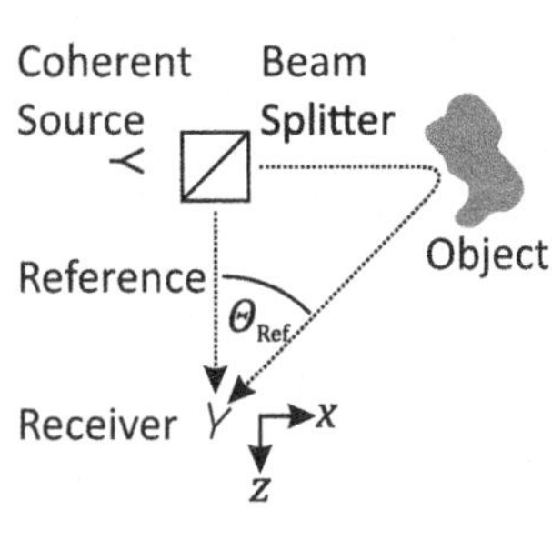

Fig. 13.8 Principal setup for off-axis holography

where $E_{\mathrm{Ref},0}$ has got as small spectral width. The measurement values recorded by a square law detector are proportional to the intensity $i(x, y)$ of the interference signal by the reference and object waves at the place of the antenna element.

$$i(x,y) \sim \left|E_{\mathrm{Ref}} + E_{\mathrm{Obj}}\right|^2 = \left|E_{\mathrm{Ref}}\right|^2 + \left|E_{\mathrm{Obj}}\right|^2 + E_{\mathrm{Ref}}E^*_{\mathrm{Obj}} + E^*_{\mathrm{Ref}}E_{\mathrm{Obj}} \quad (13.38)$$

$$\begin{aligned} I(k_x,k_y) &\sim FT_{2D}\left\{\left|E_{\mathrm{Ref}}\right|^2 + \left|E_{\mathrm{Obj}}\right|^2\right\} + FT_{2D}\left\{E_{\mathrm{Ref}}E^*_{\mathrm{Obj}}\right\} + FT_{2D}\left\{E^*_{\mathrm{Ref}}E_{\mathrm{Obj}}\right\} \\ &= FT_{2D}\left\{\left|E_{\mathrm{Ref}}\right|^2 + \left|E_{\mathrm{Obj}}\right|^2\right\} + FT_{2D}\{E_{\mathrm{Ref}}\} \otimes FT_{2D}\left\{E^*_{\mathrm{Obj}}\right\} \\ &\quad + FT_{2D}\left\{E^*_{\mathrm{Ref},0}\right\} \otimes \Phi_{\mathrm{Obj}}(k_x - k_{\mathrm{Ref}}, k_y) \end{aligned} \quad (13.39)$$

The first addend in (13.39) comprises the magnitude of the coherent background and the object wave. The last term contains $\Phi_{\mathrm{Obj}}(k_x - k_{\mathrm{Ref}}, k_y)$, which is the Fourier transform of E_{Obj}, displaced by k_{Ref} in x-direction. The $\otimes$ symbol means that it is convoluted with the spectrum of the complex conjugate reference wave, indicated by the * symbol in the upper index. The second addend includes a so-called twin image. Again, for a well-collimated reference beam, the influence of $FT_{2D}\left\{E^*_{\mathrm{Ref},0}\right\}$ on the last addend will be small because of its small bandwidth Δk_{Ref} compared to Δk_{Obj}. A sketch of the complete plane wave spectrum is depicted in Fig. 13.9, for example, case that the image is about as large as the aperture. In this figure, all three addends appearing in (13.39) can be seen. From their bandwidth and position, it follows that they will not overlap if (13.40) holds.

$$k_{\mathrm{Ref}} \geq 3\Delta k_{\mathrm{Obj}} + \Delta k_{\mathrm{Ref}} \quad (13.40)$$

In this case, the last addend can be isolated by filtering the k-space. For the derivation of the original object field, there are three possibilities. If Δk_{Ref} is small enough, the influence of the reference wave can be neglected. For that, the reference

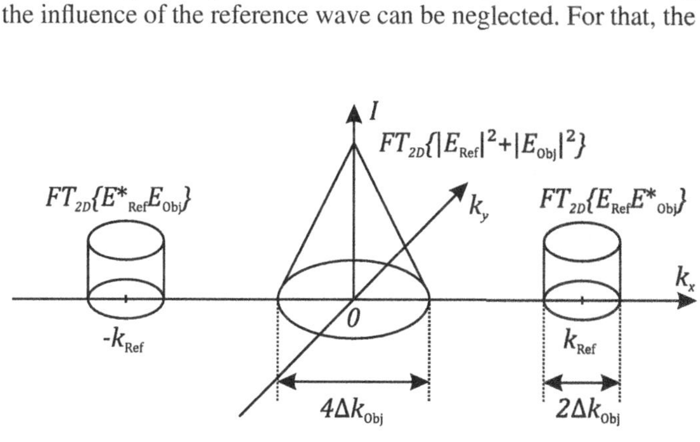

Fig. 13.9 K-space distribution for an off-axis holography setup

wave must be approximately a plane wave, which requires noticeable effort in the setup. If E_{Ref} is approximately described by a fundamental mode Gaussian beam, it acts as a taper function, affecting the resolution, but at the same time decreasing clutter by side lobes. Another way of recovering the object wave is deconvolution. As long as there is no zero of the reference wave in the aperture area, the procedure is stable. For the best signal-to-noise ratio on the other hand, matched filtering can be applied. The two latter procedures require the knowledge of the reference wave. This either has to be measured in magnitude and phase, but a simulation is usually sufficient. The final result is much more sensitive to the relative phase between two points in the aperture than to the absolute value of the phase, so small deviations in the phase center position of the simulated reference wave can be tolerated [34].

The restriction introduced by (13.40) implies that the maximum angle of incidence from the object must be much smaller than Θ_{Ref}, inherently limiting the resolution in x-direction. The resolution in y-direction is not impaired. This phenomenon has been investigated quantitatively in [35]. The maximum wavenumber $k_{\mathrm{Ref,max}} = k_0$ is obtained if $\Theta_{\mathrm{Ref}} = 90°$. Using (13.40), Δk_{Obj} is limited to $\Delta k_{\mathrm{Obj}} \leq k_0/3$.

$$\Rightarrow \delta_{x,\min} \approx 3\lambda \tag{13.41}$$

The spacing of an array in the aperture must obey a strict Nyquist criterion for the highest spatial frequency $k_{\mathrm{Ref}} + \Delta k_{\mathrm{Obj}}$.

$$\Rightarrow \Delta x \leq \frac{\pi}{k_{\mathrm{Ref}} + \Delta k_{\mathrm{Obj}}} \tag{13.42}$$

13.4.4 Inline Holography

For inline holography, the reference and the illumination beam axis are identical. In this way Gabor produced his first results. This way, the recorded intensity still contains the three addends as in (13.39), but they are not separated in the spectrum. So other methods have to be used to isolate the image information. Gabor realized that the image is in focus at a different depth than the other terms. Those therefore contribute uniformly to the clutter in the whole image, but with the correct depth, the image can be reconstructed. Hence depth resolution is crucial to the clutter level of the image. A step frequency approach increases the bandwidth of the setup, thereby improving the depth resolution and thus image clutter by background and twin image are reduced as reported in [36].

On the counterpart of the increased clutter, inline holography requires lower sampling density and less effort for the setup, because only a single beam has to be considered. Also, the required coherence length of the illumination is smaller, because the difference in propagation distance between the reference and object wave can be made small by proper setup design.

13.4.5 Modes of Detection

Many of the reported indirect microwave holograms have actually been recorded using mixers in a network analyzer instead of square law detectors, ignoring the provided phase information [34–37]. In this paragraph, those two receiving methods are compared.

First, square law detectors are considered. These also called direct detectors require an element with nonlinear characteristic, often a Schottky diode [38]. In a small range around the operating point V_0, I_0, the characteristic can be approximated by a second-degree Taylor polynomial.

$$I(V) \approx \frac{R_0}{Z_{\text{dif}}}(V - V_0)^2 + \frac{1}{Z_{\text{dif}}}(V - V_0) + I_0 \tag{13.43}$$

The choice of the coefficients R_0 and Z_{dif} seems arbitrary at the moment, but their meaning is discussed in the following. Provided with a sinusoidal voltage oscillating at an angular frequency ω with sufficiently small magnitude $\hat{V}$, a parabolic characteristic will produce three frequency components as shown in (13.44).

$$\begin{aligned} i(t) &= R_0 \frac{\hat{V}^2}{2Z_{\text{dif}}} \sin(2\omega t) + \frac{\hat{V}}{Z_{\text{dif}}} \sin(\omega t) + R_0 \frac{\hat{V}^2}{2Z_{\text{dif}}} + I_0 \\ &= R_0 P_{\text{in}} \sin(2\omega t) + \frac{\hat{V}}{Z_{\text{dif}}} \sin(\omega t) + \underbrace{R_0 P_{\text{in}}}_{I_{\text{det}}} + I_0 \end{aligned} \tag{13.44}$$

The component at ω is determined by the differential impedance Z_{dif} of the device. The DC component includes the bias current I_0 given by the operating point and additionally the detection signal I_{det}, which is proportional to R_0 and input power P_{in}. R_0 can therefore be identified as the responsivity.

$$R_0 = \frac{I''(V)}{2I'(V)} \tag{13.45}$$

In order to distinguish I_{det} from the DC bias I_0, the input signal is usually chopped and fed into a lock-in amplifier. The third spectral component is located at a frequency of 2ω and is therefore not used. R_0 can be identified as the sensitivity of the detector, which is hence determined by the first and second derivative of the characteristic. The dynamic range of a detector reaches from the noise equivalent power *NEP* to the power, where second-order Taylor approximation is no longer valid. Obviously, the *NEP* depends not only on the sensitivity of the detector, but also on the noise level. The three most relevant sources of noise for a diode detector are thermal noise, shot noise introduced by a bias current and 1/f-noise. Depending on the chopping frequency, the latter may be predominant. Otherwise, the output noise voltage of the detector is proportional to the square root of bandwidth. Therefore, the *NEP* is given in the unit $\frac{\mathrm{W}}{\sqrt{\mathrm{Hz}}}$. Accordingly, the detected *SNR* is given by (13.46).

$$SNR_{\mathrm{det}} \sim \frac{P_{\mathrm{RF}}^2}{\Delta f} \tag{13.46}$$

Unfortunately, diodes always come with parasitic components. A common equivalent circuit diagram of the parasitic components is shown in Fig. 13.10. The cutoff frequency ω_c of the diode is defined as (13.47).

$$\omega_c = \frac{1}{R_s C_J} \tag{13.47}$$

Beyond this frequency, only a fraction of the signal power reaches the junction and so sensitivity of the detector drops drastically.

Resistive mixers on the other hand often also use Schottky diodes, suffering from the same kind of cutoff frequency. The local oscillator (LO) current changes the operating point of all diodes in the mixer periodically, modulating the differential impedance and hence the radio frequency (RF) current. The created voltage component at the intermediate frequency (IF) is proportional to the RF component. A mixer can hence be considered a linear component. The IF noise power is again proportional to the bandwidth. The resulting *SNR* differs from the detector fundamentally in that the former is only proportional to the input power P_{RF} and the latter to P_{RF}^2.

$$SNR_{\mathrm{mix}} \sim \frac{P_{\mathrm{RF}}}{\Delta f} \tag{13.48}$$

As a consequence, the dynamical range of a mixer measured in dB is roundabout double the dynamical range of the detector. On the other hand, the setup of a mixer is much more complicated than the one of a square law detector. There are two signal components at high frequency, namely the RF and the LO. The device needs to be matched for both. All three ports need to be decoupled, which can be accomplished by filtering or the introduction of symmetries, coming at the cost of

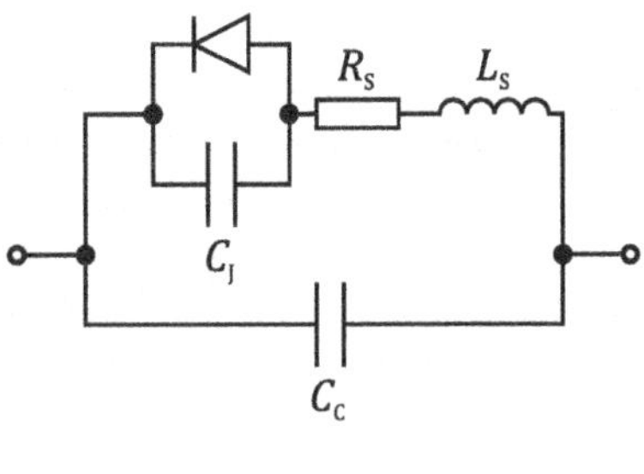

Fig. 13.10 Equivalent circuit diagram of a diode including parasitic components

more diodes with identical characteristics, as far as possible. Summarizing, mixers are costly and require considerable space impeding a dense array of receivers. To circumvent the space problem, the signals can be fed from the antenna to the mixer via a waveguide, introducing additional loss. A detector setup on the other hand is simple. Only one port carries a high frequency signal that can require a matching network. The frequency difference to the bias and converted signal is large and so filtering effort is low. In research, increasing attention is devoted to zero bias Schottky diodes that exhibit high responsivities of over $4\,\frac{\mathrm{A}}{\mathrm{W}}$ without the need for a bias circuit [39–44]. Such diodes are available with cutoff frequencies in the terahertz domain. Still, the differential impedance of these diodes is in the order of $1\,\mathrm{k}\Omega$.

13.4.6 Planar Antennas for Detector Arrays

The majority of antenna elements for an array are either horn antennas or planar antennas. For flip chip mounted zero bias Schottky diodes, a planar antenna design is appropriate. The topology of the antenna depends on the bandwidth of the application. Self-complementary antennas feature a frequency independent source impedance of about 190 Ω. Such broadband antennas are used in [41, 43] and yield a flat frequency response, while resonant antennas as in [42, 45] provide better source point matching for a small bandwidth. If the antenna elements are well matched to a high impedance, mutual coupling is reduced by an effect similar to preamplifier decoupling in MRI [46]. For high bandwidths like one waveguide band and bearing in mind the restrictions on space consumption, a good compromise is achieved by using limited size bow tie antennas [47–49]. The source point impedance of a bow tie antenna is determined by its bow angle. The smaller the bow angle the higher the input impedance. For impedances in the kΩ regime, the bow angle is too small to be considered much different from a strip line. The DC feed would be just as important to the antenna characteristic as the antenna layout itself. In order to preserve a small opening angle near the antenna feed and at the same time having a low RF coupling to the DC connections, the antenna element shown in Fig. 13.11 is considered. A zero bias Schottky diode is mounted in the center of a bow tie like structure on a quartz substrate. As depicted in Fig. 13.12, the symmetric shape is determined by two ellipses. The antenna is optimized for a frequency band from 230 to 325 GHz. The parameters for optimization include the dimensions shown in Fig. 13.12 as well as the length and width of the rectangular defected ground structure on the rear side. The latter increases the antenna source point impedance. The goal of the optimization is to increase the minimum received signal power for all angles of incidence with $\Theta < 40°$. A possible concept for a

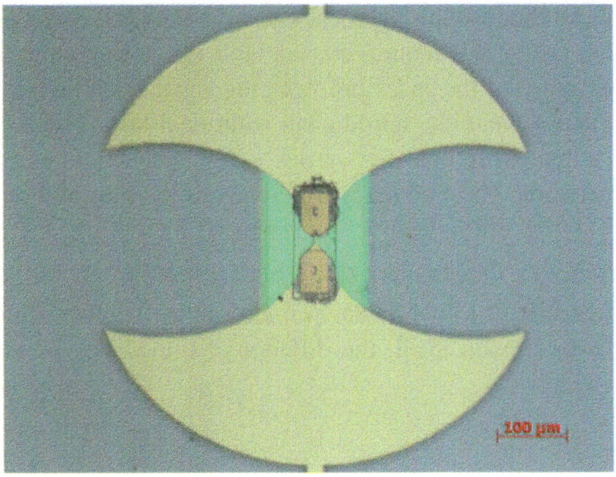

Fig. 13.11 Planar antenna element for use in 300 GHz indirect holographic imaging system

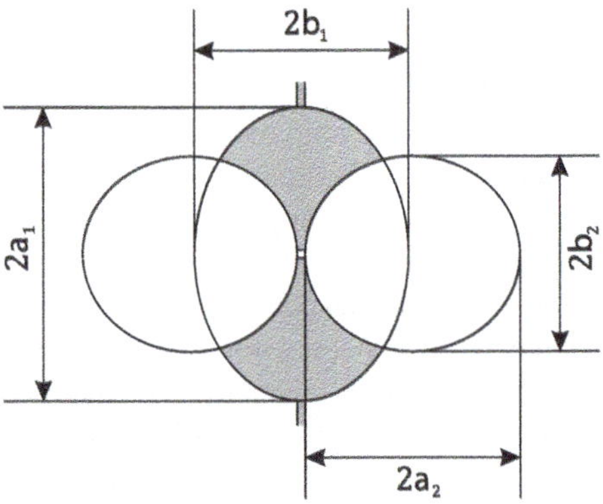

Fig. 13.12 Scheme of the antenna structure. The symmetric shape is determined by the large and half axis of the ellipses a_1, b_1, a_2 and b_2

receiver array discussed earlier is to sweep a line array across the aperture area for instance in the x-direction as indicated in Fig. 13.13. With two sparse line arrays that are shifted by n times $\lambda/2$ in x-direction and $\lambda/2$ in y-direction, the distance between two neighboring antenna elements can be enlarged, diminishing mutual coupling. After n steps in y-direction, the second array will be at the x-position of the first array, filling up the array and fulfilling the $\lambda/2$ sample criterion. Another potential improvement would be a higher level of integration. Integrating the detector diodes, the planar antennas as well as low noise amplifiers on a semiconductor waver level would allow for comparatively cheap and precise production of a large number of identical receiver units.

In the following, measurement results of a holographic setup as shown in Fig. 13.14 are discussed. The major difference to the setup in Fig. 13.8 is the position of the receive array behind the object, so transmission is recorded instead of reflection. The operating frequency is chosen to be 312 GHz. The angle Θ_{Ref} is 40° and the step size between two measurement positions is about 400 μm. The results have been recorded using the described bow tie antenna. Figure 13.15 and

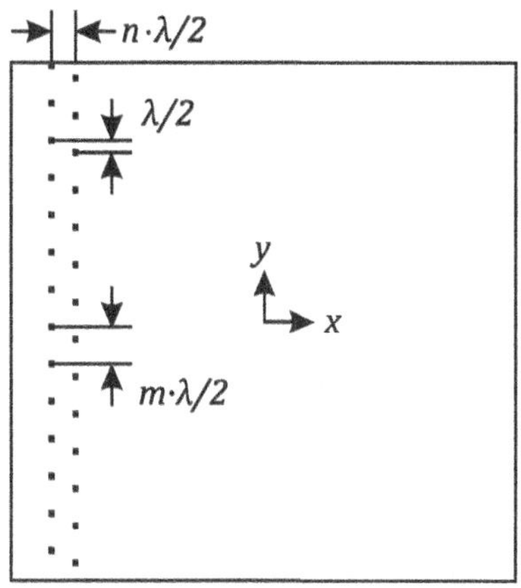

Fig. 13.13 Possible spacing of linear array of bow tie structures, fulfilling the sampling criterion and minimizing cross-coupling

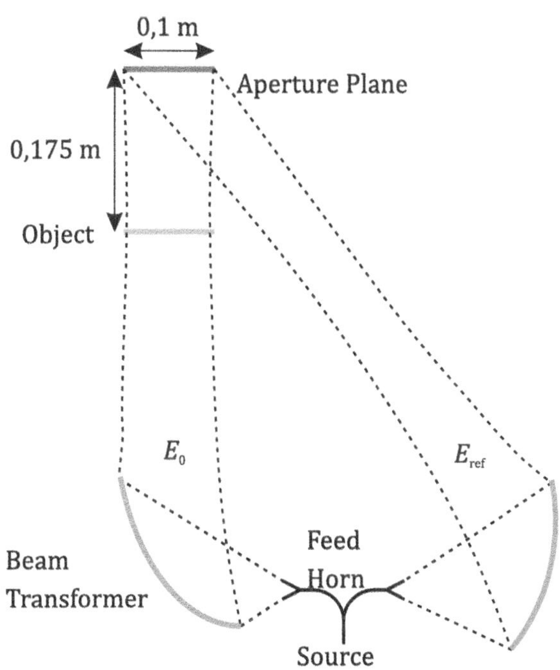

Fig. 13.14 Setup for direct holography in the millimeter wave regime

Fig. 13.16 show the recorded intensity of the reference beam and the object beam of free space, respectively. A Gaussian distribution can be noticed. Their interference pattern is shown in Fig. 13.17 and the magnitude of the respective k-space in Fig. 13.18. An object like the ones photographed in Figs. 13.19 and 13.20 perturb the received field (Figs. 13.21 and 13.22). The magnitude of the k-space is given in Figs. 13.23 and 13.24. For the imaging algorithm, a region around $k_x = -4270\,\frac{1}{\mathrm{m}}$, $k_y = 0\,\frac{1}{\mathrm{m}}$ of width $\Delta k_x = 2830\,\frac{1}{\mathrm{m}}$ and height $\Delta k_y = 5590\,\frac{1}{\mathrm{m}}$ is processed. One

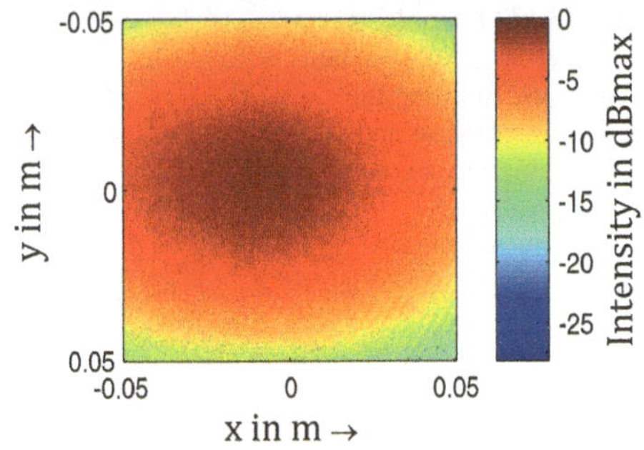

Fig. 13.15 Measured intensity of the reference beam in the aperture

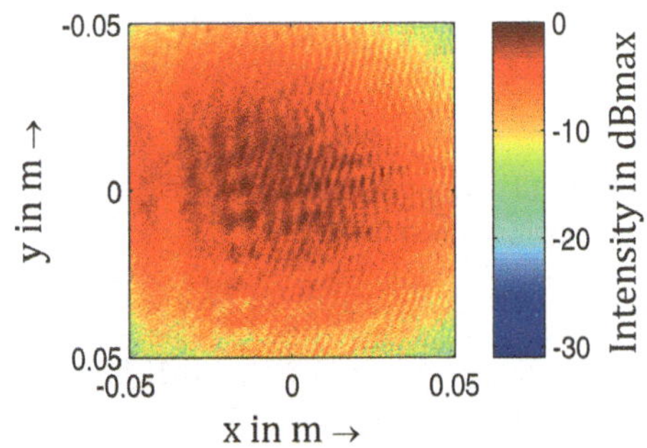

Fig. 13.16 Measured intensity of the empty space object beam in the aperture

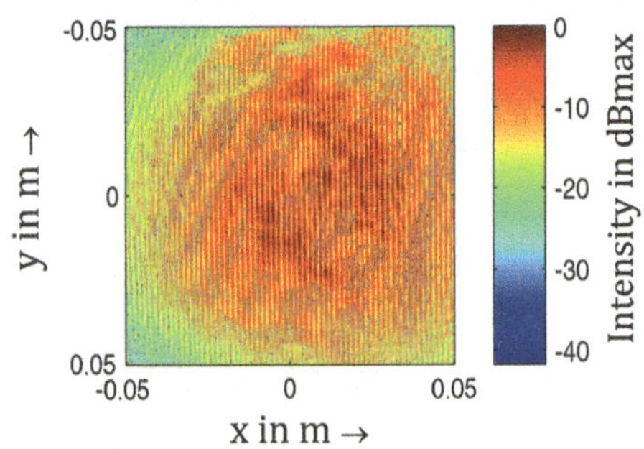

Fig. 13.17 Interference pattern of the intensities of reference beam (Fig. 13.15) and object beam (Fig. 13.16)

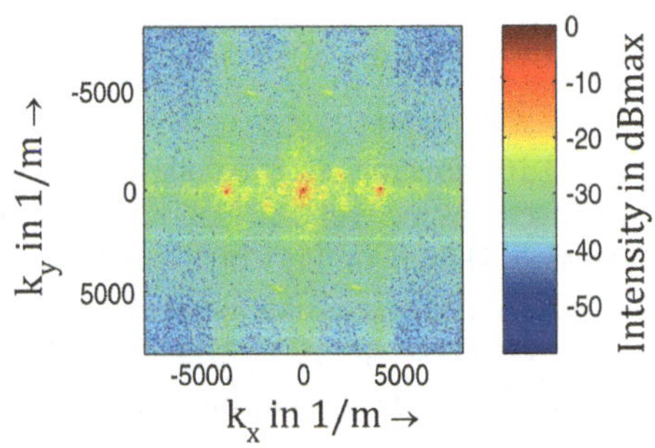

Fig. 13.18 K-space representation of Fig. 13.17

Fig. 13.19 Photograph of the first measurement object

Fig. 13.20 Photograph of the second measurement object

resolution cell measures about consequently $\delta_x = 2.2\,\text{mm}$ in width and $\delta_y = 1.1\,\text{mm}$ in hight. The reconstructed images are shown in Figs. 13.25 and 13.26. Despite considerable speckle, the prediction of the resolution is confirmed by the images.

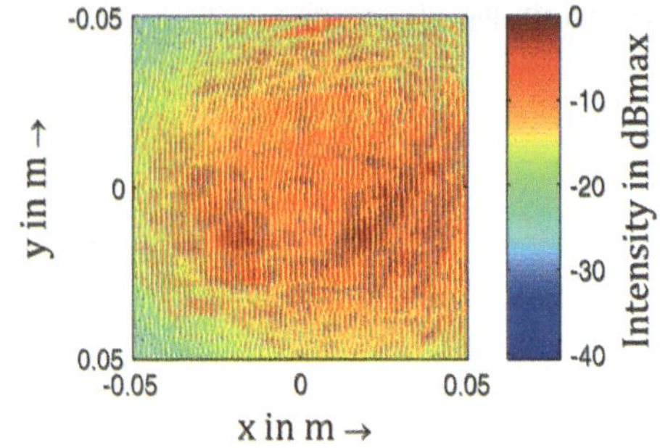

Fig. 13.21 Measured interference pattern for the first object

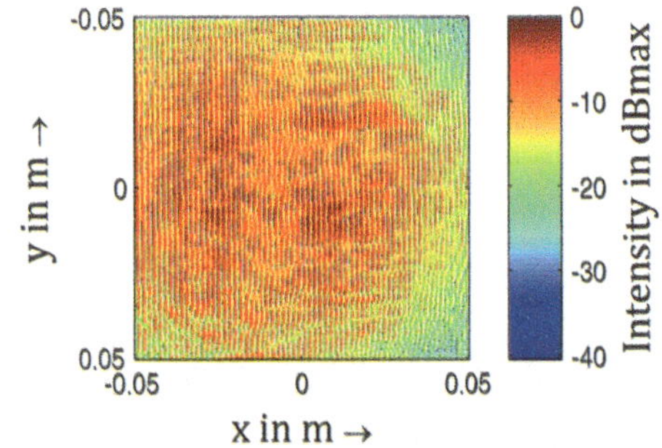

Fig. 13.22 Measured interference pattern for the second object

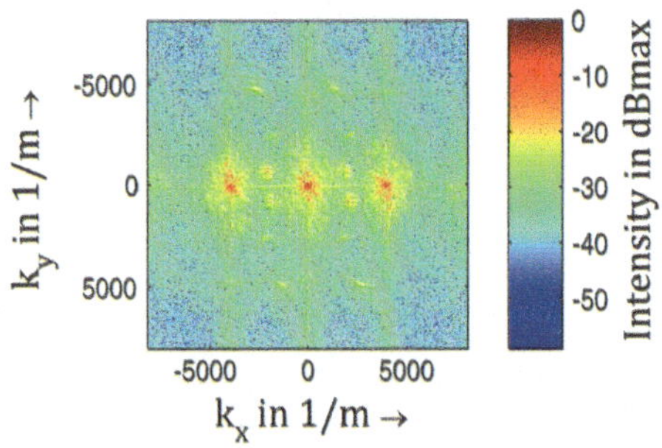

Fig. 13.23 K-space representation of Fig. 13.21

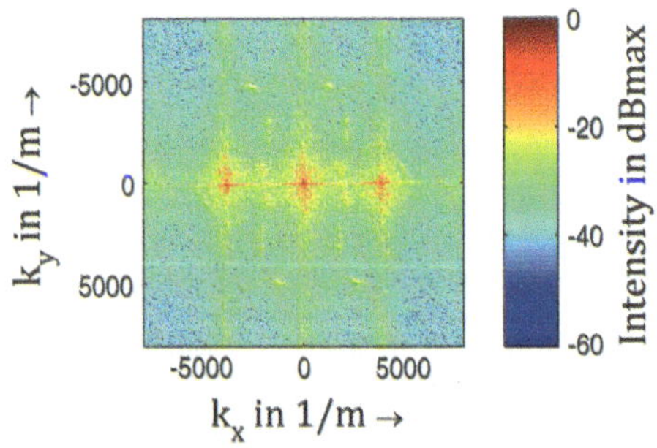

Fig. 13.24 K-space representation of Fig. 13.22

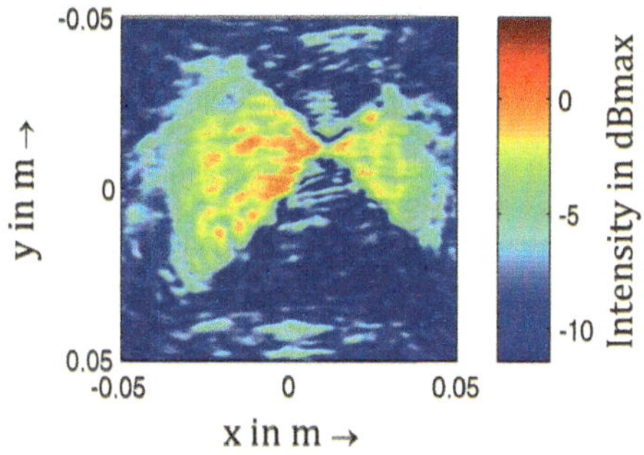

Fig. 13.25 Reconstructed image of the first object

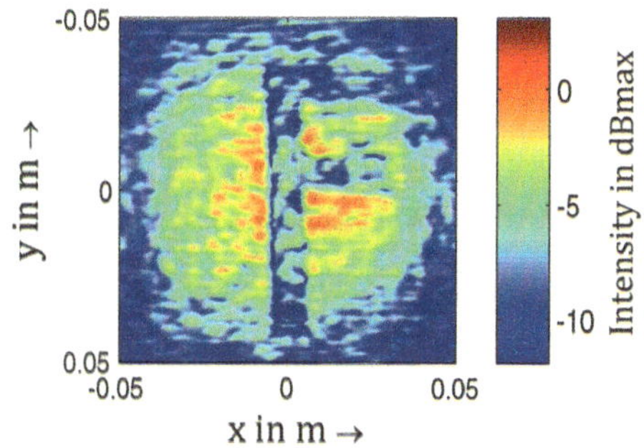

Fig. 13.26 Reconstructed image of the second object

13.5 Summary

In this chapter, different categories of holography have been compared and their specialties have been investigated. Mostly, the system effort is determined by the chosen category. For instance, a direct holographic setup using mixers will be more expensive in design and components than an indirect holographic setup using simple detector diodes. The latter on the other hand will suffer from a lower dynamic range. For off-axis holography also the smallest obtainable resolution is impaired. Finally, a specific example case has been discussed, including an off-axis indirect holographic setup. The planar antenna design has been presented and measurement data have been shown. The predicted performance agrees well with the obtained data.

References

1. D. Gabor, Holography—the reconstruction of wavefronts. Electron. Power **12**(7), 230–234 (1966)
2. D. Gabor, Holography, 1948–1971. Proc. IEEE **60**(6), 655–668 (1972)
3. J.W. Goodman, *Introduction to Fourier Optics*, 2nd edn. (McGraw-Hill, 1996)

4. E.N. Leith, Quasi-holographic techniques in the microwave region. Proc. IEEE **59**(9), 1305–1318 (1971)
5. E.N. Leith, Optical processing techniques for simultaneous pulse compression and beam sharpening. IEEE Trans. Aerosp. Electron. Syst. **AES-4**(6), 879–885 (1968)
6. L. Rayleigh, XXXI. Investigations in optics, with special reference to the spectroscope. Phil. Mag. Ser. 5 **8**(49), 261–274 (1879)
7. A.W. Jones, J. Bland-Hawthorn, P.L. Shopbell, Towards a general definition for spectroscopic resolution, in *ASP Conference on Astronomical Data Analysis Software and Systems IV* (1995), pp. 503
8. A. Sommerfeld, *Electrodynamics: Lectures on Theoretical Physics* (Academic Press, 2013)
9. J.A. Stratton, L.J. Chu, Diffraction theory of electromagnetic waves. Phys. Rev. **56**(1), 99–107 (1939)
10. S. Silver, Microwave aperture antennas and diffraction theory. J. Opt. Soc. Am. **52**(2), 131–139 (1962)
11. S. Gu, C. Li, X. Gao, Z. Sun, G. Fang, Three-dimensional image reconstruction of targets under the illumination of terahertz Gaussian beam—theory and experiment. IEEE Trans. Geosci. Remote Sens. **51**(4), 2241–2249 (2013)
12. L. Mandel, Interpretation of instantaneous frequencies. Am. J. Phys. **42**(10), 840–846 (1974), http://scitation.aip.org/content/aapt/journal/ajp/42/10/10.1119/1.1987876
13. R.L. Haupt, Thinned arrays using genetic algorithms. IEEE Trans. Antennas Propag. **42**(7), 993–999 (1994)
14. D.G. Leeper, Isophoric arrays-massively thinned phased arrays with well-controlled sidelobes. IEEE Trans. Antennas Propag. **47**(12), 1825–1835 (1999)
15. Y.T. Lo, S.W. Lee, *Antenna Handbook: Theory, Applications, and Design* (Springer Science & Business Media, 2013)
16. S. Caorsi, A. Lommi, A. Massa, M. Pastorino, Peak sidelobe level reduction with a hybrid approach based on GAs and difference sets. IEEE Trans. Antennas Propag. **52**(4), 1116–1121 (2004)
17. B.D. Steinberg, The peak sidelobe of the phased array having randomly located elements. IEEE Trans. Antennas Propag. **20**(2), 129–136 (1972)
18. B.D. Steinberg, Comparison between the peak sidelobe of the random array and algorithmically designed aperiodic arrays. IEEE Trans. Antennas Propag. **21**(3), 366–370 (1973)
19. K.C. Kerby, J.T. Bernhard, Sidelobe level and wideband behavior of arrays of random subarrays. IEEE Trans. Antennas Propag. **54**(8), 2253–2262 (2006)
20. C.A. Balanis, *Antenna Theory: Analysis and Design* (Wiley, 2005)
21. R.M. Leahy, B.D. Jeffs, On the design of maximally sparse beamforming arrays. IEEE Trans. Antennas Propag. **39**(8), 1178–1187 (1991)
22. I.J. Gupta, A.A. Ksienski, Effect of mutual coupling on the performance of adaptive arrays. IEEE Trans. Antennas Propag. **31**(5), 785–791 (1983)
23. P. Kabal, Time windows for linear prediction of speech. Technical Report (Telecomunications & Signal Processing Laboratory, Electrical & Computer Engineering, McGill University, Jan. 2003)
24. G. Tricoles, E.L. Rope, R. Hayward, Improved resolution in microwave holographic images. IEEE Trans. Antennas Propag. **29**(2), 320–326 (1981)
25. R.E. Abdel-Aal, Expansion of two-dimensional imaging apertures for resolution improvement in long-wavelength holography. IEE Proc. I (Commun. Speech Vision) **137**(3), 157–162 (1990)
26. M.D. Migliore, A simple introduction to compressed sensing/sparse recovery with applications in antenna measurements. IEEE Antennas Propag. Mag. **56**(2), 14–26 (2014)
27. G.R. Lockwood, J.R. Talman, S.S. Brunke, Real-time 3-D ultrasound imaging using sparse synthetic aperture beamforming. IEEE Trans. Ultrason. Ferroelectr. Freq. Control **45**(4), 980–988 (1998)

28. S.S. Ahmed, A. Schiessl, L. Schmidt, A novel fully electronic active real-time imager based on a planar multistatic sparse array. IEEE Trans. Microw. Theory Tech. **59**(12), 3567–3576 (2011)
29. H. Shigesawa, K. Takiyama, T. Toyonaga, O. Hirao, Microwave holography by synthetic aperture. Proc. IEEE **60**(1), 137–139 (1972)
30. W.E. Kock, Side-looking radar, holography, and doppler-free coherent radar. Proc. IEEE **56** (2), 238–239 (1968)
31. Y.K. Chan, V.C. Koo, An introduction to synthetic aperture radar (SAR). Prog. Electromagn. Res. B **2**, 27–60 (2008)
32. N. Gebert, G. Krieger, A. Moreira, Digital beamforming on receive: techniques and optimization strategies for high-resolution wide-swath SAR imaging. IEEE Trans. Aerosp. Electron. Syst. **45**(2), 564–592 (2009)
33. R.W. Larson, E.L. Johansen, J.S. Zelenka, Microwave holography. Proc. IEEE **57**(12), 2162–2164 (1969)
34. A. Tamminen, J. Ala-Laurinaho, A.V. Raisanen, Indirect holographic imaging at 310 GHz, in *2008 European Radar Conference, EuRAD-2008* (2008), pp. 168–171
35. J. Adametz, F. Gumbmann, L. Schmidt, Inherent resolution limit analysis for millimeter-wave indirect holographic imaging, in *2011 German Microwave Conference (GeMIC)* (2011), pp. 177–182
36. X. Gao, C. Li, Z. Sun, G. Fang, Implementation of step-frequency continuous-wave scheme in millimeter-wave inline holography for interferences elimination. IEEE Antennas Wirel. Propag. Lett. **12**, 1176–1179 (2013)
37. A. Enayati, A. Tamminen, J. Ala-Laurinaho, A.V. Raisanen, G.A.E. Vandenbosch, W. de Raedt, THz holographic imaging: a spatial-domain technique for phase retrieval and image reconstruction, in *2012 IEEE MTT-S International Microwave Symposium Digest (MTT)* (2012), pp. 199–201
38. B.L. Sharma, *Metal-semiconductor Schottky Barrier Junctions and Their Applications* (Springer Science & Business Media, 2013)
39. J.L. Hesler, T.W. Crowe, NEP and responsivity of THz zero-bias Schottky diode detectors, in *15th IEEE International Conference on Terahertz Electronics* (IEEE, 2007), pp. 844–845
40. M. Hoefle, A. Penirschke, O. Cojocari, R. Jakoby, Advanced RF characterization of new planar high sensitive zero-bias Schottky diodes, in *2011 European Microwave Integrated Circuits Conference (EuMIC)* (2011), pp. 89–92
41. L. Liu, JL. Hesler, H. Xu, A.W. Lichtenberger, R.M. Weikle, A broadband quasi-optical terahertz detector utilizing a zero bias Schottky diode. IEEE Microwave Wirel. Compon. Lett. **20**(9), 504–506 (2010)
42. V.I. Shashkin, Y.A. Drjagin, V.R. Zakamov, S.V. Krivov, L.M. Kukin, A.V. Murel, Y.I. Chechenin, Millimeter-wave detectors based on antenna-coupled low-barrier schottky diodes. Int. J. Infrared Millimeter Waves **28**(11), 945–952 (2007)
43. A. Semenov, O. Cojocari, H.-W. Hübers, F. Song, A. Klushin, A.-S. Müller, Application of zero-bias quasi-optical Schottky-diode detectors for monitoring short-pulse and weak terahertz radiation. IEEE Electron Device Lett. **31**(7), 674–676 (2010)
44. X. Fan, X. Pei, X. Xiong, Zero bias Schottky diodes use in high performance detection circuits, in *2011 International Conference on Electronics and Optoelectronics (ICEOE)* (2011), pp. V3-27–V3-30
45. J. Montero-de-Paz, E. Ugarte-Munoz, L.E. Garcia-Munoz, D. Segovia-Vargas, D. Schoenherr, I. Oprea, A. Amrhein, O. Cojocari, H.L. Hartnagel, Millimeter-wave receiver based on a folded dipole antenna and Schottky diode for maximum power transfer, in *2012 6th European Conference on Antennas and Propagation (EUCAP)* (2012), pp. 1259–1262
46. P.B. Roemer, W.A. Edelstein, C.E. Hayes, S.P. Souza, O.M. Mueller, The NMR phased array. Magn. Reson. Med. **16**(2), 192–225 (1990)
47. D.B. Rutledge, M.S. Muha, Imaging antenna arrays. IEEE Trans. Antennas Propag. **30**(4), 535–540 (1982)

48. C. Schildbach, J. Schür, L.-P. Schmidt, Broadband detector array concept for 3D holographic imaging at THz frequencies, in *2013 European Microwave Conference (EuMC)* (2013), pp. 1243–1246
49. V.A. Petriakov, F.F. Sizov, O. Golenkov, S.A. Dvoretskii, D.S. Krasilnikov, Direct detection MM-wave linear antenna array on the base of hemispherical lens antenna elements coupled with narrow-gap hot-carrier bolometers, in *2013 International Kharkov Symposium on Physics and Engineering of Microwaves, Millimeter and Submillimeter Waves (MSMW)*, (2013), pp. 79–81

Index

A. Boriskin and R. Sauleau (eds.), *Aperture Antennas for Millimeter and Sub-Millimeter Wave Applications*, Signals and Communication Technology,
DOI 10.1007/978-3-319-62773-1

The manufacturer's authorised representative in the EU is Springer Nature Customer Service Centre GmbH, Europaplatz 3, 69115 Heidelberg, Germany. If you have any concerns regarding our products, please contact ProductSafety@springernature.com

Printed and bound by CPI Group (UK) Ltd, Croydon, CR0 4YY
30/09/2026
02225229-0016